Ham Radio Technician + General Class License Study Guide

From Beginner to Licensed! Master the Fundamentals of Amateur Radio, Ace Both FCC Exams and Get on the Air with Confidence

Book 1: Ham Radio Technician Class License Study Guide

Book 2: Ham Radio General Class License Study Guide

Morse Code Publishing

Contents

Book 1: Ham Radio Technician Class License Study Guide

From Beginner to Licensed! Master the Fundamentals of Amateur Radio, Ace the FCC Exam and Get on the Air with Confidence

Morse Code Publishing

Introduction

AMATEUR RADIO! HAM RADIO! Welcome to a hobby or a lifestyle, as some would say, that will transform your simple curiosity into a lifetime of enjoyment, challenges, duty, and connection. Welcome to the fascinating world of ham radio.

As someone who embarked on the thrilling adventure of becoming a ham radio operator from a complete beginner, I understand the highs and lows of this journey. My passion for ham radio was sparked by a simple desire to explore a new hobby that combined my love of electronics, the thrill of a challenge, and the joy of connecting with others. This journey led me to create a resource that would make ham radio accessible to everyone, especially beginners like you and me when I first started.

This book was really created for me to help me pass my Technician Class License. It was never meant to be shared with others. It wasn't until my friend wanted to get his license and saw how well I did in preparing that I shared my notes with him. I put this together through his encouragement and am now sharing it with you. Thank you, Brian, for your encouragement and confidence in me!

This book is designed to demystify the process and equip you with everything you need to succeed. It includes a deep dive into all possible questions from the exam pool so that you feel 100% confident and ready to take your license exam.

You don't need prior experience or knowledge in ham radio to start. That's right! This guide is tailored for absolute beginners (like I was), aiming to transform daunting technical challenges into manageable, understandable concepts. I remember initially how overwhelming the technical jargon and concepts seemed, and this guide intends to break down those barriers. It combines accurate, detailed explanations with real-world applications, ensuring you grasp the theory and its application in practical scenarios.

The book includes all the questions you could see on the exam and a comprehensive breakdown of all sections of the Technician Class exam. Each chapter is structured to build your understanding progressively, turning complex regulations and technical details into explicit, actionable knowledge.

I assure you that this book will guide you from confusion to clarity. You will transition from feeling daunted by the complexity of ham radio to confidently passing the FCC exam. Through this guide, you'll prepare for the exam and be ready to join the vibrant global community of licensed amateur radio enthusiasts with a sense of accomplishment and confidence.

As we progress, remember that every expert was once a beginner—don't be fooled into thinking otherwise! So, take this first step with an open mind and an excited spirit, ready to explore the airwaves and connect with people worldwide. Your enthusiasm and open-mindedness will be your greatest assets on this journey.

To conclude with a personal touch, here's a favorite quote that has inspired me throughout my ham radio journey: "The biggest barrier to learning something new is the belief that you can't do it." Let this book be the tool that helps you break down that barrier, inviting you into a world where curiosity leads to discovery and learning becomes an exciting adventure.

Ready to start this rewarding journey? Let's dive in!

Jared Johnson, KF0RTU @ Morse Code Publishing

The Technician Class Exam

Your Gateway to Ham Radio

WELCOME TO HAM RADIO! Before you can start exploring the airwaves, you'll need to pass the Technician Exam—the first step to becoming a licensed amateur radio operator. This exam covers the basics of radio theory, regulations, and operating practices, and it's your gateway to a lifetime of learning and communication. In this section, we'll break down everything you need to know to prepare so you can approach the exam with confidence and excitement.

Chapter 1

Design of this Book

THE GOAL OF THIS book is to help you accomplish three key things:

1. Master the fundamental principles of amateur radio.

2. Prepare you to ace the FCC exam and earn your Technician license.

3. Give you the confidence to get on the air!

The exam part of this book is divided into four main sections: Rules & Regulations, Science of Radio, The Physical Radio Station, and Operating the Radio. We then end with a final section about growing your skills and knowledge in ham radio.

Each section of this book represents a quadrant in our teaching strategy. Each is interconnected and will reinforce your learning and deepen your understanding of ham radio.

Rules & Regulations	Science of Radio
The Radio Station	Operating the Radio

Science of Radio

This section delves into the technical and scientific principles behind radio communications. You'll explore concepts such as wave propagation, modulation, and frequency. We'll also discuss the different types of signals and how they travel through the atmosphere. This knowledge is the backbone of understanding radio waves and how to make the most of your communications.

The Physical Radio Station

Here, we'll focus on the hardware and setup of your radio station. From antennas and feedlines to transceivers and power supplies, this section will guide you through the essential components of a functioning amateur radio station. You'll learn the design of a station and how to troubleshoot your equipment, ensuring you're always ready to communicate.

Operating the Radio

This section is all about putting your knowledge into practice. You'll learn the procedures for making contacts, using repeaters, and operating in CW, SSB, and digital modes. We'll cover best practices for clear communication, handling interference, and staying safe. By the end of this section, you'll feel confident in your ability to operate your station effectively and responsibly.

Rules & Regulations

In this section, you'll learn about the legal aspects of amateur radio. We'll cover the Federal Communications Commission (FCC) rules, band plans, licensing requirements, and the responsibilities of being a licensed amateur radio operator. Understanding these regulations is paramount because they ensure that all operators use the radio spectrum responsibly and avoid causing interference.

Teaching Framework

Our teaching framework integrates learning across four main sections: Rules & Regulations, Science of Radio, The Physical Radio Station, and Operating the Radio. Each section interconnects, enriching your understanding as you progress. For instance, grasping radio wave propagation enhances your station setup and operation skills, while regulatory knowledge aids proper procedures.

This holistic approach ensures a well-rounded grasp of amateur radio, preparing you for the Technician exam and active participation in the radio community. At the book's end, you'll find a glossary of radio terms and a "Technician's Cheat Sheet" to reinforce key concepts and provide quick references. Let's embark on this exciting journey together!

A final word of encouragement: learning amateur radio can feel overwhelming, but it's all about simplifying and focusing on core principles. You don't need to memorize everything; instead, identify the key concepts that govern the field. Many details are simply variations or combinations of these fundamentals.

This book is designed to be concise and direct, providing the essential knowledge to succeed without getting bogged down in extraneous details. By focusing on the core principles and understanding how they interconnect, you'll find that what initially seems complex is manageable and intuitive.

Accuracy and Updates

Every effort has been made to ensure that the information in this book is accurate, up-to-date, and aligned with current standards and regulations. However, human errors can occur, and the world of amateur radio is constantly evolving. If you find any discrepancies, outdated information, or errors, please don't hesitate to reach out and let us know. Your feedback is invaluable in maintaining the quality and reliability of this resource. You can contact me at Jared@MorseCodePublishing. com, and I will do my best to address any concerns or updates in future editions. Thank you for your understanding and for contributing to this book's continued improvement.

Chapter 2

The Exam

HAVE YOU EVER STOOD at the base of a mountain, looking up at the peak, wondering how you will ever reach the top? The path might seem daunting initially, but with the correct map and tools, what seemed impossible becomes a series of manageable steps. Similarly, preparing for the Technician Class Exam might initially appear overwhelming (over 400 possible questions to know!). This chapter is your guide, mapping out the structure and nuances of the exam and equipping you with strategies to navigate this challenge effectively. From understanding the exam format to grasping the scoring system, you'll be prepared and ready to excel.

Exam Overview

The Technician Class License exam, often your first formal encounter in the ham radio community, is designed to test your understanding of basic regulations, operating practices, and electronics theory. The exam consists of 35 multiple-choice questions. Usually, the test takes about an hour. However, you will likely be given all the time you need to take the test. These questions are randomly selected from a pool of 412, covering various topics to ensure a comprehensive assessment of your beginner-level knowledge of amateur radio operations.

The Question Pool

Imagine the question pool as a well-organized library, where books are sorted into specific topics and subtopics, making it easier to find exactly what you need. Similarly, the question pool for the exam is meticulously organized into categories and subcategories, each focusing on a distinct aspect of amateur radio knowledge. Within each category, subcategories break down the material further, providing a detailed roadmap for your study. This structured setup helps you systematically approach your preparation, allowing you to tackle one area at a time and cumulatively build your knowledge.

How to Read the Exam Question Format

When preparing for the exam, you will see questions "numbered" in the following format: T5B01.

In the Technician exam, each question is labeled with a unique identifier like 'T5B01.' This identifier helps you understand the question's context:

- 'T' stands for the Technician license class. The questions are changed periodically. The current pool of questions is good through June 30, 2026.

- The first number, '5,' indicates the main topic or sub-element. In our example, 5 refers to the sub-element "Electrical Principles."

- The letter 'B' specifies a subtopic within the main topic.

- The final number '01' represents the question's specific number within that subtopic.

This structure helps you quickly locate and understand the question's context.

Sub-Element Weights

Understanding how the exam content is weighted helps prioritize your study efforts. The exam sub-elements are not weighted equally, which means that some sub-elements will have more questions on the exam than others. This insight directs your focus to areas needing a more robust review.

For instance, you will see three times as many questions about FCC rules (6 questions) as you will for antennas and feed lines (2). While having well-rounded knowledge is essential, emphasizing these heavier-weighted areas can increase your chances of a successful outcome. See below for each sub-element and the number of questions for each sub-element on the exam.

Sub-elements on the technician's exam:

1. Commission's Rules: six questions from this sub-element will be on the exam (6)

2. Operating Procedures: three questions from this sub-element will be on the exam (3)

3. Radio Wave Propagation (3)

4. Amateur Radio Practices (2)

5. Electrical Principles (4)

6. Electronic And Electrical Components (4)

7. Practical Circuits (4)

8. Signals And Emissions (4)

9. Antennas And Feed Lines (2)

10. Safety (often just written as zero) (3)

For a total of 35 exam questions.

Question Types

The questions in the Technician Class exam are structured to assess your memory and understanding. Each question offers four choices, one correct answer, and three distractors. These distractors are not random but carefully crafted to challenge common misunderstandings or computational errors. It's crucial to approach each question critically to understand why an answer is correct. This approach helps tackle distractors, which might seem correct at a surface level.

For example, a question might ask about acceptable frequency ranges for technician operators. The distractors might include frequency ranges close to correct but outside the allocated bands.

Scoring System

The scoring system of the Technician Class exam is straightforward: each correct answer earns one point. There is no benefit to leaving a question blank, which strategically implies that guessing if you're unsure is better than leaving an answer blank. This aspect of the scoring system should influence how you handle questions you find challenging. It's good practice to tackle questions you are confident about and then return to the more challenging ones, making educated guesses if needed.

By dissecting the exam structure this way, you are better prepared to allocate your study time effectively, understand what to study, and approach the exam strategically. This foundation is essential as we progress through further details and strategies in the following sections, each designed to build upon this initial groundwork, ensuring you confidently approach the exam with a clear plan.

You must score at least 74% to pass, which means getting 26 out of 35 questions correct. While this might sound straightforward, the breadth of topics requires a solid preparation strategy to ensure you cover all necessary material.

It's Time to Schedule Your Exam

Let's get you ready for the Ham Radio Technician Class exam! There are a few key steps to ensure you're fully prepared and have everything in place to take the test—think of it as your final homework before the big day. Here's what you need to do:

Step 1: FCC Registration Number (FRN)

Preparing for your amateur radio exam requires some essential items to ensure a smooth process.

First, you need an FCC Registration Number (FRN), which you can obtain by registering your Social Security Number on the FCC's website. This number is necessary for all licensing transactions.

This must be done before taking your exam. This is your homework!

Start with this website: https://www.fcc.gov/new-users-guide-getting-started-universal-licensing-system-uls

- You can also type 'New Users Guide To Getting Started With Universal Licensing System' into Google; it will be the first site to appear.

Follow the steps for a new user to register with the FCC's Commission Registration System (CORES).

- Click on 'Register.' Then follow the prompts to 'Create New Account.'
- Once you complete the steps, you will receive an email asking you to verify your account creation. Click the link in the email to confirm.
- Then click the button that says, "Go to CORES."

Next, you click on the link to 'Register New FRN."

- Follow the step-by-step instructions and register for an 'Individual FRN.'
- You will know you have completed this step when you see your new FCC Registration Number (FRN).
- Print this page and save your FRN!

Step 2: Find and Register for an Exam Session

Search for Exam Session: Visit the ARRL website and navigate to the search for exam session page.

- https://www.arrl.org/find-an-amateur-radio-license-exam-session
- Or type into Google, "Find an Amateur Radio License Exam in Your Area," and click on the ARRL site.

On the ARRL website, you can search for in-person or an online exam session. Select the type of exam you prefer. I'd recommend the in-person where possible. Especially if you are new to ham radio, it's an excellent opportunity to meet people and start networking with other hams.

Assuming you want to be in person:

- **Enter Zip Code:** Enter your zip code to get the best results and find nearby exam sessions.

- **Results:** The ARRL will provide a list of exams, including details such as the sponsoring club, location, time, and whether walk-ins are allowed.

- **Registration:** Walk-ins are usually not permitted, so you must register beforehand. For more details and contact information, click on the specific exam session to schedule your appointment.

It's that easy! You know you have been successful when you have an exam date set.

Step 3: Take the Exam

The dawn of your exam day is not just another sunrise; it marks the culmination of your diligent preparation and heralds a significant step forward in your journey as an amateur radio operator. Today, all the knowledge you've absorbed, the concepts you've untangled, and the practice tests you've navigated converge to a pivotal point.

You must also bring a legal photo ID, such as a driver's license or passport. If you don't have a photo ID, two other forms of identification, such as a birth certificate and a social security card, are required. Students and minors can present a school ID; if they are under 18, they might also need a guardian's ID. That's right! There is no age requirement to obtain your Ham radio license. Any person of any age who can pass the exam can acquire a license. But you do have to pass the exam without help.

Bring any previous Amateur Radio license or Certificates of Successful Completion of Examination (CSCE). Also, **carry two pencils, a pen, and a calculator with cleared memory**. Cell phones and other electronic devices with calculator capabilities are not allowed. Lastly, ensure you have a check, money order, or cash to cover the exam fee.

Arrive at the test location with ample time to spare. When you check-in, you'll be directed to a specific area where the exam will take place. The environment is typically set up to minimize distractions. Desks are spaced appropriately, and there may be dividers. Understanding that this setup is designed to give everyone an equal opportunity to concentrate can help ease any nerves about the testing environment.

When taking the exam, use the strategies you've practiced from our next chapter. Approach each question methodically: read the entire question and all answer choices before marking your sheet. If a question is challenging, mark it and move on, returning to it later. This helps you manage your time and avoid getting stuck on challenging problems. Keep a steady pace to ensure you cover all the questions within the allotted time, and periodically review your answers, especially those you were unsure of on your first pass.

Stay calm and focused during the exam. If you feel anxious, practice deep breathing:

- Inhale slowly through your nose. Hold for a few seconds. Exhale slowly through your mouth.

This can reduce tension and refocus your mind. Positive affirmations, like "I am prepared" or "I can do this," can help maintain your confidence.

As you sit down to take your Technician Class License exam, acknowledge the hard work you've put in and trust in your preparation. You are ready to take this important step toward joining the global community of amateur radio enthusiasts.

Now that we have an overview of the exam and how to study, let's begin our study!

Chapter 3

How to Study: Techniques That Work

WHEN MASTERING THE MATERIAL for the Technician Class exam, the effectiveness of your study techniques can make all the difference between just passing and genuinely understanding. Engaging actively with the content prepares you for the exam. It sets a solid foundation for your future as a ham radio operator. Let's explore some active learning strategies that have proven successful for many who are now licensed operators.

Active learning is a process that involves more than just reading; it requires you to engage with the material dynamically. For instance, **in this book, as you progress through it, you will want to read the questions and answers just as if they are part of the normal text—because they are! They are set up to be read as part of the teaching in the order you find them.**

You will see the exam question and the four possible answers. The correct answer will be **BOLDED** in the question. The key information you need to answer the exam question will be <u>underlined</u> in the text. So anywhere you see <u>underlined</u> text, pay attention, as that is key information you will need to know.

For example, here is an actual question from the exam. In this book, we will give you the question along with each possible answer. The **bolded** text is the correct answer—in this example, the letter 'A' is the correct answer.

T5B02: Which is equal to 1,500,000 hertz?

A. 1500 kHz
B. 1500 MHz
C. 15 GHz
D. 150 kHz

An additional active learning strategy is using flashcards. Picture this: each card holds a question on one side and the answer on the other. As you shuffle through the deck, you're testing your knowledge, reinforcing the information, and identifying areas that need more attention.

Another powerful tool is summarization. After reading a section, explain it in your own words, as if you were teaching it to someone else. This method ensures you have truly grasped the concepts rather than memorizing words.

Creating a customized study plan is helpful, as it respects the uniqueness of your schedule, learning pace, and prior knowledge. Start by assessing how many days or weeks you have until the exam, then break down the topics into manageable chunks. Allocate more time to sections with more weight or topics entirely new for you. Remember, consistency is vital in a study plan; even if it's just 20 minutes a day, it's better than cramming all the material into a few lengthy sessions.

Visual aids and mnemonics make learning more enjoyable and enhance your ability to recall the information later. Diagrams, for instance, can be invaluable for visualizing complex concepts such as signal paths or antenna designs. Mnemonics, on the other hand, are tools designed to help remember lists or sets of information. For example, to remember the order of operations in adjusting your radio settings, use a phrase where the first letter of each word stands for a step in the process. These visual and mnemonic aids transform abstract information into tangible, memorable content.

Lastly, remember the importance of feedback loops. This involves regular testing of your knowledge through practice exams, which are core in preparing for the format and time constraints of the actual test. This process helps identify weak areas and builds confidence as your scores improve.

Memorization Techniques vs. Understanding Concepts

Proper comprehension is about grasping the underlying concepts that make the facts work, enabling a more profound and enduring mastery of the subject matter. This approach prepares you for the exam and equips you with practical knowledge that enhances your participation in the amateur radio community. Let's explore how you can elevate your study methods beyond rote memorization to understand better and retain the material.

The essence of learning in any field, especially one as technical and diverse as amateur radio, lies in understanding concepts over memorizing facts. While memorization might help you recall specific answers, understanding the concepts ensures that you can apply this knowledge in various contexts, solve problems, and adapt to new situations. This is particularly relevant in amateur radio, where operating conditions and technologies constantly evolve.

For instance, understanding the concept of radio wave propagation involves more than memorizing that specific frequencies travel farther at night. It requires an appreciation of why this happens—the interaction between radio waves and the ionosphere changes characteristics based on time of day and solar activity. This deeper understanding allows you to select optimal frequencies for communication at different times, enhancing your effectiveness as a radio operator.

Integrating concepts with practical examples is another powerful tool in your study arsenal. This method involves linking theoretical knowledge to real-world applications, cementing your understanding, and making the information more relatable and easier to recall. For example, when studying the types of antennas, instead of just memorizing their names and features, consider how each type would perform under specific operating conditions. Imagine setting up a dipole antenna in your backyard and consider how its orientation and height above the ground affect your ability to communicate with other local or distant stations. Visualizing these scenarios transforms abstract information into concrete understanding that sticks with you far beyond the exam.

Analogies are particularly useful in breaking down complex technical concepts into digestible pieces. They compare a new, unfamiliar concept and something you already understand. For instance, if you're struggling to grasp the idea of electrical impedance in antennas, you might think of it as similar to water pressure in a hose. Just as water flow in a hose can be optimal at a certain pressure, radio signals transmit most effectively when the antenna impedance matches the transmitter. Such analogies not only make challenging concepts more accessible but also make the learning process more engaging and enjoyable.

Finally, practicing with purpose is about engaging with practice questions to reinforce your understanding, not just your ability to recall facts. This active practice approach prepares you for the diversity of questions you might face during the exam. It deepens your understanding, ensuring that you retain the knowledge long after you have passed the test.

By focusing on understanding concepts, integrating practical examples, utilizing analogies, and practicing with purpose, you transform your exam preparation from a task of memorization to an enriching process of learning.

Recognizing Patterns and Tricky Questions

When preparing for the Technician Class License exam, understanding the intricacies of the question pool goes beyond knowing the correct answers. It involves recognizing patterns in how questions are structured and developing strategies to handle tricky questions effectively.

Pattern Recognition

In the vast array of questions that make up the exam, specific patterns emerge that can guide you to the correct answers. Recognizing these patterns involves paying close attention to the phrasing of questions and the structure of the answers provided. For instance, examiners often use a particular phrasing style for questions that require a regulatory answer, such as those about FCC rules. These questions might include specific keywords like "regulation," "legal," or "permissible," which signal that the answer will likely involve a rule or a standard procedure.

Another typical pattern involves the use of qualifiers such as "always," "never," or "must." Questions using these terms typically test absolutes and are designed to see if you understand amateur radio operations' fundamental, non-negotiable aspects. On the other hand, options that contain qualifiers like "usually," "sometimes," or "typically" may hint at more general knowledge and understanding, which requires you to think about the most common or likely scenarios rather than absolute rules.

As you practice, reflect on why specific answers are correct and how they align with the question's phrasing. This practice will prepare you for similar questions you might face and help you feel more at ease with the exam format.

Analyzing Tricky Questions

Tricky questions are a staple of any standardized test, designed to assess your depth of understanding and ability to apply knowledge in less straightforward contexts. These questions often include extra information to distract from the core issue or subtly mislead with closely related but incorrect answers. To navigate these challenges, start by carefully reading the question to identify precisely what is being asked. Strip away any details that do not directly relate to answering the question.

Once you've isolated the main query, consider each answer choice methodically. Ask yourself why each option could be correct or incorrect based on the facts and concepts you have studied. This analytical approach helps prevent the common pitfall of rushing to select an answer that seems right at a glance but falls apart under scrutiny. Suppose an answer seems too obvious or straightforward, especially if the question is complex. In that case, it's worth taking a second look to ensure a straightforward surface-level interpretation.

Common Pitfalls and How to Overcome Them

As you prepare for the Technician Class exam, knowing and learning to navigate potential pitfalls is necessary for a smooth study experience and success. Let's explore some common challenges and how to overcome them effectively.

Overconfidence can be a subtle obstacle, often stemming from initial successes. It can lead to a false sense of readiness. The solution? Regular practice exams. They provide reality checks, highlighting areas needing more attention, like complex FCC regulations. This helps ensure you're genuinely prepared, not just feeling ready, and familiarizes you with the exam format, reducing anxiety and boosting efficiency.

Procrastination often results from the daunting nature of the material or unclear study goals. Combat this by setting small, achievable goals. Break down large topics, like "FCC regulations," into manageable chunks, such as "Frequency Allocations." This focused approach makes sessions more productive and rewarding. Additionally, create a structured study schedule with specific times dedicated to studying, making it a regular part of your day rather than an optional task.

Information overload is a significant challenge with the detailed content of the Technician Class exam. Avoid this by breaking study sessions into topic-specific chunks and allocating specific days or times to subjects. This prevents information from blurring together and helps build a solid knowledge base. Summarizing topics in your own words or creating simple mind maps can aid in consolidating information and ease recall during the exam.

Burnout is a real risk with intense studying. It can lead to decreased motivation and resentment towards the subject. Maintain a healthy study-life balance by taking regular breaks—short ones every hour—to retain information and stay focused. Engage in activities you enjoy, like hobbies or exercise, to alleviate stress and refresh your mind. Recognize signs of burnout, such as chronic fatigue or irritation, and give yourself time to rest. Remember, this journey is a marathon, not a sprint.

Adopting these strategies allows you to navigate the common pitfalls in preparing for the Technician Class exam. These approaches aid in efficient learning and make the experience enjoyable, helping you become a licensed operator with confidence and ease!

Time Management for Aspiring Hams

Balancing work, family, and other commitments while studying for your Technician Class License can feel like finding a quiet moment in a busy city—challenging but possible. Prioritizing study time means integrating it into your daily routine rather than seeing it as an interruption. Think of your time as a garden; careful planning and attention help everything flourish, including your personal, professional, and educational goals.

To prioritize study time effectively, assess your typical day or week to identify less busy periods, such as early mornings, lunch breaks, or quiet moments after dinner. Consistency is key; even 20-30 minutes daily can add up significantly over time. Communicate your goals and schedule with family or housemates to minimize interruptions and enhance productivity during these study periods.

Efficiency during study sessions is crucial. Start each session with a clear objective, such as covering a specific topic or focusing on certain question types. Use active recall techniques by reciting or writing down what you remember after reading a section, reinforcing memory, and highlighting areas needing further review. Focus on weaker areas to improve your skills. Use tools like timers to keep sessions focused and short, as prolonged periods can lead to diminishing returns.

Establishing a study routine reduces the mental effort needed to start each session. Like brushing your teeth daily, a routine eliminates the decision-making about studying. Find a distraction-free location, such as a particular desk at home, a library, or a quiet coffee shop. Consistently studying in the same spot can help trigger 'study mode.' Keep all your study materials—books, notes, flashcards—readily accessible to reduce setup time.

Setting goals is key for motivation. Begin with a broad goal, like passing the Technician Class exam by a specific date, and break it into smaller, measurable objectives, such as mastering a chapter or completing practice questions. These goals should be SMART: Specific, Measurable, Achievable, Relevant, and Time-bound. They provide clear targets and regular opportunities to celebrate your achievements, keeping motivation high.

By integrating these time management strategies into your study practices, preparing for the Technician exam becomes less about finding time and more about maximizing it. This efficient, structured approach not only prepares you for the exam but also enhances your ability to manage responsibilities across all areas of life. The goal is to seamlessly and sustainably integrate learning into your life, turning time management challenges into personal growth and success opportunities.

Practice Exams

The use of practice exams in your study routine is akin to a dress rehearsal before a major performance; it provides a vital snapshot of your current knowledge and readiness, allowing you to adjust your preparations effectively before the event. Integrating practice exams into your study schedule is essential for evaluating your memory of the material and getting comfortable with the exam format and time constraints.

We have 14 practice exams for you — yes, 14 exams! Why 14 exams? To be fully prepared for the test, you must answer each question at least once in exam format. Because section T5D has fourteen possible questions, and only one will be on the exam, we need fourteen practice tests.

Scheduling Practice Exams

As your exam date approaches, increase the frequency of the practice exams. In the last few weeks before your test, try to simulate the exam environment as closely as possible. Find a quiet, uninterrupted space and time yourself as if you were in the exam room. This not only helps with retention but also eases any anxiety you might feel about the process and timing of the actual exam. Moreover, it allows you to practice managing your time efficiently across different exam sections, ensuring you can comfortably complete all questions within the allotted time.

Access the 14 practice exams here:

14 Practice Exams!

www.MorseCodePublishing.com/TechExam

Frequency of Practice

Finding the right balance for practice exams is key to steady improvement without burnout. Start with taking exams every other week, increasing to weekly as the exam date nears. Allow time between tests to absorb the material and focus on the weaknesses identified. Avoid cramming too many practice exams before the actual test, which can lead to fatigue and diminish returns. Instead, use the final days before the exam to review material gently, concentrate on weak areas, and ensure you're well-rested and relaxed on exam day.

Analyzing Results

Analyzing practice exam results is crucial. Review incorrect answers and identify patterns in the questions you miss, whether by topic or format. This helps you adjust your study focus. Revisit your study materials for each wrong answer, ensuring you understand the concept fully. For challenging topics, seek additional resources like tutorials, extra reading, or help from a study group. Integrating practice exams into your study routine helps prepare for the exam format and pressure while deepening your material understanding, making your study time more effective. Each exam should build your knowledge and confidence, leading to success on the Technician Class License exam.

The Science of Radio

Introduction to the Science Behind Radio

UNDERSTANDING THE PRINCIPLES OF radio science is essential for mastering amateur radio operations. Radio technology is based on physics, especially electromagnetism, and encompasses electromagnetic wave generation, transmission, and reception.

By delving into the science that powers radio communication, we uncover the intricate processes that enable signals to travel vast distances through the air, allowing us to communicate across town or worldwide.

This section will explore the core concepts of electricity, radio wave propagation, frequency, modulation, and the role of antennas, providing a solid foundation of technical knowledge to ace your exam. Through this exploration, you'll better appreciate the technological marvels that make ham radio a unique and fascinating hobby.

Chapter 4

The Metric System

THE METRIC SYSTEM IS essential for mastering the technical aspects of the hobby. The metric system is a simple, standardized system of measurement based on units of ten, making it easy to learn and use.

This system's beauty lies in its simplicity and consistency. All you need to know are the base units and a few prefixes, and you can easily understand and convert between different measurements. This simplicity makes the metric system a common language in science and engineering, allowing for clear communication and accurate calculations.

So, let's dive into the metric system.

Prefixes, Values, and Symbols

Prefixes	Value	Standard form	Symbol
Tera	1 000 000 000 000	10^{12}	T
Giga	1 000 000 000	10^{9}	G
Mega	1 000 000	10^{6}	M
Kilo	1 000	10^{3}	k
Deci	0.1	10^{-1}	d
Centi	0.01	10^{-2}	c
Milli	0.001	10^{-3}	m
Micro	0.000 001	10^{-6}	μ
Nano	0.000 000 001	10^{-9}	n
Pico	0.000 000 000 001	10^{-12}	p

Here is a table highlighting the labeling of metric values. Review and memorize the prefixes, their symbols, and their values. The base value will be '1', and then you will multiply or divide by tens to get to the correct prefix.

The key to all this is memorizing the metric chart above. Now, let's help you memorize the metric units that are important for Ham Radio – from giga to pico, you can use the following mnemonic:

Giant Monsters Kick Big Mountains, Making New Paths

Giant = (Giga - G)
Monsters = (Mega - M)
Kick = (Kilo - k)
Big = (base unit)
Mountains = (Milli - m)
Making = (Micro - μ)
New = (Nano - n)
Paths = (Pico - p)

Do you want a cheat code to help you remember when the unit symbol is capitalized or not? Be a Millionaire! All unit symbols are lowercase until you get to be a Millionaire (Mega = M = Millions). At Mega (million) and up, symbols are capitalized.

No more delays. Are you ready to jump into some exam questions? Let's go!

Frequency is measured in hertz (more on this later). But the base unit of frequency is 1 hertz (Hz). Notice that in hertz, the 'H' is also capitalized. That info will help with this exam question:

T5C07: What is the abbreviation for megahertz?

A. MH
B. mh
C. Mhz
D. MHz

The chart shows that Mega means one million and is represented by the symbol 'M' (a capital M). That means the abbreviation for megahertz is MHz.

Using the metric chart again...

T5C13: What is the abbreviation for kilohertz?

A. KHZ
B. khz
C. khZ
D. kHz

Again, our base unit is hertz (Hz). Kilo means 1,000 and is represented by the 'k' (lowercase k) symbol. The abbreviation for kilohertz is kHz.

The Metric Scale

Several questions in your technician's exam are designed to ensure you understand how the metric system works. Although the questions and units of measure will change, the principle of 10s still applies.

Start with the base unit and move left (get bigger) or right (get smaller). Each unit you move is a factor of 10 (or think of it as moving one decimal place). In ham radio, we focus on the bigger prefixes. This means we move by larger units of 1,000 or 10^3 - (10^3 is just a fancy way of writing 10 x 10 x 10, which equals 1,000). In most cases, we will move our decimal place three spaces left or right based on our direction in the scale.

For example, 1,000,0000 Hz equals 1 MHz (M is for mega, meaning millions), or 1,000 Hz is equal to 1 kHz (k is for kilo, meaning thousand). We moved our units, or decimals, in groupings or multiples of 1,000s.

Metric Scale for Ham Radio

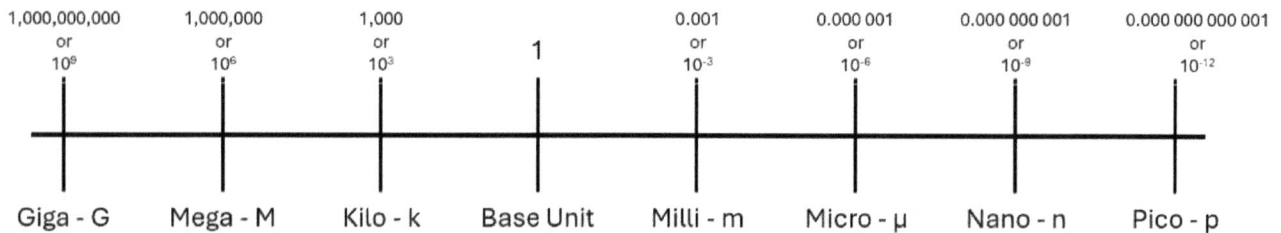

Let's practice one and then throw some exam questions at you.

T5B01: How many milliamperes is 1.5 amperes?

A. 15 milliamperes
B. 150 milliamperes
C. 1500 milliamperes
D. 15,000 milliamperes

To solve this problem, start with the base unit of amperes.

Starting at amperes, we move to the right on our number chart one space to Milli (10^3 or 1,000). <u>Moving right means the number gets bigger.</u>

On our Ham Scale, we move to the <u>right</u> one jump, which means we multiply by 1,000 (10^3 means 10 x 10 x 10 or 1,000) or think about moving the decimal three spaces.

1.5 amperes x 1,000 = 1500

But it's no longer amperes. It has now become milliamperes. So, the correct answer is 1,500 milliamperes.

How many milliamperes in 1.5 amperes?

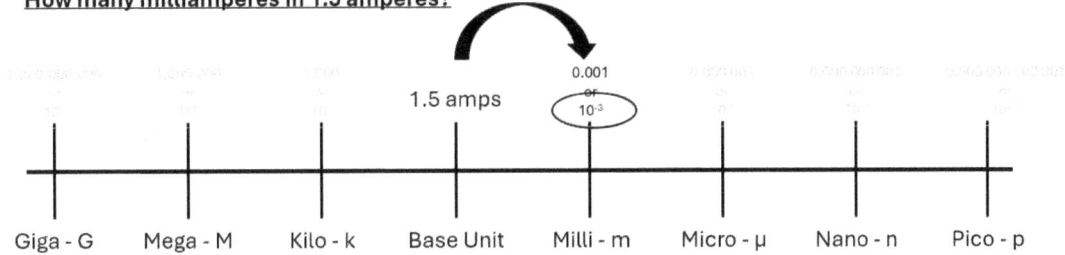

1.5 amps

0.001
or
10⁻³

| Giga - G | Mega - M | Kilo - k | Base Unit | Milli - m | Micro - µ | Nano - n | Pico - p |

1.500 = 1,500 milliamps

Moving to the Right the 'Big Number' Out Front Gets BIGGER

Let's try another one, this time moving to the left.

T5B02: Which is equal to 1,500,000 hertz?

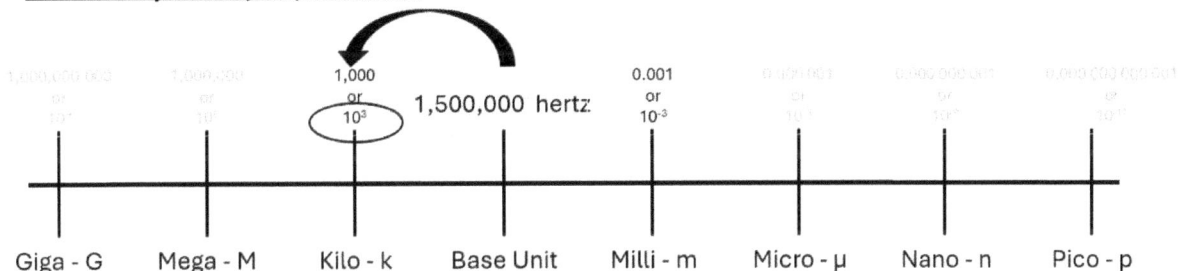

A. 1500 kHz
B. 1500 MHz
C. 15 GHz
D. 150 kHz

To solve, start with the base unit of hertz.

Starting at hertz (the base unit), we move to the left on our number chart one space to kilo. <u>Moving left means the number gets smaller.</u>

We move one "Ham Unit" to the <u>left</u> – which is equal to 10^3 or 1,000 units or 3 decimal places. They all mean the same thing!

1,500,000 hertz divided by 1,000 = 1500 (we basically dropped off three zeros)

But it's no longer hertz. It has now become kilohertz or simply kHz. So, the correct answer is 1500 kHz. If you do this for the other possible answers, none of them make sense and are incorrect.

Which is equal to 1,500,000 hertz?

1,000
or
10³

1,500,000 hertz

0.001
or
10⁻³

| Giga - G | Mega - M | Kilo - k | Base Unit | Milli - m | Micro - µ | Nano - n | Pico - p |

1,500,000 Hz= 1,500 kHz

Moving to the Left the 'Big Number' Out Front Gets SMALLER

Let's try some more.

T5B03: Which is equal to one kilovolt?

A. One one-thousandth of a volt
B. One hundred volts
C. One thousand volts
D. One million volts

Refer to the numbers table again. A kilo is a thousand, so one kilovolt is a thousand volts.

Now, a tricky one.

T5B04: Which is equal to one microvolt?

A. One one-millionth of a volt
B. One million volts
C. One thousand kilovolts
D. One one-thousandth of a volt

On the table, a micro is written as 0.000 001. Micro means millionths, which in basic terms means millions to the right of the zero. Confusing, I know!

Here is how I like to think about it. Look at the chart again. Micro is written as 0.000 001. There are, in fact, six zeros, just like a million has six zeros (a different but easy way to think about it). So, a micro is a million, but it is written in the funny metric system as a million**th.** This gives us the answer that 'one microvolt' is 'one-million**th** of a volt.'

Remember, when looking at things to the right of the decimal, we will add '**th**'.

Let's try some more.

T5B05: Which is equal to 500 milliwatts?

A. 0.02 watts
B. 0.5 watts
C. 5 watts
D. 50 watts

We start at milliwatts and move "one Ham Unit" to the left. Again, our "Ham Unit" on this metric scale is 10^3 or 1,000 or three decimal points. Moving left means the number gets smaller. We divide by 1,000. Moving the decimal three places to the right.

500.0 milliwatts divided by 1,000 = 0.5. We must update the prefix from milliwatts to watts, so 500 milliwatts equals 0.5 watts.

For reference, you will see on our metric scale 10^{-3}. That negative sign is the fancy technical way of indicating that we divide or make our big number smaller. As you become familiar with the metric system, it will become second nature. But for now, don't let it confuse you. Just remember which way to move the decimal point.

Exam Questions

The following questions are all part of the exam pool. Work your way through each one using the concepts above. The units will change, but moving right or left on the numbers line (or up and down on the table – whichever is easier for you) will remain valid.

Exam Tip: When taking the exam, draw out the metric line and use it. The visualization will help. Just remember, "Giant Monsters Kick Big Mountains Making New Paths." Draw it out and use it for the following questions.

T5B06: Which is equal to 3000 milliamperes?

A. 0.003 amperes
B. 0.3 amperes
C. 3,000,000 amperes
D. 3 amperes

Moving from milliamperes to amps (our base unit), the "number" out front gets smaller. It goes from 3,000 to 3 (as there are 1,000 or 10^3 places between milli and base unit). So, our answer is 3 amps.

T5B07: Which is equal to 3.525 MHz?

A. 0.003525 kHz
B. 35.25 kHz
C. 3525 kHz
D. 3,525,000 kHz

Refer to the scale again. Moving from megahertz to kilohertz (all answers are in kilohertz) will increase our number. There are 1,000 or 10^3 places between mega and kilo. So, we move from 3.525 to 3,525 (our decimal moved three places). We update our units from MHz to kHz and get 3,525 kHz.

T5B08: Which is equal to 1,000,000 picofarads?

A. 0.001 microfarads
B. 1 microfarad
C. 1000 microfarads
D. 1,000,000,000 microfarads

Moving from picofarads to microfarads (all answers are in microfarads) will decrease our number. There are 1,000,000 or 10^6 places between pico and micro. So, we chop off six zeros and move from 1,000,000 to 1 (our decimal moved over six zeros). We update our units from picofarads to microfarad.

T5B12: Which is equal to 28400 kHz?

A. 28.400 kHz
B. 2.800 MHz
C. 284.00 MHz
D. 28.400 MHz

Moving from kHz to MHz (all answers are in MHz, except answer 'A,' which doesn't make sense because it's the same units) will decrease our number. There are 1,000 or 10^3 places between kilo and mega. So, we move 3 decimal places and go from 28,400 to 28.400 (our decimal moved over three spaces left). We update our units from kHz to MHz.

T5B13: Which is equal to 2425 MHz?

A. 0.002425 GHz
B. 24.25 GHz
C. 2.425 GHz
D. 2425 GHz

Look back to the base scale. Moving from MHz to GHz (all answers are in GHz) will decrease our big number out front. There are 1,000 or 10^3 places between mega and giga. Again, we moved three decimal places and went from 2,425 to 2.425 (our decimal moved over three spaces left). We update our units from MHz to GHz.

Our method emphasizing the "big numbers out front" will aid you in memorization and is a technique that works with the metric system. Although somewhat unconventional, it is effective.

In conclusion, always remember, "**G**iant **M**onsters **K**ick **B**ig **M**ountains **M**aking **N**ew **P**aths!"

Chapter 5

Electrical Principles

Introduction to Electrical Principles

ELECTRICITY AND ELECTRICAL PRINCIPLES form the backbone of amateur radio. Understanding these principles helps you pass the Technician Class license exam. It equips you with the skills to troubleshoot, repair, and optimize your radio equipment. This chapter will introduce you to the fundamentals of electricity, including voltage, current, resistance, and power, Ohm's Law, and the basics of electrical circuits. By grasping these concepts, you'll be better prepared to handle the technical aspects of ham radio, ensuring your station operates efficiently and effectively. Whether building a simple antenna or setting up a sophisticated communication system, a solid foundation in electrical principles is pivotal for your success in the amateur radio world.

Current

Current is a fundamental concept in electricity. It refers to the flow of electrons in an electric circuit, like the current or flow of the ocean, but with electrons, through a conductor, such as a wire. Measured in amperes (amps - A), current represents the rate at which electrons move past a specific point in a circuit.

You will see many examples of current compared to a watering hose. Let's get creative and use the analogy of a person running through a maze. Think of current as how fast our person is running through the maze (it's a dumb analogy, but that will help it stick!).

T5A03: What is the name for the flow of electrons in an electric circuit?

A. Voltage
B. Resistance
C. Capacitance
D. Current

T5A01: Electrical current is measured in which of the following units?

A. Volts
B. Watts
C. Ohms
D. Amperes

Electrical current is driven by voltage, which provides the necessary force to push the electrons. In our fanciful analogy, voltage is the force (the electrical zap!) making our runner run through the maze. The higher the 'zap,' or voltage, the faster our runner runs.

T5A05: What is the electrical term for the force that causes electron flow?

A. Voltage
B. Ampere-hours
C. Capacitance
D. Inductance

Different Types of Current

The primary current types are direct (DC) and alternating (AC).

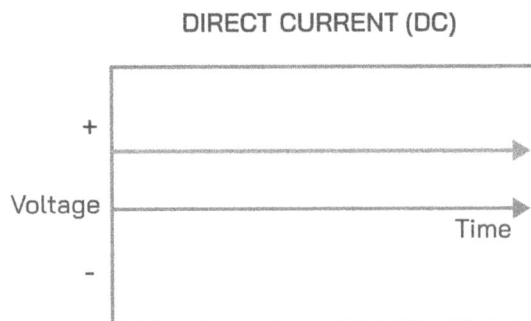

DIRECT CURRENT (DC)

The direction of the current of the voltage is always constant.

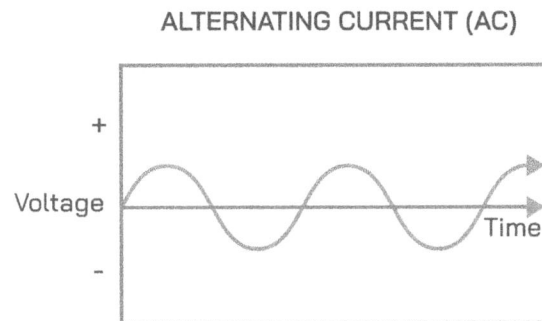

ALTERNATING CURRENT (AC)

The direction of the current is always switched periodically, and the voltage is also switched.

Direct Current (DC)

Direct current (DC) is characterized by the unidirectional flow of electric charge. In a DC circuit, electrons move in a single, consistent direction from the negative terminal to the positive terminal. This current type is typically produced by batteries, solar cells, and DC power supplies. DC is commonly used in low-voltage applications and electronic devices, including many components of amateur radio equipment. The steady voltage and current are ideal for powering transceivers, amplifiers, and other sensitive electronic circuits.

Alternating Current (AC)

Alternating current (AC) differs from DC in that the flow of electric charge periodically reverses direction. In an AC circuit, the voltage alternates between positive and negative values, causing the current to oscillate back and forth. This current type is generated by AC power sources such as power plants. It is delivered to homes and businesses through the electrical grid. AC is used for high-voltage applications, powering household appliances, and larger electronic devices. In amateur radio, AC is typically used to power base stations and other equipment that require a connection to the mains electricity supply.

T5A09: Which of the following describes alternating current?

A. current that alternates between a positive direction and zero
B. current that alternates between a negative direction and zero
C. Current that alternates between positive and negative directions
D. All these answers are correct

Electrical Resistance

Electrical resistance is <u>the property of a material that opposes the flow of electric current</u>, measured in <u>ohms (Ω)</u>. It is a crucial concept in electrical circuits, determining how much current flows for a given voltage. In our unique analogy about a runner in a maze, resistance can be thought of as material inside the maze—perhaps mud, snow, or wind. The material's resistance, the space inside our maze, slows our runner down.

Resistance can be influenced by various factors, including the material's composition, length, and cross-sectional area. Understanding resistance is essential for designing efficient circuits and selecting appropriate components in ham radio. High resistance can limit current flow, potentially leading to voltage drops and reduced performance. In contrast, low resistance ensures more efficient power transfer.

T5A04: What are the units of electrical resistance?

A. Siemens
B. Mhos
C. Ohms
D. Coulombs

Need help in remembering Ohms? I think about Ohms as a police officer saying, "Ohmmm, going to have to ask you to slow down." Which is resistance slowing our runner down.

Power

I need more power!

Power in electrical circuits is the rate at which <u>electrical energy is used</u> or transferred, measured in <u>watts (W)</u>. This concept is fundamental in ham radio, as it directly relates to the efficiency and performance of your equipment. How well can we get our runner through the maze?

Knowing your transmitter's power output in radio communications is necessary for effective signal propagation and compliance with regulatory limits.

T5A02: Electrical power is measured in which of the following units?

A. Volts
B. Watts
C. Watt-hours
D. Amperes

T5A10: Which term describes the rate at which electrical energy is used?

A. Resistance
B. Current
C. Power
D. Voltage

Decibels and the Power Ratio

Decibels (dB) are a logarithmic unit that measures the relative strength of signals, power levels, and gains in ham radio. They are invaluable for quantifying the performance of antennas, amplifiers, and other radio equipment. Using decibels, ham radio operators can easily compare signal strengths, assess the effectiveness of their setups, and make informed adjustments to improve communication quality.

Wait, what? What is logarithmic? A logarithmic scale is a way of showing numbers that grow very quickly by multiplying instead of adding. For example, on a regular scale, you count 1, 2, 3, and 4. But on a logarithmic scale, you jump by tens. If you started at 1, you would count 1, 10, 100, and 1000. This scale makes comparing very large or small numbers easier.

One common application of decibels in ham radio is measuring antenna gain. Antenna gain measures how well an antenna can focus energy in a particular direction. This measurement helps operators choose the suitable antenna for their needs and optimize its placement for better signal reception and transmission.

The Math Behind Decibels

I'm not going to lie; this is one of the most complex math parts for me. But let's work our way through it slowly.

The decibel is a logarithmic unit that expresses the ratio of two values, typically power levels. Because it is based on a logarithmic scale, a small change in decibels represents a significant change in the actual value. Memorize this formula.

1. **Power Ratio**:

$$dB = 10 \log_{10} \left(\frac{P_2}{P_1} \right)$$

In the decibel formula for power, P_1 (the initial power) is always in the denominator, and P_2 (the final power) is in the numerator. This is because decibels measure the ratio of the final power to the initial power, indicating how much the power has increased or decreased or how many times one power level is greater or smaller.

T5B09: Which decibel value most closely represents a power increase from 5 watts to 10 watts?

A. 2 dB
B. 3 dB
C. 5 dB
D. 10 dB

Solving the Decibel Increase from 5 Watts to 10 Watts

To determine the decibel (dB) increase when power goes from 5 watts to 10 watts, we use the power formula. Let's break it down step by step:

1. Identify the power values:

- Initial power (P_1): 5 watts

- Final power (P_2): 10 watts

2. Calculate the power ratio:

$$\frac{P_2}{P_1} = \frac{10}{5} = 2$$

3. Find the logarithm of the ratio:

$$\log_{10}(2) \approx 0.3$$

On your calculator, find the 'Log' key and press it. Then enter '2', and press the 'equal' key to get 0.301029957 or simply 0.3.

4. Multiply by 10 to find the decibel increase:

$$dB = 10 \times 0.3 = 3 \text{ dB}$$

So, a power increase from 5 to 10 watts is most closely represented by the rise of **3 dB**.

Quick Tip: Easy Way to Remember

A simple way to remember this is that doubling the power increases approximately 3 dB. So, when you see the power going from 5 watts to 10 watts (which is double), you can quickly recall that the increase is about 3 dB. The same is true if it's 20 watts to 40 or 25 watts to 50 watts; they are all 3 dB.

> **T5B10: Which decibel value most closely represents a power decrease from 12 watts to 3 watts?**
>
> A. -1 dB
> B. -3 dB
> **C. -6 dB**
> D. -9 dB

Work through the steps above.

1. Identify values:

   ○ Initial = 12 watts (the bottom number)

   ○ Final = 3 watts (the top number)

2. Divide the two numbers: Final power divided by initial power or (3 watts divided by 12 watts)

- ◦ Equals 0.25

3. Find the Log ratio: Press the Log key and then 0.25; press enter.

- ◦ Equals -0.602059913 or call it -0.6.

4. Multiply by 10

- ◦ -0.6 times 10 = **-6 dB**

One more time:

T5B11: Which decibel value represents a power increase from 20 watts to 200 watts?

A. 10 dB
B. 12 dB
C. 18 dB
D. 28 dB

Work through steps 1-4 again. The ending power is on top, and the starting power is on bottom. Do you get 10 dB?

Understanding and using decibels allows ham radio operators to measure and optimize their equipment and signal quality effectively. The logarithmic nature of decibels simplifies calculations and comparisons, making it an essential tool in the technical aspects of amateur radio.

You will want to memorize this formula for the exam.

Frequency

Frequency, measured in hertz (Hz), refers to the number of complete cycles an alternating current completes per second. Each cycle has one complete wave, including a positive and negative half-cycle. The frequency of an AC signal determines how many times the current changes direction in one second. The unit of frequency is hertz (Hz), with one hertz equaling one cycle per second. Standard frequencies for household AC power are 60 Hz.

Our memorization technique: our runner in the maze has to change direction, back and forth, running this way and then that way. How frequently he changes direction is measured in hertz... because it hurts to have to change direction so frequently (see how nicely both hertz and frequency get memorized together)! And remember, this is a simple and silly memorization technique, that's it. Don't take it too seriously. But silly helps things stick in memory.

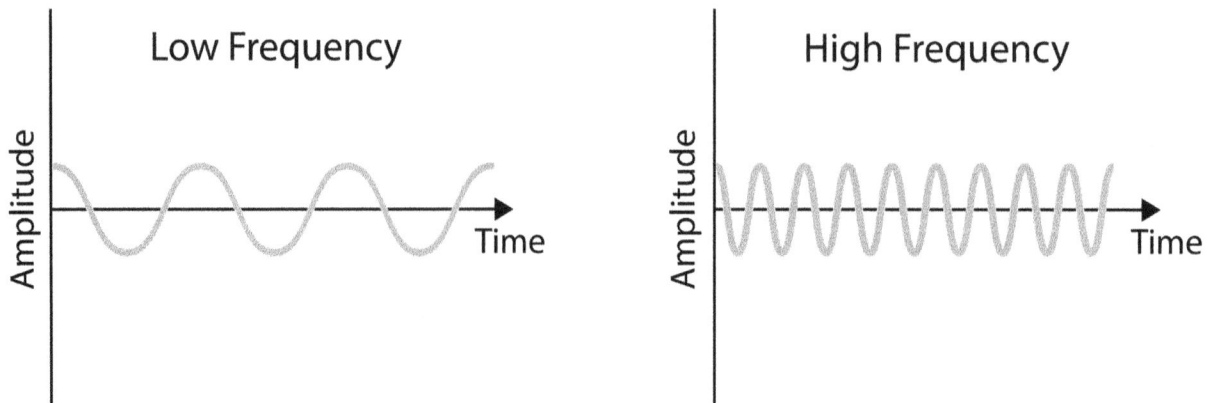

Low Frequency — Amplitude / Time

High Frequency — Amplitude / Time

T5A12: What describes the number of times per second that an alternating current makes a complete cycle?

A. Pulse rate
B. Speed
C. Wavelength
D. Frequency

T5A06: What is the unit of frequency?

A. Hertz
B. Henry
C. Farad
D. Tesla

Importance of Frequency in Ham Radio

Understanding AC frequency is fundamental in ham radio for several reasons. First, radio signals are high-frequency AC signals transmitted through the air. The frequency of these signals determines their wavelength (more on this later) and propagation characteristics. Second, many radio transceivers and other equipment use AC power from the electrical grid. Hence, knowledge of the local frequency standard is essential for proper operation and compatibility.

Understanding the terminology is one of the most essential things about ham radio. So, let's start with an easy one: RF = Radio Frequency. Easy enough!

T5C06: What does the abbreviation "RF" mean?

A. Radio frequency signals of all types
B. The resonant frequency of a tuned circuit
C. The real frequency transmitted as opposed to the apparent frequency
D. Reflective force in antenna transmission lines

Furthermore, radio communications often involve modulating a carrier AC signal to encode information. Understanding how frequency affects modulation techniques (such as AM, FM, and SSB) is essential for effective communication (again more later).

Conductors and Insulators

Through what do electrons flow?

Conductors

Conductors are materials that allow the easy flow of electric current due to their low electrical resistance. They are characterized by having free electrons that can move easily from one atom to another when an electric potential is applied. This property makes conductors ideal for use in electrical circuits and wiring.

Common examples of conductors include metals such as copper, aluminum, silver, and gold, with copper being widely used in electrical wiring due to its high conductivity and relative affordability. Graphite and saltwater are also conductors, although metals are more efficient.

T5A07: Why are metals generally good conductors of electricity?

A. They have relatively high density
B. They have many free electrons
C. They have many free protons
D. All these choices are correct

Insulators

Insulators, on the other hand, are materials that resist the flow of electric current due to their high electrical resistance. Their electrons are tightly bound to their atoms, preventing free movement. Insulators are fundamental in preventing unintended current flow and protecting against electric shocks.

Remember our previous section about the police officer saying, "Ohmmmm, going to slow you down?" Resistance opposes the flow of all current types (DC, AC, and RF).

T5A11: What type of current flow is opposed by resistance?

A. Direct current
B. Alternating current
C. RF current
D. All these choices are correct

Common examples of insulators include plastics, rubber, glass, and ceramics. Plastics are commonly used to insulate electrical wires and components. In contrast, rubber is often used in protective gear and insulating covers for electrical tools and wires. Glass and ceramics are used in high-voltage applications and as insulators in radio equipment.

In ham radio, insulators isolate different circuit parts, preventing short circuits and protecting sensitive components from high voltages. They are also used in antenna systems to separate the conductive elements from supporting structures, ensuring efficient signal transmission without losses due to unintended grounding.

T5A08: Which of the following is a good electrical insulator?

A. Copper
B. Glass
C. Aluminum
D. Mercury

Proper selection and use of these materials ensures the reliability and safety of radio equipment. Conductors create efficient pathways for electrical signals, while insulators prevent unwanted current flow and protect the operator and the equipment.

Circuit Types: Parallel vs. Series

Circuits are the pathways through which electric current flows. Understanding the different types of circuits—particularly parallel and series circuits—is a fundamental electrical concept. Each type of circuit has its unique characteristics and applications.

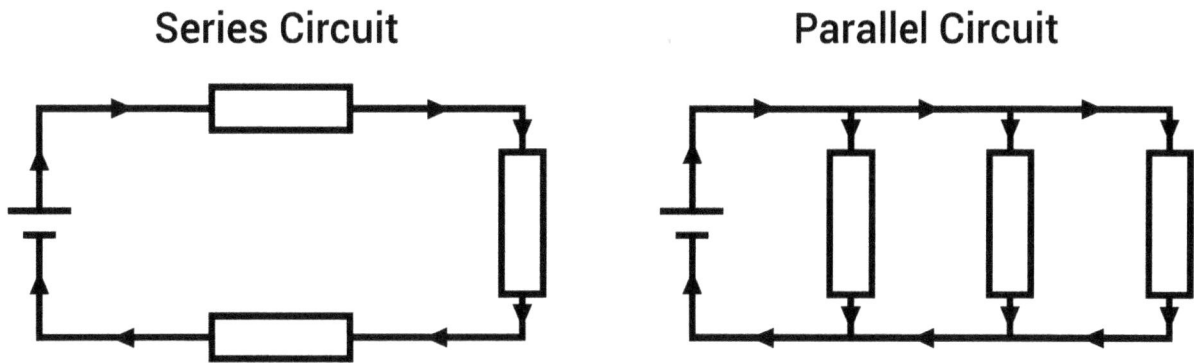

Series Circuit   **Parallel Circuit**

Series Circuits

In a series circuit, components are connected end-to-end in a single path for the current to flow. This means that the same current passes through each component one after another. Here are some points about series circuits:

1. Current: The current (I) is the same through all components in series. Because there is only one path to take.

2. Voltage: The total voltage (V) is the sum of the voltages across each component. If you have three resistors in series, the total voltage is the sum of the voltage drops across each resistor.

3. Resistance: The total resistance (R) is the sum of the individual resistances. Adding more resistors in series increases the total resistance.

T5D13: In which type of circuit is DC current the same through all components?

A. Series
B. Parallel
C. Resonant
D. Branch

Don't let the word 'DC current' in the question throw you off. It's an example of a "filler word" that we talked about back in chapter 3.

Parallel Circuits

In a parallel circuit, components are connected across the same two points, creating multiple paths for the current to flow. Here are some points about parallel circuits:

1. Current: The total current (I) is the sum of the currents through each parallel branch. Different branches can have different currents depending on their resistance.

2. Voltage: The voltage (V) across each component is the same and equal to the source's total voltage. Everywhere gets "zapped" the same amount.

3. Resistance: The total resistance (R) of a parallel circuit is <u>less than the resistance of the smallest</u> individual resistor.

T5D14: In which type of circuit is voltage the same across all components?

A. Series
B. Parallel
C. Resonant
D. Branch

Practical Applications

Series Circuits: Series circuits are often used in applications where the same current must pass through all components, such as string lights or certain types of sensors.

Parallel Circuits: Parallel circuits are used in applications where components must operate independently, such as in-home electrical wiring, where each appliance gets the same voltage but can draw different currents.

Understanding the differences between series and parallel circuits is key to designing effective and efficient electronic systems. Mastering these concepts will better equip you to build, troubleshoot, and optimize your ham radio equipment.

Ohm's Law

Ohm's Law is one of the fundamental principles of electronics and ham radio, and it forms the foundation for understanding how electrical circuits work. This Law describes the relationship between voltage (V), current (I), and resistance (R) in a simple and intuitive way. Think of it as the "rule of three" for electricity. You can easily calculate the third if you know two of these quantities.

The Formula

Ohm's Law is expressed by the formula:

$$E = I \times R$$

Where:

- **E** (voltage) is the electrical potential difference measured in volts (V).

  - 'E' what? It goes back to when and where things were named. Skipping the history lesson, remember that E = voltage.

- **I** (current) is the flow of electric charge measured in amperes (A).

  - 'I'? Again, it is a fascinating historical story, but it is unnecessary for the exam. Just remember that I = current.

- **R** (resistance) opposes the current flow measured in ohms (Ω).

  - Finally, an easy one to remember!

Breaking It Down

To understand this better, let's use the ever-present water analogy. Imagine a garden hose:

- **Voltage (E)** is like the water pressure in the hose. Higher pressure pushes more water through.

- **Current (I)** is the amount of water flowing through the hose. More water flow means a higher current.

- **Resistance (R)** is like the width of the hose. A narrow hose (higher resistance) makes water flow harder, while a wide hose (lower resistance) allows more water to pass through easily.

The Ohm's Law Triangle is a simple tool for visualizing how the three interact and is useful for memorization on the exam. Start with the "E" on top and then go alphabetically counterclockwise- "I", then "R".

Using the Triangle: If you want to find voltage (E), cover up the "E" in the triangle. You're left with "I" next to "R," which means Voltage = Current × Resistance (E = I × R).

- To find current (I), cover up the "I". You'll see "E" over "R," which means Current = Voltage ÷ Resistance (I = E ÷ R).

- To find resistance (R), cover up the "R". You'll see "E" over "I", which means Resistance = Voltage ÷ Current (R = E ÷ I).

This simple triangle allows you to quickly solve problems involving voltage, current, and resistance without having to memorize each formula separately.

Write out the triangle on a piece of paper. This will help with memorization. Then, answer the three exam questions, which involve rewriting the formula to solve for each measurement.

T5D01: What formula is used to calculate current in a circuit?

A. I = E * R
B. I = E / R
C. I = E + R
D. I = E - R

Draw the triangle. To find current (I), cover up the "I". You'll see "E" over "R", which means Current = Voltage ÷ Resistance (I = E ÷ R).

T5D02: What formula is used to calculate voltage in a circuit?

A. E= I x R
B. E= I / R
C. E= I + R
D. E = I - R

If you want to find voltage (E), cover up the "E" in the triangle. You're left with "I" next to "R", which means Voltage = Current × Resistance (E = I × R).

T5D03: What formula is used to calculate resistance in a circuit?

A. R = E x I
B. R = E / I
C. R = E + I
D. R = E - I

To find resistance (R), cover up the "R". You'll see "E" over "I", which means Resistance = Voltage ÷ Current (R = E ÷ I).

Now, let's try using some numbers.

T5D04: What is the resistance of a circuit in which a current of 3 amperes flows when connected to 90 volts?

A. 3 ohms
B. 30 ohms
C. 93 ohms
D. 270 ohms

The Easy 4-Step Approach

Step 1: look at the values we have been given. 90 volts and 3 amps.

Step 2: fill in the triangle.

- E = 90 volts

- I = 3 amps

Step 3: do the math.

- 90 divided by 3 = 30

- Since we are solving for resistance (R), we have 30 ohms!

The following problems are part of the exam pool. Use the triangle and solve for the missing measurement.

Remember, what are they asking for? What measurement: resistance (R), current (I), or voltage (E)?

Create your triangle on a piece of paper and then fill in the numbers from the question. With your triangle, you should easily complete the following exam questions.

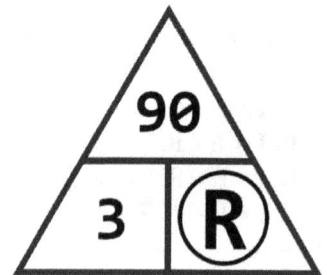

Voltage	Current	Resistance
E = I.R	I = E/R	R = E /I

T5D05: What is the resistance of a circuit for which the applied voltage is 12 volts, and the current flow is 1.5 amperes?

A. 18 ohms
B. 0.125 ohms
C. 8 ohms
D. 13.5 ohms

Resistance = Voltage ÷ Current (R = E ÷ I) = 12 volts divided by 1.5 amps = 8 ohms of resistance

T5D06: What is the resistance of a circuit that draws 4 amperes from a 12-volt source?

A. 3 ohms
B. 16 ohms
C. 48 ohms
D. 8 ohms

Resistance = Voltage ÷ Current (R = E ÷ I) = 12 volts divided by 4 amps = 3 ohms

T5D07: What is the current in a circuit with an applied voltage of 120 volts and a resistance of 80 ohms?

A. 9600 amperes
B. 200 amperes
C. 0.667 amperes
D. 1.5 amperes

Current = Voltage ÷ Resistance (I = E ÷ R) = 120 volts divided by 80 ohms = 1.5 amps of current

T5D08: What is the current through a 100-ohm resistor connected across 200 volts?

A. 20,000 amperes
B. 0.5 amperes
C. 2 amperes
D. 100 amperes

Current = Voltage ÷ Resistance (I = E ÷ R) = 200 volts divided by 100 ohms = 2 amps

T5D09: What is the current through a 24-ohm resistor connected across 240 volts?

A. 24,000 amperes
B. 0.1 amperes
C. 10 amperes
D. 216 amperes

Current = Voltage ÷ Resistance (I = E ÷ R) = 240 volts divided by 24 ohms = 10 amps

T5D10: What is the voltage across a 2-ohm resistor if a current of 0.5 amperes flows through it?

A. 1 volt
B. 0.25 volts
C. 2.5 volts
D. 1.5 volts

Voltage = Current × Resistance (E = I × R) = 0.5 amps multiplied by 2 ohms = 1 volt

T5D11: What is the voltage across a 10-ohm resistor if a current of 1 ampere flows through it?

A. 1 volt
B. 10 volts
C. 11 volts
D. 9 volts

Voltage = Current × Resistance (E = I × R) = 1 amp multiplied by 10 ohms = 10 volts

T5D12: What is the voltage across a 10-ohm resistor if a current of 2 amperes flows through it?

A. 8 volts
B. 0.2 volts
C. 12 volts
D. 20 volts

Voltage = Current × Resistance (E = I × R) = 2 amps multiplied by 10 ohms = 20 volts

That's a lot of math! Just remember the Ohm Triangle, and you will be able to move between Voltage (E), Current (I), and Resistance (R).

The Power Pyramid

Just when you thought it was safe to move on, we have one more triangle for you. The triangle of power!

In addition to Ohm's Law, the power formula is another essential formula. This formula helps you calculate the electrical power in a circuit, which is key to understanding how much energy your devices use or generate. The power formula relates power (P) to current (I) and voltage (E).

The Formula

The power formula is expressed as:

$$P = I \times E$$

Where:

- **P** (power) is the amount of electrical energy transferred or consumed per unit of time, measured in watts (W). And yes! P for power is easy!

- **I** (current) is the flow of electric charge, measured in amperes (A).

- **E** (voltage) is the electrical potential difference, measured in volts (V).

Remember, there is Power in PIE!

Why It Matters in Ham Radio

Understanding the power formula is essential for ham radio operators because it helps manage your equipment's energy consumption and output. Knowing how much power your transmitter uses or how much power your antenna can handle ensures that your setup operates efficiently and safely. It also helps you make informed decisions when choosing power supplies and other components.

To remember the order of the pyramid, spell P-I-E counterclockwise—the power of pie.

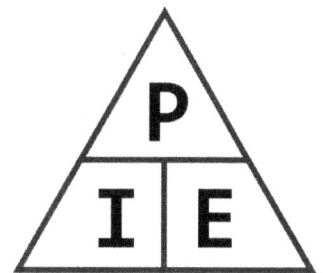

T5C08: What is the formula used to calculate electrical power (P) in a DC circuit?

A. P = E * I
B. P = E / I
C. P = E - I
D. P = E + I

The Power Triangle works the same way. Simply know which values you have and the ones for which you are solving. Plug them into the pyramid and solve. Let us practice!

Power	Current	Voltage
P = E × I	I = P / E	E = P / I

T5C09: How much power is delivered by a voltage of 13.8 volts DC and a current of 10 amperes?

A. 138 watts
B. 0.7 watts
C. 23.8 watts
D. 3.8 watts

Power = Voltage × Current (P = V × I) = 13.8 volts multiplied by 10 amps = 138 watts of power. Here is another example of DC just being filler for this question.

T5C10: How much power is delivered by a voltage of 12 volts DC and a current of 2.5 amperes?

A. 4.8 watts
B. 30 watts
C. 14.5 watts
D. 0.208 watts

Power = Voltage × Current (P = V × I) = 12 volts multiplied by 2.5 amps = 30 watts of power

T5C11: How much current is required to deliver 120 watts at a voltage of 12 volts DC?

A. 0.1 amperes
B. 10 amperes
C. 12 amperes
D. 132 amperes

Current = Power ÷ Voltage (I = P ÷ V) = 120 watts of power divided by 12 volts = 10 amps

Having any trouble getting the correct answer? Go back to the Easy 4-Step Approach and follow each one. The only difference between the Ohm and Power pyramids is the units of measurement. The formula and math work the same.

Chapter 6

The Control of Electrical Principles

Capacitance

CAPACITANCE IS A FUNDAMENTAL electrical property that describes a component or circuit's ability to store and release electrical energy. It is measured in farads (F). Capacitance is typically found in capacitors, essential components in many electronic circuits, including ham radios. Think about, "I have the capacity to store energy in an ELECTRIC field!" That electric piece is important.

A capacitor consists of two conductive plates separated by an insulating material called a dielectric (more on this in the physical radio station section). When a voltage is applied across the plates, an electric field is created, allowing the capacitor to store energy as an electrostatic charge.

This stored energy can be released when the circuit requires it, helping to smooth out fluctuations in voltage and maintain stable operation. In ham radio, capacitors are used for various purposes, such as filtering signals, tuning antennas, and coupling or decoupling different circuit stages.

T5C01: What describes the ability to store energy in an electric field?

A. Inductance
B. Resistance
C. Tolerance
D. Capacitance

I don't have a good way to remember this one, so just memorize it!

T5C02: What is the unit of capacitance?

A. The farad
B. The ohm
C. The volt
D. The henry

Inductance

Inductance is the electrical property that describes the ability of a component or circuit to store energy in a magnetic field when an electric current flows through it. That seems like capacitance! So, the key here is that inductance refers to the MAGNETIC field.

Inductance is measured in Henry's (H). It is primarily associated with inductors, which are components made by winding a coil of wire around a core. When current flows through the coil, it creates a magnetic field, storing energy within the inductor. This stored energy can then be released when the current changes, providing a means to regulate current flow and filter signals.

In ham radio, inductors play an important role in various applications, such as tuning circuits, filtering out unwanted frequencies, and forming part of oscillators and transformers. Understanding inductance is essential for designing and troubleshooting radio equipment, as it helps ensure that circuits operate efficiently and effectively. By mastering the principles of inductance, amateur radio operators can enhance their ability to create and maintain stable, high-performing communication systems.

T5C03: What describes the ability to store energy in a magnetic field?

A. Admittance
B. Capacitance
C. Resistance
D. Inductance

T5C04: What is the unit of inductance?

A. The coulomb
B. The farad
C. The henry
D. The ohm

So, who is still confused? One more look at the difference between capacitance and inductance.

Capacitance and inductance are fundamental electrical properties but operate based on different principles and serve distinct circuit functions.

Capacitance is storing and releasing electrical energy in an electrostatic field. Capacitance, measured in farads (F), helps smooth out voltage fluctuations, filter signals, and couple or decouple circuit stages.

Inductance is the ability to store and release energy in a magnetic field. Measured in henrys (H), it regulates current flow, filters unwanted frequencies, and creates oscillators and transformers.

In essence, capacitors oppose changes in voltage by storing and releasing energy quickly. In contrast, inductors oppose changes in current by generating a counteracting voltage. Understanding these differences is crucial for designing and optimizing circuits, giving us control over electricity in our radios.

Memorization technique.

We have four ideas here: 1) **C**apacitance, 2) **I**nductance, 3) **E**lectric Field, and 4) **M**agnetic Field. Lucky for us, if we put the words in alphabetical order, partners line up. **C – E – I – M**

- **C**apacitance = **E**lectric Field

- **I**nductance = **M**agnetic Field

Impedance

Impedance is <u>a circuit's total opposition to the flow of alternating current</u> (AC).

It includes resistance (which opposes the flow of direct current) and reactance (which comes from capacitors and inductors in the circuit).

What was that?

Impedance affects how much current flows for a given AC voltage. <u>Like resistance, it's measured in ohms (Ω).</u> Any time our cop slows us down, it's an ohm.

Matching impedance between components, like a transmitter and antenna, is crucial in ham radio. Proper impedance matching ensures efficient power transfer and reduces signal loss, improving radio equipment performance. Understanding impedance helps design, tune, and troubleshoot your ham radio setup.

T5C12: What is impedance?

A. The opposition to AC current flow
B. The inverse of resistance
C. The Q or Quality Factor of a component
D. The power handling capability of a component

T5C05: What is the unit of impedance?

A. The volt
B. The ampere
C. The coulomb
D. The ohm

Understanding the electrical principles of current, resistance, power, capacitance, inductance, and impedance is fundamental for any aspiring ham radio operator. These concepts form the backbone of radio equipment operating and interacting with the surrounding environment.

By grasping these principles, you can effectively design, build, and troubleshoot circuits, ensuring your equipment performs at its best. Whether you're smoothing out voltage fluctuations with capacitors, managing current flow with inductors, or ensuring optimal power transfer through impedance matching, a solid foundation in these electrical principles will enhance your skills and confidence as a ham radio operator.

Mastery of these topics further prepares you for the Technician Class license exam.

Chapter 7

Radio Waves

IMAGINE STANDING ON A cliff overlooking the ocean, where the waves travel tirelessly from the horizon to the shore. This relentless movement of water is much like how radio waves traverse through space, an invisible force carrying voices and data across vast distances. In this chapter, we'll decode the mystique of radio waves and their propagation, transforming the seemingly complex into something understandable. By grasping these fundamentals, you'll be equipped to pass your exam and engage effectively in amateur radio, making connections that were once beyond reach.

Physics of Radio Waves

Radio waves are electromagnetic radiation, akin to the light you see with your eyes but with much longer wavelengths and lower frequencies. They are part of the electromagnetic spectrum, which includes other types of waves such as microwaves, infrared radiation, ultraviolet light, X-rays, and gamma rays.

These waves are generated by moving electric charges, typically in a piece of metal called an antenna. When an electric current flows through an antenna, it creates an electromagnetic field that radiates outwards as radio waves. Depending on their frequency, the length of these waves can range from about one millimeter to over 100 kilometers.

The frequency of radio waves, measured in hertz (Hz), is the number of times the wave oscillates per second. Radio frequencies used in ham radio range from 30 kHz (kilohertz) to 300 GHz (gigahertz), covering a broad spectrum of wavelengths.

When you speak into a ham radio's microphone, your voice is converted into electrical signals. These signals are then used to alter (modulate) a radio wave generated by the transmitter. The modulated wave travels through the air (and even the vacuum of space) at the <u>speed of light – 300,000 kilometers per second</u>, carrying your message.

> **T3B04: What is the velocity of a radio wave traveling through free space?**
>
> **A. speed of light**
> B. speed of sound
> C. Speed inversely proportional to its wavelength
> D. Speed that increases as the frequency increases

For comparison, the Earth's circumference at the equator is 40,075 kilometers. Radio waves travel at the speed of light, which is why communication is instantaneous anywhere in the world.

T3B11: What is the approximate velocity of a radio wave in free space?

A. 150,000 meters per second
B. 300,000,000 meters per second
C. 300,000,000 miles per hour
D. 150,000 miles per hour

For this question, remember that we are in the metric system, which rules out a few answers. Also, we saw the speed of light written in kilometers above. Now, we have to convert kilometers to meters. Remembering our chapter on the metric system, we move to the right (the big number out front gets bigger) by 1,000 or three decimal places.

The speed of light changes from 300,000 kilometers per second to 300,000,000 meters per second.

Components of Radio Waves:

- **Electric Field (E-field)**: This component oscillates <u>perpendicular (at a right angle)</u> to the wave's direction of travel. <u>The electric field's strength and orientation determine the radio wave's polarization.</u>

- **Magnetic Field (H-field)**: This component oscillates <u>perpendicular (at a right angle)</u> to both the direction of travel and the electric field. The magnetic field is always in phase with the electric field.

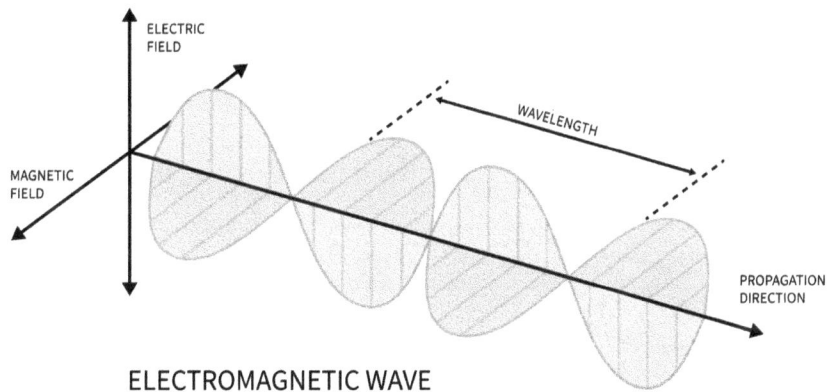

ELECTROMAGNETIC WAVE

T3B03: What are the two components of a radio wave?

A. impedance and reactance
B. voltage and current
C. Electric and magnetic fields
D. Ionizing and non-ionizing radiation

T3B01: What is the relationship between the electric and magnetic fields of an electromagnetic wave?

A. They travel at different speeds
B. They are in parallel
C. They revolve in opposite directions
D. They are at right angles

The property of a radio wave that defines its polarization is the <u>orientation of its electric field (E-field).</u> Polarization refers to the direction in which the electric field oscillates as the radio wave travels through space.

T3B02: What property of a radio wave defines its polarization?

A. The orientation of the electric field
B. The orientation of the magnetic field
C. The ratio of the energy in the magnetic field to the energy in the electric field
D. The ratio of the velocity to the wavelength

Polarization

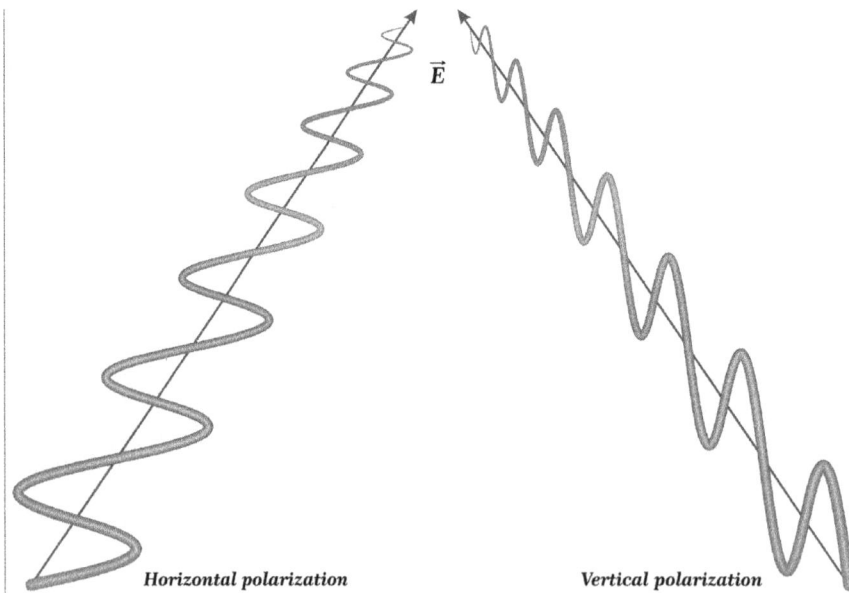

Horizontal polarization   *Vertical polarization*

- **Vertical Polarization**:

  - In vertically polarized waves, the electric field oscillates in a vertical direction.

  - These waves are commonly used in mobile and handheld radio communications because vertically oriented antennas are easy to mount and use.

- **Horizontal Polarization**:

  - In horizontally polarized waves, the electric field oscillates in a horizontal direction.

  - This type of polarization is often used in television broadcasting and some forms of long-distance communication, where horizontal antennas can be more practical.

- **Circular Polarization**:

  - Circularly polarized waves have an electric field that rotates in a circular motion as they propagate. This can be either right-hand circular polarization (RHCP) or left-hand circular polarization (LHCP).

  - Circular polarization is used in satellite and space communications because it can reduce the effects of signal fading due to rotation or orientation changes of the transmitting and receiving antennas.

- **Elliptical Polarization**:

  - Elliptically polarized waves have an electric field that traces an ellipse as it propagates. This more general form of polarization includes linear and circular polarization.

Matching the polarization of the transmitting and receiving antennas is crucial for efficient communication. If the polarizations are mismatched, signal strength can be significantly reduced, leading to poor reception and communication quality. For example, a vertically polarized antenna will not effectively receive a horizontally polarized signal.

The orientation of a radio wave's electric field defines its polarization. Understanding and matching the polarization of antennas in radio communication systems is essential for maximizing signal strength and communication efficiency.

Memorization technique.

Think about holding a "walkie-talkie." I know, I know, but don't hurt me. It's just for memorization. You hold a walkie-talkie straight up and down... or vertically. The antenna is vertical. Meaning it is vertically polarized. Suppose you turned the walkie-talkie sideways. The antenna is now parallel to the ground and so horizontally polarized.

Remember that, and you have this polarization stuff down!

Chapter 8

Frequency Bands

IN REAL-WORLD APPLICATIONS, THE choice of frequency band can dramatically affect the success and scope of communication.

Each band, with its unique propagation characteristics, opens up different aspects of the hobby to explore, from local emergency preparedness networks to worldwide digital communication networks, each offering its own set of challenges and rewards.

Look at some of those characteristics and relate them to real-world usage.

Short-Hand Name	VLF	LF	MF	HF	VHF	UHF	SHF	EHF
Name	Very Low Frequency	Low Frequency	Medium Frequency	High Frequency	Very High Frequency	Ultra High Frequency	Super High Frequency	Extremely High Frequency
Frequency	3 – 30 kHz	30 – 300 kHz	300 – 3,000 kHz	3 MHz – 30 MHz	30 – 300 MHz	300 – 3,000 MHz	3 GHz – 30 GHz	30 – 300 GHz
Wavelength	100 – 10 km	10 – 1 km	1,000 m – 100 m	100 – 10 m	10 – 1 m	1 – 0.1 m	10 cm – 1 cm	1 – 0.1 cm

Frequency Band Table

Characteristics of Radio Bands

VLF (Very Low Frequency)

- Propagation: Ground waves capable of long-distance communication.

- Uses: Submarine communication, navigation beacons.

- Penetration: Good penetration through water and Earth.

- Antenna Size: Huge antennas are needed due to long wavelengths.

LF (Low Frequency)

- Propagation: Ground waves are reliable over long distances.

- Uses: Long-range navigation, time signals, AM broadcasting.

- Antenna Size: Large antennas are required.

- Penetration: Good ground penetration, useful for geophysical surveys.

MF (Medium Frequency)

- Propagation: Ground waves during the day, skywaves at night.

- Uses: AM radio broadcasting, maritime communication.

- Antenna Size: Medium-sized antennas.

- Night-time Reach: Better long-distance propagation at night due to skywave reflections.

HF (High Frequency)

- Propagation: Skywave (ionospheric) propagation is capable of global communication.

- Uses: Shortwave radio, amateur radio, international broadcasting, military communication.

- Antenna Size: Practical antenna sizes for home and portable use.

- Day/Night Variability: Propagation conditions vary with time of day and solar activity.

VHF (Very High Frequency)

- Propagation: Line-of-sight, limited by the horizon.

- Uses: FM radio, television broadcasting, air traffic control, amateur radio.

- Antenna Size: Relatively small antennas.

- Penetration: Good penetration through buildings, suitable for mobile communication.

UHF (Ultra High Frequency)

- Propagation: Line-of-sight, affected by obstacles.

- Uses: Television broadcasting, mobile phones, Wi-Fi, Bluetooth, amateur radio.

- Antenna Size: Very small antennas.

- Penetration: Limited building penetration, but suitable for urban environments with many repeaters.

SHF (Super High Frequency)

- Propagation: Line-of-sight, very sensitive to obstacles and atmospheric conditions.

- Uses: Radar, satellite communication, microwave links, Wi-Fi.

- Antenna Size: Tiny, often parabolic dishes.

- Atmospheric Effects: Affected by rain, fog, and other atmospheric conditions.

EHF (Extremely High Frequency)

- Propagation: Line-of-sight, very short range.

- Uses: Advanced radar, experimental communication systems, scientific research.

- Antenna Size: Tiny antennas.

- Atmospheric Effects: Highly affected by atmospheric absorption, limited practical use in open air.

Technician Band Access

As a Technician class licensee, you will have full access to the UHF (Ultra High Frequency) and VHF (Very High Frequency) bands allocated for amateur radio use in the United States. These bands are ideal for local and regional communication.

In this section, we will explore the characteristics of VHF and UHF bands, including their propagation properties, typical uses, and practical tips to help you make the most of your operating privileges. Understanding these aspects will enhance your ability to communicate effectively, enjoy the full benefits of your technician license, and pass your exam.

Insights into VHF and UHF for Technician Licensees

VHF (Very High Frequency)

- **Frequency Range**: 30 to 300 MHz

- **Wavelength**: 10 meters to 1 meter

- **Characteristics**:

  - **Propagation**: VHF signals primarily travel by line-of-sight, which means they usually do not bend around obstacles or follow the curvature of the Earth. However, they can reflect off buildings, hills, and other large objects, extending their range slightly beyond the horizon.

  - **Penetration**: VHF signals can penetrate through foliage and buildings better than UHF signals, making them suitable for mobile and portable communication.

  - **Typical Uses**: FM radio, television broadcasting (channels 2-13), public service communications (police and fire departments), marine communications, and amateur radio.

- **Advantages**:

  - **Range**: This is effective for medium-range communication, typically up to 30 miles, depending on the terrain and antenna height.

  - **Less Interference**: VHF is less affected by man-made electrical noise than lower frequency bands.

  - **Antenna Size**: VHF antennas are generally manageable, making them suitable for home use and mobile installations.

- **Disadvantages**:

  - **Obstructions**: Large buildings, hills, and other obstructions can significantly reduce signal strength.

○ **Bandwidth**: Less bandwidth compared to UHF, which can limit the number of available channels.

UHF (Ultra High Frequency)

- **Frequency Range:** 300 MHz to 3 GHz

- **Wavelength:** 1 meter to 0.1 meter

- **Characteristics:**

  ○ **Propagation:** UHF signals also travel by line-of-sight but are more affected by physical obstructions than VHF signals. They tend to have a shorter range but can penetrate through buildings better than VHF signals.

  ○ **Penetration:** UHF signals can pass through walls and buildings more effectively than VHF, making them ideal for indoor communication and urban environments.

  ○ **Typical Uses:** Television broadcasting (channels 14-83), mobile phones, Wi-Fi, Bluetooth, GPS, and amateur radio.

- **Advantages:**

  ○ **Bandwidth:** UHF offers more bandwidth, allowing for more channels and higher data transmission rates.

  ○ **Antenna Size:** UHF antennas are smaller and more compact, making them easy to install in confined spaces.

  ○ **Urban Use:** More effective for use in densely populated urban areas with many buildings and obstacles.

- **Disadvantages:**

  ○ **Range:** UHF typically has a shorter range than VHF due to its higher frequencies and greater susceptibility to obstacles.

  ○ **Interference:** More prone to interference from other electronic devices and appliances.

Practical Applications for Technicians

VHF in Ham Radio:

- **2-Meter Band**: This is the most popular VHF band for amateur radio, ranging from 144 to 148 MHz. It is widely used for local communication, repeater operation, and emergency services.

- **Simplex and Repeater Operation**: Technician licensees can use VHF frequencies for simplex (direct radio-to-radio) communication and through repeaters, which extend the communication range.

UHF in Ham Radio:

- **70-centimeter Band**: This band ranges from 420 to 450 MHz and is another popular amateur radio band. It is ideal for local communication, especially in urban areas where building penetration is essential.

- **Linked Repeaters**: UHF repeaters are often linked to other repeaters via the internet or other means, allowing for long-distance communication through a network of repeaters.

Tips for Using VHF and UHF:

- **Antennas**: Use high-gain antennas to improve signal strength and range. Vertical antennas are common for VHF for mobile and base stations. In contrast, smaller and more directional antennas are effective for UHF.

- **Line-of-Sight**: Ensure that antennas are placed as high as possible and in a clear line of sight to minimize obstructions and maximize range.

- **Repeaters**: Utilize local repeaters to extend your communication range. Familiarize yourself with the location and frequencies of repeaters in your area.

Understanding the characteristics and best practices for using VHF and UHF bands will help Technician licensees maximize their privileges and enhance their communication capabilities in the amateur radio community.

Let's tackle some exam questions.

T3A02: What is the effect of vegetation on UHF and microwave signals?

A. Knife-edge diffraction
B. Absorption
C. Amplification
D. Polarization rotation

When dealing with UHF (Ultra-High Frequency) and microwave signals, one key factor to consider is how vegetation, like trees and shrubs, can impact these signals. The main effect here is absorption. Unlike lower-frequency signals, UHF and microwave signals have shorter wavelengths, making them more susceptible to being absorbed by water-containing materials, like the leaves and branches of plants.

Think of it like this: when UHF and microwave signals pass through vegetation, the plants absorb part of the signal's energy, weakening it. This is why you might experience reduced signal strength or even signal loss when operating in areas with dense foliage.

T3A12: What is the effect of fog and rain on signals in the 10-meter and 6-meter bands?

A. Absorption
B. There is little effect
C. Deflection
D. Range increase

Unlike higher frequency signals, which can be heavily impacted by moisture in the air, the signals in the 10-meter and 6-meter bands are largely unaffected by fog and rain.

This minimal impact occurs because the wavelengths of these bands are relatively long compared to the size of raindrops and fog droplets. These water particles are simply too small to significantly scatter or absorb the radio waves in these bands. So, while you might notice some interference in other situations, you can generally expect clear communication on these frequencies, regardless of the weather.

But:

T3A07: What weather condition might decrease range at microwave frequencies?

A. High winds
B. Low barometric pressure
C. Precipitation
D. Colder temperatures

When working with microwave frequencies, one important factor to keep in mind is how weather can affect your signal's range. Precipitation, such as rain, snow, or sleet, can significantly reduce the range of microwave signals. Because of their high frequency and short wavelength, microwave signals are particularly sensitive to weather conditions. When it rains or snows, the tiny water droplets or ice particles in the air interact with these signals in two main ways: absorption and scattering.

Summary

In summary, for ham radio operators with a Technician license, several bands are particularly popular due to their accessibility and the range of activities they support. The 2-meter and 70-centimeter bands are the most popular for everyday use, while the 6-meter and 10-meter bands offer opportunities for more specialized and long-distance communications.

10-Meter Band (HF)

- **Wavelength**: Approximately 10 meters

- **Frequency Range**: 28.0–28.5 MHz (Technician operators have limited privileges in the 28.0–28.5 MHz range. See Privileges Cheat Sheet for exact details.)

- **Usage**: This band can support long-distance (DX) communications, especially during periods of high solar activity. It offers a great introduction to HF operating and is a favorite among Technicians interested in experimenting with digital modes or single-sideband (SSB) phone operations.

6-Meter Band (VHF)

- **Wavelength**: Approximately 6 meters

- **Frequency Range**: 50–54 MHz

- **Usage**: Known as "The Magic Band," the 6-meter band can provide both local and long-distance communications, depending on atmospheric conditions. It can exhibit characteristics of both HF and VHF bands, making it a versatile choice for Technician operators.

2-Meter Band (VHF)

- **Wavelength**: Approximately 2 meters

- **Frequency Range**: 144–148 MHz

- **Usage**: This is the most popular VHF band for Technician operators. It is widely used for local communications, especially through repeaters. It is also popular for simplex (direct, station-to-station) communications and emergency services.

70-Centimeter Band (UHF)

- **Wavelength**: Approximately 70 centimeters

- **Frequency Range**: 420–450 MHz

- **Usage**: This band is popular for local communications, often using repeaters. The shorter wavelength allows for better performance in urban environments, where signals can penetrate buildings and other obstacles more effectively than VHF signals.

Memorization technique: repetition. We saw a lot of repetition in this chapter. These radio bands are a core fundamental part of radio. Through repetition, you will begin to remember and know these bands until one day, this knowledge is just second nature.

Chapter 9

Propagation: How Your Signal Travels

PROPAGATION REFERS TO RADIO waves traveling through the atmosphere from a transmitter to a receiver. This is a critical concept in ham radio, as understanding how different propagation mechanisms work can help operators optimize their communication range and reliability.

The environment through which radio waves travel can influence their behavior. Like sound waves, which can echo off walls or be muffled by cushions, radio waves interact with objects and atmospheric conditions in various ways. Understanding these interactions is imperative for effective transmission and reception.

Propagation Modes

Different environments require different propagation modes to transmit signals effectively. As a Technician licensee, you will primarily encounter three modes:

1. **Ground Wave Propagation**: This occurs when radio waves travel along the surface of the Earth. Ground waves are predominantly used for local communications within about 50 to 100 miles, ideal for chatting with fellow hams in your region. These waves tend to follow the Earth's curvature and can be reliable during the day and night.

2. **Skywave Propagation**: At higher frequencies, radio waves can be reflected to Earth by the ionosphere, an ionized layer of the atmosphere. This mode allows for international communication, as the waves can 'bounce' between the ionosphere and the Earth, covering vast distances. However, this type of propagation is highly dependent on solar and atmospheric conditions, and thus, it can be pretty variable.

3. **Line-of-Sight Propagation**: As the name suggests, this mode requires no physical obstruction between the transmitting and receiving antennas. Line-of-sight propagation is perfect for applications like FM radio and television broadcasts, where clarity and consistency are essential. The range can be extended beyond the visual line of sight by the slight bending of waves in the atmosphere, known as tropospheric ducting.

Factors Affecting Propagation

Several environmental factors can influence how well your radio signals travel. Solar activity, such as sunspots and solar flares, can dramatically affect the ionization levels of the ionosphere, altering its ability to reflect radio waves. Higher solar activity generally enhances skywave propagation conditions but can also lead to increased levels of disruptive solar noise.

The Earth's atmosphere, with its varying layers of temperature and ionization, plays a significant role in radio wave propagation. Weather conditions, particularly storms and heavy cloud cover, can absorb or scatter radio waves, reducing signal strength and clarity. Understanding these factors can help you choose the best times and frequencies for your radio activities.

Knife-edge diffraction

Knife-edge diffraction is a phenomenon that allows radio signals to bend around sharp edges, such as the tops of mountains or buildings, enabling them to travel beyond obstacles that would typically block direct line-of-sight communication. When a radio wave encounters a sharp edge (like hitting the edge of a building in the image below), part of the wave is bent, or diffracted, around the obstacle, continuing on the other side. This effect helps signals reach areas that would otherwise be in a shadow zone due to the obstruction.

Remembering knife-edge diffraction is easy if you think of a knife slicing through the air, creating a path for the radio wave to follow. This effect is particularly useful in urban environments or mountainous regions where obstacles frequently interfere with line-of-sight communication.

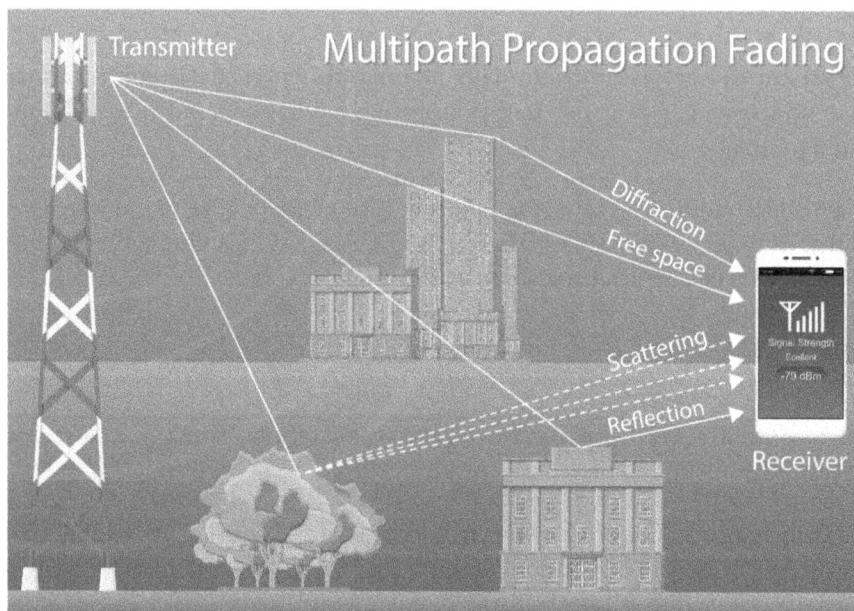

T3C05: Which of the following effects may allow radio signals to travel beyond obstructions between the transmitting and receiving stations?

A. Knife-edge diffraction
B. Faraday rotation
C. Quantum tunneling
D. Doppler shift

Multipath Propagation

Multipath propagation is a phenomenon where radio signals travel from a transmitter to a receiver along multiple paths. These paths include direct line-of-sight, reflections off buildings, mountains, or other obstacles, and even refraction through the atmosphere.

As a result, the signal arriving at the receiver combines these multiple paths, which can cause constructive or destructive interference. This can lead to signal strength and quality variations, sometimes resulting in fading or distortion.

Look back at the image above on propagation. A radio wave can take many paths from its transmitter to its intended receiver.

When these multiple signal paths converge at the receiving antenna, they can cancel or reinforce each other depending on their phase relationships.

In Phase and Out of Phase

Phase relationships describe the relative positions of waves in time, specifically how the peaks and troughs of one wave align with those of another. When discussing radio signals, phase relationships become crucial, especially in multipath propagation.

Phase – Waves

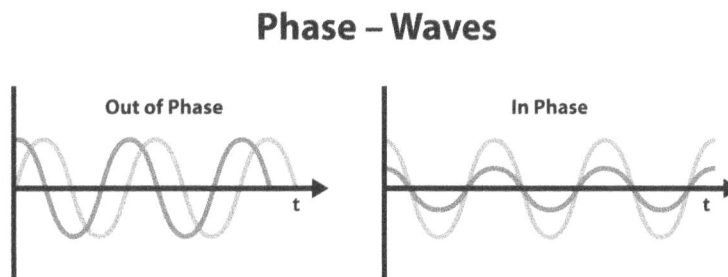

Out of Phase   In Phase

In Phase

When two waves are in phase, their peaks (highest points) and troughs (lowest points) align perfectly. This means that the waves reach their maximum and minimum values at the same time. When in-phase waves combine, they reinforce each other, producing a more robust overall signal. This is known as constructive interference.

For example, suppose two VHF signals arrive at an antenna in phase. In that case, their combined effect will be stronger than either of the individual signals alone. This can improve the clarity and strength of the received signal.

Out of Phase

When two waves are out of phase, their peaks and troughs do not align. Instead, the peak of one wave aligns with the trough of another. If two waves are entirely out of phase (180 degrees out of phase), one wave's peak aligns perfectly with the other wave's trough. When out-of-phase waves combine, they can cancel each other out, resulting in a weaker or even null signal. This is known as destructive interference.

For instance, if two VHF signals arrive at an antenna out of phase, their combined effect may be a much weaker signal or, in some cases, no signal. This can lead to signal fading or dead spots.

Importance in Radio Communication

Understanding phase relationships is key for optimizing radio communication. In multipath propagation, signals often arrive at the receiver through different paths, each with a phase relationship. Small changes in the position of the receiving antenna can cause significant changes in the phase relationships of these signals, leading to variations in signal strength.

Understanding that VHF signals are especially prone to these effects—due to their relatively short wavelengths (1 to 10 meters), VHF signals can easily reflect off buildings, terrain, and other obstacles —helps clarify why signal strength can vary so dramatically with minor changes in antenna position.

T3A01: Why do VHF signal strengths sometimes vary greatly when the antenna is moved only a few feet?

A. The signal path encounters different concentrations of water vapor
B. VHF ionospheric propagation is very sensitive to path length
C. Multipath propagation cancels or reinforces signals
D. All these choices are correct

By understanding how in-phase and out-of-phase signals interact, radio operators can better position their antennas to maximize constructive interference and minimize destructive interference, ensuring clearer and more reliable communication.

Picket Fencing

"Picket fencing" refers to the rapid fluttering or fading of mobile radio signals caused by multipath propagation. When a vehicle with a mobile radio moves through an area, the transmitted signal can reflect off various objects such as buildings, trees, and other structures. These reflections create multiple paths for the signal to reach the receiver. As the vehicle moves, the relative phases of these numerous signals constantly change, causing the received signal to rapidly fluctuate in strength.

This rapid variation in signal strength resembles the visual effect of looking through the gaps in a picket fence while moving past it, hence the name "picket fencing."

T3A06: What is the meaning of the term "picket fencing"?

A. Alternating transmissions during a net operation
B. Rapid flutter on mobile signals due to multipath propagation
C. A type of ground system used with vertical antennas
D. Local vs long-distance communications

The irregular fading of signals is often caused by the random combination of signals that arrive via different paths. Transmitted radio signals can travel to the receiver through multiple routes, such as direct paths, reflections off the ionosphere, or even numerous reflections. Each path may vary in length and the conditions it encounters, causing the signals to arrive at slightly different times and with different phases.

T3A08: What is a likely cause of irregular fading of signals propagated by the ionosphere?

A. Frequency shift due to Faraday rotation
B. Interference from thunderstorms
C. Intermodulation distortion
D. Random combining of signals arriving via different paths

So, what effect does all this multipath propagation have? What's the real problem with it?

T3A10: What effect does multipath propagation have on data transmissions?

A. Transmission rates must be increased by a factor equal to the number of separate paths observed
B. Transmission rates must be decreased by a factor equal to the number of separate paths observed
C. No significant changes will occur if the signals are transmitted using FM
D. Error rates are likely to increase

And that is the ultimate problem. <u>Error rates are likely to increase</u>, making our transmissions less reliable.

Chapter 10

The Atmosphere

THE ATMOSPHERE PLAYS AN interesting role in radio wave propagation and significantly impacts radio communications. Understanding these layers helps operators predict and optimize their transmission ranges and reception quality, making the atmosphere important enough for an overview and then a deep dive into it.

Understanding how radio waves travel through the atmosphere is essential for optimizing ham radio communication. The atmosphere, with its various layers and properties, significantly influences the behavior and propagation of radio waves. This section will explore the complex interactions between radio waves and the atmosphere, including the roles of the troposphere, stratosphere, mesosphere, and thermosphere.

By delving into these atmospheric relationships, you will gain valuable insights into how different frequencies are affected by atmospheric conditions, how to predict and utilize favorable propagation paths, and how to overcome environmental challenges. This knowledge will enhance your ability to make reliable long-distance contacts and improve communication effectiveness.

T3C06: What type of propagation is responsible for allowing over-the-horizon VHF and UHF communications to ranges of approximately 300 miles on a regular basis?

A. Tropospheric ducting
B. D region refraction
C. F2 region refraction
D. Faraday rotation

The troposphere is the part we live in. VHF and UHF work within the troposphere. Answers 'B' and 'C' relate to the ionosphere. Faraday rotation is the phenomenon where the polarization plane of an electromagnetic wave, such as a radio wave, rotates as it passes through a magnetic field in a medium like the ionosphere (key word here: ionosphere).

So, the only answer that makes sense is 'A' tropospheric ducting.

Tropospheric ducting is a phenomenon that allows radio waves to travel much greater distances than usual by being trapped in a layer of the atmosphere called the troposphere. This occurs mainly during temperature inversions, where a layer of warmer air lies above a layer of cooler air, reversing the normal temperature gradient – colder air tends to be on top because things get really cold as you move closer to space.

The inversion creates a refractive index gradient, bending radio waves downward and trapping them between the layers. As the waves reach the lower boundary, they are refracted back upwards, effectively guiding them along the curvature of the Earth.

This guided path enables radio waves, especially VHF and UHF signals, to propagate hundreds of miles beyond their typical line-of-sight range. Tropospheric ducting often provides stable signal paths, which can be advantageous for long-distance communication.

Ham radio operators can leverage this phenomenon to make contacts far beyond their usual range, particularly during specific weather conditions that favor temperature inversions, such as clear, calm nights or over large bodies of water. This extended range capability enhances the effectiveness and reliability of VHF and UHF communications.

T3C08: What causes tropospheric ducting?

A. Discharges of lightning during electrical storms
B. Sunspots and solar flares
C. Updrafts from hurricanes and tornadoes
D. Temperature inversions in the atmosphere

T3C11: Why is the radio horizon for VHF and UHF signals more distant than the visual horizon?

A. Radio signals move somewhat faster than the speed of light
B. Radio waves are not blocked by dust particles
C. The atmosphere refracts radio waves slightly
D. Radio waves are blocked by dust particles

Radio waves can travel over the visual horizon with help from the atmosphere. However, the atmosphere doesn't really impact light waves, so radio waves get the boost and travel a little further.

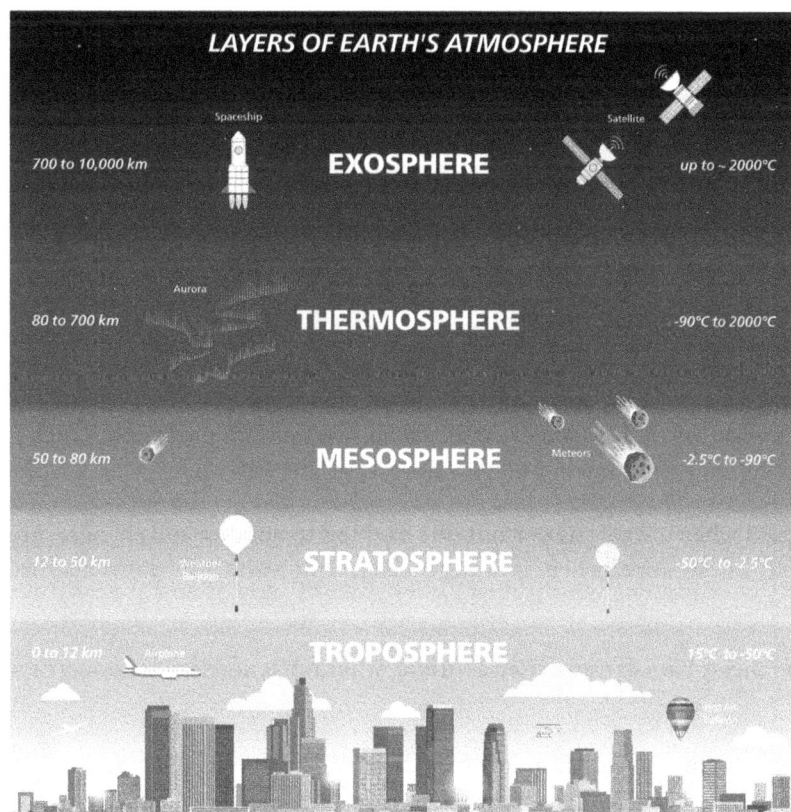

LAYERS OF EARTH'S ATMOSPHERE

Layers of the Atmosphere

1. **Troposphere**:

   ○ **Location**: Extends from the Earth's surface to about 12 kilometers (~8 miles).

   ○ **Characteristics**: Contains most of the atmosphere's water vapor and weather phenomena. We live here!

   ○ **Impact on Radio Waves**: Primarily affects VHF and UHF signals through tropospheric refraction, which can extend their range. Weather conditions like temperature inversions can also enhance or hinder propagation.

2. **Stratosphere**:

   ○ **Location**: Ranges from about 12 to 50 kilometers (31 miles) above the Earth's surface.

   ○ **Characteristics**: It contains the ozone layer, which absorbs and scatters ultraviolet solar radiation.

   ○ **Impact on Radio Waves**: Generally, it has minimal direct effect on radio wave propagation.

3. **Mesosphere**:

   ○ **Location**: Extends about 50 to 80 kilometers (~50 miles) above the Earth's surface.

   ○ **Characteristics**: The middle layer of the atmosphere where temperatures decrease with altitude.

   ○ **Impact on Radio Waves**: This layer plays a minor role in radio wave propagation, but meteors burn up, occasionally affecting radio signals.

4. **Thermosphere**:

   ○ **Location**: Ranges from about 80 to 700 kilometers (435 miles) above the Earth's surface.

   ○ **Characteristics**: Temperatures rise significantly with altitude as high-energy solar radiation heats the sparse gas molecules present, increasing their energy and raising the temperature with height. Contains the ionosphere.

   ○ **Impact on Radio Waves**: The <u>ionosphere within the thermosphere is central for HF radio wave propagation.</u>

The Ionosphere

The ionosphere <u>reflects or refracts radio waves</u> primarily due to its unique composition and the presence of ionized particles—a quick detour about the difference between reflecting and refracting.

- **Reflection** occurs when a wave, like light or a radio signal, hits a surface and bounces back at the same angle it arrived. This resembles seeing your reflection in a mirror, where the light waves bounce back to your eyes.

- **Refraction** happens when a wave passes from one medium to another and changes direction due to a change in speed. For example, a straw appears bent in water because light waves slow down and bend when they move from air to water.

In short, reflection is the bouncing back of waves from a surface, while refraction is the bending of waves as they pass through different materials.

This region of the Earth's upper atmosphere, extending from about 60 kilometers (37 miles) to 1,000 kilometers (620 miles) above the Earth's surface, contains a high concentration of ions and free electrons. These ions and electrons are created by ionizing atmospheric gases due to solar radiation, particularly ultraviolet (UV) light and X-rays from the Sun.

When solar radiation hits the molecules and atoms in the upper atmosphere, it strips away electrons, creating ions and free electrons. An ion is an atom or molecule that has gained or lost one or more electrons, giving it a net electrical charge. Atoms typically have an equal number of protons (positively charged) and electrons (negatively charged), making them neutral. However, when an atom gains extra electrons, it becomes negatively charged and is called an anion. Conversely, it becomes positively charged when it loses electrons and is called a cation.

Ions are core in many chemical reactions and processes, such as in batteries, where the flow of ions generates an electric current.

This ionization process creates layers within the ionosphere that can reflect or refract radio waves. Radio waves, being electromagnetic waves, interact with these charged particles in the ionosphere. This interaction affects the speed and direction of the radio waves. At specific frequencies, the density of ions in the ionosphere is high enough to reflect radio waves back toward the Earth, acting like a mirror. This reflection is more effective at lower frequencies within the HF band (3 to 30 MHz).

The ability of the ionosphere to reflect or refract radio waves is influenced by several factors, including frequency, solar activity, time of day, and seasonal changes. While lower-frequency radio waves are more likely to be reflected (HF), in contrast, higher-frequency waves (UHF) usually pass through or are refracted. Increased solar activity enhances ionization, improving reflective and refractive properties, while low solar activity reduces ionization. The ionosphere's properties also change between day and night, with higher solar radiation during the day increasing ionization. Seasonal variations further impact ionization levels, with higher levels typically occurring in the summer due to increased solar exposure.

Layers of the Ionosphere

1. **D Layer**:

   - **Location**: Approximately 60 to 90 kilometers (37 to 56 miles) above the Earth's surface.

   - **Characteristics**: Exists only during the daytime and absorbs lower-frequency HF signals, causing attenuation. Attenuation refers to the gradual weakening of a signal as it travels.

   - **Impact on Radio Waves**: Absorbs and weakens HF signals during the day, especially at frequencies below 10 MHz.

2. **E Layer**:

   - **Location**: Approximately 90 to 120 kilometers (56 to 75 miles) above the Earth's surface.

   - **Characteristics**: Can reflect HF waves and is more pronounced during the daytime.

   - **Impact on Radio Waves**: Provides helpful reflection for HF signals in the daytime, aiding short to medium-range communication (up to 2,000 kilometers or 1,200 miles).

3. **F Layers (F1 and F2)**:

   - **Location**: The F1 layer is about 150 to 220 kilometers (93 to 137 miles) above the Earth's surface during the daytime and merges into the F2 layer at night. The F2 layer extends from about 220 to 800 kilometers (137 to 500 miles).

   - **Characteristics**: The most important layers for long-distance HF propagation. The F1 layer disappears at night, while the F2 layer remains, providing reliable reflection for HF waves.

   - **Impact on Radio Waves**: The F2 layer supports long-distance HF communication (up to 3,000 kilometers or 1,900 miles per hop). This layer is highly variable and influenced by solar activity.

Practical Applications for Ham Radio Operators

Understanding the ionosphere and atmospheric layers allows ham radio operators to predict and optimize communication strategies. By selecting the appropriate frequencies and times for transmission, operators can take advantage of favorable ionospheric conditions to achieve longer and more reliable contacts.

T3A11: Which region of the atmosphere can refract or bend HF and VHF radio waves?

A. The stratosphere
B. The troposphere
C. The ionosphere
D. The mesosphere

T3C09: What is generally the best time for long-distance 10-meter band propagation via the F region?

A. From dawn to shortly after sunset during periods of high sunspot activity
B. From shortly after sunset to dawn during periods of high sunspot activity
C. From dawn to shortly after sunset during periods of low sunspot activity
D. From shortly after sunset to dawn during periods of low sunspot activity

As underlined in the 'F-Region,' this layer disappears at night, meaning we need to use it from dawn to dusk.

Sporadic E Propagation

Sporadic E propagation is a fascinating and somewhat unpredictable phenomenon. It allows occasional strong signals on the 10 (HF), 6, and 2-meter VHF bands beyond the typical radio horizon.

This type of propagation occurs when patches of highly ionized gas, known as Sporadic E layers, form in the lower part of the ionosphere (the E layer). These ionized patches can reflect VHF and upper HF signals, enabling communication over distances far greater than usual line-of-sight limits.

Sporadic E is most commonly associated with strong, intermittent signals on the 10, 6, and 2-meter bands, making it an exciting opportunity for ham radio operators to make long-distance contacts. These events typically occur during late spring and summer but can happen at any time of the year.

T3C04: Which of the following types of propagation is most commonly associated with occasional strong signals on the 10, 6, and 2-meter bands from beyond the radio horizon?

A. Backscatter
B. Sporadic E
C. D region absorption
D. Gray-line propagation

The key phrases to remember in this question are occasional, propagation, and beyond the horizon!

Auroral Backscatter

"Auroral" refers to phenomena related to the aurora, which are natural light displays in the Earth's sky typically seen in high-latitude regions around the Arctic and Antarctic. These displays, known as the Aurora Borealis in the Northern Hemisphere and the Aurora Australis in the Southern Hemisphere, are caused by the interaction of solar wind particles with the Earth's magnetic field and atmosphere.

Auroral backscatter occurs when VHF signals are reflected off the ionized particles in the aurora.

This type of propagation is fascinating but can also present some unique challenges. VHF signals received via auroral backscatter are typically distorted and exhibit considerable variations in signal strength. This happens because the reflecting auroral particles are in constant motion, creating a turbulent and fluctuating medium that affects the stability and clarity of the radio waves.

When you encounter auroral backscatter, you'll notice that the signal may sound "fluttery" or "hollow," which can rapidly change in strength. These characteristics make auroral backscatter a challenging and exciting mode of communication for operators.

> **T3C03: What is a characteristic of VHF signals received via auroral backscatter?**
>
> A. They are often received from 10,000 miles or more
> **B. They are distorted, and signal strength varies considerably**
> C. They occur only during winter nighttime hours
> D. They are generally strongest when your antenna is aimed west

Sunspots

> **T3C10: Which of the following bands may provide long-distance communications via the ionosphere's F region during the peak of the sunspot cycle?**
>
> **A. 6 and 10 meters**
> B. 23 centimeters
> C. 70 centimeters and 1.25 meters
> D. All these choices are correct

Recall the previous section. One key concept to remember for the exam is that whenever the exam talks about the atmosphere, it mainly refers to HF and VHF.

During the peak of the sunspot cycle, increased solar activity enhances the ionization of the ionosphere's F region, significantly improving its ability to reflect higher-frequency radio waves. This heightened ionization allows the 6-meter (50 to 54 MHz – VHF) and 10-meter (28 to 29.7 MHz – HF) bands to achieve long-distance communication.

The 6-meter band, typically used for line-of-sight communication, can exhibit HF-like propagation characteristics during this period, enabling contacts over hundreds or thousands of miles.

Similarly, the 10-meter band, already known for its long-distance capabilities, becomes even more effective, allowing global communication with relatively low power and simple antennas.

The F region's enhanced ionization during the sunspot peak allows these bands to refract signals back to Earth, facilitating long-distance propagation. This phenomenon enables amateur radio operators to take advantage of optimal conditions for communication, making the 6-meter and 10-meter bands particularly valuable for achieving extended-range contacts during periods of high solar activity.

Meteor Scatter

Meteor scatter communication (yes, there is such a thing as communicating by bouncing signals off parts of meteors!!!) takes advantage of the ionized trails left by meteors as they enter the Earth's atmosphere. When a meteor burns up, it leaves behind a trail of ionized particles that can reflect radio waves, allowing for brief periods of long-distance communication. The 6-meter VHF band (50 to 54 MHz) is best suited for this type of communication.

Again, we have a theme on which bands and frequencies to use with things in space.

The 6-meter band is ideal for meteor scatter because its frequency is high enough to take advantage of meteors' short, intense bursts of ionization. These ionized trails can reflect the 6-meter signals, allowing radio operators to make contacts over hundreds of miles for a few seconds to a few minutes. This makes the 6-meter band, often called the "magic band," remarkably effective for this unique mode of communication.

T3C07: What band is best suited for communicating via meteor scatter?

A. 33 centimeters
B. 6 meters
C. 2 meters
D. 70 centimeters

The Curvature of the Earth

Now that we have covered bands, frequencies, wavelengths, and the atmosphere, let's apply those concepts, starting with radio waves and the curvature of the Earth.

The ionosphere's ability to reflect and refract radio waves is critically vital for communication over long distances, especially due to the curvature of the Earth. Because the Earth is round, radio waves traveling in a straight line from a transmitter would quickly leave the surface and travel into space, making long-distance terrestrial communication impossible without some means of bending or reflecting the waves back toward the Earth's surface.

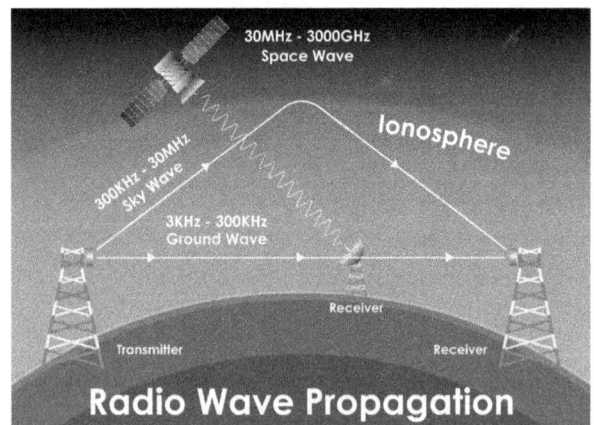
Radio Wave Propagation

The ionosphere acts as a natural reflector or refractor for radio waves. When radio waves are transmitted upwards, they encounter the ionized layers of the ionosphere. Depending on the frequency and ionization density of the waves, they can be reflected back to the Earth or refracted at an angle, allowing them to follow the curvature of the Earth. This process enables signals to cover distances far beyond the line-of-sight horizon, effectively "bouncing" between the ionosphere and the Earth's surface in what is known as skywave propagation.

Long-distance radio communication would be severely limited without the ionosphere's ability to bend radio waves back toward the Earth. The curvature of the Earth would cause radio signals to dissipate into space, restricting effective communication to line-of-sight distances, which are typically at most 30-40 miles for ground-based transmissions. Thus, the ionosphere's role in radio wave propagation is crucial for overcoming the Earth's curvature and enabling global radio communication.

T3C02: What is a characteristic of HF communication compared with communications on VHF and higher frequencies?

A. HF antennas are generally smaller
B. HF accommodates wider bandwidth signals
C. Long-distance ionospheric propagation is far more common on HF
D. There is less atmospheric interference (static) on HF

This section is all about the ionosphere and its relationship with HF. That is one key concept separating HF from higher-frequency VHF or UHF bands.

T3C01: Why are simplex UHF signals rarely heard beyond their radio horizon?

A. They are too weak to go very far
B. FCC regulations prohibit them from going more than 50 miles
C. UHF signals are usually not propagated by the ionosphere
D. UHF signals are absorbed by the ionospheric D region

For this question, they used 'simplex,' a concept we haven't covered yet, to emphasize that this communication occurs directly between two radio stations without using intermediate relays such as repeaters or satellites. In simplex operation, both transmitting and receiving occur on the same frequency.

Because we are using UHF with no other resources (repeaters or satellites) and the question states that we are going 'beyond the horizon,' we understand that we would need to bounce signals off the ionosphere. However, UHF frequencies are generally too high to be reflected.

Chapter 11

The Electromagnetic Spectrum and the Radio Bands

You NEED TO CHOOSE the right frequency to maximize the reach and clarity of your transmissions. Depending on the time of day and atmospheric conditions, specific frequencies will be more effective than others.

Understanding the relationship between frequency and wavelength is akin to learning the notes on a piano; each note has a specific pitch and tone, essential for creating music. In radio communication, frequency refers to the number of times a wave oscillates per second, measured in hertz (Hz), and wavelength is the physical distance over which the wave's shape repeats.

These two concepts are inversely related: <u>the higher the frequency, the shorter the wavelength</u>. This fundamental relationship shapes all aspects of radio transmission and reception.

T3B05: What is the relationship between wavelength and frequency?

A. Wavelength gets longer as frequency increases
B. Wavelength gets shorter as frequency increases
C. Wavelength and frequency are unrelated
D. Wavelength and frequency increase as path length increases

---

ELECTROMAGNETIC RADIATION (FREQUENCY)

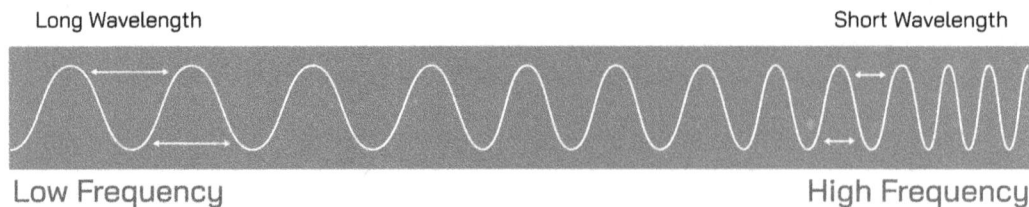

You will need to know a mathematical relationship between wavelength and frequency for the test. The relationship between frequency and wavelength is given by the formula:

Where:

- λ (pronounced lambda) is the wavelength in <u>meters</u>.

- c is the speed of light (approximately 300,000,000 <u>meters</u> per second).

- f is the frequency in hertz (Hz).

Now, on the test, the question is:

$$\lambda = \frac{c}{f}$$

Wavelength in Meters

T3B06: What is the formula for converting frequency to approximate wavelength in meters?

A. wavelength in meters equals frequency in hertz multiplied by 300
B. wavelength in meters equals frequency in hertz divided by 300
C. Wavelength in meters equals frequency in megahertz divided by 300
D. Wavelength in meters equals 300 divided by frequency in megahertz

---

Here, you can see they are being a little sneaky. Whereas our formula is in hertz, the question is looking at <u>megahertz</u>. So, let's adjust (plus, it makes the formula a little easier; that's why they do it, less about being sneaky and more about simplicity).

We must adjust the formula when using megahertz (MHz) for frequency. Since 1 MHz equals 1,000,000 Hz, the formula can be rewritten. Now:

- λ (pronounced lambda) is the wavelength in <u>meters</u>.

- f is the frequency in megahertz (MHz).

This new formula makes calculating the wavelength for any given frequency in megahertz easier.

$$\text{meters} = \frac{300}{\text{Frequency (MHz)}}$$

Memorize this formula, and remember that it has a frequency of megahertz (MHz).

Electromagnetic Spectrum

Radio makes up just one part of the overall electromagnetic spectrum. The electromagnetic spectrum also includes things like the light we see, the visible part of the spectrum, and x-rays (and gamma rays for any Hulk fans!).

THE ELECTROMAGNETIC SPECTRUM

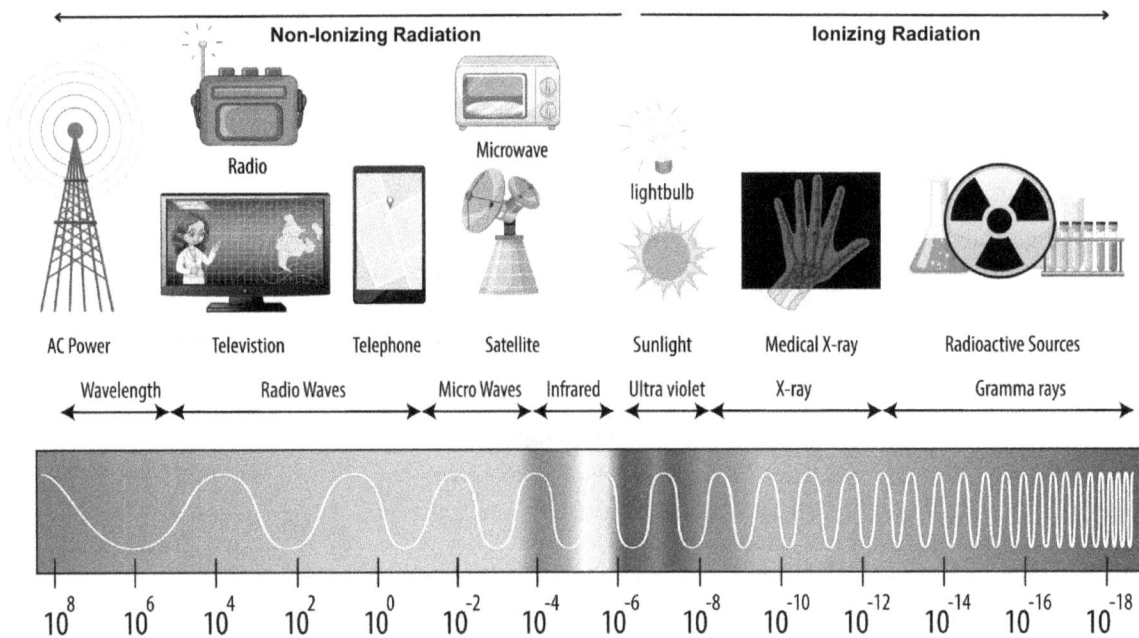

The radio spectrum encompasses all frequencies used for communication, which are meticulously organized into bands – think of them as sections or pieces of the electromagnetic spectrum.

Each band is allocated for specific use. And within particular bands, we amateur radio licensed operators have been given privileges to use parts of them for different types of communication and experimentation.

Radio Bands

Radio bands are broad categories that describe specific ranges of frequencies. Here, we name a few common ones. Yes! More repetition. Have you memorized them all yet?

- **VLF (Very Low Frequency)**

- **LF (Low Frequency)**

- **MF (Medium Frequency)**: includes AM radio

- **HF (High Frequency)**: includes shortwave radio and many ham bands

- **VHF (Very High Frequency)**: includes FM radio and TV channels

- **UHF (Ultra High Frequency)**: includes TV channels and mobile phones

- **SHF (Super High Frequency)**: used for radar and satellite communication

- **EHF (Extremely High Frequency)**

These names help quickly identify the general frequency range and the type of communication typically found there.

Short-Hand Name	**VLF**	**LF**	**MF**	**HF**	**VHF**	**UHF**	**SHF**	**EHF**
Name	Very Low Frequency	Low Frequency	Medium Frequency	High Frequency	Very High Frequency	Ultra High Frequency	Super High Frequency	Extremely High Frequency
Frequency	3 – 30 kHz	30 – 300 kHz	300 – 3,000 kHz	3 MHz – 30 MHz	30 – 300 MHz	300 – 3,000 MHz	3 GHz – 30 GHz	30 – 300 GHz
Wavelength	100 – 10 km	10 – 1 km	1,000 m – 100 m	100 – 10 m	10 – 1 m	1 – 0.1 m	10 cm – 1 cm	1 – 0.1 cm

Wavelength

Wavelength is the distance a radio wave travels in one cycle. Let's now add wavelengths to our names.

VLF (Very Low Frequency)

- **Wavelength**: 100 km to 10 km

  - For anyone new to the metric scale, 10 km is over 6 miles! This means in the time it takes for this wave to travel through its entire cycle, it travels over 6 miles.

  - Even though I'm adding in some miles, feet, and inches to help a beginner understand the metric system, ham radio uses the metric system.

LF (Low Frequency)

- **Wavelength**: 10 km to 1 km

  - 1 km is about 0.6 miles

MF (Medium Frequency)

- **Wavelength**: 1,000 meters (which is the same as 1 km) to 100 meters

  - 100 meters is about a football field (slightly longer but roughly)

HF (High Frequency)

- **Wavelength**: 100 meters to 10 meters

  - 10 meters is about 33 feet

VHF (Very High Frequency)

- **Wavelength**: 10 meters to 1 meter

  - 1 meter is about 3.28 feet

UHF (Ultra High Frequency)

- **Wavelength**: 1 meter to 0.1 meter (10 centimeters)

  - 10 cm is about 4 inches

SHF (Super High Frequency)

- **Wavelength**: 10 centimeters to 1 centimeter

  - 1 cm is a little under half an inch

EHF (Extremely High Frequency)

- **Wavelength**: 1 centimeter to 1 millimeter

At this point, it's helpful to start memorizing bands and wavelengths. Or, do what I do: print off a band chart. As Einstein said, use your mind to think, not memorize.

Printable band chart: **www.MorseCodePublishing.com/TechExam**

Band Chart

You will need to memorize part of this for the exam. Focus on the exam questions and memorize them. Otherwise, print off the included chart and hang it on your wall. Over time, as you become a wiser and older ham, it will sink into memory.

Frequency

Frequency describes how often a radio wave oscillates per second, measured in hertz (Hz). It indicates how many cycles occur in one second:

- **30 MHz**: The radio wave oscillates 30 million times per second. Remember, M = Mega, and Mega equals 1,000,000 (back from our metric system section).

- **144 MHz**: This means the wave oscillates 144 million times per second.

Higher frequencies correspond to shorter wavelengths, and lower frequencies correspond to longer wavelengths.

In the two examples above, 30 MHz has a lower frequency than 144 MHz, so a lower frequency means that 30 MHz has a longer wavelength than 144 MHz. A longer wavelength means it's easier to go farther, so if we wanted to consider a long-distance radio call, 30 MHz would be a better frequency than 144 MHz. There will be more on all this, but you can start to see how things are coming together!

Continuing to Connect the Concepts, Let's Add in Frequency

Remember, each term—whether a band name, wavelength size, or frequency —offers a different perspective on the same radio spectrum, making it easier to navigate and utilize. A few sections have been specifically called out as they are important to technicians.

VLF (Very Low Frequency)

- **Frequency**: 3 to 30 kHz (very low frequency)

- **Wavelength**: 100,000 meters to 10,000 meters (very long wavelengths)

LF (Low Frequency)

- **Frequency**: 30 to 300 kHz

- **Wavelength**: 10,000 meters to 1,000 meters

MF (Medium Frequency)

- **Frequency**: 300 kHz to 3 MHz (includes AM radio)

- **Wavelength**: 1,000 meters to 100 meters

Let's now highlight the ones important to use as Technicians.

HF (High Frequency)
- **Frequency**: 3 to 30 MHz (includes shortwave radio and many ham bands)

- **Wavelength**: 100 meters to 10 meters

VHF (Very High Frequency)
- **Frequency**: 30 to 300 MHz (includes FM radio and TV channels)

- **Wavelength**: 10 meters to 1 meter

UHF (Ultra High Frequency)
- **Frequency**: 300 MHz to 3 GHz (includes TV channels and mobile phones)

- **Wavelength**: 1 meter to 0.1 meter (10 centimeters)

SHF (Super High Frequency)

- **Frequency**: 3 to 30 GHz (used for radar and satellite communication)

- **Wavelength**: 10 centimeters to 1 centimeter

EHF (Extremely High Frequency)

- **Frequency**: 30 to 300 GHz (very, very fast frequency)

- **Wavelength**: 1 centimeter to 1 millimeter (very, very short wavelength)

Now, based on what we've learned, let's answer some questions. Refer specifically to the called-out bands above:

T3B07: In addition to frequency, which of the following is used to identify amateur radio bands?

A. The approximate wavelength in meters
B. Traditional letter/number designators
C. Channel numbers
D. All these choices are correct

Radio bands are often identified by wavelength in addition to frequency.

T3B08: What frequency range is referred to as VHF?

A. 30 kHz to 300 kHz
B. 30 MHz to 300 MHz
C. 300 kHz to 3000 kHz
D. 300 MHz to 3000 MHz

Stay tuned for the next section if you haven't memorized this yet.

T3B09: What frequency range is referred to as UHF?

A. 30 to 300 kHz
B. 30 to 300 MHz
C. 300 to 3000 kHz
D. 300 to 3000 MHz

T3B10: What frequency range is referred to as HF?

A. 300 to 3000 MHz
B. 30 to 300 MHz
C. 3 to 30 MHz
D. 300 to 3000 kHz

Do you need additional help with those last three questions or help remembering the frequencies? Here is my memory trick for this part. It's idiotic, but it works.

For Frequency:

1. I always start with High Frequency (HF) as my anchor. Then, High Frequency advances to '*Very*' High Frequency and then to '*Ultra*' High Frequency. Easy enough! Write it down. We now have the names in order.

2. Next, there are 3 ways to identify a band: the name, the wavelength, and the frequency—three different methods (3). Remember the #3.

3. Anchoring to High Frequency and the number 3, I now remember that the frequency for HF starts at 3 MHz. Nice and convenient!

4. HF begins at 3 MHz of frequency. A good start! Write it down.

5. Finally, the metric system is based on 10s. So, each band starts at a level and expands 10 times.

6. Putting this all together, HF starts at 3 MHz, and multiplied by 10, it ends at 30 MHz. Which is where VHF begins.

   ◦ If VHF begins at 30 MHz, multiplying by 10, it must end around 300 MHz.

   ◦ This is where UHF would start (300 MHz) and end around 3,000 MHz (or, using the metric system to simplify, it could be written as 3 GHz).

HF	VHF	UHF
High Frequency	Very High Frequency	Ultra High Frequency
3 MHz – 30 MHz	30 – 300 MHz	300 – 3,000 MHz
100 m – 10 m	10 – 1 m	1 – 0.1 m

This exercise even works for wavelength.

For Wavelength:

1. Again, I will start with high-frequency (HF) as my anchor. Then, High Frequency advances to '*Very*' High Frequency and then to '*Ultra*' High Frequency. Easy enough! Write it down.

2. Next, we use the metric system, which is based on multiples of 10. So, we use 10s in the following steps.

3. Anchoring to High Frequency and using multiples of 10, I now remember that the wavelength for HF starts at 100 Meters. That's a nice, convenient round number!

4. HF begins at 100 meters of wavelength. Write it down.

5. Finally, the metric system is based on 10s. So, to move to the next band, we DIVIDE by 10.

6. Putting this all together, HF starts at 100 meters, and divided by 10, it ends at 10 meters. Which is where VHF begins.

   ◦ If VHF begins at 10 meters, divided by 10, it must end at 1 meter.

   ◦ This is where UHF would start at 1 meter and end at around 0.1 meter (10 cm).

Exam questions:

T1B03: Which frequency is in the 6-meter amateur band?

A. 49.00 MHz
B. 52.525 MHz
C. 28.50 MHz
D. 222.15 MHz

When we have either frequency or wavelength, and we want to know the other, we use our simplified formula.

$$\text{Frequency} = \frac{300}{\text{Wavelength in Meters}}$$

We want frequency. We have wavelength. We divide 300 by 6 meters to equal 50 MHz. But none of the answers equal 50!?!?

But we can at least eliminate 'C' and 'D,' which are far from 50. But we are still left with 'A' and 'B.' Remember the section about bands and the "magic 6-meter band." In that summary, we shared that the 6-meter band is from 50 to 54 MHz. Based on our math and memorization, we know the answer is 52.525 MHz, which is in the 6 m band.

T1B04: Which amateur band includes 146.52 MHz?

A. 6 meters
B. 20 meters
C. 70 centimeters
D. 2 meters

Our math comes close to answering this question, and that counts! 300 divided by 146.52 equals 2.047 m, which is close enough to know that the answer is 'D.' The way to estimate this is that 146.52 MHz is very close to 150 MHz. Doing that math, 300 divided by 150 equals 2 meters.

$$\text{meters} = \frac{300}{\text{Frequency (MHz)}}$$

Radio Bands Summary

For a ham operator, knowing which bands are available and their specific characteristics is crucial for both regulatory compliance and effective communication. Not all frequencies are accessible for use, and frequency allocation is managed by national and international bodies to prevent interference. The Federal Communications Commission (FCC) oversees these regulations in the United States, assigning specific bands to different services, including amateur radio. Following these band plans is legally required and essential for ensuring clear and organized communication. These band plans provide a structured framework that supports efficient amateur radio operations.

Each band is further divided into segments dedicated to specific communication modes, such as Morse code (CW), digital modes, or voice communication, with recommended usage guidelines. More detailed rules and regulations will be covered in a later section.

It can be overwhelming to navigate terms like bands, wavelengths, frequencies, and the electromagnetic spectrum, along with UHF, VHF, and HF. Let's break it down: radio bands can be identified by different names, wavelength sizes (like 2 meters), or frequencies (like 30 MHz). These terms are interconnected and will become clearer as you delve deeper into amateur radio.

Chapter 12

Modulation and Bandwidth

Modulation

MODULATION IS AKIN TO understanding how different musical instruments can play the same note but sound distinct. Each instrument alters the sound wave in its unique way.

In amateur radio, modulation refers to how the basic form of a radio wave is changed to carry information, be it voice, data, or video. It involves combining speech with a radio frequency (RF) carrier signal.

T7A08: Which of the following describes combining speech with an RF carrier signal?

A. Impedance matching
B. Oscillation
C. Modulation
D. Low-pass filtering

There are several types of wave modulation, including Amplitude Modulation (AM), Frequency Modulation (FM), and Phase Modulation (PM). Each alters the carrier signal in different ways to encode the audio information.

AM (Amplitude Modulation)

AM (Amplitude Modulation) changes the height or strength of the radio wave to match the sound being sent. It was one of the first ways to send radio signals. It is still used today, especially in aviation and AM radio stations. AM is more susceptible to interference because changes in amplitude can be caused by various forms of electrical noise, such as thunderstorms and machinery.

Single Sideband (SSB) Modulation

Single Sideband (SSB) is a highly efficient type of amplitude modulation (AM) used extensively in ham radio. In traditional AM, a carrier signal is modulated with audio information, producing two identical sidebands (upper and lower) and a carrier, which consumes a lot of bandwidth and power. SSB improves on this by eliminating the carrier and one of the sidebands, leaving just one sideband (either upper or lower) to carry the information. This makes SSB much more efficient in terms of bandwidth and power usage.

Modulation

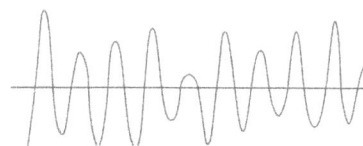

AM - Amplitude Modulation

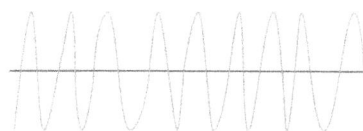

FM - Frequency Modulation

SSB is a streamlined version of AM. Instead of transmitting redundant information, SSB focuses all the power into a single sideband, making it more robust for long-distance communication. This efficiency is why SSB is so popular among amateur radio operators for voice communication on HF bands.

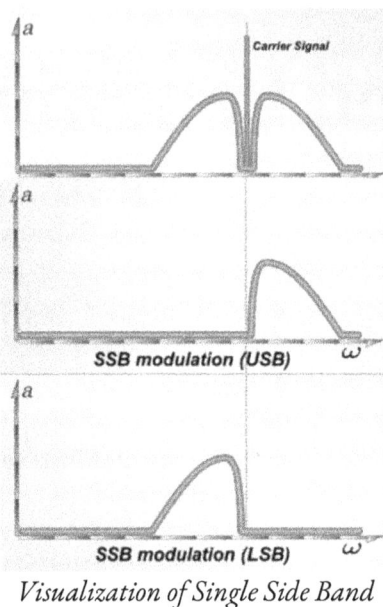

Visualization of Single Side Band

T8A01: Which of the following is a form of amplitude modulation?

A. Spread spectrum
B. Packet radio
C. Single sideband
D. Phase shift keying (PSK)

SSB for Long-Distance Contacts on VHF and UHF

Single-sideband (SSB) voice modulation is often used for long-distance, weak signal contacts on the VHF and UHF bands. SSB is the go-to mode for reaching long distances with low signal strength. Its efficiency in utilizing power and bandwidth allows for clearer communication even when the signal is faint.

SSB's efficiency is particularly suitable for long-distance communication, especially on the VHF and UHF bands, where conserving bandwidth and power is crucial.

T8A03: Which type of voice mode is often used for long-distance (weak signal) contacts on the VHF and UHF bands?

A. FM
B. DRM
C. SSB
D. PM

The upper sideband (USB) is typically used for SSB communications for these bands. This convention helps standardize operations and ensures compatibility between different stations. Operators can achieve better and more reliable communication over longer distances using the upper sideband, even when signals are weak.

Most modern ham radios have a mode selection button or dial. To use it, you will switch your radio to the appropriate SSB mode: USB (Upper Sideband) for 10-meter HF, VHF, and UHF bands or LSB (Lower Sideband) for other HF bands below 10 meters.

T8A06: Which sideband is normally used for 10-meter HF, VHF, and UHF single-sideband communications?

A. Upper sideband
B. Lower sideband
C. Suppressed sideband
D. Inverted sideband

FM (Frequency Modulation)

FM (Frequency Modulation) changes the speed of the radio wave to send information. FM is great for sending clear sound without much interference. It is perfect for radio stations playing music and public service broadcasts. FM is less susceptible to interference because it relies on changes in frequency rather than amplitude, making it better at rejecting noise and providing better sound quality.

Frequency Modulation (FM) or Phase Modulation (PM) is commonly used for VHF packet radio transmissions. Packet radio is a digital mode where data is transmitted in small packets, which requires a robust and reliable transmission method. FM is favored because it provides clear and consistent signal quality, which is central for accurately transmitting digital data. FM works by varying the carrier wave frequency to encode the data, making it less susceptible to noise and interference.

T8A02: What type of modulation is commonly used for VHF packet radio transmissions?

A. FM or PM
B. SSB
C. AM
D. PSK

VHF packet radio transmissions are a method of digital communication used by amateur radio operators to send data over Very High Frequency (VHF) bands. This method involves dividing data into small packets, transmitting them using Frequency Modulation (FM) or Phase Modulation (PM), and reassembling them at the receiving end. FM and PM are preferred due to their robust signal quality, which ensures accurate data transfer. VHF packet radio is commonly used for text messaging, weather data, and position reporting applications. It is especially valuable for emergency communications due to its reliability and efficiency in maintaining clear and accurate data transmission.

Similarly, Phase Modulation (PM) can also be used for packet radio. PM encodes data by varying the carrier wave phase, which, like FM, ensures that the digital signals are transmitted clearly and efficiently. Both FM and PM are well-suited for VHF packet radio because they maintain signal integrity over the distances typically covered by VHF frequencies.

Modulation for VHF and UHF Voice Repeaters

Frequency Modulation (FM) and Phase Modulation (PM) are also commonly used for VHF and UHF voice repeaters. These modulation types are preferred because they provide clear, reliable, high-quality voice communication, essential for effective repeater operation. FM works by varying the carrier wave frequency according to the voice signal, ensuring the transmitted audio is clear and less susceptible to noise and interference.

Similarly, Phase Modulation (PM) encodes the voice signal by varying the phase of the carrier wave. Both FM and PM are well-suited for the VHF (30-300 MHz) and UHF (300 MHz-3 GHz) bands because they can maintain signal integrity over longer distances and through various obstacles.

T8A04: Which type of modulation is commonly used for VHF and UHF voice repeaters?

A. AM
B. SSB
C. PSK
D. FM or PM

Digital Modulation

Digital modes, however, encode data into a series of digital signals, which are then used to modulate the carrier wave. These modes are increasingly popular in amateur radio because they provide efficient and reliable communication even under challenging conditions where traditional analog signals might fail.

More on digital later.

Bandwidth

Each modulation type inherently requires a certain amount of bandwidth, the range of frequencies the signal occupies on the radio spectrum. Bandwidth is pivotal in radio communications as it determines how much information can be transmitted and how susceptible the transmission might be to interference.

Said another way, bandwidth is the range of frequencies that a radio signal occupies. Imagine you're tuning a radio to a station. Each station has its own slice of frequencies. For example, an FM radio station might broadcast on 101.1 MHz. Still, it uses a small range of frequencies around that number to send its signal. This range is the station's bandwidth.

The wider the bandwidth, the more information the signal can carry. For instance, FM radio uses more bandwidth (about 200 kHz) than AM radio (10 kHz), so FM can deliver higher-quality sound. Each type of radio service (like AM, FM, or digital) uses a different amount of bandwidth to fit its needs.

Following this line of thinking and remembering what we learned about SSB (Single-Side Band), being part of AM, you should now be able to answer this question:

T8A07: What is a characteristic of single sideband (SSB) compared to FM?

A. SSB signals are easier to tune in correctly
B. SSB signals are less susceptible to interference
C. SSB signals have narrower bandwidth
D. All these choices are correct

SSB typically requires a bandwidth of about 2.5 to 3 kHz for effective transmission. This is much narrower than traditional amplitude modulation (AM), which requires twice the bandwidth due to transmitting both the upper and lower sidebands and the carrier.

T8A08: What is the approximate bandwidth of a typical single sideband (SSB) voice signal?

A. 1 kHz
B. 3 kHz
C. 6 kHz
D. 15 kHz

Typical Bandwidth by Mode

Mode	Bandwidth	Usage / Application
Digital Modes (e.g., PSK31)	31 Hz	Low bandwidth digital communication.
CW (Continuous Wave)	150-500 Hz	Morse code transmission.
SSB (Single Sideband)	2.5-3 kHz	Voice communication, HF ham radio.
AM (Amplitude Modulation)	10 kHz	Standard for AM broadcasting, voice communication.
NBFM (Narrowband FM)	10-15 kHz	Two-way radio, VHF repeaters, public safety.
WBFM (Wideband FM)	150-200 kHz	Commercial FM broadcasting (music and voice).
NTSC (Analog TV)	6 MHz	Analog television broadcasting.
ATSC (Digital TV)	6 MHz	Digital television broadcasting.

FM vs Single Sideband (SSB)

One of the main disadvantages of frequency modulation (FM) compared with single-sided band (SSB) is that FM can only receive one signal at a time at a given frequency. This limitation arises because FM signals occupy a wider bandwidth and are designed to be strong and clear, effectively blocking out other signals on the same frequency. In contrast, SSB signals use narrower bandwidth, allowing multiple signals to coexist on nearby frequencies and making it possible to tune in to and separate individual signals.

Remembering this fundamental difference can help you understand why SSB is often preferred for long-distance and crowded band conditions, where multiple stations may operate simultaneously. Knowing that FM allows only one signal per frequency will help you recall this concept for your exam, highlighting the trade-offs between signal clarity and bandwidth efficiency in different modulation modes.

T8A12: Which of the following is a disadvantage of FM compared with single sideband?

A. Voice quality is poorer
B. Only one signal can be received at a time
C. FM signals are harder to tune
D. All these choices are correct

Digital modes can also be designed to use very narrow bandwidths, which makes them less prone to interference and allows more simultaneous transmissions within a given part of the band.

Efficient spectrum use is fundamental to successful amateur radio operations. Managing bandwidth effectively ensures that multiple communications can co-occur without interfering with each other. This is not only a technical necessity but also a regulatory requirement, as the FCC allocates specific portions of the spectrum for different uses to prevent chaos on the airwaves.

Bandwidth of a VHF Repeater FM Voice Signal

Let's continue to learn more about ham radio by looking at actual questions on the exam.

The approximate bandwidth of a VHF repeater FM voice signal is between 10 and 15 kHz. This bandwidth range is necessary to accommodate the frequency variations required for transmitting clear and intelligible voice audio. Frequency Modulation (FM) works by varying the carrier wave frequency per the audio signal, and a bandwidth of 10 to 15 kHz ensures that the transmitted voice is clear and free from distortion.

This bandwidth is optimized to balance audio quality and spectrum efficiency, allowing multiple repeaters to operate within the VHF band without interfering with each other.

T8A09: What is the approximate bandwidth of a VHF repeater FM voice signal?

A. Less than 500 Hz
B. About 150 kHz
C. Between 10 and 15 kHz
D. Between 50 and 125 kHz

AM Fast-Scan TV Transmissions

Another question about bandwidths caused issues regarding where to place it within this book. We put it inside the bandwidth section because it was related to the exam question.

AM fast-scan TV, commonly called amateur television (ATV), transmits video and audio signals using amplitude modulation (AM) over radio frequencies. This form of television broadcasting is similar to commercial analog TV. Still, amateur radio operators use it for personal and experimental purposes.

In AM fast-scan TV, the video signal modulates the carrier wave's amplitude, meaning the carrier wave's strength varies per the brightness and detail of the picture. Audio is typically transmitted alongside the video by frequency modulation (FM) on a subcarrier or by using a separate carrier. The term "fast-scan" refers to the high frame rate, similar to standard broadcast television, providing smooth and continuous video, which contrasts with slow-scan television (SSTV), which transmits still images much slower.

AM fast-scan TV transmissions require substantial bandwidth, typically around 6 MHz, to carry the detailed video information and ensure high-quality image and sound. Amateur radio operators use AM fast-scan TV for various purposes, including personal communication, experimentation, and public service broadcasting. It allows them to share real-time video content, conduct remote surveillance, or provide visual information during emergency operations.

The relatively wide bandwidth (remember it was 6 MHz) is necessary to transmit the high-resolution video and accompanying audio signals that make up a television broadcast. Unlike voice transmissions, which require much narrower bandwidth, TV signals contain much more information and, therefore, need more space in the frequency spectrum.

The 6 MHz bandwidth allows for the detailed video data to be transmitted alongside the audio, ensuring that the received picture is clear and the sound is synchronized and of high quality.

T8A10: What is the approximate bandwidth of AM fast-scan TV transmissions?

A. More than 10 MHz
B. About 6 MHz
C. About 3 MHz
D. About 1 MHz

Morse Code: Continuous Wave (CW)

Continuous Wave (CW) signals, commonly used for Morse code communication, have the narrowest bandwidth among various types of signals. In CW transmission, a constant-amplitude radio wave is turned on and off to create a series of short and long signals corresponding to the dots and dashes of Morse code.

Morse code comprises dots and dashes representing letters, numbers, and punctuation. CW is the method by which this language is transmitted over the airwaves.

T8D09: What is CW?

A. A type of electromagnetic propagation
B. A digital mode used primarily on 2-meter FM
C. A technique for coil winding
D. Another name for a Morse code transmission

CW signals typically occupy only about 150 Hz of bandwidth, which is significantly narrower than other modes like Single Sideband (SSB) and Frequency Modulation (FM), which are in kilohertz. This narrow bandwidth makes CW very efficient for long-distance communication, especially in crowded or noisy bands where minimizing interference is vital.

Remembering that CW has the narrowest bandwidth is easy if you consider it the most "compact" form of communication. It allows operators to fit more signals into the same frequency space without causing interference.

T8A05: Which of the following types of signal has the narrowest bandwidth?

A. FM voice
B. SSB voice
C. CW
D. Slow-scan TV

T8A11: What is the approximate bandwidth required to transmit a CW signal?

A. 2.4 kHz
B. 150 Hz
C. 1000 Hz
D. 15 kHz

Morse Code

Morse code is one of the oldest and most enduring methods of communication in radio and telegraphy. It was developed in the early 1830s by Samuel Morse and Alfred Vail. This system uses a series of dots (short signals) and dashes (long signals) to represent letters, numbers, and punctuation marks.

A short pause separates each character. In contrast, words are separated by longer pauses, allowing for the transmission of complex messages using simple, binary-like sequences. Initially used for telegraph communication, operators manually key in the dots and dashes to send messages over long distances via telegraph lines. With the advent of radio in the early 20th century, Morse code became a vital tool for ship-to-shore and ship-to-ship communication, military operations, and amateur radio, valued for its simplicity and effectiveness in poor signal conditions or limited bandwidth scenarios.

The simplicity of the equipment needed to send and receive Morse code makes it accessible and cost-effective. Furthermore, Morse code's reliability, even with weak or partially obstructed signals, has made it a popular choice for emergency communication.

Although less commonly used in commercial and military communication today, Morse code remains an essential skill for amateur radio operators. It continues to be used in specific aviation and maritime contexts. It is appreciated for its historical significance and effectiveness in low-signal environments.

Many amateur radio enthusiasts enjoy learning and using Morse code as part of their hobby, maintaining a link to the rich history of radio communication. However, you will not need to know Morse Code for any test.

Bandwidth Summary

Adjusting bandwidth in radio communications involves changing the range of frequencies used to transmit a signal. Depending on the equipment and mode of communication, various methods can be used to accomplish this. Most modern radios have adjustable settings labeled "IF Bandwidth" or "Filter," allowing users to narrow or widen the bandwidth.

Different communication modes, such as Single Sideband (SSB), Frequency Modulation (FM), and Morse code (CW), naturally use different bandwidths, so switching between these modes can also adjust the bandwidth. External filters can be used to restrict or broaden the range of frequencies the radio can handle.

Properly adjusting bandwidth is necessary for minimizing interference and ensuring clear communication, especially in crowded frequency bands.

The Physical Radio Station

SETTING UP YOUR FIRST station is an exciting endeavor. Start with the basics: a transceiver (a device that can transmit and receive communications), a power supply, and an antenna.

The type of equipment you choose should match the operations you intend to perform – and, most importantly, your license. For instance, a simple VHF or UHF setup might suffice if you're interested in local communications. For global communications, you might look into HF radios.

And remember, the antenna is as crucial as the radio itself, if not more so. A well-installed antenna will significantly enhance your ability to send and receive signals.

In this section, we will walk through the physical components of your radio and then discuss its actual use in the next section.

Chapter 13

Electronic And Electrical Components

WELCOME TO THE SECTION on electrical components, where we delve into the fundamental building blocks of radio communication systems. Understanding these components is pivotal for a beginner to troubleshoot, repair, and optimize their equipment. We'll explore essential elements like resistors, capacitors, inductors, and more, explaining their functions, characteristics, and interactions within a circuit.

Electrical Schematic

A schematic is a wiring diagram that represents the elements of a system using abstract, graphic symbols rather than realistic pictures. It illustrates a circuit's electrical connections and functions.

T6C01: What is the name of an electrical wiring diagram that uses standard component symbols?

A. Bill of materials
B. Connector pinout
C. Schematic
D. Flow chart

Electrical schematics focus on accurately representing component connections. These diagrams use standardized symbols to depict various electrical components, such as resistors, capacitors, inductors, and transistors, and show how they are interconnected by lines representing wires. The schematic provides a clear and detailed map of the circuit, illustrating how current flows from one component to another, which is essential for understanding the circuit's functionality and troubleshooting.

Think of an electrical schematic as a blueprint for building or analyzing an electronic device. Each symbol corresponds to a specific component; the lines connecting them indicate electrical current paths. Understanding that the purpose of a schematic is to show these connections will help you recall this concept for your exam and appreciate the importance of schematics in designing and maintaining electronic circuits.

T6C12: Which of the following is accurately represented in electrical schematics?

A. Wire lengths
B. Physical appearance of components
C. Component connections
D. All these choices are correct

Here is an example of an electrical schematic, which, lucky for us, is very similar to the one you will see on the exam.

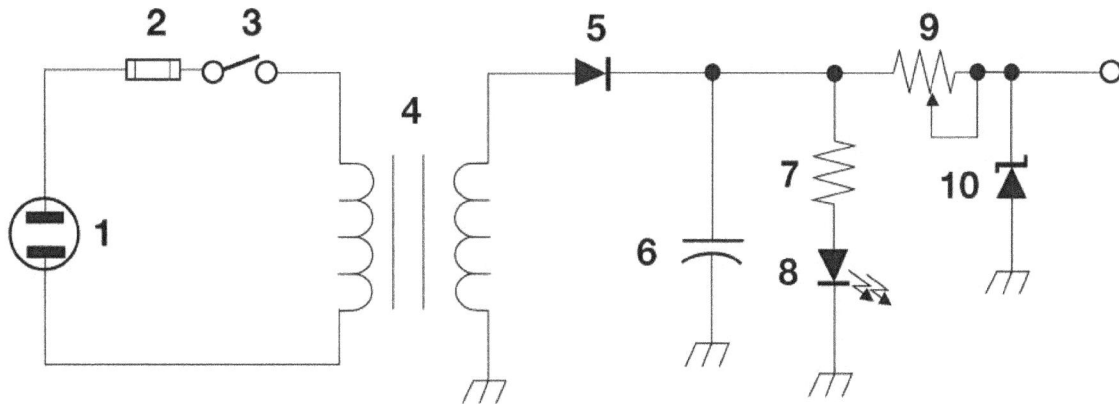

Schematic Components

Here is the identification of each numbered part in the schematic:

1. **AC Power Source**: Provides the electrical energy to the circuit from a wall plug.

2. **Fuse**: Protects the circuit by breaking the connection if the current exceeds a certain level.

3. **Single-Pole Single-Throw Switch**: Controls the flow of electricity, allowing the circuit to be opened or closed.

4. **Transformer**: Changes the voltage level with primary and secondary windings.

5. **Diode**: Allows current to flow in one direction only, used for rectification (the process of converting alternating current (AC) to direct current (DC)).

6. **Capacitor**: Stores electrical energy and releases it when needed.

7. **Resistor**: Limits or opposes the current flow and adjusts signal levels in a circuit.

8. **Light Emitting Diode (LED)**: Emits light when current flows through it.

9. **Variable Resistor (Potentiometer)**: Manually adjusts resistance to control signal levels or other parameters.

10. **Zener Diode**: This device allows current to flow in the reverse direction when a specific voltage is reached, and it is used for voltage regulation.

Together, these components form a circuit that can transform, control, and utilize electrical energy for various purposes. Let's examine each one and prepare for the test. Take some time and begin to memorize the components and symbols.

Resistor

> ### T6A01: What electrical component opposes the flow of current in a DC circuit?
>
> A. Inductor
> **B. Resistor**
> C. Inverter
> D. Transformer

Don't be fooled by them throwing in the word 'DC circuit.' A resistor (component #7 in the schematic above) is an electrical component that opposes, limits, regulates, or "resists" the flow of electrical current in a circuit by providing resistance (in an

AC or DC circuit). It converts electrical energy into heat, helping to control voltage and current levels within the circuit. I remember the resistor as the zig-zag as it's throwing off heat while limiting the voltage passing through it.

Capacitor

A capacitor is an electrical component that stores energy in an electric field. It consists of two conductive plates separated by an insulating material known as the dielectric. When voltage is applied across these plates, an electric field forms, causing one plate to accumulate a positive charge and another to accumulate a negative charge. This stored energy can be quickly released when needed, making capacitors essential for various applications in electronic circuits.

T6A04: What electrical component stores energy in an electric field?

A. Varistor
B. Capacitor
C. Inductor
D. Diode

And then:

T6C06: What is component 6 in the figure?

A. Resistor
B. Capacitor
C. Regulator IC
D. Transistor

A capacitor (component #6) is an electrical component that stores and releases electrical energy by accumulating charge on its plates when connected to a power source. It is used to smooth out voltage fluctuations, filter signals, and store energy in electronic circuits. I remember this as "storing energy" on the horizontal line and then shooting it out on the curved line. Again, memorize the schematic however you want, but that's what I do.

T6A05: What type of electrical component consists of conductive surfaces separated by an insulator?

A. Resistor
B. Potentiometer
C. Oscillator
D. Capacitor

A capacitor is constructed with two conductive plates separated by an insulating material called the dielectric. The plates can be made of metal, and the dielectric can be composed of ceramic, plastic, or electrolytic substances. When a voltage is applied, electric charge accumulates on the plates, creating an electric field in the dielectric, which allows the capacitor to store energy.

Fuse

T6A09: What electrical component is used to protect other circuit components from current over-loads?

A. Fuse
B. Thyratron
C. Varactor
D. All these choices are correct

A fuse (component #2) is a safety device used in electrical circuits to protect against overcurrents and overloads, which can cause damage to equipment or even start a fire. It consists of a metal wire or strip that melts when too much current flows through it, breaking the circuit and stopping the flow of electricity. Once a fuse has blown, it must be replaced to restore the circuit, ensuring that any issues causing the overcurrent are addressed before resuming operation. This one is easy to remember because it looks like a fuse!

Switches

Switches are fundamental components in electrical and electronic circuits. They act as devices that can open or close a circuit, thereby controlling the flow of electrical current. They come in various forms, including toggle, push-button, rotary, and slide switches, each designed for specific applications. By providing a simple means to start or stop current flow, switches are used in virtually every type of electronic equipment, from household appliances to complex industrial machinery.

Understanding how switches work and their different types is key for anyone using electronics. Switches can range from simple single-pole single-throw (SPST) types, which control a single circuit, to more complex configurations like double-pole double-throw (DPDT) switches, which can control multiple circuits.

The Single-Pole Single-Throw (SPST) (component #3) switch is the simplest type of switch, featuring one input (pole) and one output (throw). It acts as an on-off switch that controls a single circuit, allowing current to flow when in the "on" position and interrupting the flow when in the "off" position. Think of this as looking down on two poles. But the switch only touches one – a single pole.

T6A12: What type of switch is represented by component #3 in the schematic?

A. Single-pole single-throw
B. Single-pole double-throw
C. Double-pole single-throw
D. Double-pole double-throw

A Single-Pole Double-Throw (SPDT) switch is a versatile component that allows a single input circuit to be switched between one of two output circuits. It has one common terminal and two output terminals, enabling the user to route the input signal to either of the two outputs. This functionality is helpful in applications where you must toggle between two different circuits or pathways, such as selecting between two audio sources or switching operational modes in a device.

T6A08: What is the function of an SPDT switch?

A. A single circuit is opened or closed
B. Two circuits are opened or closed
C. A single circuit is switched between one of two other circuits
D. Two circuits are each switched between one of two other circuits

It's helpful to think of an SPDT switch as a bridge connecting one path to either of two destinations. When the switch is in one position, the common terminal is connected to the first output terminal; when flipped to the other position, the common terminal connects to the second output terminal.

Relays: Electrically Controlled Switches

A relay is an electrically controlled switch that allows a low-power signal to control a higher-power circuit. It consists of an electromagnet, a set of contacts, and a switching mechanism. When a small current flows through the electromagnet, it generates a magnetic field that either opens or closes the contacts, switching the higher-power circuit on or off. This ability to control large electrical loads with a small input signal makes relays invaluable in various applications, from automotive systems to industrial machinery and home automation.

For beginners, it's helpful to think of a relay as a remote control for electrical circuits. By operating the relay using a small, safe electrical signal, you can manage high-voltage or high-current devices without direct human intervention, ensuring safety and efficiency.

T6D02: What is a relay?

A. An electrically-controlled switch
B. A current controlled amplifier
C. An inverting amplifier
D. A pass transistor

Diode

T6B02: What electronic component allows current to flow in only one direction?

A. Resistor
B. Fuse
C. Diode
D. Driven element

A diode (component #5) is an electronic component that allows current to flow in one direction only, acting as a one-way valve for electric current. It has two terminals: the anode (positive) and the cathode (negative). When a voltage is applied across the diode in the correct direction (forward bias), it conducts electricity; when the voltage is reversed (reverse bias), it blocks the current, providing essential functionality in circuits such as rectification, signal demodulation, and protection against reverse polarity. This one should be easy to remember, as it's an arrow and a bar. Traffic goes only one way.

T6B09: What are the names for the electrodes of a diode?

A. Plus and minus
B. Source and drain
C. Anode and cathode
D. Gate and base

Marking the Cathode Lead of a Semiconductor Diode

A semiconductor diode's (negative) cathode lead is often marked on the package with a distinctive stripe or band. This band is typically located near the cathode end of the diode, making it easy to identify the polarity. The other end of the diode, which is the anode, does not have this marking.

Why does this matter? Recognizing this marking ensures the diode (positive lead) is correctly oriented in the circuit. Installing a diode backward can prevent the circuit from functioning correctly or even damage the diode. Remember that the stripe indicates the cathode will help you correctly install diodes and avoid common mistakes.

T6B06: How is the cathode lead of a semiconductor diode often marked on the package?

A. With the word "cathode"
B. With a stripe
C. With the letter C
D. With the letter K

Understanding Forward Voltage Drop in Diodes

T6B01: Which is true about forward voltage drop in a diode?

A. It is lower in some diode types than in others
B. It is proportional to peak inverse voltage
C. It indicates that the diode is defective
D. It has no impact on the voltage delivered to the load

The forward voltage drop of a diode is the minimum voltage required to allow current to flow through the diode in the forward direction. When the applied voltage exceeds this threshold, the diode conducts electricity, allowing current to pass.

For example, a typical silicon diode has a forward voltage drop of about 0.7 volts, meaning that it needs at least 0.7 volts to start conducting. This voltage drop is due to the energy required to overcome the barrier potential of the diode's semiconductor material, which is essential for its operation in controlling the direction of current flow in electronic circuits.

This voltage drop varies among different types of diodes. For example, a standard silicon diode typically has a forward voltage drop of about 0.7 volts. In contrast, a Schottky diode has a lower forward voltage drop, around 0.2 to 0.3 volts. This lower voltage drop makes Schottky diodes more efficient in applications with low power loss.

For beginners, it's helpful to remember that the forward voltage drop is a characteristic to consider when selecting a diode for a specific application. Knowing that some diodes, like Schottky diodes, have a lower forward voltage drop can help you choose the suitable diode to minimize power loss and improve efficiency in your circuits.

LED (Light Emitting Diode)

T6C07: What is component 8 in Figure T-2?

A. Resistor
B. Inductor
C. Regulator IC
D. Light-emitting diode

Are you thinking, why does a diode sound so familiar? It does because of the 'light emitting diode' we see in component #8 ... yes, the LED, just like your TV screen.

A Light-Emitting Diode (LED) is a semiconductor device that emits light when an electrical current passes through it (otherwise known as a forward current—just like it sounds, current is moving forward through the diode). Unlike traditional incandescent bulbs, which produce light through filament heating, LEDs generate light through electroluminescence, a process in which electrons recombine with holes within the semiconductor material, releasing energy in the form of photons.

T6B07: What causes a light-emitting diode (LED) to emit light?

A. Forward current
B. Reverse current
C. Capacitively-coupled RF signal
D. Inductively-coupled RF signal

LEDs are highly efficient, consuming less power and offering longer lifespans than conventional light sources. They are widely used in various applications, from indicator lights and displays to general illumination. In the schematic, the LED is a visual indicator that shows when the circuit is active.

T6D07: Which of the following is commonly used as a visual indicator?

A. LED
B. FET
C. Zener diode
D. Bipolar transistor

This symbol is precisely like a diode, but it emits light in the form of lightning bolts.

Variable Resistor

T6C08: What is component 9 in the figure?

A. Variable capacitor
B. Variable inductor
C. Variable resistor
D. Variable transformer

This one has the same zig-zag as a resistor but with an extra line arrow, meaning it has some variability. It should now be memorized for you!

A variable resistor, also known as a <u>potentiometer</u> or rheostat, is an adjustable resistor used to <u>control the current or voltage in a circuit</u>. It consists of a <u>resistive</u> element and a <u>sliding or rotating</u> contact (wiper) that moves along the component, changing the resistance value. By adjusting the position of the wiper, you can increase or decrease the resistance, allowing for precise control over the electrical parameters in the circuit. Variable resistors are commonly used for tuning and calibration in electronic devices, such as <u>adjusting the volume on a radio</u> or the brightness of a light.

What was that again? If I listen to my car radio and want to adjust the volume?

Potentiometer

T6A02: What type of component is often used as an adjustable volume control?

A. Fixed resistor
B. Power resistor
C. Potentiometer
D. Transformer

Who knew it had such a fancy name? And what exactly does a potentiometer control?

T6A03: What electrical parameter is controlled by a potentiometer?

A. Inductance
B. Resistance
C. Capacitance
D. Field strength

Hopefully, you remembered that a potentiometer controls current, which is done with a resistor. So, it's resistance we are controlling here (fun how it all comes around full circle).

Transformer

T6C09: What is component 4 in the figure?

A. Variable inductor
B. Double-pole switch
C. Potentiometer
D. Transformer

One of the cooler-looking symbols (component #4), a transformer is an electrical device that transfers electrical energy <u>between two or more circuits</u> through electromagnetic induction.

It consists of two or more wire windings, called coils, wrapped around a common magnetic core. The primary coil receives an alternating current (AC) voltage, which creates a varying magnetic field in the core. This magnetic field induces a voltage in the secondary coil, which can be higher or lower than the primary voltage depending on the ratio of the number of turns in the coils.

Transformers are essential in power distribution. They allow electricity to be transmitted efficiently over long distances by <u>stepping up</u> the voltage to reduce energy loss and then <u>stepping it down</u> for safe use in homes and businesses.

T6D06: What component changes 120 V AC power to a lower AC voltage for other uses?

A. Variable capacitor
B. Transformer
C. Transistor
D. Diode

Here is a practical example. A transformer is essential to convert 120 V AC power to a lower AC voltage suitable for various applications. For beginners, it's helpful to think of a transformer as a device that adapts electrical energy to meet the needs of different devices safely. When you plug in a device that requires less voltage than the standard 120 V AC supply, the transformer ensures that the device receives the appropriate voltage, protecting it from damage and ensuring efficient operation.

Another Schematic to Work Through

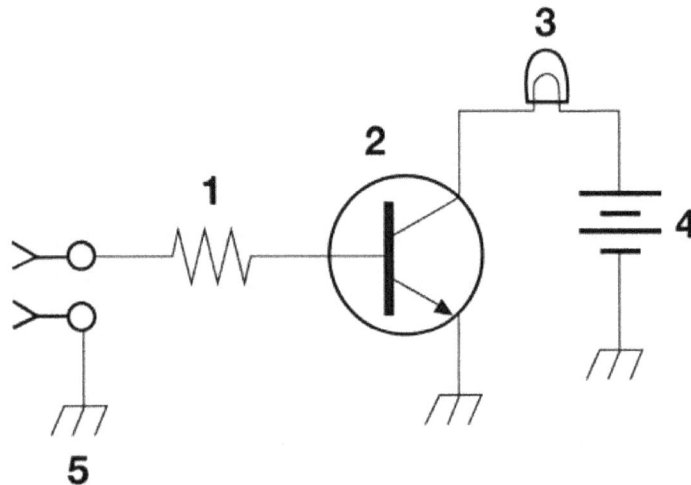

Here is the identification of each numbered part in the schematic:

1. **Resistor**: Limits the current flowing into the base of the transistor.

2. **Transistor (NPN)**: Acts as a switch or amplifier, controlling the current flow through the circuit.

3. **Lamp**: An indicator that lights up when current flows through, showing the circuit is active and functioning correctly.

4. **Battery**: Provides the electrical power for the circuit.

5. **Ground**: Indicates a reference point in the circuit where the voltage is considered to be zero. Ground symbols are essential in circuits to provide a common return path for electric current, prevent electrical shock, and maintain circuit stability.

Let's practice:

T6C02: What is component 1 in the figure?

A. Resistor
B. Transistor
C. Battery
D. Connector

Hopefully, you remember this zig-zag from the last section, too.

Transistor

T6C03: What is component 2 in the figure?

A. Resistor
B. Transistor
C. Indicator lamp
D. Connector

A transistor (component #2) is a semiconductor device that amplifies or switches electronic signals and electrical power. It has three regions of semiconductor material, which are important for its function as an amplifier or switch in electronic circuits: the base, the collector, and the emitter.

Applying a small current or voltage to the base allows the transistor to control a larger current flowing between the collector and the emitter. This capability makes transistors essential components in many electronic devices, enabling them to perform signal amplification, switching, and voltage regulation functions.

Transistors are essential—-, so they get a fancy symbol (again, how I remember it).

T6D10: What is the function of component 2?

A. Give off light when current flows through it
B. Supply electrical energy
C. Control the flow of current
D. Convert electrical energy into radio waves

Because it can switch electronic signals:

T6B03: Which of these components can be used as an electronic switch?

A. Varistor
B. Potentiometer
C. Transistor
D. Thermistor

And another one about transistors:

T6B04: Which of the following components can consist of three regions of semiconductor material?

A. Alternator
B. Transistor
C. Triode
D. Pentagrid converter

Transistors come in various types, the most common being bipolar junction transistors (BJTs) and field-effect transistors (FETs).

In BJTs (bipolar junction transistors), the current flowing into the base controls the current between the collector and emitter.

T6B12: What are the names of the electrodes of a bipolar junction transistor?

A. signal, bias, power
B. emitter, base, collector
C. Input, output, supply
D. Pole one, pole two, output

While in FETs (field effect transistors), a voltage applied to the gate controls the current between the drain and source.

T6B05: What type of transistor has a gate, drain, and source?

A. Varistor
B. Field-effect
C. Tesla-effect
D. Bipolar junction

To check if you remember what you read above:

T6B08: What does the abbreviation FET stand for?

A. Frequency Emission Transmitter
B. Fast Electron Transistor
C. Free Electron Transmitter
D. Field Effect Transistor

Power Gain in Transistors

A transistor can provide power gain, which means it can amplify the power of an input signal. This is achieved by controlling a large current flowing through the transistor with a smaller input current or voltage. In a standard configuration like the common-emitter arrangement in bipolar junction transistors (BJTs), a small current at the base controls a much larger current flowing from the collector to the emitter, significantly amplifying the signal's power.

For beginners, it is helpful to think of a transistor as a powerful amplifier. By applying a small input signal to one of its terminals, the transistor can produce a much larger output signal, making it an essential component in audio amplifiers, radio transmitters, and many other electronic devices.

Understanding that a transistor can provide power gain will help you recall this essential function for your exam and recognize its critical role in enhancing signal strength in various applications.

T6B10: Which of the following can provide power gain?-

A. Transformer
B. Transistor
C. Reactor
D. Resistor

The Relationship Between Gain and Transistors

Gain describes a device's ability to amplify a signal, representing the ratio of the output signal to the input signal in an electronic circuit. It is a critical parameter in amplifiers, indicating how much the signal's power, voltage, or current increases.

For beginners, gain measures how much an amplifier boosts a signal's strength. If an amplifier has a high gain, a small input signal will significantly increase the output, making the sound louder in audio systems or boosting weak radio signals.

Gain and transistors are closely related. Transistors are key components in amplifying signals, and their primary function is to provide gain. In a transistor, gain refers to the ratio of the output signal to the input signal, indicating how effectively the transistor can amplify the current, voltage, or power. Transistors, whether they are bipolar junction transistors (BJTs) or field-effect transistors (FETs), utilize their internal properties to control a larger current or voltage with a minor input, achieving amplification.

T6B11: What is the term that describes a device's ability to amplify a signal?

A. Gain
B. Forward resistance
C. Forward voltage drop
D. On resistance

Lamp

T6C04: What is component 3?

A. Resistor
B. Transistor
C. Lamp
D. Ground symbol

Hopefully, this one is easy to memorize, as it looks like a lamp or at least a small light bulb you screw in.

In the context of an electrical circuit, a lamp (component #3) is a device that produces light when electrical current flows through it. It typically consists of a filament or a light-emitting diode (LED) that glows when energized by an electrical current. In the schematic provided, the lamp acts as an indicator, illuminating to show that the circuit is active and the current is flowing correctly. This visual feedback helps confirm the circuit's operation and diagnose any issues with the current flow.

Battery

> **T6C05: What is component 4 in the figure?**
>
> A. Resistor
> B. Transistor
> C. Ground symbol
> **D. Battery**

A battery (component #4) is a device that stores chemical energy and converts it into electrical energy to provide a steady supply of direct current (DC) to a circuit. It consists of one or more electrochemical cells, each containing positive and negative electrodes immersed in an electrolyte.

When connected to a circuit, a chemical reaction occurs within the battery, generating a flow of electrons from the negative terminal to the positive terminal, providing the necessary power to operate electronic devices. In the schematic, the battery acts as the primary power source, supplying the electrical energy needed for the components, such as the lamp and transistor, to function. In this case, the battery component looks like a battery having multiple polls or cells (again, it's just a way to memorize what it is on a schematic). More on batteries in the power chapter.

One More Schematic to Work Through

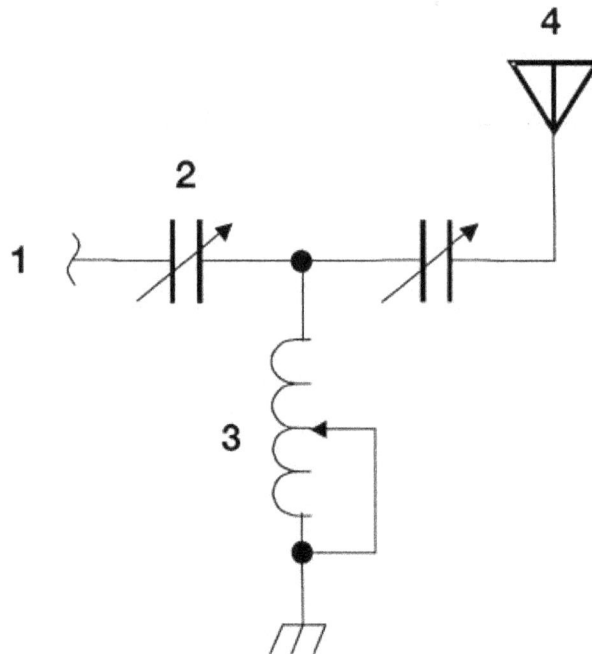

Here is the identification of each numbered part in the schematic:

1. **AC Power Source**: Carries the power to our equipment.

2. **Variable Capacitor**: Allows for capacitance adjustment to tune the circuit for resonance.

3. **Variable Inductor**: Works with the capacitors to form a resonant circuit, which can be tuned to match the antenna's impedance.

4. **Antenna**: Radiates the radio frequency signal into the air or receives signals from the air.

This circuit is an antenna tuner used to match the impedance of the transmission line to the antenna for efficient signal transmission and reception. The chapter on antennas explains this more.

Inductor

An inductor is an electrical component that <u>stores energy in a magnetic field</u> when electric current flows through it. It consists of a <u>coil of wire, and as current passes through it, it generates a magnetic field</u> around it. When the current changes, the magnetic field changes, inducing a voltage that opposes the change in current. This property of inductors makes them helpful in filtering signals, temporarily storing energy, and managing the flow of AC and DC currents in circuits.

For beginners, think of an inductor as a device that resists changes in current, storing energy in the form of a magnetic field when current flows through it. This ability to store and release energy helps <u>smooth out fluctuations in current</u>, making inductors essential in power supplies, radio transmitters, and various other electronic applications.

> **T6A06: What type of electrical component stores energy in a magnetic field?**
>
> A. Varistor
> B. Capacitor
> **C. Inductor**
> D. Diode

> **T6A07: What electrical component is typically constructed as a coil of wire?**
>
> A. Switch
> B. Capacitor
> C. Diode
> **D. Inductor**

Variable Inductor

> **T6C10: What is component 3 in the figure?**
>
> A. Connector
> B. Meter
> C. Variable capacitor
> **D. Variable inductor**

A variable inductor is an electrical component whose inductance can be adjusted. Inductance is the property of a component that opposes changes in current flow and is measured in henries (H).

A variable inductor typically <u>consists of a coil of wire</u> (like the component image in #3), and its inductance can be changed by altering the position of a movable core (often made of ferrite) within the coil or by adjusting the spacing between turns of the coil.

Key Characteristics:

1. **Adjustability**: The ability to fine-tune the inductance allows for precise control of the circuit's characteristics, making variable inductors useful in applications requiring tuning, such as radio frequency (RF) circuits, filters, and oscillators.

2. **Construction**: Variable inductors are usually built with a screw mechanism that moves the core in and out of the coil or changes the coil's geometry. This adjustability affects the magnetic field and, thus, the inductance.

3. **Applications**: They are commonly used in RF tuning circuits, antenna matching networks, and any application where the inductance needs to be varied to achieve optimal performance.

Practical Example:

In an antenna tuning circuit, a variable inductor can be adjusted to match the antenna's impedance with the transmitter's. This matching ensures maximum power transfer and efficient signal transmission.

Understanding the function and use of a variable inductor helps design and optimize circuits requiring precise inductance control, ensuring better performance and flexibility in various electronic applications.

Antenna

T6C11: What is component 4 in the figure?

A. Antenna
B. Transmitter
C. Dummy load
D. Ground

A lot more on antennas later. Just remember this "antenna-looking" thing is meant to illustrate an antenna on a schematic. Makes sense...right!

Other Components

Rectifiers

Converting AC to DC

A rectifier is a device or circuit that <u>converts alternating current (AC) into varying direct current (DC).</u> This conversion is essential in power supplies for electronic devices, which typically require DC power. Rectifiers allow current to flow only in one direction, using components such as diodes. When AC voltage is applied to a rectifier, the diodes block the negative portions of the AC cycle, resulting in a pulsating DC output.

Think of a rectifier as a one-way valve for electrical current. It takes the back-and-forth motion of AC and turns it into a steady flow of DC, which is necessary for powering most electronic circuits.

T6D01: Which of the following devices or circuits changes an alternating current into a varying direct current signal?

A. Transformer
B. Rectifier
C. Amplifier
D. Reflector

Meters

Displaying Electrical Quantities

A meter is a device that displays an electrical quantity as a numeric value, allowing for precise measurement and monitoring of various parameters within a circuit. Common types of meters include voltmeters, which measure voltage; ammeters, which measure current; and ohmmeters, which measure resistance. These devices provide accurate readings essential for diagnosing and troubleshooting electrical issues and ensuring that circuits operate within their specified parameters.

For beginners, think of a meter as the instrument panel for your electrical circuit, much like a speedometer in a car. It provides real-time, numerical data that helps you understand the electrical system's behavior. Whether you are checking the voltage of a power supply, measuring the current draw of a device, or verifying the resistance of a component, meters are indispensable tools for anyone working with electronics.

T6D04: Which of the following displays an electrical quantity as a numeric value?

A. Potentiometer
B. Transistor
C. Meter
D. Relay

Regulators (Voltage)

Controlling Voltage in Circuits

A voltage regulator is a circuit designed to maintain a constant output voltage from a power supply, regardless of variations in input voltage or load conditions. It "regulates" the voltage. This ensures that electronic devices receive a stable and reliable voltage, which is crucial for their proper functioning and longevity. Voltage regulators can be found in various forms, such as linear and switching regulators, each with its own method of controlling and stabilizing the voltage.

Think of a voltage regulator as a manager that ensures the voltage supplied to your electronic components remains steady, much like a thermostat maintains a constant temperature. By keeping the voltage within a specific range, voltage regulators prevent damage to sensitive components and ensure consistent performance of electronic devices.

T6D05: What type of circuit controls the amount of voltage from a power supply?

A. Regulator
B. Oscillator
C. Filter
D. Phase inverter

Circuits

Components like resistors, capacitors, and inductors have unique properties and functions in electronics. However, these components' true power emerges when combined in various configurations.

By strategically connecting different components, we can create circuits with new capabilities and behaviors, such as filtering signals, storing energy, and generating oscillations.

Combining these basic building blocks opens up many possibilities for designing more complex and functional electronic systems. This section will delve into the principles and techniques of combining components, providing a foundation for creating sophisticated circuits with a wide range of practical applications.

Resonant Circuit

A resonant circuit, also known as a tuned circuit, is an electrical circuit that combines inductors and capacitors to create a system capable of oscillating at a specific frequency known as the resonant frequency. At this frequency, the inductive and capacitive reactance are equal in magnitude but opposite in phase, causing them to cancel each other out. This results in the circuit having a high impedance at the resonant frequency, allowing it to store and transfer energy efficiently between the inductor and capacitor.

T6D08: Which of the following is combined with an inductor to make a resonant circuit?

A. Resistor
B. Zener diode
C. Potentiometer
D. Capacitor

At resonance, energy oscillates back and forth between the inductor's magnetic field and the capacitor's electric field, enabling the circuit to store energy effectively.

Resonant circuits are widely used in radio frequency applications, such as tuning radios to specific frequencies, filtering signals, and generating stable frequencies in oscillators.

T6D11: Which of the following is a resonant or tuned circuit?

A. An inductor and a capacitor in series or parallel
B. A linear voltage regulator
C. A resistor circuit used for reducing standing wave ratio
D. A circuit designed to provide high-fidelity audio

Don't let the series or parallel fool you. An inductor and a capacitor is a resonant circuit.

Integrated Circuit

An integrated circuit (IC) is a miniature electronic device comprising a complex assembly of interconnected electronic components, such as transistors, diodes, resistors, and capacitors, all fabricated onto a single piece of semiconductor material, typically silicon. These components are embedded in a tiny chip, which can perform various functions depending on the circuit's design and configuration.

T6D09: What is the name of a device that combines several semiconductors and other components into one package?

A. Transducer
B. Multi-pole relay
C. Integrated circuit
D. Transformer

Integrated circuits have revolutionized electronics by significantly reducing their size, cost, and power consumption while increasing reliability and performance. They are used in virtually all modern electronic equipment, from simple calculators and digital watches to complex computers and communication systems.

Understanding integrated circuits is crucial for anyone involved in electronics. They form the backbone of modern electronic technology and enable the creation of sophisticated and compact devices.

Go back through one more time. Study the schematics and ensure you can name each component.

Chapter 14

Essential Tools and Techniques for Electronics

HAVING THE RIGHT TOOLS and mastering fundamental techniques are crucial for success in electronics. Whether you are a beginner just starting out or an experienced enthusiast, understanding how to use basic test instruments and perform essential tasks like soldering is fundamental. This chapter will guide you through the use of voltmeters, ammeters, and ohmmeters, explaining how these tools help measure and diagnose electrical quantities. Additionally, we will delve into soldering, a critical skill for assembling and repairing electronic circuits.

By the end of this chapter, you will be equipped with the knowledge to accurately measure voltage, current, and resistance and understand the fundamentals of joining components together using soldering techniques. These skills are the building blocks of practical electronic work.

Voltmeter

A voltmeter is a measuring instrument used to measure the electrical potential difference, or voltage, between two points in an electrical circuit. It is an essential tool for anyone working with electronics, as it helps ensure that circuits operate correctly and safely. Voltmeters can be analog or digital, with digital voltmeters providing numerical readings on a display and analog voltmeters using a needle to indicate the voltage on a scale.

T7D01: Which instrument would you use to measure electric potential?

A. An ammeter
B. A voltmeter
C. A wavemeter
D. An ohmmeter

Voltmeters are typically connected in parallel with the component or section of the circuit being measured. This parallel connection allows the voltmeter to measure the voltage without significantly affecting the circuit's operation. Using a voltmeter, you can diagnose issues, verify circuit functionality, and ensure that your electronic devices receive the correct voltage. Understanding how to use a voltmeter is fundamental for troubleshooting and maintaining electronic systems.

T7D02: How is a voltmeter connected to a component to measure applied voltage?

A. In series
B. In parallel
C. In quadrature
D. In phase

Precautions When Measuring High Voltages

When measuring high voltages with a voltmeter, it is vital to <u>ensure that the voltmeter and its leads are rated for the voltages you intend to measure.</u> Using equipment not rated for high voltage can lead to inaccurate readings, damage to the equipment, or even personal injury. High voltage can cause arcs, which might damage the voltmeter or create hazardous conditions for the operator.

T0A12: Which of the following precautions should be taken when measuring high voltages with a voltmeter?

A. Ensure that the voltmeter has very low impedance
B. Ensure that the voltmeter and leads are rated for use at the voltages to be measured
C. Ensure that the circuit is grounded through the voltmeter
D. Ensure that the voltmeter is set to the correct frequency

Always check the voltmeter's maximum voltage rating and probes to measure high voltages safely. This rating should be clearly marked and must exceed the voltage you plan to measure. Additionally, appropriate personal protective equipment, such as insulated gloves and eye protection, should be used to reduce the risk of electric shock. Understanding and following these precautions is essential for ensuring accurate measurements and safety during high-voltage testing.

Ammeter

An ammeter is a measuring instrument used to <u>measure the electric current flowing</u> through a circuit. It is a tool for anyone working with electronics, as it allows for monitoring and troubleshooting of current flow to ensure that circuits operate within their intended parameters. Ammeters can be analog or digital.

T7D04: Which instrument is used to measure electric current?

A. An ohmmeter
B. An electrometer
C. A voltmeter
D. An ammeter

To measure current accurately, an <u>ammeter is connected in series</u> with the component or section of the circuit being measured. <u>This series connection allows the entire current flowing through the circuit to pass through the ammeter.</u> Using an ammeter, you can detect issues such as excessive current that might indicate a short circuit or insufficient current that could suggest a poor connection or high resistance.

Connected in Series

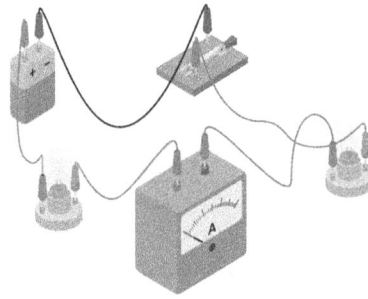

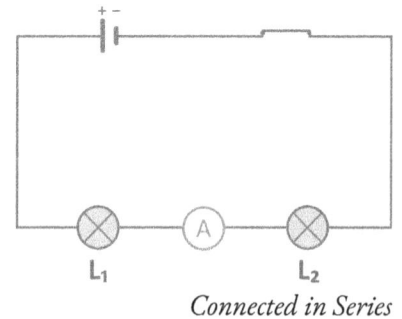

T7D03: When configured to measure current, how is a multimeter connected to a component?

A. In series
B. In parallel
C. In quadrature
D. In phase

Ohmmeter

An ohmmeter is an instrument used to <u>measure the electrical resistance</u> of a component or a circuit. Resistance measures how much a component opposes the flow of electric current and is typically measured in ohms (Ω).

To measure resistance, an ohmmeter is connected across the component or section of the circuit being tested, with the <u>circuit power turned off</u> to avoid damaging the meter or the circuit.

The <u>ohmmeter applies a small known voltage to the component</u>. It measures the resulting current flow, using Ohm's Law to calculate the resistance. Using an ohmmeter, you can check for open and short circuits and ensure that components have the correct resistance values.

T7D11: Which of the following precautions should be taken when measuring in-circuit resistance with an ohmmeter?

A. Ensure that the applied voltages are correct
B. Ensure that the circuit is not powered
C. Ensure that the circuit is grounded
D. Ensure that the circuit is operating at the correct frequency

---

Understanding Ohmmeter Readings Across a Discharged Capacitor

When an ohmmeter is connected across a large, discharged capacitor, <u>the reading will show an increasing resistance with time.</u> This occurs because the <u>capacitor starts to charge due to the small voltage supplied by the ohmmeter.</u> Initially, the ohmmeter measures a low resistance as the capacitor is uncharged and allows current to flow freely. As the capacitor charges, the current flow decreases, causing the resistance reading on the ohmmeter to increase gradually.

For beginners, it's helpful to visualize this process as the capacitor filling up with electric charge. When fully discharged, the capacitor behaves almost like a short circuit with minimal resistance. As it charges, it starts to resist the current flow more and more, leading to an increasing resistance reading. Understanding this behavior helps identify and verify capacitors in a circuit, ensuring they function correctly. Remembering that a rising resistance reading indicates a charging capacitor will help you recall this concept for your exam and practical applications. Be safe!

T7D10: What reading indicates that an ohmmeter is connected across a large, discharged capacitor?

A. Increasing resistance with time
B. Decreasing resistance with time
C. Steady full-scale reading
D. Alternating between open and short circuit

Multimeter

A multimeter combines the functions of several measurement tools into a single device, typically including the ability to measure voltage (voltmeter), current (ammeter), and resistance (ohmmeter). Multimeters come in analog and digital forms, with digital multimeters (DMMs) being more common due to their ease of use and precision.

Very rarely, in today's world, will you see a separate voltmeter, ammeter, or ohmmeter. Today, they are all combined into one tool – a multimeter.

Functions of a Multimeter

1. **Voltage Measurement (Voltmeter)**: This allows you to measure the electrical potential difference between two points in a circuit. You can measure AC (alternating current) and DC (direct current) voltages.

2. **Current Measurement (Ammeter)**: This function enables you to measure the flow of electric current through a component or section of a circuit. It helps diagnose issues related to current flow.

3. **Resistance Measurement (Ohmmeter)**: This measures the opposition to current flow within a component or circuit. It helps check the integrity of resistors and ensure there are no short circuits or open circuits.

In ham radio, multimeters are indispensable tools for building, testing, and maintaining radio equipment. They help ensure that circuits function correctly, identify faults, and verify the proper operation of various components. For instance, you can use a multimeter to check a power supply's voltage, measure the transmitter's current draw, or test the resistance of a connection. By mastering a multimeter, ham radio operators can ensure their equipment operates efficiently and effectively, leading to better communication and more reliable performance.

T7D07: Which of the following measurements are made using a multimeter?

A. Signal strength and noise
B. Impedance and reactance
C. Voltage and resistance
D. All these choices are correct

Because it's an all-in-one tool, a multimeter can be easily damaged if you attempt to mismeasure components.

For example, suppose you attempt to <u>measure voltage</u> while it is set to <u>measure resistance.</u> In that case, the multimeter applies a small voltage to the circuit in resistance mode to measure the current flow and calculate the resistance. When you connect the multimeter to a voltage source while in this mode, the external voltage can overload the multimeter's internal components, potentially causing permanent damage.

It's wise always to double-check that your multimeter is set to the correct measurement mode before taking any readings. This simple habit can prevent costly mistakes and ensure your multimeter remains in good working condition. Remember, using the wrong setting risks damaging your equipment and can lead to incorrect readings and faulty diagnostics.

T7D06: Which of the following can damage a multimeter?

A. Attempting to measure resistance using the voltage setting
B. Failing to connect one of the probes to ground
C. Attempting to measure voltage when using the resistance setting
D. Not allowing it to warm up properly

One note here. This is one of those tricky questions that you need to make sure you are paying attention to. Answer 'A' might appear valid. It's the opposite of 'C,' the correct answer.

But why is this statement wrong? When you set a multimeter to measure voltage, it is designed to measure the potential difference between two points without sending any current into the circuit. If you attempt to measure resistance while in the voltage setting, the multimeter will not send the necessary small current through the component to measure its resistance, leading to a failed or inaccurate measurement. However, this action does not usually harm the multimeter itself.

Solder

Understanding Solder and Its Importance in Ham Radio

Solder is a fusible metal alloy that creates a permanent bond between metal workpieces. It is commonly composed of tin and lead, although lead-free varieties are also available for health and environmental reasons. The soldering process involves melting the solder and applying it to the joint between two metal surfaces, such as electronic components and circuit board pads, where it cools and solidifies, forming a solid electrical and mechanical connection.

In ham radio, soldering is used to assemble and maintain electronic equipment. Proper soldering ensures reliable electrical connections, which are vital for the performance and longevity of radio circuits. Here are some key reasons why solder is vital in ham radio:

1. **Component Assembly**: Soldering attaches electronic components to printed circuit boards (PCBs), ensuring they stay in place and function correctly.

2. **Signal Integrity**: Good solder joints provide low-resistance connections essential for maintaining signal quality and preventing signal loss or interference.

3. **Durability**: Soldered connections are mechanically robust, which is vital for withstanding the physical stresses that can occur in portable or mobile radio equipment.

4. **Repairs and Modifications**: Soldering skills are necessary for repairing faulty equipment and modifying or up-grading radio gear to improve performance or add new features.

Mastering soldering techniques is essential for any ham radio operator. It enables the construction, maintenance, and repair of radio equipment, ensuring reliable communication and optimal performance. Practice makes perfect. This is a skill you master over time.

The Importance of Using the Right Solder in Electronics

For radio and electronic applications, avoiding acid-core solder is imperative. Acid-core solder contains a flux with an acid-based cleaning agent designed for plumbing and metalwork, not for delicate electronic components. The acid flux can corrode and damage electronic parts, leading to poor electrical connections and potential failure of the circuit over time.

Instead, rosin-core solder should be used for radio and electronic work. Rosin-core solder contains a safe flux for electronic components, providing good cleaning action without the risk of corrosion. Using the correct solder type ensures reliable and long-lasting connections for electronic devices' proper functioning and durability.

T7D08: Which of the following types of solder should not be used for radio and electronic applications?

A. Acid-core solder
B. Lead-tin solder
C. Rosin-core solder
D. Tin-copper solder

Read the question carefully and pick up on the "NOT to be used in radio."

Cold Tin-Lead Solder Joints

A cold tin-lead solder joint is characterized by a rough or lumpy surface, indicating that the solder did not properly melt and flow during the soldering process. This typically occurs when the soldering iron does not reach a high enough temperature or when the joint is disturbed before the solder has thoroughly cooled and solidified. Cold solder joints are weak and unreliable, leading to poor electrical conductivity and potential connection failure.

For beginners, it's essential to recognize that a proper solder joint should have a smooth, shiny, and uniform appearance, ensuring a solid mechanical and electrical connection. Understanding the appearance of cold solder joints helps diagnose and fix soldering issues, ensuring the reliability and performance of your electronic circuits.

T7D09: What is the characteristic appearance of a cold tin-lead solder joint?

A. Dark black spots
B. A bright or shiny surface
C. A rough or lumpy surface
D. Excessive solder

Understanding and mastering essential tools like voltmeters, ammeters, and ohmmeters, along with proper soldering techniques, are foundational skills in electronics. These skills ensure accurate measurements and reliable connections and enhance the overall performance and durability of electronic equipment. As you continue your journey in electronics, remember that these fundamental practices are the building blocks for more advanced projects and successful repairs.

Chapter 15

The Ham Radio Station

WHEN YOU'RE READY TO set up your first ham radio station, it's like preparing to plant a garden. It would help to have the right tools, a good piece of land, and knowledge about what you're about to grow. Similarly, setting up a ham radio station requires a thoughtful selection of equipment, careful setup planning, and a keen eye for safety and operational efficiency. Let's walk through these steps individually to ensure your first foray into setting up a ham radio station is successful and enjoyable.

Transmitter + Receiver = Transceiver

Selecting your first transceiver— a device that transmits and receives radio signals—is akin to picking the seed you will plant. It must suit your environment and the communication you wish to engage in. Start with a clear understanding of your goals. Are you interested in local communications or aiming to reach across continents? This decision will influence whether you should opt for a VHF/UHF transceiver, mainly used for local and regional communication, or an HF transceiver capable of global communication.

T7A02: What is a transceiver?

A. A device that combines a receiver and transmitter
B. A device for matching feed line impedance to 50 ohms
C. A device for automatically sending and decoding Morse code
D. A device for converting receiver and transmitter frequencies to another band

---

Your budget is another factor. Transceivers vary in price from relatively inexpensive to quite costly, depending on their features and capabilities. As a beginner, it might be wise to start with something more fundamental; many entry-level transceivers offer a solid foundation to learn without overwhelming you with too many advanced features. Space is also a consideration—ensure the equipment size matches your available area. A compact, all-in-one transceiver might be ideal if space is limited.

The Theoretical and Practical

Let's recap and start to bring the theoretical and practical together. The science of ham radio begins with radio waves, forms of electromagnetic energy that travel through the air carrying information (remember we covered this already! You do remember it, right?). At its core, the process involves a transmitter converting information into a radio wave, then sending it across distances where a receiver captures it and converts it back into usable information.

Depending on what was initially sent, this might be voice, data, or Morse code. The transmitter begins by taking electrical signals from the microphone or other input device and, using a component called a modulator, alters these signals into a form suitable for transmission - the oscillator generates the radio frequency (RF) carrier wave necessary for carrying the information. This modulation (remember modulation?) can vary; it might involve adjusting the base signal's amplitude

(AM), frequency (FM), or phase (PM) - the carrier wave - to encode the information. Once modulated, the signal is sent to an antenna, converted into a radio wave, and broadcast over the airwaves.

The receiver's job is to capture these radio waves using its antenna and then demodulate them, reversing the modulation process to extract the original information from the carrier wave. The receiver features a mixer, which takes the incoming RF signal and combines it with a locally generated signal to produce an intermediate frequency (IF) signal that is easier to process. The IF stage helps improve selectivity—the ability to discriminate between multiple signals—and sensitivity, the ability to detect weak signals. Following this stage, the signal undergoes filtering and amplification, preparing it for the demodulation stage, where the original audio or data signal is recovered.

This is achieved through a component called the demodulator. After demodulation, the signal is usually weak and needs to be amplified; this is where the receiver's amplifier steps in, boosting the signal to a level suitable for output to speakers, a headset, or other devices. The precision with which these tasks are performed directly affects the quality and clarity of the received transmission, making the design and condition of transmitters and receivers indispensable for effective communication.

Features

Reverse Function

The "reverse" function on a VHF/UHF transceiver listens to a repeater's input frequency rather than its output frequency. Typically, when using a repeater, you transmit on the repeater's input frequency and receive on its output frequency. Your transceiver temporarily switches to listening on the input frequency by activating the reverse function. This lets you hear the station transmitting directly to the repeater, bypassing the repeater's output signal.

T2B01: How is a VHF/UHF transceiver's "reverse" function used?

A. To reduce power output
B. To increase power output
C. To listen on a repeater's input frequency
D. To listen on a repeater's output frequency

This feature is handy for troubleshooting and improving communication. For example, suppose you hear a weak or distorted signal on the repeater's output. In that case, you can use the reverse function to check the input signal's quality. The issue might be with the repeater if the input signal is strong and clear. The reverse function can also help you establish direct contact with other stations during emergencies or high-traffic times, ensuring more reliable communication.

Utilizing Memory Channels and Scanning Functions on FM Transceivers

Many operators utilize memory channels and scanning functions to enhance the convenience and efficiency of operating an FM transceiver. Storing a favorite frequency or channel in a memory channel allows quick and easy access without manually tuning each time. This feature is useful for frequently used frequencies (funny turn of phrase), such as local repeaters, common simplex channels, or emergency communication frequencies. By storing these frequencies in memory, you can instantly switch to them with 'the push of a button,' saving time and ensuring you are always ready to communicate.

T4B04: What is a way to enable quick access to a favorite frequency or channel on your transceiver?

A. Enable the frequency offset
B. Store it in a memory channel
C. Enable the VOX
D. Use the scan mode to select the desired frequency

This is no different than setting your favorite radio stations in your car.

Additionally, the scanning function of an FM transceiver enhances the ability to monitor a range of frequencies. When the scanning function is activated, the transceiver <u>automatically tunes through a preset range of frequencies, pausing whenever it detects activity</u>. This lets operators quickly find active channels without manually tuning through each frequency. Scanning helps identify active conversations, monitor emergency frequencies, or check for open channels in a busy band. Together, memory channels and scanning functions streamline the operation of FM transceivers, making it easier for amateur radio operators to stay connected and informed.

T4B05: What does the scanning function of an FM transceiver do?

A. Checks incoming signal deviation
B. Prevents interference to nearby repeaters
C. Tunes through a range of frequencies to check for activity
D. Checks for messages left on a digital bulletin board

Multimode

<u>Having multiple receive bandwidth choices on a multimode transceiver provides a significant advantage in reducing noise and interference.</u> Each transmission mode, such as SSB, CW, or FM, benefits from a specific bandwidth that optimizes signal clarity and reception quality. For instance, SSB typically uses a narrower bandwidth (around 3 kHz) compared to FM, which uses a wider bandwidth (around 200 kHz). By selecting the appropriate bandwidth for the mode you are operating, you can minimize unwanted noise and interference, improving the overall quality of the received signal.

T4B08: What is the advantage of having multiple receive bandwidth choices on a multimode transceiver?

A. Permits monitoring several modes at once by selecting a separate filter for each mode
B. Permits noise or interference reduction by selecting a bandwidth matching the mode
C. Increases the number of frequencies that can be stored in memory
D. Increases the amount of offset between receive and transmit frequencies

This flexibility is particularly beneficial in crowded or noisy environments, where signals from adjacent channels or other sources can cause interference. By adjusting the receive bandwidth to match the mode, you can effectively filter out these unwanted signals, enhancing your ability to hear the intended transmission clearly.

The Transmitter

<u>The PTT (Push-To-Talk) input on a transceiver is a control that switches the device from receive mode to transmit mode when grounded.</u> When the operator presses the PTT button, it completes a circuit that grounds the PTT input, signaling the transceiver to stop receiving and start transmitting. This function facilitates communication, allowing operators to switch between listening and talking with a simple button push.

Remembering the function of the PTT input is straightforward if you remember it as the "talk button" on your radio. You activate the transmitter by grounding the PTT input, enabling your voice or data to be sent out over the airwaves.

T7A07: What is the function of a transceiver's PTT input?

A. Input for a key used to send CW
B. Switches transceiver from receive to transmit when grounded
C. Provides a transmit tuning tone when grounded
D. Input for a preamplifier tuning tone

An oscillator is a circuit that generates a signal at a specific frequency. It produces a continuous, periodic waveform—typically a sine or square wave—essential for various radio communication and electronics functions. Oscillators are the heart of many devices, providing the stable frequencies needed for signal generation, clock timing in digital circuits, and frequency synthesis in transceivers.

By generating precise and stable frequencies, oscillators ensure that communication systems operate accurately and reliably.

T7A05: What is the name of a circuit that generates a signal at a specific frequency?

A. Reactance modulator
B. Phase modulator
C. Low-pass filter
D. Oscillator

A mixer converts a signal from one frequency to another. It combines two input signals—typically, a signal from the oscillator and the incoming RF signal—to produce new signals at the sum and difference of the original frequencies. This process is known as frequency mixing or heterodyning.

Remember, a mixer's function is to "mix" frequencies to create new ones. This frequency conversion is vital for tuning and demodulating signals in communication systems.

T7A03: Which of the following is used to convert a signal from one frequency to another?

A. Phase splitter
B. Mixer
C. Inverter
D. Amplifier

A transverter is a device that converts the RF (radio frequency) input and output of a transceiver to another band. This allows a transceiver that operates on one frequency band to communicate on different frequency bands. Essentially, the transverter extends the frequency range of a transceiver, enabling it to access bands that it was not originally designed to cover. For example, a transverter can convert a 28 MHz (10-meter band) signal to 144 MHz (2-meter band) and vice versa.

Remembering the term transverter helps understand how amateur radio operators can expand their communication capabilities without needing multiple transceivers. Operators can use a transverter to explore new frequency bands and enhance flexibility in various communication scenarios. This concept is particularly valuable for operators who want to experiment with different bands or who need to adapt their equipment for specific operating conditions.

T7A06: What device converts the RF input and output of a transceiver to another band?

A. High-pass filter
B. Low-pass filter
C. Transverter
D. Phase converter

To enter a transceiver's operating frequency, you can use either the keypad or the VFO (Variable Frequency Oscillator) knob. The keypad allows for direct frequency input by typing in the desired frequency digits, which is particularly useful for quickly setting a specific frequency. This method provides precision and speed, making it easy to switch between frequencies without manually tuning through the band.

T4B02: Which of the following can be used to enter a transceiver's operating frequency?

A. The keypad or VFO knob
B. The CTCSS or DTMF encoder
C. The Automatic Frequency Control
D. All these choices are correct

Alternatively, the VFO knob allows you to adjust the frequency incrementally. By turning the knob, you can fine-tune the operating frequency up or down, which helps find the exact signal or make minor adjustments. The VFO knob is handy during live communication when you need to alter the frequency to improve reception or avoid interference slightly. Understanding how to use both the keypad and VFO knob ensures efficient and accurate frequency management on your transceiver.

The Receiver

Receiver sensitivity refers to a radio receiver's ability to detect the presence of weak signals. It is a critical parameter that determines how well a receiver can pick up signals, especially those faint or distant. A high-sensitivity receiver can detect lower-power signals, making it more effective in various communication scenarios, including long-distance or weak signal operations.

The term sensitivity is essential to remember because it directly impacts the performance of your radio equipment. This highlights the importance of having a sensitive receiver to ensure reliable communication, even under challenging conditions where signals might be weak or subject to interference.

T7A01: Which term describes the ability of a receiver to detect the presence of a signal?

A. Linearity
B. Sensitivity
C. Selectivity
D. Total Harmonic Distortion

Selectivity refers to the ability of a radio receiver to discriminate between multiple signals that are close in frequency. This means that a receiver with high selectivity can effectively isolate and process the desired signal while rejecting nearby unwanted signals or interference. Selectivity is core in crowded radio environments, where many signals are present within a narrow frequency range, such as during contests or in urban areas with numerous transmissions.

A highly selective receiver ensures clear and accurate reception by filtering out adjacent channel interference, allowing operators to focus on the intended communication.

T7A04: Which term describes the ability of a receiver to discriminate between multiple signals?

A. Discrimination ratio
B. Sensitivity
C. Selectivity
D. Harmonic distortion

Electronic Keyer

An electronic keyer is a device designed to assist amateur radio operators in manually sending Morse code. This device simplifies the process of creating the precise timing needed for dots and dashes in Morse code transmissions. When an operator presses the keyer paddles, the electronic keyer generates consistent and accurate Morse code signals, making sending clear and readable messages easier.

T4A12: What is an electronic keyer?

A. A device for switching antennas from transmit to receive
B. A device for voice-activated switching from receive to transmit
C. A device that assists in manual sending of Morse code
D. An interlock to prevent unauthorized use of a radio

Using an electronic keyer can significantly improve the efficiency and accuracy of Morse code communication. It helps to reduce the fatigue associated with manual keying. It ensures that the transmitted code is uniform and easy to decipher. For beginners and experienced operators alike, an electronic keyer is a valuable tool for enhancing Morse code proficiency and ensuring high-quality transmissions.

Power Amplifier

An RF power amplifier is a device that increases the transmitted output power from a transceiver. By boosting the strength of the radio frequency signal generated by the transceiver, an RF power amplifier enables the signal to travel farther and penetrate obstacles more effectively, thereby improving communication range and signal quality. This is particularly useful when higher power is needed to maintain a clear and reliable connection, such as long-distance communication or in environments with significant signal attenuation.

Remembering the role of an RF power amplifier is simple if you think of it as a "signal booster" for your transceiver. It takes the initial signal and amplifies it to a higher power level, making it more robust and capable of overcoming the challenges of distance and interference.

T7A10: What device increases the transmitted output power from a transceiver?

A. A voltage divider
B. An RF power amplifier
C. An impedance network
D. All these choices are correct

An RF preamplifier device will be installed between the antenna and the receiver to amplify weak signals before they enter the receiver. This amplification boosts the signal strength, making it easier for the receiver to process and decode the incoming transmission. By enhancing weak signals, an RF preamplifier improves the overall sensitivity and performance of the receiving system, allowing for clearer and more reliable reception of distant or faint signals.

To remember where an RF preamplifier is installed, consider it the "first stage" in the signal path from the antenna to the receiver. Placing the preamplifier right after the antenna ensures the signal is amplified at the earliest possible stage, minimizing noise and signal degradation.

T7A11: Where is an RF preamplifier installed?

A. Between the antenna and receiver
B. At the output of the transmitter power amplifier
C. Between the transmitter and the antenna tuner
D. At the output of the receiver audio amplifier

At this point, the technician's exam will ask particular questions about VHF.

Understanding the SSB/CW-FM Switch on a VHF Power Amplifier

The SSB/CW-FM switch on a VHF power amplifier allows the user to set the amplifier for proper operation according to the selected mode of transmission: Single Sideband (SSB), Continuous Wave (CW), or Frequency Modulation (FM).

Each mode has different characteristics and power requirements, and this switch ensures that the amplifier adjusts its operating parameters to match the mode being used. For instance, SSB signals require linear amplification to preserve the integrity of the waveform. In contrast, FM signals can benefit from higher power efficiency without distortion.

Think of the SSB/CW-FM switch as a mode selector that optimizes the amplifier's performance for the specific type of signal you are transmitting. Setting the amplifier correctly helps prevent signal distortion and ensures efficient and reliable operation.

T7A09: What is the function of the SSB/CW-FM switch on a VHF power amplifier?

A. Change the mode of the transmitted signal
B. Set the amplifier for proper operation in the selected mode
C. Change the frequency range of the amplifier to operate in the proper segment of the band
D. Reduce the received signal noise

Wiring it Together

Wires for Transceiver's DC Power Connection

Using short, heavy-gauge wires for a transceiver's DC power connection is best for minimizing voltage drop when transmitting. Voltage drop occurs when there is resistance in the wires connecting the power source to the transceiver. This resistance causes a reduction in voltage, which can lead to decreased performance or even malfunction of the transceiver, especially under high current demands during transmission. Heavy-gauge wires have lower resistance than thinner wires, which helps maintain the voltage levels at the transceiver.

T4A03: Why are short, heavy-gauge wires used for a transceiver's DC power connection?

A. To minimize voltage drop when transmitting
B. To provide a good counterpoise for the antenna
C. To avoid RF interference
D. All these choices are correct

In general, when wiring components together in a radio setup, it's essential to consider the quality and specifications of the wiring. Shorter wires reduce resistance and minimize potential interference and signal loss. Additionally, ensuring secure and solid connections between components helps maintain stable and reliable power delivery. Proper wiring practices, such as using suitable wire gauges and lengths, ensure optimal performance and safety in amateur radio operations.

Chapter 16

Antenna Basics

ANTENNAS ARE THE UNSUNG heroes of ham radio; they are the interface that bridges the gap between the electronic world inside your radio and the vast expanse of the airwaves. Understanding how antennas work is akin to understanding how a plant turns sunlight into energy—both are about converting one form of energy into another more usable form.

The Role of Antennas

Antennas are key pieces of equipment in the transmission and reception of radio waves. They come in various shapes and sizes, each designed for specific frequencies and types of operations.

For transmitting, the antenna converts the RF electrical signals into electromagnetic waves, efficiently radiating them into the atmosphere. For receiving, the process reverses, with the antenna capturing electromagnetic waves and converting them back into RF electrical signals for the receiver to process.

The effectiveness of an antenna is often measured by its gain and directivity; gain refers to its ability to focus energy in a particular direction, and directivity is its ability to receive energy from a specific direction. Understanding these properties helps you select a suitable antenna for your needs, whether for local communications on the VHF and UHF bands or for chasing distant contacts on the HF bands.

Types of Antennas

The most common antennas you might encounter or consider as a Technician licensee include dipoles, verticals, and beam antennas. Each type has characteristics suitable for different radio activities and environments.

Several factors should guide your decision when choosing an antenna. The space you have available is one of the most vital constraints. A large beam antenna might offer excellent long-distance capabilities. Still, having only a tiny rooftop or yard will make it impractical. Similarly, your budget is an essential consideration. While spending much on sophisticated antenna systems is possible, many practical antennas can be constructed with inexpensive materials. Your desired communication range also influences your choice; a beam antenna might be necessary for worldwide communications, but a simple dipole might suffice for chatting with fellow hams in your region.

Dipoles

Dipoles are perhaps the most widely used type of antenna in amateur radio. They consist of two conductive elements, usually metal rods or wires, that are equal in length and oriented in a straight line, with the feedline connected to the center. The term "dipole" means "two poles," referring to the two elements of the antenna.

You will also commonly hear the term 'half-wave dipole.' A half-wave dipole is a specific type of dipole antenna where the total length of the antenna is half the wavelength of the frequency it is designed to transmit or receive. "Half-wave" refers to this specific length relative to the wavelength.

For example, a half-wave dipole for a 10-meter band would be approximately 5 meters long, with each element being about 2.5 meters.

In summary, all half-wave dipoles are dipole antennas, but not all dipole antennas are half-wave dipoles.

Dipole antennas are generally popular because of their straightforward design, ease of construction (you need two pieces of wire, some basic tools, and a place to set them up), and effective performance.

They produce a broadside radiation pattern, meaning they radiate most of the signal energy perpendicular to the length of the antenna. This pattern provides good coverage and is ideal for many applications.

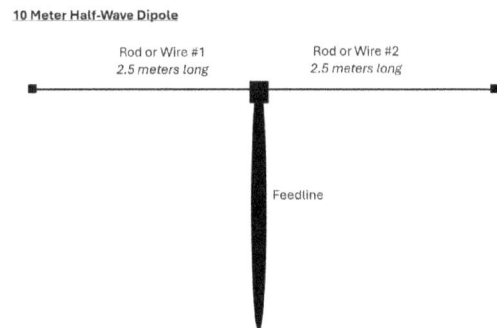

10 Meter Half-Wave Dipole

Rod or Wire #1
2.5 meters long

Rod or Wire #2
2.5 meters long

Feedline

Understanding the Radiation Pattern of Dipole Antennas

Dipole antennas produce a broadside radiation pattern, which means they radiate most of the signal energy perpendicular to the length of the antenna. Imagine the dipole antenna as a straight line; the most substantial signal radiation occurs at right angles (90 degrees) to this line, both above and below the antenna, rather than off the ends. This directional pattern is often described as doughnut-shaped when visualized in three dimensions, with the dipole antenna at the center of the doughnut hole.

A simple dipole antenna oriented parallel to the Earth's surface is known as a horizontally polarized antenna. Polarization is a key concept in antenna theory that refers to the orientation of the electric field of the radio waves transmitted or received by an antenna. In the case of a horizontally polarized dipole antenna, the electric field oscillates horizontally, meaning that the electric field oscillation is aligned with the horizon. This orientation is achieved when the dipole elements—usually metal rods or wires—are placed horizontally.

The horizontal polarization of an antenna has specific implications for signal transmission and reception. The most important aspect is the antenna's radiation pattern. A horizontally polarized dipole antenna exhibits a broadside radiation pattern, where the strongest signals are radiated in directions perpendicular to the antenna. This pattern creates two primary lobes of radiation, one on each side of the antenna, with minimal radiation along the axis of the antenna itself. This characteristic makes horizontally polarized dipole antennas highly effective for general communication needs, as they can provide a wide coverage area.

For example, when a dipole antenna is mounted horizontally, it radiates efficiently in all directions perpendicular to its length. This is particularly advantageous for amateur radio operators, broadcasters, and receiving antennas. In amateur radio, horizontal polarization is often used for long-distance communication, as it tends to interact more favorably with the Earth's ionosphere, enabling signals to travel farther distances through a process known as skywave propagation. Additionally, many amateur radio operators prefer horizontally polarized antennas for their ability to reject vertically polarized noise, such as that from electrical equipment, which can interfere with clear communication.

Moreover, understanding the concept of polarization is crucial when designing or selecting an antenna for a specific application. Matching the polarization of the transmitting and receiving antennas maximizes signal strength and clarity, as mismatched polarizations can lead to signal loss. For instance, if a horizontally polarized antenna transmits a signal, a horizontally polarized receiving antenna will pick up that signal more effectively than a vertically polarized one.

To summarize, a horizontally polarized dipole antenna is a versatile and widely used design in radio communications. Its horizontal polarization and broadside radiation pattern make it ideal for various applications, from amateur radio to broadcasting, providing robust signal coverage and enhanced communication effectiveness. By understanding these principles, users can optimize their antenna setups to achieve the best possible performance in their specific use cases.

T9A03: Which of the following describes a simple dipole oriented parallel to the Earth's surface?

A. A ground-wave antenna
B. A horizontally polarized antenna
C. A traveling-wave antenna
D. A vertically polarized antenna

T9A10: In which direction does a half-wave dipole antenna radiate the strongest signal?

A. Equally in all directions
B. Off the ends of the antenna
C. In the direction of the feed line
D. Broadside to the antenna

Resonant Frequency and Dipole Antenna Length

The resonant frequency of a dipole antenna is the frequency at which it naturally resonates and efficiently radiates signals (basically, the frequency at which it works best).

This means an antenna can transmit and receive radio waves most effectively at this frequency, with minimal energy loss. When an antenna operates at its resonant frequency, its electrical length matches the signal's wavelength, allowing optimal energy transfer between the antenna and the air.

In practical terms, if an antenna is designed to have a resonant frequency of 14 MHz, it will perform best when transmitting or receiving signals at that frequency. This concept is fundamental for ensuring effective communication, as using an antenna at its resonant frequency maximizes signal strength and clarity. Understanding resonant frequency helps design and tune antennas to operate efficiently for the desired frequency band.

This frequency is inversely related to the length of the antenna. Remember back: higher frequency = shorter wavelength.

Therefore, shortening a dipole antenna increases its resonant frequency. This happens because the shorter the antenna, the higher the frequency it will resonate, as its physical length becomes more in tune with the shorter wavelength of higher frequencies.

To remember this, think of a guitar string: a shorter string produces a higher pitch. Similarly, a shorter dipole antenna resonates at a higher frequency. Understanding that shortening a dipole antenna increases its resonant frequency will help you recall this concept for your exam, highlighting the importance of antenna length in determining the operating frequency range of your antenna system.

T9A05: Which of the following increases the resonant frequency of a dipole antenna?

A. Lengthening it
B. Inserting coils in series with radiating wires
C. Shortening it
D. Adding capacitive loading to the ends of the radiating wires

Dipole Math Time:

> **T9A09: What is the approximate length, in inches, of a half-wavelength 6-meter dipole antenna?**
>
> A. 6
> B. 50
> **C. 112**
> D. 236

To calculate the length of a half-wavelength dipole antenna, you can use the formula:

$$Length\ (feet) = \frac{468}{Frequency\ (MHz)}$$

For the curious, the constant 468 is derived from the speed of light and the relationship between wavelength and frequency in free space.

For a 6-meter band, the frequency is typically around 50 MHz. Plugging this into the formula gives:

$$Length\ (feet) = \frac{468}{50} = 9.36\ feet$$

To convert this length to inches:

$$Length\ (feet) = 9.36 \times 12 = 112.32\ inches$$

Therefore, the approximate length of a half-wavelength 6-meter dipole antenna is about 112.32 inches. This calculation helps ensure that the antenna is adequately sized to resonate at the desired frequency, providing optimal performance for communication on the 6-meter band.

Ok, but wait. We can do this easier, too. Let's estimate.

First, a 6-meter dipole, and we need "half-wavelength." Half of 6 meters is 3 meters.

Next, we know that 1 meter is approximately equal to 1 yard, 3 feet, or 36 inches.

The question asks for inches, so we multiply 3 meters by 36 inches to get 108.

But 108 isn't the correct answer or even a possible answer on the test. Right, but remembering our study strategy, we look at all the possible answers and eliminate the wrong ones. No other answer is anywhere close to 108 except 112. There you go! There is no hard math involved. But you have also seen the actual formula to figure this out.

Vertical

Vertical antennas are a popular choice in radio communication due to their omnidirectional radiation pattern and compact design. As their name suggests, these antennas stand upright (like a flagpole) and radiate radio waves equally in all horizontal directions. This makes them highly effective for mobile and base station use, especially in areas with limited space, like small backyards or city apartments.

Have you ever seen a skyscraper with protruding vertical antennas on the top? It's that simple!

A common example is the quarter-wavelength vertical antenna. For the 146 MHz frequency band, a quarter-wavelength vertical antenna is approximately 19.2 inches tall (an interesting example; let's look at the math below). These antennas are easy to install and require minimal space, making them ideal for vehicles or locations with limited space. The ground plane or surface on which the antenna is mounted acts as a reflective surface, enhancing the radiation pattern and efficiency of the antenna.

Vertical antennas are favored for their simplicity, efficiency, and ability to provide reliable communication over a wide area. Understanding their operation and applications can help you choose a suitable antenna, ensuring optimal performance and coverage.

Quarter-Wavelength Vertical Antenna Math

T9A08: What is the approximate length, in inches, of a quarter-wavelength vertical antenna for 146 MHz?

A. 112
B. 50
C. 19
D. 12

Start with our favorite formula. It's our favorite formula because it has frequency and length. The test gives us frequency and wanting length (in inches) in the question.

$$\text{meters} = \frac{300}{\text{Frequency (MHz)}}$$

- Step 1) 300 divided by 146 MHZ = 2.05 (let's call it 2.0).

  - We could have also estimated it again by saying 300 divided by 150 MHz (close to 146 MHz), which gives us an estimated answer of 2.

- Step 2) We have 2 meters now, and the question concerns a 'quarter-wavelength' vertical antenna.

  - 2 meters divided by 4 (a quarter equals 4) = 0.5 meters.

- Step 3) We take our 0.5 meters and convert it into inches (because that's what the question wants).

  - 0.5 meters multiplied by 36 inches (1 meter is approximately 1 yard or 36 inches) = 18 inches (half of 36 is 18).

- Step 4) Again, thinking critically about the possible answers, only one is close to our estimate of 18.

  - The correct answer is 'C', 19 inches.

This height ensures the antenna resonates effectively at 146 MHz, allowing for optimal signal transmission and reception. Remembering this calculation helps you design and tune antennas for specific frequencies, ensuring efficient communication.

In summary, the main dimension of concern for a quarter-wave vertical antenna is its height, which is one-quarter of the wavelength at the operating frequency. The width and length are typically not significant factors, as the antenna's functionality is primarily determined by its vertical length.

Beam / Yagi (Directional)

A beam antenna is a directional antenna that concentrates signals in one specific direction, significantly enhancing radio waves' transmission and reception. Compared to omnidirectional antennas, which radiate signals equally in all directions, a beam antenna can achieve higher gain by focusing the signal. This directional focus allows for better long-distance communication and reduced interference from unwanted sources, making beam antennas particularly useful in amateur radio operations, television broadcasting, and other applications where signal clarity and range are critical.

Like the popular Yagi antenna, beam antennas have a central part that sends and receives signals, called the driven element. They also have extra parts that help focus the energy in specific directions. Looking at the image, you see the energy is focused in one direction. This makes them more powerful and able to reach farther distances than simpler antennas. These antennas are highly directional, transmitting and receiving signals in one direction. This focus allows them to go farther than non-directional antennas like dipoles or verticals. However, they are more complex to build and require more space and a rotator to turn them toward the signal you are trying to receive or transmit.

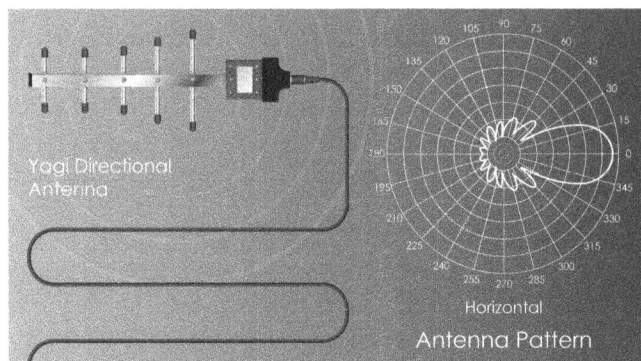

Focused in One Direction

T9A01: What is a beam antenna?

A. An antenna built from aluminum I-beams
B. An omnidirectional antenna invented by Clarence Beam
C. An antenna that concentrates signals in one direction
D. An antenna that reverses the phase of received signals

And because of the greater concentration...

T9A06: Which of the following types of antenna offers the greatest gain?

A. 5/8 wave vertical
B. Isotropic
C. J pole
D. Yagi

Remember that **gain = effectiveness.**

And one more time...

Antenna gain refers to the increase in signal strength that an antenna provides in a specified direction compared to a reference antenna, typically an isotropic radiator or a simple dipole antenna.

This measurement indicates how effective an antenna is at directing radio waves in a particular direction, enhancing both the transmission and reception of signals. Higher gain antennas focus the radio energy more narrowly, similar to how a flashlight concentrates light, which results in stronger signals in the desired direction and improved communication range.

Remembering the concept of antenna gain is simple if you consider it the antenna's ability to "boost" the signal strength in a specific direction. This directed focus makes the antenna more efficient at sending and receiving signals where they are needed most, reducing interference from unwanted directions.

T9A11: What is antenna gain?

A. The additional power that is added to the transmitter power
B. The additional power that is required in the antenna when transmitting on a higher frequency
C. The increase in signal strength in a specified direction compared to a reference antenna
D. The increase in impedance on receive or transmit compared to a reference antenna

Handheld Radios

The short, flexible antennas supplied with most handheld radio transceivers, often referred to as "rubber duck" antennas, have the disadvantage of low efficiency compared to full-sized quarter-wave antennas.

Due to their compact size, these short antennas cannot radiate signals as effectively as longer, quarter-wave antennas, which are better matched to the wavelength of the transmitted signals. This reduced efficiency means less transmitted power is effectively radiated into the air, leading to weaker signal strength and reduced communication range.

Remembering this disadvantage is straightforward if you consider the trade-off between convenience and performance. While the short, flexible antenna is portable and easy to carry, its low efficiency can limit the effectiveness of your handheld radio, especially in challenging communication environments.

T9A04: What is a disadvantage of the short, flexible antenna supplied with most handheld radio transceivers, compared to a full-sized quarter-wave antenna?

A. It has low efficiency
B. It transmits only circularly polarized signals
C. It is mechanically fragile
D. All these choices are correct

For this next question, they are really throwing you a specific situation. But think about the general principles involved. In this exam question, you have a handheld VHF radio and are sitting in a metal or near-metal vehicle.

T9A07: What is a disadvantage of using a handheld VHF transceiver with a flexible antenna inside a vehicle?

A. Signal strength is reduced due to the shielding effect of the vehicle
B. The bandwidth of the antenna will decrease, increasing SWR
C. The SWR might decrease, decreasing the signal strength
D. All these choices are correct

Using a handheld VHF transceiver with a flexible antenna inside a vehicle has the disadvantage of reduced signal strength. This reduction occurs due to the shielding effect of the vehicle's metal body, which can block or reflect radio waves. The vehicle acts like a partial Faraday cage, preventing radio signals from reaching the antenna efficiently. This results in weaker transmissions and reception, making communication less reliable when using the transceiver inside the vehicle.

That's why we use car-mounted, usually magnet, antennas on the outside of our vehicle!

Here is another weirdly specific question. Think about the overall concepts being asked. Compare each answer to the question. What in the world are they getting at?

T9A12: What is an advantage of a 5/8 wavelength whip antenna for VHF or UHF mobile service?

A. It has more gain than a 1/4-wavelength antenna
B. It radiates at a very high angle
C. It eliminates distortion caused by reflected signals
D. It has 10 times the power gain of a 1/4 wavelength whip

Answers B, C, and D make little sense because they say nothing specific about a 5/8 wavelength whip antenna.

Answer 'A' discusses gain. Gain refers to the antenna's ability to focus the radio signal in a particular direction, which enhances the effective radiated power and improves signal strength.

A 5/8 wavelength antenna provides more gain than a 1/4 wavelength antenna because its longer length allows it to focus the radiated energy more effectively in the horizontal plane. This increased focus directs more signal energy outward rather than upward or downward, resulting in a stronger and more efficient signal over longer distances.

In contrast, a 1/4 wavelength antenna has a broader radiation pattern that disperses energy more evenly in all directions, including upward and downward, leading to lower gain and reduced range.

This makes the 5/8 wavelength antenna particularly advantageous for mobile and terrestrial communication, where a robust and horizontally directed signal is necessary for maintaining reliable connections over varying distances and terrain.

Antenna Loading

Antenna loading is a technique for making antennas resonate at desired frequencies without increasing their physical length. This is key when space constraints prevent the use of full-sized antennas. By inserting inductors or capacitors, the electrical properties of the antenna can be modified to achieve resonance at specific frequencies, enhancing performance in limited spaces.

In many practical applications, especially in mobile and portable setups, full-sized antennas are impractical due to their length. "For instance, a half-wave dipole for the 80-meter band must be approximately 40 meters long (using easy math for approximation – as a Technician, you have access to the 80-meter band only for CW from 3.525 to 3.600 MHz). Antenna loading solves this by electrically lengthening the antenna, allowing it to resonate at the desired frequency without needing excessive physical length.

One standard method is inductive loading, where inductors (coils) are inserted into the antenna's radiating elements. This creates reactance, making the antenna behave as if it were longer. Another method is capacitive loading, adding capacitive hats or elements to increase capacitance and effectively lengthen the antenna electrically. These techniques benefit mobile and portable antennas, ensuring they operate efficiently on lower frequencies without requiring large physical sizes.

T9A02: Which of the following describes a type of antenna loading?

A. Electrically lengthening by inserting inductors in radiating elements
B. Inserting a resistor in the radiating portion of the antenna to make it resonant
C. Installing a spring in the base of a mobile vertical antenna to make it more flexible
D. Strengthening the radiating elements of a beam antenna to better resist wind damage

Be careful when answering. You might look at 'B' and consider it.

Inserting a resistor in the radiating portion of an antenna to make it resonant is incorrect because resistors do not create resonance. Resonance in antennas is achieved through reactive components, such as inductors (which provide inductance) or capacitors (which provide capacitance). These reactive components store and release energy to enhance the antenna's ability to resonate at specific frequencies.

Resistors, however, dissipate energy as heat and do not contribute to the reactive properties needed for resonance. Adding a resistor to an antenna would reduce efficiency by converting part of the signal energy into heat, weakening the transmitted or received signal. This would degrade the antenna's overall performance rather than improve it.

In summary, while inductors and capacitors can tune an antenna to the desired resonant frequency, resistors cannot. They serve a different purpose by limiting current and dissipating power, which does not help achieve resonance. Understanding this distinction is key to designing efficient and effective antennas.

Antenna Polarization

Antenna polarization is a fundamental concept in radio communication that refers to the orientation of the electric field of the radio wave radiated by the antenna. An antenna's polarization can significantly affect the performance and clarity of the transmitted and received signals. Understanding and selecting the appropriate polarization for your antenna setup is important for optimizing communication efficiency and reducing signal losses.

This section will explore the different types of antenna polarization, their applications, and how they impact radio communication. Whether setting up a simple dipole antenna or a complex array, knowing how to manage antenna polarization will enhance your ability to establish precise and reliable connections.

Understanding Antenna Polarization

Antenna polarization refers to the orientation of the electric field of the radiated radio wave relative to the Earth's surface. The two primary types of polarization are vertical and horizontal.

- **Vertical Polarization**: The electric field is perpendicular to the Earth's surface. This type of polarization is commonly used for mobile and handheld radio communications because vertically polarized antennas are often more practical for vehicles and portable devices. This type of antenna is straight "up and down" compared to the Earth. So, it's vertically polarized.

- **Horizontal Polarization**: The electric field is parallel to the Earth's surface. This polarization is typically used for fixed-station communications, such as television broadcasting and point-to-point radio links. Horizontal antennas (for example, a dipole) are antennas that are "sideways" compared to the Earth. So, they are horizontally polarized.

    - Horizontal polarization is usually used for long-distance CW (Continuous Wave) and SSB (Single Sideband) contacts on the VHF and UHF bands. This preference arises because horizontally polarized signals tend to suffer less from ground reflections and multipath interference, which can degrade signal quality and reliability over long distances. Horizontal polarization is also advantageous for reducing man-made, vertically polarized noise.

T3A03: What antenna polarization is normally used for long-distance CW and SSB contacts on the VHF and UHF bands?

A. Right-hand circular
B. Left-hand circular
C. Horizontal
D. Vertical

In practical terms, horizontal polarization helps ensure clearer and more stable communication for long-distance contacts, making it the standard choice for serious VHF and UHF operators aiming for maximum range and signal integrity.

Mismatched Polarizations

Choosing the correct polarization for your antenna matters because <u>mismatched polarization between transmitting and receiving antennas can significantly reduce the strength or signal loss.</u> For instance, if a vertically polarized antenna is transmitted to a horizontally polarized receiving antenna, the signal strength can be reduced considerably. Therefore, ensuring that both the transmitting and receiving antennas share the same polarization can greatly enhance signal clarity and strength.

T3A04: What happens when antennas at opposite ends of a VHF or UHF line of sight radio link are not using the same polarization?

A. The modulation sidebands might become inverted
B. Received signal strength is reduced
C. Signals have an echo effect
D. Nothing significant will happen

By understanding and properly implementing antenna polarization, you can optimize your radio communications, ensuring that your signals are transmitted and received with maximum efficiency and minimal interference.

Elliptical Polarization

Elliptical Polarization and Its Impact on Transmission

Elliptical polarization occurs when radio waves are propagated by the ionosphere. The electric field of the waves rotates and traces an elliptical shape as they travel, resulting in a signal that is not purely vertically or horizontally polarized but a combination of both. The key advantage of elliptically polarized signals is their versatility in terms of antenna orientation.

<u>Due to elliptical polarization, vertically or horizontally polarized antennas can be used effectively for transmission or reception.</u> This flexibility simplifies the setup and improves the chances of establishing reliable communication, as the signal can be received by antennas with different polarizations. For amateur radio operators, understanding elliptical polarization helps explain why transmissions via the ionosphere are less dependent on matching the exact polarization between the transmitting and receiving antennas.

T3A09: Which of the following results from the fact that signals propagated by the ionosphere are elliptically polarized?

A. Digital modes are unusable
B. Either vertically or horizontally polarized antennas may be used for transmission or reception
C. FM voice is unusable
D. Both the transmitting and receiving antennas must be of the same polarization

Final Thoughts on Antennas

For those inclined towards DIY projects, building your antenna can be a rewarding endeavor that deepens your understanding of radio physics and propagation. Simple projects like constructing a dipole or a vertical can be accomplished with basic materials like wire, insulators, and some coaxial cables.

These projects save money and offer the satisfaction of communicating through the equipment you have built yourself. Online forums and amateur radio clubs often provide plans and guidance, making these projects accessible even to those with minimal experience building radio equipment.

Chapter 17

Feedline Basics

A FEEDLINE IS A cable or transmission line that connects a radio transmitter or receiver to an antenna.

Its primary function is to transfer radio frequency (RF) signals between the radio equipment and the antenna with minimal power loss. Feedlines are essential in any radio communication system, ensuring that the transmitter's maximum signal power reaches the antenna and that signals received are efficiently delivered to the receiver.

Feedlines come in various types, including coaxial cables (coax), twin-lead, and ladder lines, each with specific characteristics and suitable applications. Coaxial cables, for example, are widely used due to their ease of installation and good shielding properties, which help prevent signal loss and interference.

Standing Waves

A standing wave occurs when two radio waves of the same frequency and amplitude travel in opposite directions along a transmission line, such as a feedline connected to an antenna, and interfere with each other.

These waves interfere with each other, creating points of constructive interference (where the waves add together) and points of destructive interference (where the waves cancel each other out), resulting in a pattern of fixed nodes (points of no movement) and antinodes (points of maximum movement) along the line. This pattern appears stationary, hence the name "standing wave."

Standing Wave Ratio (SWR)

Standing Wave Ratio (SWR) is a measure used to evaluate power transfer efficiency from a radio transmitter to an antenna. It indicates how well the antenna is matched to the transmission line (feedline) and transmitter, which is important for effective communication.

When a radio transmitter sends a signal to an antenna, the goal is to radiate all the power out as radio waves. However, suppose there is a mismatch between the impedance (opposition to the current flow) of the transmission line (feedline) and the antenna. In that case, some of the power gets reflected back toward the transmitter, creating standing waves along the transmission line.

The measurement for standing waves is the standing wave ratio, which measures how well a load is matched to a transmission line.

T9B12: What is the standing wave ratio (SWR)?

A. A measure of how well a load is matched to a transmission line
B. The ratio of amplifier power output to input
C. The transmitter efficiency ratio
D. An indication of the quality of your station's ground connection

A high Standing Wave Ratio (SWR) means that a lot of power is reflected back toward the transmitter instead of being radiated by the antenna, leading to inefficient operation and potential equipment damage. Understanding and minimizing standing waves helps ensure efficient and effective radio communication.

- **Ideal SWR (1:1)**: This means that there is perfect impedance matching, and all the power is transferred to the antenna without reflection.

  ○ **Now, there are ideals and reality**: any measurement under 2.0:1 is ok to use (you can try to improve it, but it's ok). And if you ever do get a 1:1 reading, take a picture of it and post it online. It's a rare site!

- **High SWR (e.g., 2:1 or higher)**: This indicates an impedance mismatch, meaning some power is being reflected back, resulting in less efficient transmission and potential damage to the transmitter over time.

T7C04: What reading on an SWR meter indicates a perfect impedance match between the antenna and the feed line?

A. 50:50
B. Zero
C. 1:1
D. Full Scale

T7C06: What does an SWR reading of 4:1 indicate?

A. loss of -4 dB
B. Good impedance match
C. gain of +4 dB
D. Impedance mismatch

Measuring the Standing Wave Ratio (SWR) involves using a directional wattmeter. A directional wattmeter measures the power of radio frequency (RF) signals traveling through a transmission line.

T7C08: Which instrument can be used to determine SWR?

A. Voltmeter
B. Ohmmeter
C. Iambic pentameter
D. Directional wattmeter

Unlike a standard wattmeter, a directional wattmeter can differentiate between forward power (the power sent from the transmitter to the antenna) and reflected power (the power that bounces back from the antenna due to impedance mismatches).

Why It Matters

Understanding SWR is vital for beginners because it helps tune and adjust the antenna for optimal performance. A low SWR ensures the antenna radiates most of the transmitted power, improving communication effectiveness (reduced signal loss). Regularly checking and maintaining a good SWR can also prevent potential damage to the radio equipment due to excessive reflected power and heat.

T9B01: What is a benefit of low SWR?

A. Reduced television interference
B. Reduced signal loss
C. Less antenna wear
D. All these choices are correct

T7C07: What happens to power lost in a feed line?

A. It increases the SWR
B. It is radiated as harmonics
C. It is converted into heat
D. It distorts the signal

Simple Explanation

Think of SWR as a health check for your antenna system. A perfect SWR of 1:1 means your antenna system is in great shape, efficiently transmitting all the power it receives. A higher SWR value indicates that your system needs adjustment to minimize power loss, which results in signal loss.

You can maintain an efficient and reliable antenna system by monitoring the SWR and making necessary adjustments.

SWR Meter

When selecting an accessory SWR meter, it's essential to consider the frequency and power level at which the measurements will be made. Different SWR meters are designed to operate optimally within specific frequency ranges, such as HF, VHF, or UHF bands. Using an SWR meter outside its intended frequency range can lead to inaccurate readings, which can affect the efficiency of your antenna system and potentially harm your equipment.

T4A02: Which of the following should be considered when selecting an accessory SWR meter?

A. The frequency and power level at which the measurements will be made
B. The distance that the meter will be located from the antenna
C. The types of modulation being used at the station
D. All these choices are correct

Your transmitter's power level should match the SWR meter's power handling capability. Suppose the SWR meter is not rated for the power level of your transmissions. In that case, it can be damaged or provide unreliable measurements. Therefore, ensuring that your SWR meter is suitable for the frequencies and power levels you intend to use will help maintain accurate readings and protect your SWR meter and radio equipment.

Erratic Changes in SWR

Erratic changes in your Standing Wave Ratio (SWR) <u>are often caused by loose connections in the antenna or feed line</u> (those pesky things!). When connections are not secure, they can introduce intermittent contact or variable resistance, leading to fluctuations in the impedance seen by the transmitter. This causes the SWR to change unpredictably as the match between the antenna system and the transmission line becomes unstable.

For beginners, it's important to remember that maintaining tight and secure connections in your antenna setup is vital for stable and efficient operation. Loose connections can lead to poor performance and potential damage to your equipment due to reflected power. Regularly inspecting and ensuring all connections are secure can prevent erratic SWR changes and provide reliable communication. This concept is important for both practical operation and the exam.

T9B09: What can cause erratic changes in SWR?

A. Local thunderstorm
B. Loose connection in the antenna or feed line
C. Over-modulation
D. Overload from a strong local station

Solid State Transmitters

A solid-state transmitter uses electronic components, such as transistors, diodes, and integrated circuits, instead of vacuum tubes to amplify and generate radio frequency signals. These modern components offer advantages, including greater reliability, efficiency, and compactness.

Key Features of Solid-State Transmitters:

1. **Transistor-Based Amplification**: Solid-state transmitters use transistors for signal amplification. Transistors are more durable and energy-efficient than vacuum tubes. They can handle high power levels and are less prone to failure, making them ideal for continuous operation.

2. **Compact and Lightweight**: Because they rely on solid-state components, these transmitters are generally more compact and lighter than their vacuum tube counterparts. This makes them suitable for mobile and portable applications and environments with limited space.

3. **Efficiency and Performance**: Solid-state transmitters are known for their high efficiency in converting electrical power into radio frequency power, resulting in less heat generation and lower cooling requirements. They also provide consistent performance with fewer maintenance needs.

4. **Protective Features**: These transmitters often come with built-in protective features, such as automatic power reduction when SWR increases, to safeguard the internal components from damage due to impedance mismatches. <u>Most solid-state transmitters automatically reduce the output power when the SWR (Standing Wave Ratio) increases beyond a certain level to protect the output amplifier transistors.</u>

   ○ High SWR indicates a significant impedance mismatch between the transmitter and the antenna, causing a large portion of the transmitted power to be reflected back into the transmitter. This reflected power can generate excessive heat and stress the transistors, potentially leading to damage or failure.

○ The transmitter limits the reflected power amount by reducing output power, safeguarding its internal components.

A solid-state transmitter uses modern electronic components to generate and amplify radio signals, offering reliability, efficiency, and size advantages. Understanding the benefits and features of solid-state transmitters highlights their importance in contemporary radio communication, ensuring efficient and dependable operation.

T7C05: Why do most solid-state transmitters reduce output power as SWR increases beyond a certain level?

A. To protect the output amplifier transistors
B. To comply with FCC rules on spectral purity
C. Because power supplies cannot supply enough current at high SWR
D. To lower the SWR on the transmission line

Coaxial Cable

Coaxial cable, commonly called "coax," is a transmission line that carries radio frequency (RF) signals from a transmitter to an antenna or from an antenna to a receiver. Due to its efficiency and ease of use, it is widely used in various communication systems, including amateur radio, television broadcasting, and internet connections.

Structure of Coaxial Cable

1. **Inner Conductor**: The central wire that carries the RF signal. It can be solid or stranded and is typically made of copper for good conductivity.

2. **Dielectric Insulator** surrounds the inner conductor, providing insulation and maintaining the spacing between it and the outer shield.

3. **Outer Conductor (Shield)**: A braided or solid metallic shield surrounding the dielectric insulator. It serves two primary purposes: providing a signal return path and shielding the inner conductor from external electromagnetic interference (EMI).

4. **Outer Jacket**: The protective outer layer that covers the shield. It protects the cable from physical damage and environmental factors.

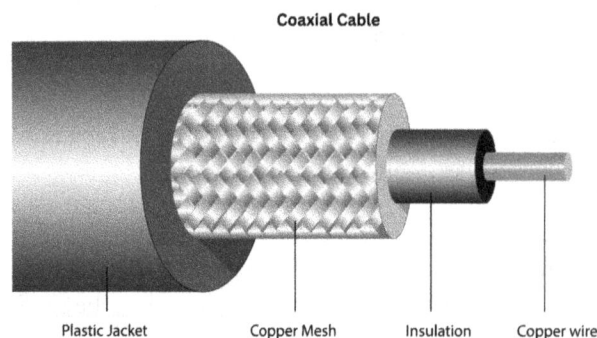

Coaxial Cable

Plastic Jacket   Copper Mesh   Insulation   Copper wire

Advantages of Coaxial Cable

• **Good Shielding**: The outer conductor effectively shields the inner conductor from external interference, ensuring a clean signal.

- ○ Shielded Wire: <u>Shielded wire is used in electrical and electronic systems to prevent the coupling of unwanted signals to or from the wire</u>. This shielding is typically made of a conductive layer surrounding the inner conductor, such as braided copper or aluminum foil. The shield is a barrier against electromagnetic interference (EMI) and radio frequency interference (RFI), which can distort or disrupt the wires' signals. For beginners, consider shielded wire a protective shield for the signals traveling through the wire.

T6D03: Which of the following is a reason to use shielded wire?

A. To decrease the resistance of DC power connections
B. To increase the current carrying capability of the wire
C. To prevent coupling of unwanted signals to or from the wire
D. To couple the wire to other signals

- • **Ease of Installation**: <u>Coax is flexible and relatively easy to install</u>, making it a convenient choice for many applications.

- • **Impedance Consistency**: Standard coaxial cables have a consistent impedance (typically 50 ohms), for minimizing signal reflections and ensuring efficient power transfer.

  - ○ The <u>most common coaxial cable impedance used in amateur radio is 50 ohms.</u> This standard impedance is chosen because it balances power handling capability and low signal loss, making it ideal for various radio frequencies and applications. Using a 50-ohm coaxial cable ensures that the transmission line impedance matches the typical impedance of most radio transmitters and antennas. This is vital for minimizing signal reflections and maximizing power transfer.

  - ○ For beginners, think of impedance as the "resistance" the cable offers to the flow of the radio signal. Matching this impedance correctly helps to ensure that the maximum amount of power reaches the antenna, leading to more efficient transmission and clearer communication.

Understanding coaxial cable is essential for anyone involved in radio communications. It plays a vital role in connecting radios to antennas and ensuring efficient signal transmission.

T9B03: Why is coaxial cable the most common feed line for amateur radio antenna systems?

A. It is easy to use and requires few special installation considerations
B. It has less loss than any other type of feed line
C. It can handle more power than any other type of feed line
D. It is less expensive than any other type of feed line

T9B02: What is the most common impedance of coaxial cables used in amateur radio?

A. 8 ohms
B. 50 ohms
C. 600 ohms
D. 12 ohms

Signal Loss in Coaxial Cables at Higher Frequencies

As the frequency of a signal in coaxial cable increases, the loss also increases. This means that higher frequency signals experience more significant attenuation (the reduction in the strength or amplitude of a signal as it travels through a wire) as they travel through the cable.

The primary reason for this increased loss is the skin effect, where higher-frequency currents tend to flow near the conductor's surface, effectively reducing the cross-sectional area available for current flow and increasing resistance. Dielectric losses in the insulating material between the conductors also contribute to the overall signal loss at higher frequencies.

For beginners, it's helpful to remember that while coaxial cables are efficient for transmitting signals, they are not perfect. As you move to higher frequencies, more of the signal's power is lost as heat within the cable, leading to weaker signals at the other end.

This concept is integral for ensuring efficient signal transmission, especially in applications requiring high-frequency operations, such as VHF and UHF bands.

T9B05: What happens as the frequency of a signal in coaxial cable is increased?

A. The characteristic impedance decreases
B. The loss decreases
C. The characteristic impedance increases
D. The loss increases

Sources of Loss in Coaxial Feed Lines

Several factors can cause loss in coaxial feed lines, including water intrusion into coaxial connectors, high SWR (Standing Wave Ratio), and the use of multiple connectors in the line.

Water intrusion can severely degrade the performance of a coaxial cable by increasing its resistance and causing signal attenuation. Moisture in the connectors can lead to corrosion and create additional pathways for signal loss, causing failure. Water and electricity don't mix! Keep moisture out.

High SWR indicates a poor match between the antenna and the transmission line, resulting in significant power being reflected back toward the transmitter instead of radiating. This reflection not only reduces the efficiency of the transmission but also increases losses in the coaxial cable.

Additionally, having multiple connectors in the line introduces more points where signal degradation can occur due to imperfect connections and increased resistance. Understanding these sources of loss is core to maintaining an efficient and effective communication system, ensuring minimal signal loss and optimal performance.

T9B08: Which of the following is a source of loss in coaxial feed line?

A. Water intrusion into coaxial connectors
B. High SWR
C. Multiple connectors in the line
D. All these choices are correct

T7C09: Which of the following causes failure of coaxial cables?

A. Moisture contamination
B. Solder flux contamination
C. Rapid fluctuation in transmitter output power
D. Operation at 100% duty cycle for an extended period

Coaxial Construction

UV-Resistant Outer Jackets on Coaxial Cables

The outer jacket of the coaxial cable should be resistant to ultraviolet (UV) light because prolonged exposure to UV radiation can degrade the material, making it brittle and prone to cracking. When the outer jacket is damaged, it can no longer effectively protect the inner components of the cable from environmental factors. This damage can allow water to enter the cable, leading to increased attenuation, corrosion of the conductors, and overall signal degradation.

For beginners, it's essential to understand that a UV-resistant jacket ensures the coaxial cable's longevity and reliability, especially when used outdoors. The cable maintains its integrity and performance by preventing UV damage and subsequent water intrusion, ensuring efficient signal transmission, and protecting your communication setup from potential failures.

Simply put, most feedlines go outside. The sun is outside, and UV light damages things over time. It rains outside, and moisture destroys our cables. Protect your feedlines!

T7C10: Why should the outer jacket of coaxial cable be resistant to ultraviolet light?

A. Ultraviolet resistant jackets prevent harmonic radiation
B. Ultraviolet light can increase losses in the cable's jacket
C. Ultraviolet and RF signals can mix, causing interference
D. Ultraviolet light can damage the jacket and allow water to enter the cable

Air Core Coaxial Cable

Air core coaxial cable is a transmission line that uses air as the primary dielectric medium between the inner conductor and the outer shield. This design minimizes signal loss because air has a very low dielectric constant compared to solid or foam dielectrics. The reduced dielectric constant allows for lower capacitance per unit length, resulting in lower signal attenuation, making air core coaxial cables highly efficient for high-frequency and long-distance transmissions.

However, the primary disadvantage of air core coaxial cables is their vulnerability to moisture intrusion. Unlike foam or solid dielectric cables with a more substantial insulating material, air core cables depend on maintaining an air gap. If moisture penetrates this air space, it can drastically degrade the cable's performance by increasing signal loss and causing corrosion of the conductors. Special installation techniques are required to mitigate this, such as using moisture-blocking connectors and ensuring all seals are airtight. These precautions add complexity and cost to the installation process but are necessary to maintain the cable's high performance. Understanding these characteristics of air core coaxial cables highlights the trade-offs between efficiency and environmental vulnerability.

T7C11: What is a disadvantage of air core coaxial cable when compared to foam or solid dielectric types?

A. It has more loss per foot
B. It cannot be used for VHF or UHF antennas
C. It requires special techniques to prevent moisture in the cable
D. It cannot be used at below freezing temperatures

Understanding Air Core Coax and Air-Insulated Hardline

While air core coax and air-insulated hardline both use air as the primary dielectric to reduce signal loss, they are different in their construction and typical applications.

Air Core Coax

- **Construction**: Air core coaxial cable uses air as the dielectric medium between the inner conductor and the outer shield, often supported by spacers or other structures to maintain the air gap.

- **Flexibility**: These cables are generally more flexible than hardlines, making them easier to install in applications where the cable needs to bend or route through tight spaces.

- **Use**: Commonly used in situations where flexibility and lower cost are more important, but they still require careful handling to prevent moisture ingress.

Air-Insulated Hardline

- **Construction**: An air-insulated hardline is a rigid coaxial cable with a solid or semi-rigid outer conductor and an air dielectric. It often includes structural elements to support the inner conductor and maintain the air gap.

- **Durability**: Hardlines are more robust, offer better shielding, and have lower loss than flexible air core coax. They are typically used in professional settings where maximum signal integrity is needed.

- **Use**: It is ideal for long-distance and high-frequency applications, such as VHF and UHF communications, where low signal loss and high durability are essential.

Key Differences

- **Flexibility**: Air core coax is more flexible, while air-insulated hardline is rigid and less flexible.

- **Performance**: Air-insulated hardline offers lower signal loss and better shielding than air core coax.

- **Applications**: Air core coax is used in applications requiring flexibility. In contrast, air-insulated hardline is used in high-performance applications needing minimal signal loss.

Understanding these differences helps select the correct type of feed line for specific needs, ensuring optimal performance in various radio communication scenarios.

Air-insulated hardline feed lines have the lowest loss at VHF and UHF frequencies compared to other feed lines. These cables are designed with an air core as the primary dielectric, significantly reducing signal attenuation. These typically have a robust construction with solid outer conductors, further reducing resistive losses and providing excellent shielding against external interference.

For beginners, it's helpful to remember that air-insulated hardlines are ideal for high-frequency applications because they maintain signal strength over long distances, making them particularly efficient for VHF and UHF communications. While they can be more expensive and require careful installation to prevent moisture intrusion, their superior performance in terms of low signal loss makes them a preferred choice for professional and high-performance amateur radio setups.

T9B11: Which of the following types of feed line has the lowest loss at VHF and UHF?

A. 50-ohm flexible coax
B. Multi-conductor unbalanced cable
C. Air-insulated hardline
D. 75-ohm flexible coax

Understanding "RG" in Coaxial Cables

"RG" stands for "Radio Guide," a designation initially used by the military to classify different types of coaxial cables. Each RG number corresponds to a specific type of coaxial cable with particular characteristics, such as impedance, diameter, and shielding. This system helps users identify the appropriate cable based on standardized specifications.

Key Characteristics of RG Coaxial Cables

- **Impedance**: RG cables typically come with standard impedance values, such as 50 ohms or 75 ohms, which are critical for matching the cable to the transmitter, receiver, and antenna to minimize signal reflection and loss.

- **Diameter**: The physical size of the cable, which affects its flexibility and suitability for different applications.

- **Shielding**: The type and amount of shielding (e.g., braided, foil) that protects the signal from external electromagnetic interference (EMI).

Examples of Common RG Cables

- **RG-58**: A 50-ohm cable commonly used in amateur radio and networking applications.

- **RG-59**: A 75-ohm cable often used for cable television and other video applications.

- **RG-213**: A 50-ohm cable with a larger diameter and better shielding, suitable for higher power applications.

Understanding the "RG" designation helps select the correct coaxial cable for specific requirements, ensuring efficient and reliable signal transmission in various communication systems.

T9B10: What is the electrical difference between RG-58 and RG-213 coaxial cable?

A. There is no significant difference between the two types
B. RG-58 cable has two shields
C. RG-213 cable has less loss at a given frequency
D. RG-58 cable can handle higher power levels

The lower loss in RG-213 is due to its larger diameter and superior shielding, which reduce resistance and minimize the signal's attenuation as it travels through the cable.

For beginners, it's helpful to think of RG-213 as a thicker and more robust cable that maintains signal strength better than the thinner RG-58. This makes RG-213 a preferred choice for applications requiring <u>longer cable runs</u> or <u>higher power transmission,</u> such as in professional and high-performance amateur radio setups.

Plug (PL) Connectors

Whereas the RG designation stands for "Radio Guide" and is used to classify different types of coaxial cables, **PL (plug)** is a type of connector used to terminate coaxial cables. It is commonly used in amateur radio and other communication applications. The PL-259 is designed for connecting coaxial cables to radio equipment and typically mates with an SO-239 socket. It is known for its <u>durability and reliability, particularly in HF (High Frequency) and VHF (Very High Frequency) applications.</u>

The PL-259 connector features a threaded coupling that ensures a secure and stable connection, essential for maintaining signal integrity, especially at higher power levels and frequencies.

The PL-259 connectors are commonly found in many radio setups because they balance mechanical strength and electrical performance well. Their design helps minimize signal loss and reflections, making them suitable for various frequencies typically used in amateur radio.

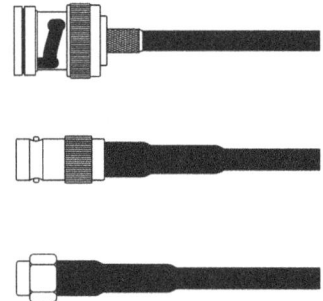

T9B07: Which of the following is true of PL-259 type coax connectors?

A. They are preferred for microwave operation
B. They are watertight
C. They are commonly used at HF and VHF frequencies
D. They are a bayonet-type connector

Type N (RF Connectors)

Type N connectors are designed for high-frequency applications, typically above 400 MHz. They provide excellent performance with low signal loss and good shielding, making them suitable for UHF and microwave frequencies.

They feature a threaded coupling mechanism that ensures a secure and stable connection, which maintains signal integrity at higher frequencies.

- **Compatibility**: Depending on the application and frequency requirements, Type N and PL-259 connectors can be used with various RG coaxial cables. For example, RG-213 is a thicker, lower-loss cable that can be terminated with either type N or PL-259 connectors, depending on whether the application is at higher frequencies (Type N) or lower frequencies (PL-259).

- **Application**: The right combination of RG cable and connector type depends on the communication system's specific requirements, such as frequency range, power handling, and environmental conditions. <u>Type N connectors are preferred for high-frequency applications,</u> while PL-259 connectors are more commonly used in HF and VHF setups.

- **Signal Integrity**: Correctly matched and installed connectors and cables are crucial for maintaining signal integrity and minimizing loss. For instance, using Type N connectors with RG cables in UHF applications helps achieve better performance and reliability.

Understanding the roles and compatibility of Type N, PL-259 connectors, and RG cables is essential for building effective and efficient coaxial communication systems.

T9B06: Which of the following RF connector types is most suitable for frequencies above 400 MHz?

A. UHF (PL-259/SO-239)
B. Type N
C. RS-213
D. DB-25

In this chapter, we explored the critical role of feedlines in radio communication systems, focusing on their types, functionality, and the importance of minimizing signal loss. Understanding the use of different feedlines, such as coaxial cables and ladder lines, along with concepts like standing wave ratio (SWR), is essential for optimizing signal transmission and protecting equipment. Proper selection and maintenance of feedlines ensure efficient and reliable communication, making it a foundational knowledge area for any radio enthusiast.

Chapter 18

Power Sources and Electrical Safety

SAFETY, A FUNDAMENTAL CORE of amateur radio, can often be overlooked. In these next two chapters, we'll delve into the best practices and precautions necessary to ensure the safe installation, maintenance, and operation of antennas and feedlines. From understanding electrical hazards and proper grounding techniques to mitigating risks associated with high-frequency RF exposure and structural safety, this chapter will provide you with the knowledge needed to protect yourself, your equipment, and those around you. Prioritizing safety ensures compliance with regulations and enhances the overall reliability and longevity of your radio setup.

Power Safety

When setting up your radio station, choosing a power source is as important as selecting the transceiver or antenna. It influences not just the functionality but also the safety and portability of your operations. This section will review power sources and, most importantly, the safety of working with electricity.

The station licensee is responsible for ensuring the safety of equipment and power sources and that no person is exposed to RF energy above the FCC exposure limits. Yes, your responsibility is to ensure your setup is safe!

T0C13: Who is responsible for ensuring that no person is exposed to RF energy above the FCC exposure limits?

A. The FCC
B. The station licensee
C. Anyone who is near an antenna
D. The local zoning board

This individual who holds the license for the amateur radio station must ensure that their station operates within the safety guidelines set by the FCC. These guidelines protect both the licensee and the public from potential health risks associated with excessive RF exposure.

To comply with these regulations, the station licensee must evaluate their RF emissions, considering frequency, power level, antenna type, and placement factors. They must take appropriate measures to reduce exposure if it exceeds the permissible limits, such as adjusting power levels, relocating antennas, or implementing barriers to restrict access to high RF areas.

Understanding and adhering to these responsibilities is essential for maintaining a safe operating environment and regulatory compliance.

Power Sources

The three primary power sources you can consider are mains electricity, batteries, and solar power. Each has its unique advantages and considerations.

Mains Electricity

Mains electricity - also known as household power, utility power, or grid power - is the standard electrical power supplied to homes and businesses by the electric utility company; it is the electricity that comes through the outlets in your walls. It is the home-based amateur radio station's most common power source. It's reliable and can handle high-power outputs needed for long-range communications. However, using mains power requires careful handling to avoid electrical hazards. It's key to ensure that all equipment is correctly rated for the voltage and current it will handle and that all connections are secure and insulated. Installing a dedicated circuit for your radio equipment can also prevent circuit overloads and reduce the risk of electrical fires.

Batteries

Batteries are an excellent choice for those who prefer operating their stations in remote locations or prioritize emergency preparedness. They provide a portable and relatively safe power source, making them ideal for amateur radio applications. The most commonly used batteries in amateur radio are sealed lead-acid and lithium-ion batteries. Lead-acid batteries are cheaper and more robust but heavier and require regular maintenance to prevent degradation. Lithium-ion batteries, while more expensive, offer higher energy density and longer life cycles with minimal maintenance. Regularly checking the charge level and ensuring correct recharging is essential to avoid damaging the battery.

Several rechargeable battery chemistries are commonly used in amateur radio, each offering unique advantages. <u>Lithium-ion (Li-ion), Nickel-Metal Hydride (NiMH), and Lead-Acid are the most notable rechargeable battery chemistries</u>. Lithium-ion batteries are known for their high energy density, lightweight design, and long cycle life, making them ideal for portable devices and handheld radios. Nickel-metal hydride batteries are valued for their safety and environmental friendliness, and they are often used in household electronics and portable equipment. Although heavier and with lower energy density, lead-acid batteries are robust, cost-effective, and frequently used in backup power systems and automotive applications. Selecting the right rechargeable battery type ensures reliable power for your equipment and operations.

> **T6A10: Which of the following battery chemistries is rechargeable?**
>
> A. Nickel-metal hydride
> B. Lithium-ion
> C. Lead-acid
> **D. All these choices are correct**

In contrast, non-rechargeable batteries, also known as primary batteries, are designed for single use and cannot be recharged once depleted. <u>Carbon-zinc batteries are a common type of non-rechargeable battery</u>. They are among the oldest battery chemistries and are often used in low-drain devices such as remote controls, clocks, and flashlights. Carbon-zinc batteries are inexpensive and readily available but typically have a shorter lifespan and lower energy density than alkaline batteries.

T6A11: Which of the following battery chemistries is not rechargeable?

A. Nickel-cadmium
B. Carbon-zinc
C. Lead-acid
D. Lithium-ion

Understanding the difference between rechargeable and non-rechargeable batteries is important for amateur radio operators, as it impacts their equipment's choice of power sources. While rechargeable batteries are suitable for repeated use and are more cost-effective over time, non-rechargeable batteries like carbon zinc offer a convenient, ready-to-use power solution for low-drain or infrequently used devices.

Determining Battery Life for Your Equipment

To determine the length of time equipment can be powered from a battery, you need to divide the battery's ampere-hour (Ah) rating by the average current draw of the equipment. A battery's ampere-hour rating indicates the current it can supply over a specific period. For example, a battery with a 10 Ah rating can theoretically supply 10 amps for one hour or 1 amp for ten hours.

T4A09: How can you determine the length of time that equipment can be powered from a battery?

A. Divide the watt-hour rating of the battery by the peak power consumption of the equipment
B. Divide the battery ampere-hour rating by the average current draw of the equipment
C. Multiply the watts per hour consumed by the equipment by the battery power rating
D. Multiply the square of the current rating of the battery by the input resistance of the equipment

To calculate how long your equipment will run on a fully charged battery, first find the average current draw of your equipment in amps. Then, divide the battery's ampere-hour rating by this current draw. For instance, if your radio equipment draws an average of 2 amps and your battery is rated at 20 Ah, the equipment can be powered for approximately 10 hours (20 Ah / 2 A = 10 hours). This calculation helps amateur radio operators plan their operations, especially in the field or during emergencies, ensuring they have adequate power for their needs.

Solar

Solar power's appeal as a renewable energy source is becoming increasingly popular among amateur radio operators, especially those looking to build sustainable and independent stations. Setting up a solar-powered station involves investing in solar panels, charge controllers, and batteries to store the harvested energy. While the initial setup cost can be high, the long-term benefits of reduced energy bills and the ability to operate during power outages make it a worthwhile investment. Solar power systems require minimal maintenance once installed and can provide a reliable power supply, given sufficient sunlight.

Mobile Power

Ham Radios in Vehicles: Power Considerations

Operating ham radios in vehicles involves unique power considerations to ensure reliable performance and safety. Mobile transceivers typically require a stable power supply to function effectively, and vehicles' 12-volt electrical systems provide an ideal power source. However, proper connection and power management are vital to avoiding issues such as voltage drops, interference, or even damage to the radio equipment.

Connecting the Negative Power Return

For a <u>mobile transceiver, the negative power return should be connected to the 12-volt battery chassis ground</u>. This connection ensures a solid and stable ground reference, reducing the risk of electrical noise and interference affecting the transceiver's performance. Connecting directly to the battery chassis ground helps maintain a low-resistance path for the return current, which is essential for reliable operation, especially during high-current conditions such as transmitting.

T4A11: Where should the negative power return of a mobile transceiver be connected in a vehicle?

A. At the 12-volt battery chassis ground
B. At the antenna mount
C. To any metal part of the vehicle
D. Through the transceiver's mounting bracket

Finding the Chassis Ground in a Vehicle

Finding the chassis ground in your vehicle is relatively straightforward. Here's a step-by-step guide to help you locate it:

1. **Locate the Battery**: Find your vehicle's battery, usually under the hood. Identify the negative terminal, typically marked with a minus sign (-) and connected to a black cable.

2. **Follow the Negative Cable**: Trace the black cable from the battery's negative terminal. This cable is usually connected to a large metal part of the car, such as the engine block or frame.

3. **Identify the Ground Point**: The spot where the negative cable attaches to the metal part of the car is your chassis ground. This connection point often has a bolt or clamp securing the cable to the car's body or frame.

4. **Alternative Ground Points**: If you need an additional grounding point, look for other substantial metal parts of the car's frame or body that are clean and free of paint or rust. If they connect well to the vehicle's metal structure, they can serve as effective grounding points.

5. **Check for Existing Grounds**: Many vehicles have designated grounding points for various electrical systems. These are often marked and can be found in the vehicle's manual. They are typically in the engine compartment or near the battery.

Connecting your mobile transceiver's negative power return to this chassis ground ensures a stable and reliable grounding connection, reducing electrical noise and maintaining optimal performance.

Appropriate Power Supply Rating for a Mobile Transceiver

A typical <u>50-watt output mobile FM transceiver requires a power supply rating of 13.8 volts at 12 amperes.</u> This rating is key because it matches the voltage and current requirements of the transceiver under typical operating conditions. <u>The 13.8 volts is the standard voltage for vehicle electrical systems</u>, ensuring compatibility with the transceiver. The current rating of 12 amperes is necessary to support the power demands during transmission, where the current draw can be substantial. Providing a power supply with these specifications ensures that the transceiver operates efficiently, avoiding problems like voltage drops or overheating that could occur with an inadequate power supply.

T4A01: Which of the following is an appropriate power supply rating for a typical 50-watt output mobile FM transceiver?

A. 24.0 volts at 4 amperes
B. 13.8 volts at 4 amperes
C. 24.0 volts at 12 amperes
D. 13.8 volts at 12 amperes

This can be tricky on its surface, especially when you are new to ham and perhaps electrical equipment.

To understand the appropriate power supply rating for a typical 50-watt output mobile FM transceiver, you can use a basic formula derived from our Power Formula, which we reviewed earlier. Do you remember the Power in Pie? In the question and answers, we have watts (P- Power), volts (E- Voltage) and amperes (I- Current).

Power (P) = Current (I) x Voltage (E)

For a 50-watt transceiver, we want to calculate the required current (I) when the voltage (E) is 13.8 volts, which is a standard operating voltage for mobile transceivers.

Rearrange the formula to solve for current (I):

I = P / E

Substitute the values:

I = 50 watts / 13.8 volts ≈ 3.62 amperes

However, this calculation only considers the power output, not the transceiver's circuitry's efficiency and additional power requirements. Reread and think about the question carefully. It says this mobile FM transceiver has an output of 50 watts (a net or final output of 50 watts). A typical efficiency rating for a mobile transceiver might be around 50-60%, meaning the actual current draw would be higher than this ideal calculation.

Considering inefficiencies and safe operating margins, a general rule of thumb is to double or even triple the calculated current to estimate the required power supply rating. This is why a power supply rating of 13.8 volts at 12 amperes is recommended. It accounts for the additional current needed due to inefficiencies and ensures stable operation under various conditions.

So, while the basic formula can help understand the concept, the reader should remember that the final recommended power supply rating (13.8V at 12A) considers practical factors beyond just the power output calculation. This value is typically provided by the manufacturer or can be found in the transceiver's specifications.

Radio Frequency (RF) Safety

As an amateur radio operator, you need to know and understand the guidelines for minimizing RF exposure. Radio Frequency (RF) radiation, at high levels, can pose health risks, which is why the FCC has established specific exposure limits. These guidelines are designed to protect both operators and the general public from potentially harmful RF radiation effects.

Radio signals are a form of non-ionizing radiation. Unlike ionizing radiation, which has enough energy to remove tightly bound electrons from atoms, non-ionizing radiation lacks sufficient energy to ionize atoms or molecules, which could result in chemical changes in cells and even damage DNA. Instead, it primarily causes atoms and molecules to vibrate, which can produce heat but does not cause chemical changes or damage to biological tissues at the atomic level.

T0C01: What type of radiation are radio signals?

A. Gamma radiation
B. Ionizing radiation
C. Alpha radiation
D. Non-ionizing radiation

T0C12: How does RF radiation differ from ionizing radiation (radioactivity)?

A. RF radiation does not have sufficient energy to cause chemical changes in cells and damage DNA
B. RF radiation can only be detected with an RF dosimeter
C. RF radiation is limited in range to a few feet
D. RF radiation is perfectly safe

In amateur radio, understanding that radio signals are non-ionizing radiation is essential for safety and regulatory reasons. This type of radiation is generally considered safe for everyday use, as it does not have the harmful effects of ionizing radiation, such as X-rays or gamma rays.

Factors Affecting RF Exposure

The RF exposure of people near an amateur station antenna is influenced by several key factors: the frequency and power level of the RF field, the distance from the antenna to the person, and the antenna's radiation pattern. Higher frequencies and power levels result in stronger RF fields, which can increase the potential for exposure. For example, antennas transmitting at higher power levels or operating at higher frequencies can generate more intense RF fields, potentially leading to greater exposure.

Distance from the antenna is also important. The intensity of RF radiation decreases rapidly with increased distance from the source. Thus, maintaining a safe distance from the antenna can significantly reduce exposure. Additionally, the antenna's radiation pattern determines the direction and spread of the RF energy.

Directional antennas focus energy in specific directions, leading to higher exposure in those areas. In contrast, omnidirectional antennas distribute energy evenly. Understanding these factors helps operators design their stations to minimize RF exposure and comply with safety regulations, ensuring a safe operating environment for both themselves and the public.

T0C04: What factors affect the RF exposure of people near an amateur station antenna?

A. frequency and power level of the RF field
B. Distance from the antenna to a person
C. Radiation pattern of the antenna
D. All these choices are correct

Duty Cycle

The duty cycle refers to the percentage of time during which a device or signal is active and transmitting, compared to the total time. It is usually expressed as a percentage. For example, suppose a transmitter is on for 1 second and off for 1 second. In that case, the duty cycle is 50% because the transmitter is active for half the total time.

T0C11: What is the definition of duty cycle during the averaging time for RF exposure?

A. The difference between the lowest power output and the highest power output of a transmitter
B. The difference between the PEP and average power output of a transmitter
C. The percentage of time that a transmitter is transmitting
D. The percentage of time that a transmitter is not transmitting

The duty cycle is a parameter in determining safe RF radiation exposure levels because it directly <u>affects the average radiation</u> exposure. Remember, it says average, not necessarily total. A higher duty cycle means the transmitter is on more frequently, increasing the average RF exposure. Conversely, a lower duty cycle reduces the average exposure, as the transmitter is off for a more significant portion of the time. By considering the duty cycle, regulators and operators can assess and manage the potential risks associated with prolonged RF radiation exposure, ensuring that safety guidelines are followed.

T0C10: Why is duty cycle one of the factors used to determine safe RF radiation exposure levels?

A. It affects the average exposure to radiation
B. It affects the peak exposure to radiation
C. It takes into account the antenna feed line loss
D. It takes into account the thermal effects of the final amplifier

During the averaging time for RF exposure, the duty cycle is defined as the percentage of time that a transmitter is transmitting. This metric is used to calculate the overall exposure levels to RF radiation. For instance, if a transmitter operates with a duty cycle of 25%, it means that the transmitter is active and emitting RF signals for 25% of the time and inactive for the remaining 75%. This averaging approach helps evaluate the cumulative exposure and ensure that it stays within safe limits set by regulatory bodies.

In addition to RF exposure, the duty cycle is essential for electronic devices' thermal management and power consumption. A higher duty cycle means the equipment is active for extended periods in high-power applications like radio transmitters. If not properly managed, this leads to increased heat generation and potential overheating. Conversely, a lower duty cycle can help reduce heat build-up and extend the life of the equipment by allowing it to cool down during off periods.

Practical Example

The allowable power density for RF safety depends on the duty cycle of a transmitter. <u>The allowable power density can be doubled (increased by a factor of 2) if the duty cycle decreases from 100 percent to 50 percent</u>. This is because the duty cycle indicates how long the transmitter is actively emitting RF signals. The transmitter is always on at a 100 percent duty cycle, resulting in continuous exposure. However, at a 50 percent duty cycle, the transmitter is only active half the time, effectively halving the average exposure.

By halving the duty cycle, the average power density over time decreases, allowing for a higher instantaneous power density while maintaining safe exposure levels.

T0C03: How does the allowable power density for RF safety change if duty cycle changes from 100 percent to 50 percent?

A. It increases by a factor of 3
B. It decreases by 50 percent
C. It increases by a factor of 2
D. There is no adjustment allowed for lower duty cycle

Frequency

Maximum Permissible Exposure

RF exposure refers to the electromagnetic energy emitted by radio transmitters and its potential effects on human health. Exposure can vary depending on several factors, including frequency.

The maximum permissible exposure (MPE) has the lowest value at 50 MHz. This frequency is particularly significant because it falls within the range where the human body can absorb RF energy most efficiently. The absorption of RF energy by the body is influenced by the wavelength of the signal, and at around 50 MHz, the wavelength is such that it maximizes energy absorption. As a result, the exposure limits are set lower at this frequency to protect against potential health risks.

T0C02: At which of the following frequencies does maximum permissible exposure have the lowest value?

A. 3.5 MHz
B. 50 MHz
C. 440 MHz
D. 1296 MHz

Repetition helps build memorization. 50 MHz lies within the "6-meter band" and is a segment in the VHF (Very High Frequency) spectrum. This band spans from 50 to 54 MHz and is known for its unique propagation characteristics. The 6-meter band is particularly valued by amateur radio operators because it bridges the gap between HF (High Frequency) and VHF, offering local and long-distance communication opportunities under favorable conditions.

Exposure Limits

Exposure limits vary with frequency because the human body absorbs RF energy differently across the spectrum. The body's tissues absorb specific frequencies more efficiently, creating a higher potential for thermal and non-thermal effects. For instance, frequencies around 30-300 MHz are more readily absorbed, which can cause heating of body tissues and potentially lead to harmful biological effects. To mitigate these risks, exposure limits are set lower for these frequencies. By varying the limits according to frequency, regulatory guidelines ensure that RF exposure remains safe, protecting operators and the public from excessive RF energy absorption.

T0C05: Why do exposure limits vary with frequency?

A. Lower-frequency RF fields have more energy than higher-frequency fields
B. Lower frequency RF fields do not penetrate the human body
C. Higher frequency RF fields are transient in nature
D. The human body absorbs more RF energy at some frequencies than at others

Antenna RF Safety

The antenna may look like only a big dumb metal pole, but touching an antenna during transmission can result in an RF burn to the skin. RF energy generated during transmission can induce high currents on the surface of the antenna. When someone touches the antenna, these currents can flow through their skin, causing localized heating and potentially painful burns known as RF burns. This is because the human body can act as a conductor, allowing RF energy to pass through and generate heat at the point of contact.

T0C07: What hazard is created by touching an antenna during a transmission?

A. Electrocution
B. RF burn to skin
C. Radiation poisoning
D. All these choices are correct

To prevent RF burns and ensure safety, avoiding touching any part of the antenna system while transmitting is essential. This includes maintaining a safe distance from the antenna and ensuring that any necessary adjustments or repairs are made while the transmitter is off.

One practical action to reduce exposure to RF radiation is to relocate antennas. Positioning antennas farther away from living spaces, work areas, or places where people frequently gather can significantly lower the risk of exposure. By increasing the distance between the antenna and individuals, the intensity of the RF field decreases due to the inverse square law, which states that the strength of the signal diminishes rapidly with distance.

T0C08: Which of the following actions can reduce exposure to RF radiation?

A. Relocate antennas
B. Relocate the transmitter
C. Increase the duty cycle
D. All these choices are correct

Additionally, placing antennas in elevated positions, such as on rooftops or towers, can help direct the RF radiation away from populated areas. This enhances safety by reducing ground-level exposure and improves the antenna's performance by minimizing obstructions and increasing the effective range.

Bonding for Safety

Bonding in electrical systems refers to connecting all metallic parts of an electrical installation to create a continuous conductive path. This process ensures that all components have the same electrical potential, which minimizes the risk of electrical shock and helps effectively manage electrical currents. Bonding is paramount for both safety and the efficient operation of electrical systems, including those used in amateur radio setups.

Importance of Bonding in RF Systems

In RF (radio frequency) systems, bonding is vital for managing RF currents and minimizing interference. Proper bonding helps create a low-impedance path to the ground, which is essential for dissipating unwanted RF energy. This reduces the potential for RF interference that can affect the performance of radio equipment. Additionally, adequate bonding helps in protecting the equipment and the operator from electrical faults and static build-up. Using materials like flat copper straps for bonding ensures efficient current flow. It enhances the overall stability and reliability of the radio station. Understanding the importance of bonding is essential for maintaining a safe and effective amateur radio operation.

Preferred Conductor for RF Bonding: Flat Copper Strap

Flat copper straps are the preferred conductor for bonding at RF (radio frequency). They are chosen because of their low inductance and high surface area, which are key characteristics for effective RF bonding. At high frequencies, RF currents flow along the surface of conductors, a phenomenon known as the skin effect. Flat copper straps provide a larger surface area than round wires, thus minimizing resistance and inductance, ensuring efficient current flow.

T4A08: Which of the following conductors is preferred for bonding at RF?

A. Copper braid removed from coaxial cable
B. Steel wire
C. Twisted-pair cable
D. Flat copper strap

Using flat copper straps for RF bonding helps create a low-impedance path to the ground, which is essential for reducing RF noise and interference. This improves the overall performance of the radio equipment by ensuring stable and reliable connections. Flat copper straps are also flexible and easy to install, making them a practical choice for various bonding and grounding applications in an amateur radio station.

Compliance

You can use several acceptable methods to determine whether your amateur radio station complies with FCC RF exposure regulations. One method is by calculation based on FCC OET Bulletin 65, which provides detailed guidelines and formulas for evaluating RF exposure levels. This bulletin helps operators calculate the potential exposure from their equipment and ensure it falls within safe limits. If you love math and reading government regulations, Google "FCC OET Bulletin 65" and open up the PDF. Enjoy!

Another method is computer modeling. Various software tools can simulate RF exposure levels based on your station's specific parameters, such as antenna type, power output, and operating frequency. These models can provide a detailed analysis of the RF fields around your station, helping you verify compliance with FCC regulations.

Lastly, you can measure field strength using calibrated equipment. This involves using specialized RF measurement devices to measure the RF fields in and around your station directly. Accurate measurements can assess the actual RF exposure levels and compare them to the FCC's safety limits. Understanding and applying these methods ensures that your station operates safely and within regulatory guidelines, an essential aspect of practical operation and exam preparation in amateur radio.

T0C06: Which of the following is an acceptable method to determine whether your station complies with FCC RF exposure regulations?

A. By calculation based on FCC OET Bulletin 65
B. By calculation based on computer modeling
C. By measurement of field strength using calibrated equipment
D. All these choices are correct

Ensuring that your station complies with FCC RF exposure regulations can seem complex. Still, there are straightforward methods and guidelines to help beginners navigate this critical aspect of station operation. Here are some key points and guidance to simplify the process:

Practical Tips for Beginners:

- **Start with FCC OET Bulletin 65**:
  - **Guidance**: Begin by familiarizing yourself with FCC OET Bulletin 65. It provides a solid foundation and step-by-step instructions on how to perform calculations. It's a trusted source directly from the FCC. It's technical, but you should at least look it over once.

- **Use Simple Software Tools**:

  - **Guidance**: Many user-friendly software tools are available for amateur radio operators. These tools can automate much of the calculation process. Look for software that is specifically designed for amateur radio RF exposure assessment.

  - **Start here**: Google 'ARRL RF Exposure Calculator' and it's the first link that comes up. This is a trusted source and relatively simple to use. Just plug in your numbers and the calculator will do the rest.

- **Seek Help from Experienced Operators**:

  - **Guidance**: Don't hesitate to ask for help from more experienced operators or join local amateur radio clubs. Many clubs offer resources and mentorship programs to assist beginners in understanding and applying these regulations.

  - **Recommended**: As a newly minted ham or soon-to-be, this is a great place to start and get some hands-on experience by working with trusted peers.

- **Regularly Review and Update**:

  - **Guidance**: Regularly review your station setup and any changes you make to ensure ongoing compliance. Modifications to your equipment or operating practices can change RF exposure levels.

By following these methods and practical tips, you can confidently ensure your station complies with FCC RF exposure regulations while maintaining safety for yourself and others.

Power Safety

Power safety is a vital aspect of radio operations, encompassing the safe handling and management of electrical power to prevent accidents, equipment damage, and personal injury. Understanding and following proper safety protocols is critical, whether working with low-voltage batteries or high-voltage power supplies. This section will cover key safety practices, including safe measurement techniques, proper grounding, handling of high-voltage equipment, and the importance of appropriately rated tools and protective gear. By adhering to these guidelines, you can ensure a safe and effective radio setup.

Electrical Safety

Electrical current flowing through the body can pose serious health hazards. One significant risk is tissue injury caused by heating, as the electrical energy can generate considerable heat, potentially leading to burns or internal damage. This heating effect is hazardous as it can affect the skin and deeper tissues.

In addition to heating, electrical current can disrupt the body's electrical functions at a cellular level. This disruption can interfere with the regular operation of cells, potentially leading to serious health issues such as heart arrhythmias or neurological damage. Furthermore, electrical currents can cause involuntary muscle contractions. These contractions can be severe enough to cause physical injuries or make it impossible to let go of the source of the current, prolonging exposure and increasing the risk of serious harm.

T0A02: What health hazard is presented by electrical current flowing through the body?

A. It may cause injury by heating tissue
B. It may disrupt the electrical functions of cells
C. It may cause involuntary muscle contractions
D. All these choices are correct

Fuses

The primary purpose of a fuse in an electrical circuit is to <u>remove power in case of an overload,</u> protecting the circuit and connected devices from damage. A fuse is a safety device that contains a thin wire or filament that melts when the current flowing through it exceeds a specific threshold. This melting action breaks the circuit, stopping the flow of electricity and preventing potential damage caused by excessive current.

T0A04: What is the purpose of a fuse in an electrical circuit?

A. To prevent power supply ripple from damaging a component
B. To remove power in case of overload
C. To limit current to prevent shocks
D. All these choices are correct

Fuses are key components in ensuring electrical safety. They help prevent overheating, fires, and damage to sensitive electronic components by automatically disconnecting the power during an overload condition. Fuses safeguard both the equipment and the users by ensuring that circuits operate within safe limits.

Correct Fuse Rating

Using a fuse with the correct rating is for preventing excessive current from causing damage or posing a fire hazard. A fuse ensures that the electrical system operates within safe limits. Exceeding these limits risks serious damage to your equipment and creates significant safety hazards.

For example, a 5-ampere fuse should never be replaced with a 20-ampere fuse because it compromises the safety of the electrical circuit. Fuses are designed to protect circuits by breaking the connection when the current exceeds a specific limit, in this case, 5 amperes. Replacing a 5-ampere fuse with a 20-ampere fuse allows a much higher current to flow through the circuit than it was designed to handle. <u>This can lead to overheating of the wires and components, potentially causing them to melt or catch fire.</u>

T0A05: Why should a 5-ampere fuse never be replaced with a 20-ampere fuse?

A. The larger fuse would be likely to blow because it is rated for higher current
B. The power supply ripple would greatly increase
C. Excessive current could cause a fire
D. All these choices are correct

Standardization of Electrical Components

Understanding Wire Color Coding in the United States

<u>In the United States, the black wire insulation in a three-wire 120 V cable indicates the "hot" or live wire.</u> This wire carries the electrical current from the power source to the load, which is essential for the operation of electrical devices. The hot wire delivers the voltage needed to power appliances, lights, and other equipment.

T0A03: In the United States, what circuit does black wire insulation indicate in a three-wire 120 V cable?

A. Neutral
B. Hot
C. Equipment ground
D. Black insulation is never used

Understanding wire color coding is important for safely working with electrical systems. In a typical three-wire 120 V setup, besides the black hot wire, there is usually a white wire that serves as the neutral conductor and a green or bare wire that acts as the ground. The neutral wire completes the circuit by returning current to the power source. In contrast, the ground wire provides a safe path for electricity in case of a fault. Identifying and using these wires ensures electrical circuits' safe and efficient operation, a fundamental concept for practical electrical work and exam preparation.

Placement of Fuses and Circuit Breakers in a 120V AC Power Circuit

In a 120V AC power circuit, a fuse or circuit breaker <u>should only be installed in series with the hot conductor.</u> The hot conductor is the wire that carries the live current from the power source to the load. Installing the protective device in series with the hot conductor ensures that the circuit will be interrupted in the event of an overload or short circuit, effectively cutting off the flow of electricity and preventing potential damage or fire.

T0A08: Where should a fuse or circuit breaker be installed in a 120V AC power circuit?

A. In series with the hot conductor only
B. In series with the hot and neutral conductors
C. In parallel with the hot conductor only
D. In parallel with the hot and neutral conductors

Placing the fuse or circuit breaker in the hot conductor is crucial because it directly controls the current flow into the circuit. If the protective device were installed in the neutral or ground wire instead, it would not provide the same level of protection. The hot wire is where the electrical energy enters the circuit, so interrupting it with a fuse or circuit breaker ensures that the entire circuit is safely disconnected from the power source in case of a fault.

Guarding Against Electrical Shock at Your Station

Guarding against electrical shock is vital for maintaining a safe amateur radio station. One effective method is to <u>use three-wire cords and plugs</u> for all AC-powered equipment. The three-wire system includes a hot wire, a neutral wire, and a ground wire. The ground wire provides a path for electrical current to return to the ground in case of a fault, thereby reducing the risk of electric shock.

Additionally, <u>connecting all AC-powered station equipment to a common safety ground</u> ensures that any stray electrical currents are safely directed away from the equipment and operator. This grounding practice helps to equalize potential differences between devices, preventing electrical shocks. Another safety measure is to <u>install mechanical interlocks in high-voltage circuits</u>. Mechanical interlocks ensure that high-voltage circuits cannot be accessed; at the same time, they are energized, providing an additional layer of protection.

T0A06: What is a good way to guard against electrical shock at your station?

A. Use three-wire cords and plugs for all AC-powered equipment
B. Connect all AC powered station equipment to a common safety ground
C. Install mechanical interlocks in high-voltage circuits
D. All these choices are correct

Hazard of Stored Charge in Filter Capacitors

A significant hazard in a power supply immediately after turning it off is the charge stored in filter capacitors. Filter capacitors are designed to smooth out fluctuations in the power supply by storing electrical energy. However, even after the power supply is turned off, these capacitors can retain a substantial charge for some time. This stored energy can pose a severe risk of electric shock if someone touches the components connected to the capacitors.

T0A11: What hazard exists in a power supply immediately after turning it off?

A. Circulating currents in the dc filter
B. Leakage flux in the power transformer
C. Voltage transients from kickback diodes
D. Charge stored in filter capacitors

To mitigate this risk, it's essential to safely discharge the capacitors before handling the power supply components. This can be done using a resistor to slowly bleed off the stored charge, ensuring the capacitors are fully discharged. Understanding this hazard and taking appropriate precautions will help prevent accidental shocks and injuries.

Battery Safety

Battery safety is fundamental to operating an amateur radio station, especially when using 12-volt storage batteries. These batteries are commonly used for their reliability and capacity to provide ample power. However, they come with specific safety hazards that operators must be aware of to prevent accidents and ensure safe operation.

One significant safety hazard of a 12-volt storage battery is the risk of shorting the terminals. Suppose the positive and negative terminals are accidentally connected with a conductive material. In that case, a rapid flow of current can cause intense heat. This can result in burns, fire, or even an explosion, posing a severe danger to the operator and surrounding equipment. Always handle batteries carefully to prevent this, ensuring the terminals are covered or insulated when not used.

T0A01: Which of the following is a safety hazard of a 12-volt storage battery?

A. Touching both terminals with the hands can cause electrical shock
B. Shorting the terminals can cause burns, fire, or an explosion
C. RF emissions from a nearby transmitter can cause the electrolyte to emit poison gas
D. All these choices are correct

Another significant hazard is caused by charging or discharging a battery too quickly. Rapid charging or discharging can lead to overheating of the battery. This excessive heat can damage the battery's internal components and reduce lifespan. Additionally, fast charging can cause out-gassing, where the battery releases potentially harmful gases. These gases can create an explosive atmosphere if not adequately ventilated. To mitigate these risks, use appropriate charging equipment designed for your specific battery type and follow the manufacturer's guidelines for safe charging and discharging rates.

T0A10: What hazard is caused by charging or discharging a battery too quickly?

A. Overheating or out-gassing
B. Excess output ripple
C. Half-wave rectification
D. Inverse memory effect

Understanding these hazards and implementing proper safety measures ensures the safe use of batteries in your amateur radio operations.

Grounding

Grounding is fundamental to setting up a safe and efficient amateur radio station. Proper grounding ensures protection against electrical faults, lightning strikes, and static build-up, safeguarding both the operator and the equipment. Grounding panels are an essential part of this system, providing a centralized location for all grounding connections. By using a grounding panel, operators can maintain a low-impedance path to the earth, which is needed for adequate grounding.

Bonding External Ground Rods

All external ground rods or earth connections (these are metal rods, typically made of copper or galvanized steel, driven into the ground to provide a direct path to the earth for electrical currents) should be <u>bonded together with heavy wire</u> or a conductive strap to ensure an effective and cohesive grounding system. Bonding these ground connections ensures that they have the same electrical potential, preventing voltage differences that could cause dangerous arcing or reduce the effectiveness of the grounding system. By connecting all ground rods, operators create a more reliable and robust grounding network that enhances overall safety and performance.

T0A09: What should be done to all external ground rods or earth connections?

A. Waterproof them with silicone caulk or electrical tape
B. Keep them as far apart as possible
C. Bond them together with heavy wire or conductive strap
D. Tune them for resonance on the lowest frequency of operation

Where to Install a Lightning Arrester

A lightning arrester is a device used to protect electrical equipment from damage caused by lightning strikes or other high-voltage surges. It functions by providing a pathway for the excess electrical energy to be safely diverted to the ground, preventing it from reaching and damaging sensitive equipment. It should be installed in the coaxial feed line <u>on a grounded panel near where the feed lines enter the building</u>. This strategic placement ensures that any surge caused by a lightning strike is safely diverted to the ground before entering the building and damaging the equipment. Installing the arrester on a grounded panel provides a direct path for the electrical surge to follow, minimizing the risk of it traveling through the internal wiring and causing extensive damage. This practice is key for protecting the radio equipment and the operator from the potentially devastating effects of lightning strikes.

T0A07: Where should a lightning arrester be installed in a coaxial feed line?

A. At the output connector of a transceiver
B. At the antenna feed point
C. At the ac power service panel
D. On a grounded panel near where feed lines enter the building

Ensuring Compliance

To ensure your station complies with RF safety regulations, it is essential to reevaluate it whenever there are transmitter or antenna system changes. This includes adjusting the power output, changing the antenna type or location, or altering the transmission frequency. Each of these changes can affect the RF exposure levels around your station, potentially bringing it out of compliance with the FCC's safety limits.

T0C09: How can you make sure your station stays in compliance with RF safety regulations?

A. By informing the FCC of any changes made in your station
B. By reevaluating the station whenever an item in the transmitter or antenna system is changed
C. By making sure your antennas have low SWR
D. All these choices are correct

Regular re-evaluation involves measuring or calculating the RF field strength to ensure exposure levels are within the permissible limits. This process may include using tools like RF exposure calculators, conducting field measurements, and ensuring that any adjustments do not result in excessive RF exposure. By diligently reevaluating your station after any changes, you can maintain a safe operating environment and adhere to regulatory requirements, thus protecting yourself and the public from potential RF hazards.

Understanding and implementing these safety measures ensures your compliance with legal requirements and secures your well-being and that of others. Whether experimenting with different power sources or fine-tuning your station's setup, keeping safety at the forefront of your activities is essential. This approach prevents accidents and ensures your radio experience is enjoyable and sustainable.

Chapter 19

Antennas, Feedlines and Safety

Antenna Tuner

THE PRIMARY FUNCTION OF an antenna tuner, also known as an antenna coupler, <u>is to match the impedance of the antenna system to the transceiver's output impedance</u>, typically 50 ohms. This matching is important because it ensures that the transceiver's maximum power is transferred to the antenna for efficient radiation.

When the impedances are not matched, significant power can be reflected back to the transceiver, resulting in poor transmission efficiency and potential damage to the equipment.

The tuner then uses adjustable inductors and capacitors to create a matching network. By altering the inductance and capacitance, the tuner changes the impedance seen by the transceiver.

For beginners, it's helpful to think of an antenna tuner as a device that "tunes" the antenna system to make it compatible with the transceiver. By adjusting the impedance, the tuner minimizes signal reflection and maximizes power transfer, leading to clearer and stronger signals.

T9B04: What is the major function of an antenna tuner (antenna coupler)?

A. It matches the antenna system impedance to the transceiver's output impedance
B. It helps a receiver automatically tune in weak stations
C. It allows an antenna to be used on both transmit and receive
D. It automatically selects the proper antenna for the frequency band being used

---

Antenna Analyzer

An antenna analyzer is a specialized device radio operators use to evaluate and optimize the performance of their antennas. It measures impedance, SWR (Standing Wave Ratio), and resonant frequency, providing detailed insights into how well an antenna system functions. This information helps tune and adjust the antenna for optimal performance.

Essential Functions of an Antenna Analyzer:

- **Impedance Measurement**: The analyzer measures the antenna system's impedance across a range of frequencies, helping to identify how closely the antenna's impedance matches the desired value (typically 50 ohms).

- **SWR Measurement**: It provides real-time SWR readings, indicating the power transfer efficiency from the transmitter to the antenna. A low SWR indicates good impedance matching and minimal power loss.

- **Resonant Frequency Identification**: The analyzer identifies the antenna's resonant frequency, which is the frequency at which the antenna operates most efficiently. This helps in tuning the antenna to the desired frequency band and determines if it is resonant at the desired operating frequency.

- **Graphical Displays**: Many modern antenna analyzers include graphical displays that show SWR curves, impedance plots, and other useful data, making it easier to visualize the antenna system's performance.

Think of an antenna analyzer as a diagnostic tool that helps you tune your antenna to its "sweet spot." This ensures that your radio system operates efficiently, with maximum signal strength and clarity.

T7C02: Which of the following is used to determine if an antenna is resonant at the desired operating frequency?

A. A VTVM
B. An antenna analyzer
C. A Q meter
D. A frequency counter

Dummy Loads

A dummy load is an essential tool for radio operators. It is a substitute for an antenna during the testing and calibration of radio equipment.

It provides a non-radiating, resistive load that safely absorbs the transmitter's power, preventing interference with actual broadcasts. By matching the impedance of the radio's output (typically 50 ohms), a dummy load allows operators to test and adjust their equipment under realistic conditions without radiating signals into the air.

This helps ensure transmitters are properly tuned and functioning before being connected to an antenna for real-world communication.

T7C01: What is the primary purpose of a dummy load?

A. To prevent transmitting signals over the air when making tests
B. To prevent over-modulation of a transmitter
C. To improve the efficiency of an antenna
D. To improve the signal-to-noise ratio of a receiver

A dummy load consists of a non-inductive resistor mounted on a heat sink. The non-inductive resistor provides a purely resistive load without introducing inductance, which could affect the accuracy of the testing. This resistor matches the typical impedance of the radio equipment, ensuring that the transmitter operates under normal conditions.

The heat sink is essential as it dissipates the heat the resistor generates when it absorbs the transmitter's power. Without proper heat dissipation, the resistor could overheat and fail.

By safely absorbing the transmitted power and converting it into heat, the dummy load allows radio operators to test and calibrate their equipment without broadcasting signals, ensuring that their transmitters are properly tuned and functioning correctly.

T7C03: What does a dummy load consist of?

A. A high-gain amplifier and a TR switch
B. A non-inductive resistor mounted on a heat sink
C. A low-voltage power supply and a DC relay
D. A 50-ohm reactance used to terminate a transmission line

From the intricate details of antenna resonance and impedance matching to the practical applications of dummy loads and coaxial cables, each component plays a key role in maintaining signal integrity and optimizing performance. Mastering these concepts ensures your radio equipment operates reliably, providing clear and consistent communication in any scenario.

RF Power Meter

An RF power meter is a device used to measure the power level of radio frequency (RF) signals. It is an essential tool for amateur radio operators and professionals working with RF equipment. RF power meters provide accurate measurements of the power being transmitted or received by an antenna, helping operators ensure that their equipment functions correctly and efficiently.

Importance of an RF Power Meter

Using an RF power meter, operators can verify that their transmitters are producing the expected power output and that their antennas are radiating this power effectively. This helps diagnose and troubleshoot issues such as power loss, signal distortion, or mismatched impedance in the transmission line. Accurate power measurements are imperative for maintaining optimal performance, ensuring compliance with regulatory power limits, and protecting the equipment from potential damage due to over-powering or inefficient operation.

Proper Installation of an RF Power Meter

An RF power meter should be <u>installed between the transmitter and the antenna in the feed line.</u> This placement allows the meter to accurately measure the transmitter's power output before it reaches the antenna. By positioning the RF power meter in this location, operators can monitor the transmitted power and ensure it is within the desired range.

T4A05: Where should an RF power meter be installed?

A. In the feed line, between the transmitter and antenna
B. At the power supply output
C. In parallel with the push-to-talk line and the antenna
D. In the power supply cable, as close as possible to the radio

Installing the RF power meter in the transmission line is needed to diagnose and troubleshoot potential transmitter, feed line, or antenna issues. It helps verify that the transmitter is functioning correctly and that the power is effectively delivered to the antenna. Proper installation ensures accurate readings, which is essential for optimizing performance, maintaining regulatory compliance, and protecting the equipment from potential damage.

Antenna Safety

Safety around and with antennas is vital to setting up and maintaining your radio station. While optimizing performance and adhering to RF exposure guidelines are essential, it is equally important to consider the physical safety aspects associated with antenna installation and maintenance. This includes preventing accidents during installation, avoiding hazards related to high structures, and ensuring the antenna system is secure and stable under various weather conditions. In this section, we will explore safety practices and precautions that every amateur radio operator should follow to protect themselves and others from potential physical hazards associated with antennas.

Key Considerations When Setting Up an Antenna

Setting up an antenna involves several fundamental considerations to ensure optimal performance, safety, and regulation compliance. Here are the main factors to take into account:

1. **Location**: Choose an area that provides the best line-of-sight for the antenna, free from obstructions like buildings, trees, and other structures. Elevating the antenna can significantly improve signal reception and transmission.

2. **Height**: The height of the antenna affects its range and performance. Higher placement typically results in better coverage and reduced interference. However, it is essential to comply with local zoning laws and regulations regarding maximum allowable heights.

3. **Safety**: Ensure the installation is safe for both people and property. This includes:

   ◦ **Grounding**: Proper grounding protects against lightning strikes and electrical surges.

   ◦ **Guy Lines**: Use tension guy lines to stabilize tall masts and towers, especially in windy areas.

   ◦ **Clearance**: Maintain a safe distance from power lines and other electrical infrastructure to prevent electrical hazards.

4. **RF Exposure**: To comply with safety regulations, evaluate and minimize RF exposure. This includes ensuring the antenna is placed far enough from areas where people spend time and adjusting power levels as needed.

5. **Structural Integrity**: Use durable materials and mounting hardware suitable for the environmental conditions. Regularly inspect and maintain the antenna and its supporting structures to ensure continued safety and performance.

6. **Regulatory Compliance**: Adhere to all relevant laws, regulations, and guidelines from authorities like the FCC. This includes obtaining necessary permits and ensuring the installation does not interfere with other services.

7. **Environmental Factors**: Consider the local climate and weather conditions. Antennas should withstand wind, rain, snow, and other environmental stresses.

Considering these factors, you can ensure your antenna setup is safe, efficient, and compliant with all relevant standards. This approach not only enhances the performance of your amateur radio station but also safeguards against potential hazards and legal issues.

Constructing the Antenna

One of the most critical safety precautions to observe when installing an antenna tower is <u>to look for and stay clear of any overhead electrical wires.</u> Contact with electrical wires can result in severe injury or even fatality due to electric shock. **<u>It is essential to thoroughly survey the installation site for any nearby power lines</u>** and ensure that the tower and its components are kept at a safe distance.

T0B04: Which of the following is an important safety precaution to observe when putting up an antenna tower?

A. Wear a ground strap connected to your wrist at all times
B. Insulate the base of the tower to avoid lightning strikes
C. Look for and stay clear of any overhead electrical wires
D. All these choices are correct

To ensure safety, maintain a minimum clearance distance as recommended by local electrical codes; typically, though, you **should ensure that if your antenna falls over, no part will ever be closer than 10 feet away from a power line.** Additionally, always assume that power lines are live and potentially dangerous. By taking these precautions, you can prevent accidental contact with electrical wires, safeguarding yourself and others involved in the installation process.

This part is both underlined and bolded – it's that important.

T0B06: What is the minimum safe distance from a power line to allow when installing an antenna?

A. Add the height of the antenna to the height of the power line and multiply by a factor of 1.5
B. The height of the power line above ground
C. 1/2 wavelength at the operating frequency
D. Enough so that if the antenna falls, no part of it can come closer than 10 feet to the power wires

One more piece to add to this. If the above wasn't enough of a warning, the FCC wants to stress one more piece from the test.

Attaching an antenna to a utility pole is highly discouraged because of the significant risk that the antenna or its supporting structures could contact high-voltage power lines. Utility poles often carry wires that transmit electricity at high voltages, posing a severe risk of electric shock or electrocution if touched. Even indirect contact, such as a wire or antenna element coming close to a power line, can lead to dangerous arcing or electrical discharge.

T0B09: Why should you avoid attaching an antenna to a utility pole?

A. The antenna will not work properly because of induced voltages
B. The 60 Hz radiations from the feed line may increase the SWR
C. The antenna could contact high-voltage power lines
D. All these choices are correct

Always choose an installation site well away from utility poles and overhead power lines to ensure safety. This precaution helps prevent accidental contact with high-voltage lines, which can cause severe injury, fatality, or damage to equipment.

Securing the Antenna

Tension Guy Lines

Tension guy lines are essential to stabilize and support tall structures such as antenna masts and towers. They are typically made of strong, durable materials like steel cable or synthetic fiber and designed to withstand significant tension. You've seen them. They are anchored at multiple points around the structure's base and connected to the tower at various heights, forming a supportive framework that resists wind and gravity forces.

- **Stability**: Tension guy lines help maintain the vertical position of the mast or tower by distributing the load and providing counteracting forces, preventing it from swaying or toppling over.

- **Wind Resistance**: Guy lines are crucial in withstanding wind loads, particularly in regions prone to strong winds. They help dissipate the wind force across multiple points, reducing the risk of structural failure.

- **Safety**: Properly installed guy lines ensure the safety of the antenna system and surrounding areas, preventing accidents caused by collapsing structures.

Understanding the importance and correct installation of tension guy lines is vital for ensuring the physical safety and longevity of your antenna setup.

Turnbuckles

A turnbuckle is a device that adjusts the tension or length of cables, ropes, or guy lines in various applications, including antenna installations. It consists of two threaded eye bolts screwed into either end of a metal frame, one with right-hand threads and the other with left-hand threads. The turnbuckle either tightens or loosens by rotating the frame, allowing for precise adjustment of the tension in the guy lines.

Key Features and Uses of Turnbuckles

- **Tension Adjustment**: Turnbuckles are primarily used to adjust the tension in guy lines, ensuring that structures like antenna masts and towers remain stable and secure. Proper tension helps prevent swaying and structural failure, particularly in windy conditions.

- **Versatility**: They are used in various applications, from construction and marine settings to amateur radio installations, where precise line tension control is essential.

- **Safety Enhancement**: Turnbuckles help maintain the integrity of the support system. They allow you to easily adjust and maintain the necessary tension in the guy lines, contributing to the overall safety and reliability of the installation.

Example of Use

Turnbuckles are attached to the guy lines that stabilize the antenna mast in an antenna installation. By tightening or loosening the turnbuckles, the operator can ensure that each guy line is evenly tensioned, maintaining the mast's vertical position and stability. This is critical for preventing the mast from leaning or collapsing, especially in adverse weather conditions.

For additional safety, a wire is put through the turnbuckle. <u>The purpose of a safety wire through a turnbuckle used to tension guy lines is to prevent the turnbuckle from loosening inadvertently</u>. Over time, vibrations from wind, weather changes, or other environmental factors can cause turnbuckles to rotate and loosen, potentially compromising the stability of the antenna structure.

By threading a safety wire through the turnbuckle and securing it, you ensure it remains adjusted, maintaining the necessary tension on the guy lines. This added security helps prevent structural failure and enhances the overall safety and reliability of the antenna installation.

T0B05: What is the purpose of a safety wire through a turnbuckle used to tension guy lines?

A. Secure the guy line if the turnbuckle breaks
B. Prevent loosening of the turnbuckle from vibration
C. Provide a ground path for lightning strikes
D. Provide an ability to measure for proper tensioning

Crank-Up Tower

A crank-up tower is a telescoping tower used in amateur radio and other applications to support antennas at varying heights. The tower consists of multiple nested sections that can be extended and retracted using a manual or motorized crank mechanism. This design allows for flexibility in raising the antenna to a desired height for optimal performance while enabling the tower to be lowered for maintenance, transport, or to reduce exposure to severe weather.

Key Features and Uses of Crank-Up Towers

- **Adjustable Height**: The primary feature of a crank-up tower is its ability to adjust the height of the antenna. Operators can elevate their antennas by cranking the tower up to improve signal reception and transmission. Conversely, the tower can be lowered to protect the antenna during storms or for easy maintenance access.

- **Ease of Use**: Crank-up towers are relatively easy to operate. The crank mechanism, which can be manual or motorized, allows for smooth raising and lowering of the tower sections. This makes it convenient for operators to adjust the antenna height as needed without requiring extensive effort or equipment.

- **Safety Considerations**: While crank-up towers offer convenience and flexibility, they also have specific safety considerations. <u>The tower should only be climbed if fully retracted or equipped with mechanical safety locks to prevent unintended movement.</u> Ensuring the tower is stable and secure is vital to avoid accidents and injuries.

Practical Applications

Crank-up towers are widely used by amateur radio enthusiasts who need to optimize their antenna performance by adjusting their height. They are also beneficial for other applications requiring temporary or adjustable equipment elevation, such as temporary broadcasting setups or emergency communication systems.

T0B07: Which of the following is an important safety rule to remember when using a crank-up tower?

A. This type of tower must never be painted
B. This type of tower must never be grounded
C. This type of tower must not be climbed unless it is retracted, or mechanical safety locking devices have been installed
D. All these choices are correct

Climbing an Antenna Tower

Climbing a tower is often necessary for amateur radio operators and technicians who install, maintain, or adjust antennas and other equipment. However, it is also one of the most hazardous activities associated with radio operations. Ensuring safety while climbing towers requires a thorough understanding of proper techniques, appropriate safety equipment, and adherence to established protocols. This section will cover the essential safety practices and precautions to follow when climbing towers, helping you minimize risks and perform your tasks effectively and safely. Whether you are a seasoned climber or new to tower work, these guidelines are to protect you and ensure a secure working environment.

Essential Requirements for Climbing an Antenna Tower

Climbing an antenna tower safely requires adhering to several safety measures. First and foremost, <u>climbers must have sufficient training in safe tower climbing techniques</u>. This is not training you get from a book but real-life training that ensures climbers understand the risks involved and are proficient in using climbing equipment and techniques to minimize these risks. Proper training is essential for developing the skills and knowledge necessary to navigate the complexities of tower climbing safely.

In addition to training, using an approved climbing harness and <u>appropriate tie-off methods</u> are mandatory. An <u>approved climbing harness</u> is designed to provide support and security, preventing falls and reducing the risk of injury. The harness must be properly fitted and secured before climbing. Furthermore, climbers must use appropriate tie-off techniques always to remain connected to the tower. This continuous connection ensures that the harness and tie-off system prevent a fall if a climber loses their grip or footing.

T0B02: What is required when climbing an antenna tower?

A. Have sufficient training on safe tower climbing techniques
B. Use appropriate tie-off to the tower at all times
C. Always wear an approved climbing harness
D. All these choices are correct

<u>Climbing a tower without a helper or observer is never safe</u>. Having a second person present is essential for several reasons. A helper or observer can immediately assist in an emergency, such as a fall or medical issue. They can also help manage climbing equipment, ensure safety lines are correctly secured, and communicate with the climber to monitor their status throughout the climb.

Additionally, a helper or observer acts as an extra set of eyes to spot potential hazards the climber might overlook. This added vigilance enhances overall safety by ensuring all safety protocols are followed, and any issues are promptly addressed.

T0B03: Under what circumstances is it safe to climb a tower without a helper or observer?

A. When no electrical work is being performed
B. When no mechanical work is being performed
C. When the work being done is not more than 20 feet above the ground
D. Never

Understand the importance of never climbing a tower alone. This will help maintain safety and prevent accidents. This principle is vital for both practical tower work and exam preparation, emphasizing the necessity of teamwork and proper safety measures in tower climbing activities.

Grounding the Antenna

Grounding the antenna is fundamental to setting up and maintaining a safe and effective amateur radio station. Proper grounding provides several benefits, including protection from lightning strikes, reduced electrical noise, and enhanced overall signal performance.

Establishing a reliable grounding system can safeguard your equipment from damage, improve communication clarity, and ensure compliance with safety regulations. This section will explore the principles and practices of grounding antennas, offering essential guidelines to help you implement a robust and effective grounding solution for your radio setup.

Grounding Requirements

Local electrical codes establish ground requirements for an amateur radio tower or antenna. These codes provide specific guidelines and regulations designed to ensure the safety and effectiveness of the grounding system. Adhering to local electrical codes is important because they are tailored to address the area's unique environmental and infrastructural conditions, such as soil conductivity and regional weather patterns.

Following local electrical codes helps prevent electrical hazards, such as lightning strikes and electrical surges, by properly designing and installing the grounding system. Compliance with these codes protects your equipment, enhances its performance, and ensures that your installation meets legal and safety standards.

T0B11: Which of the following establishes grounding requirements for an amateur radio tower or antenna?

A. FCC Part 97 rules
B. Local electrical codes
C. FAA tower lighting regulations
D. UL recommended practices

Proper Grounding Method for a Tower

A proper grounding method for a tower involves using separate eight-foot ground rods for each tower leg, which are then bonded to the tower and each other. This setup provides a robust grounding system that effectively disperses electrical currents, such as those from lightning strikes, into the earth. By installing a ground rod at each tower leg and connecting these rods together, you create a low-resistance path to the ground that helps protect both the tower and any connected equipment from electrical surges.

T0B08: Which is a proper grounding method for a tower?

A. A single four-foot ground rod, driven into the ground no more than 12 inches from the base
B. A ferrite-core RF choke connected between the tower and ground
C. A connection between the tower base and a cold water pipe
D. Separate eight-foot ground rods for each tower leg, bonded to the tower and each other

---

Bonding the ground rods to the tower and each other ensures the grounding system works as a unified structure, providing consistent protection throughout the installation. This method not only enhances safety by reducing the risk of electrical shocks and equipment damage but also helps improve the overall performance of the antenna system by minimizing electrical noise and interference.

Best Practices for Installing Ground Wires for Lightning Protection

When installing ground wires on a tower for lightning protection, it is good practice to ensure that the connections are short and direct. Short, direct ground connections provide the most efficient path for lightning-induced electrical currents to travel into the ground, minimizing the potential for damage to the tower and connected equipment. Long or convoluted ground paths can increase resistance and the likelihood of side-flash, where the lightning current jumps to nearby objects, potentially causing damage or injury.

T0B01: Which of the following is good practice when installing ground wires on a tower for lightning protection?

A. Put a drip loop in the ground connection to prevent water damage to the ground system
B. Make sure all ground wire bends are right angles
C. Ensure that connections are short and direct
D. All these choices are correct

Direct connections help maintain a low-impedance path to the ground, which is essential for quickly dissipating the high-energy surge from a lightning strike. This practice enhances the installation's safety and the grounding system's overall effectiveness.

Importance of Avoiding Sharp Bends in Grounding Conductors

Sharp bends must be avoided when installing grounding conductors used for lightning protection. Sharp bends can create points of high impedance, which hinder the efficient flow of lightning-induced electrical currents to the ground. This impedance can cause the electrical energy to slow or even arc at the bend, potentially damaging the grounding system, the tower, or connected equipment.

Ensuring that grounding conductors have smooth, gradual bends instead of sharp angles provides a more direct and low-resistance path for the electrical currents. This practice enhances the overall effectiveness of the grounding system, allowing it to dissipate the energy from a lightning strike quickly and safely into the earth.

T0B10: Which of the following is true when installing grounding conductors used for lightning protection?

A. Use only non-insulated wire
B. Wires must be carefully routed with precise right-angle bends
C. Sharp bends must be avoided
D. Common grounds must be avoided

As we conclude our discussion on safety, it should be clear that a thorough understanding of these principles will not only keep your equipment safe and in top operating condition but will also ensure the safety of you and others from the potential harm of operating an amateur radio station.

Operating Your Radio

Now that you have a solid understanding of the science behind radio and have familiarized yourself with the physical setup and components, it's time to start operating your radio. This section will guide you through the basics of getting on the air, from tuning and selecting frequencies to making your first contact. Whether using a handheld transceiver, a mobile unit in your vehicle, or a base station at home, learning how to operate your equipment effectively is fundamental to enjoying all that amateur radio offers. We'll cover essential skills such as setting up your radio, choosing the right modes and bands, and following proper operating procedures. With these foundational skills, you'll be ready to communicate with other hams, participate in nets, and explore exciting amateur radio communication.

Chapter 20

Decoding Ham Radio Jargon

In HAM RADIO, EVERY term and abbreviation carries weight, serving as a bridge to more efficient communication. For instance, a "QSO," one of the most commonly heard terms, refers simply to a conversation between two or more amateur radio operators.

Each interaction you have on the air is a QSO, the basic building block of your ham radio experience. Another term, "DX," stands for "distance" and is a nod to the thrill of long-distance communication, often crossing international borders. "CQ," pronounced in a way like "'seek you," is an invitation for any operator who hears the call to respond, typically used when you want to initiate a new conversation without a specific individual in mind. Understanding these terms is your first step in feeling at home among seasoned operators.

Are you already feeling overwhelmed and concerned about keeping all these terms straight? Don't worry. An extensive glossary is provided at the end of the book.

Phonetic Alphabet

The phonetic alphabet is a standardized set of words used to represent each letter of the alphabet in voice communication. This system ensures clear and precise communication, especially when audio quality may be poor or misunderstandings due to similar-sounding letters are to be avoided. Each word in the phonetic alphabet uniquely corresponds to a letter, making it easier to spell words and prevent confusion.

Here is the NATO phonetic alphabet, commonly used in aviation, military, and amateur radio communications:

Phonetic Alphabet

A- Alfa	**G**- Golf	**M**- Mike	**S**- Sierra	**Y**- Yankee
B- Bravo	**H**- Hotel	**N**- November	**T**- Tango	**Z**- Zulu
C- Charlie	**I**- India	**O**- Oscar	**U**- Uniform	
D- Delta	**J**- Juliet	**P**- Papa	**V**- Victor	
E- Echo	**K**- Kilo	**Q**- Quebec	**W**- Whiskey	
F- Foxtrot	**L**- Lima	**R**- Romeo	**X**- X Ray	

Importance in Ham Radio

In ham radio, the phonetic alphabet is essential for ensuring clear communication, especially when conveying call signs, coordinates, or other critical information. It minimizes the risk of errors caused by mishearing. It helps operators exchange information accurately and efficiently, even in noisy or challenging conditions.

Although helpful and a common practice, officially, the FCC rules state that using a phonetic alphabet for station identification in the Amateur Radio Service is encouraged—not mandated, not required, but encouraged.

T1A03: What do the FCC rules state regarding the use of a phonetic alphabet for station identification in the Amateur Radio Service?

A. It is required when transmitting emergency messages
B. It is encouraged
C. It is required when in contact with foreign stations
D. All these choices are correct

Q-Codes

Q-codes are a standardized set of three-letter codes used in radio communications to convey complex information quickly and efficiently. Developed originally for commercial radiotelegraph communication and later adopted by amateur radio operators, Q codes simplify and standardize the communication process, particularly when language barriers or signal conditions might make verbal communication difficult.

Learning these codes is not just about efficiency; it's about fluency in a language that speeds up exchanges and builds camaraderie among operators who might be continents apart. The use of Q-codes dates back to the early 20th century, making them a tradition carried forward by today's digital-age hams.

Each Q code represents a specific question or statement, allowing operators to convey detailed information using short, universally understood codes. Here are a few examples of commonly used Q codes in amateur radio:

- **QRM**: Are you experiencing interference? / I am experiencing interference.

- **QRN**: Are you troubled by static? / I am troubled by static.

- **QRO**: Should I increase power? / Increase power.

- **QRP**: Should I decrease power? / Decrease power.

- **QRS**: Should I send more slowly? / Send more slowly.

- **QRT**: Should I stop sending? / Stop sending.

- **QRZ**: Who is calling me? / You are being called by _____.

- **QSB**: Are my signals fading? / Your signals are fading.

- **QSL**: Can you acknowledge receipt? / I acknowledge receipt.

- **QSY**: Should I change the frequency? / Change frequency to _____.

- **QTH**: What is your location? / My location is _____.

A printable version of Q-codes and the Phonetic Alphabet can be found here:
www.MorseCodePublishing.com/TechExam

Printable Copy

Scan the QR code or click the link to receive 'Your Ham Packet,' which includes a formatted and printable copy of Q-Codes and the Phonetic Alphabet. Print and keep close to your radio.

A couple of these will be on the test. So let's practice.

T2B10: Which Q signal indicates that you are receiving interference from other stations?

A. QRM
B. QRN
C. QTH
D. QSB

QRM: Are you experiencing interference? / I am experiencing interference.

T2B11: Which Q signal indicates that you are changing frequency?

A. QRU
B. QSY
C. QSL
D. QRZ

QSY: Should I change the frequency? / Change frequency to _____.

My recommendation: You will learn these as you gain experience. Go get the printable Q-Codes linked above and print them out. Keep them by your radio. Over time, you will memorize them all.

For now, for today, just memorize the two for the test.

Common Words and Terms

Beyond the structured codes and official terms, ham radio is rich in slang and common phrases that color the conversations. "Rag chew" refers to a long, casual conversation, usually without any specific agenda other than enjoying good company over the airwaves. "Elmer," another endearing term, describes an experienced ham who mentors newer enthusiasts, helping them navigate their initial amateur radio adventures.

As you become more familiar with these terms, abbreviations, and phrases, you'll become an integral part of the ham radio community, ready to share your stories and experiences across the airwaves. Each term learned is a step closer to confident transmissions and enriched interactions, ensuring that each QSO you engage in is as rewarding as it is educational.

See the glossary for more common terms!

Operational Terms

Understanding operational terms is important for effective communication and smooth operation in amateur radio. These terms describe various modes, methods, and protocols used during radio operations, providing a common language that all operators can understand and use.

This section will introduce you to essential operational terms, such as simplex, duplex, and more, that you will encounter frequently in your ham radio activities. By familiarizing yourself with these terms, you will be better equipped to navigate and participate in radio communications, ensuring you can successfully operate your equipment and confidently engage with other operators.

Simplex

The term "simplex" describes an amateur station <u>transmitting and receiving on the same frequency.</u> In simplex communication, both stations <u>take turns transmitting and receiving</u> on a single frequency, allowing for a straightforward exchange of information. This method is commonly used for local line-of-sight communication, where both stations can hear each other clearly without additional infrastructure.

Simplex operation is easy to set up and use, making it ideal for casual conversations, field operations, and emergency communication scenarios. Using the same frequency for transmission and reception, simplex communication ensures that all participating stations are on the same channel, simplifying the process of making contact and maintaining a conversation.

T2A11: What term describes an amateur station that is transmitting and receiving on the same frequency?

A. Full duplex
B. Diplex
C. Simplex
D. Multiplex

Duplex

Duplex communication is a method where <u>transmission and reception occur on different frequencies</u>. This setup allows simultaneous two-way communication, as each station can transmit and receive simultaneously without interference. There are two types of duplex communication: half-duplex and full-duplex.

1. **Half-Duplex**: In half-duplex systems, communication can occur in both directions but not simultaneously. One station transmits while the other receives, and then they switch roles. An example of half-duplex communication is using a repeater, where the repeater receives a signal on one frequency and retransmits it on another.

2. **Full-Duplex**: In full-duplex systems, both stations transmit and receive simultaneously at different frequencies. This type of communication is commonly used in telephone networks and some advanced radio systems, allowing for continuous and uninterrupted conversation.

Duplex communication is essential when continuous and simultaneous two-way communication is needed, such as in extended-range repeaters or complex communication systems.

Squelch

Squelch is a feature in radio receivers <u>that mutes the audio output when the received signal is below a certain threshold or not present.</u> This helps eliminate background noise and static, ensuring that only signals strong enough to be intelligible are heard. By adjusting the squelch control, users can set the minimum signal strength required to open the audio path, allowing for a better listening experience.

T2B13: What is the purpose of a squelch function?

A. Reduce a CW transmitter's key clicks
B. Mute the receiver audio when a signal is not present
C. Eliminate parasitic oscillations in an RF amplifier
D. Reduce interference from impulse noise

For beginners, think of squelch as a noise gate for your radio. Without squelch, you would constantly hear static and weak signals, making it difficult to discern important communications. Setting the squelch level appropriately allows you to filter out these unwanted sounds, ensuring that only strong, clear signals come through.

Squelch in Action

Squelch is a feature in FM receivers that mutes the audio output when the received signal falls below a certain threshold, effectively reducing background noise when no signal is present. The squelch threshold must be adjusted appropriately to hear a weak FM signal. The squelch control can be set so that the <u>receiver's audio output is always on</u>, even when the signal is weak.

To adjust the squelch for weak signals, you turn the squelch knob until the background noise stops. This setting allows the receiver to produce audio output for even the faintest signals, ensuring you can hear weak transmissions. Setting the squelch threshold low ensures that the receiver remains open to any incoming signal, no matter how weak, which is essential for receiving distant or low-power stations.

T4B03: How is squelch adjusted so that a weak FM signal can be heard?

A. Set the squelch threshold so that receiver output audio is on all the time
B. Turn up the audio level until it overcomes the squelch threshold
C. Turn on the anti-squelch function
D. Enable squelch enhancement

Continuous Tone-Coded Squelch System (CTCSS)

CTCSS, which stands for Continuous Tone-Coded Squelch System, is a method used in radio communication to <u>control a receiver's squelch function. It involves transmitting a sub-audible tone,</u> typically between 67 and 254 Hz, along with the regular voice audio signal. The receiver is set to open its squelch and allow the audio to be heard only when it detects the correct tone, effectively reducing interference from unwanted signals on the same frequency.

For beginners, think of CTCSS as a selective gatekeeper for your radio communications. This system ensures that your receiver only picks up transmissions intended for you, filtering out background noise and transmissions from other users, not sending the specific tone. This is particularly useful in environments with heavy radio traffic or shared frequency scenarios.

T2B02: What term describes the use of a sub-audible tone transmitted along with normal voice audio to open the squelch of a receiver?

A. Carrier squelch
B. Tone burst
C. DTMF
D. CTCSS

Digital Coded Squelch (DCS)

DCS, which stands for Digital Coded Squelch, is an advanced method of controlling the squelch function in radio communication systems. Unlike CTCSS, which uses sub-audible analog tones, DCS employs digital codes to achieve the same purpose. When a radio transmits a signal, it includes a specific digital code and the voice or data. The receiving radio will only open its squelch and allow the audio to be heard if it detects the correct digital code.

Key Features of DCS:

- **Digital Encoding**: DCS uses a series of digital bits as the squelch code, offering a larger number of unique codes compared to CTCSS tones. This reduces the chances of interference from other users who do not use the correct DCS code at the same frequency.

- **Interference Reduction**: Using digital codes, DCS provides more precise control over which signals are allowed through the squelch, further minimizing unwanted noise and enhancing communication clarity.

- **Flexibility**: DCS is widely used in commercial and amateur radio systems, offering flexibility and enhanced privacy for users needing reliable communication in busy radio environments.

Consider DCS as a digital key unlocking your radio's audio only when the correct digital code is received. This ensures that you only hear transmissions intended for you.

The 'Your Ham Packet' link above also has a full list of all the CTCSS and DCS codes in printable format.

Now that you know some of the terms and common language. Let's try putting it into action.

Chapter 21

Making Your First Contact

Choice of Frequencies and Time of Day

SELECTING THE CORRECT FREQUENCY and time for making contacts is much like choosing the right time and place for planting seeds in a garden. It would help if you considered the conditions to maximize the chance of success. For example, trying to connect with people in the United States at 10am on a Tuesday might be more difficult because a lot of us work during the day. But at 8:30 pm, when the routine housework is done, the kids are in bed, and now it's time to relax, offers you a greater chance of success.

National Calling Frequency for FM Simplex on the 2-Meter band

The national calling frequency for FM simplex operations in the 2-meter band is **146.520 MHz**. This frequency is designated for initial contact between amateur radio operators before moving to another frequency to continue their conversation. It is a standard frequency where hams can reach out and establish communication, especially in unfamiliar areas or trying to contact new stations.

For beginners, it's important to remember 146.520 MHz as the primary frequency for calling and making initial contact on the 2-meter band (VHF). This helps make connecting with other operators and participating in the ham radio community easier.

T2A02: What is the national calling frequency for FM simplex operations in the 2 meter band?

A. 146.520 MHz
B. 145.000 MHz
C. 432.100 MHz
D. 446.000 MHz

Handling Nerves

It's natural to feel nervous when making your first few contacts. The key to overcoming this is preparation and practice. Begin by listening to other exchanges to get a feel for the typical flow of a conversation. To build your confidence, you might also try simulated QSOs with a friend or mentor. Remember, every ham was once a beginner, and the community is known for being welcoming and helpful. Most operators will be patient and supportive, understanding the jitters of being new to the hobby.

Steps to Take Before Calling CQ

Before calling CQ on a frequency, it is essential to ensure that you are not interfering with ongoing communications and that you are authorized to use that frequency. The first step is to listen carefully to the frequency for a short period to ensure it is clear and not already used by another station. This helps avoid interrupting any ongoing conversations or activities.

Next, it is good practice to ask if the frequency is in use by transmitting a brief message such as, "Is this frequency in use?" and waiting for a response. This courteous step ensures that you are not inadvertently causing interference. Finally, verify that you can use the frequency based on your license class and the regulations governing that band. Following these steps, you help maintain orderly and respectful use of the airwaves. Understanding and remembering these practices will help you prepare for your exam and ensure proper etiquette in real-world communications.

T2A12: What should you do before calling CQ?

A. Listen first to be sure that no one else is using the frequency
B. Ask if the frequency is in use
C. Make sure you are authorized to use that frequency
D. All these choices are correct

QSO Etiquette

Just as there are rules of etiquette in social settings, there are unwritten rules that govern the conduct of QSOs. These include identifying yourself correctly with your call sign, listening more than you talk to ensure you're not interrupting ongoing communications, and being clear and concise in your transmissions. Always be polite and considerate, thank the other operator for the contact, and sign off appropriately with your call sign.

QSO Basics

A QSO, or a radio contact, is the fundamental interaction in the ham radio community. It begins with the simple yet profound act of reaching out through the ether and connecting with another person. This process involves several key steps: initiating the contact, conducting the conversation, and logging the interaction details.

To initiate a QSO, you first need to find a clear frequency that is available for use. This involves tuning your radio while listening for ongoing conversations to avoid interference. Once you've found a clear frequency - you can always start at 146.520 MHz - you call "CQ," followed by your call sign, inviting others to respond.

The procedural signal "CQ" is used in amateur radio to indicate that a station is calling for any station that might be listening. When an operator transmits "CQ," they are essentially sending out an open invitation for a conversation with anyone who can hear the signal. It's a way for operators to initiate contact and make new connections, whether for casual chatting, exchanging information, or seeking specific communication partners.

T2A08: What is the meaning of the procedural signal "CQ"?

A. Call on the quarter hour
B. Test transmission, no reply expected
C. Only the called station should transmit
D. Calling any station

Managing a QSO requires a delicate balance of listening and speaking. It's like a dance, where participants move in sync, respecting each other's space and pace. This exchange serves the practical purpose of testing and enjoying your radio setup and fosters a sense of community and shared interest.

After the exchange, log the details of your QSO. This log should include the call sign of the station contacted and the frequency, mode, date, and time of the QSO. Logging serves as a personal record and is also a requirement in many aspects of the hobby, especially if you wish to apply for awards or participate in contests.

Transmit Frequency Placement

When tuning an FM receiver, accurately set it to the signal's exact frequency. <u>The audio signal will become distorted if the receiver is tuned slightly above or below the intended frequency</u>. This distortion occurs because FM (Frequency Modulation) receivers are designed to demodulate signals at a specific frequency. Deviating from this frequency disrupts the proper demodulation process, resulting in poor audio quality, including static, hissing, or muffled sounds.

> **T4B12: What is the result of tuning an FM receiver above or below a signal's frequency?**
>
> A. Change in audio pitch
> B. Sideband inversion
> C. Generation of a heterodyne tone
> **D. Distortion of the signal's audio**

One note about setting a frequency. Setting your transmit frequency precisely at the edge of an amateur band or sub-band can lead to several issues that might result in unintentional non-compliance with FCC regulations. One primary reason to avoid this practice is to account for <u>calibration errors in your transmitter's frequency</u> display. Even minor inaccuracies in calibration can cause your signal to extend beyond the permissible frequency range, leading to potential violations.

Additionally, when transmitting, the <u>modulation sidebands</u> of your signal can extend beyond your set frequency. If your transmit frequency is right at the edge of a band, these sidebands might spill over into frequencies that are not allocated for your use, causing interference with other services. Lastly, transmitter <u>frequency drift</u>, which can occur due to temperature changes or component aging, can shift your signal outside the intended band. Setting your transmit frequency slightly away from the band edge ensures that your entire signal, including any sidebands, remains within the authorized frequency range, thus maintaining compliance and avoiding interference with adjacent frequencies.

> **T1B09: Why should you not set your transmit frequency to be exactly at the edge of an amateur band or sub-band?**
>
> A. To allow for calibration error in the transmitter frequency display
> B. So that modulation sidebands do not extend beyond the band edge
> C. To allow for transmitter frequency drift
> **D. All these choices are correct**

Example First Contact

Let's look at an example of a typical QSO (contact) between two amateur radio operators. Here's a simplified version of how it might go:

Operator 1: "CQ CQ CQ, this is November Alpha One Charlie Charlie, NA1CC, calling CQ and standing by."

Operator 2: "November Alpha One Charlie Charlie, this is Kilo Bravo Nine Romeo Echo Mike, KB9REM responding. Your signal is 5 and 9 into Chicago."

Operator 1: "Kilo Bravo Nine Romeo Echo Mike, thanks for the 5 and 9. You're also 5 and 9 into Boston. The name here is John. How do you copy?"

Operator 2: "Roger, John. Good copy. You're loud and clear. My name is Mike. It's a pleasure to make the QSO with you today. What's the weather like over there?"

Operator 1: "Thanks, Mike. The weather here is sunny and warm, and it is a beautiful day for radio. How about there?"

Operator 2: "It's pretty much the same here, John. Great conditions for a chat. Are you running any special equipment today?"

Operator 1: "I'm using an Icom 7300 with a dipole antenna about 20 feet up. Power is set to 100 watts. How about yourself, Mike?"

Operator 2: "I'm on a Yaesu FT-991 with a vertical antenna. Also running about 100 watts. It's working well, as you can hear!"

Operator 1: "Sounds great, Mike. Your setup is doing a fine job. I won't hold it too long since I know others want to make contact. Thanks for the QSO and 73!"

Operator 2: "73, John. Thanks for the contact, and have a great day. KB9REM clear."

Operator 1: "NA1CC clear."

In this example, "CQ" means a general call to any operators who might be listening. "5 and 9" refers to the signal report: "5" means perfectly readable, and "9" indicates a strong signal. "73" is a standard ham radio shorthand for "best regards" or "goodbye."

Now, let's pretend you are 'operator 2.'

Responding to a Station Calling CQ

When you hear a station calling CQ, it indicates they are looking to contact any listening station. To respond correctly, you should <u>transmit the calling station's call sign followed by your call sign.</u>

From the example above, if you hear NA1CC calling CQ, you would respond by saying, " NA1CC, this is KB9REM," where NA1CC is the station calling CQ and KB9REM is your call sign.

This method ensures the calling station knows precisely who is responding and can acknowledge your call. Keeping your response clear and concise helps maintain efficient communication on the airwaves, allowing both stations to establish contact quickly and begin their conversation.

T2A05: How should you respond to a station calling CQ?

A. Transmit "CQ" followed by the other station's call sign
B. Transmit your call sign followed by the other station's call sign
C. Transmit the other station's call sign followed by your call sign
D. Transmit a signal report followed by your call sign

As you begin making your first contacts in amateur radio, remember that the choice of frequency and time, coupled with proper etiquette and technical knowledge, plays a pivotal role in your enjoyment.

Now that you've learned the basics of initiating QSOs and ensuring clear communication, you're ready to explore the next exciting aspect of amateur radio: repeaters. In the upcoming chapter, we'll delve into how repeaters work, their benefits, and how to use them to extend your communication range and connect with other operators.

Chapter 22

Repeaters

IN AMATEUR RADIO, REPEATERS play a vital role in extending the range and reliability of communications. These powerful systems are strategically placed to receive and retransmit signals, allowing operators to communicate over much greater distances than possible with direct radio-to-radio contact. Understanding how repeaters work and how to use them is essential for any ham radio enthusiast.

A repeater is an electronic device that receives a radio signal and then retransmits it at a higher power level, extending the signal's range. Repeaters are typically located on high towers, buildings, or mountains, giving them a wide coverage area. They consist of a receiver, a transmitter, and an antenna, all tuned to specific frequencies. The components of a repeater include:

1. **Receiver**: Captures the incoming signal on one frequency.

2. **Transmitter**: Amplifies and rebroadcasts the signal on a different frequency.

3. **Duplexer**: A single antenna can receive and transmit without interference.

4. **Antenna**: Positioned at a high location to maximize coverage.

How Repeaters Work

When a radio operator transmits a signal, the repeater's receiver picks it up. It immediately <u>retransmits it at a higher power on a different frequency</u>. This process effectively boosts the original signal, allowing it to cover a much larger area. The standard frequency offset used in repeaters ensures that the transmitted and received signals do not interfere with each other.

> **T1F09: What type of amateur station simultaneously retransmits the signal of another amateur station on a different channel or channels?**
>
> A. Beacon station
> B. Earth station
> **C. Repeater station**
> D. Message forwarding station

Repeater Offset

Repeater offset is <u>the difference between a repeater's transmit and receive frequencies</u>. Repeaters extend the communication range by receiving a signal at one frequency and retransmitting it at another. This frequency separation, known as the offset, prevents the repeater from interfering with its own signal. For example, in the 2-meter band, a standard repeater offset is 600 kHz. If the repeater receives 146.240 MHz, it might transmit 146.340 MHz. Offsets can be either positive or negative.

T2A07: What is meant by "repeater offset"?

A. The difference between a repeater's transmit and receive frequencies
B. The repeater has a time delay to prevent interference
C. The repeater station identification is done on a separate frequency
D. The number of simultaneous transmit frequencies used by a repeater

---

Benefits of Using Repeaters

Repeaters offer several advantages:

- **Extended Range**: They significantly increase the communication range, allowing contacts over distances that would be impossible with direct communication.

- **Improved Signal Quality**: Repeaters improve signal clarity and reduce interference by retransmitting signals with higher power and from elevated locations.

- **Emergency Communication**: Repeaters are crucial in emergencies, providing reliable communication channels for coordinating rescue and relief efforts.

Understanding repeaters and how to effectively use them is a fundamental aspect of ham radio operation. Whether for everyday communication, participating in local nets, or handling emergency traffic, repeaters enhance the functionality and reach of amateur radio systems.

Repeaters in Action

2 Meter Band (VHF Repeater)

To be repetitive, because that is how we learn, the 2-meter band is also commonly called the "VHF (Very High Frequency) band." This band (2-meter) covers frequencies from 144 MHz to 148 MHz. It is a popular range for amateur radio operators due to its favorable propagation characteristics and the availability of repeaters.

The VHF band allows for reliable communication over relatively long distances, especially when using repeaters to extend the signal range.

A standard repeater frequency offset in the 2-meter band is 600 kHz. This offset means that the repeater's transmit frequency is either 600 kHz higher or 600 kHz lower than its receive frequency. For example, if a repeater receives signals on 146.940 MHz, it might transmit on 146.340 MHz (600 kHz lower) or 147.540 MHz (600 kHz higher). This separation of transmit and receive frequencies allows the repeater to listen for incoming signals and retransmit them without interference simultaneously.

T2A01: What is a common repeater frequency offset in the 2-meter band?

A. Plus or minus 5 MHz
B. Plus or minus 600 kHz
C. Plus or minus 500 kHz
D. Plus or minus 1 MHz

---

70 cm Band (UHF Repeater)

The 70-centimeter band is called the "UHF (Ultra High Frequency) band." This band covers frequencies from 420 MHz to 450 MHz. Amateur radio operators widely use it for various types of communication, including local and regional contacts and repeater operations. The UHF band is known for penetrating buildings and other obstacles more effectively than lower frequencies, making it ideal for urban environments and mobile operations.

A standard repeater frequency offset in the 70-centimeter band is 5 MHz. This means the repeater's transmit frequency is either 5 MHz higher or 5 MHz lower than its receive frequency. For instance, if a repeater receives signals on 444.000 MHz, it might transmit on 449.000 MHz (5 MHz higher) or 439.000 MHz (5 MHz lower). This offset allows the repeater to receive and transmit signals simultaneously without interference, facilitating clear and effective communication over extended distances.

T2A03: What is a common repeater frequency offset in the 70 cm band?

A. Plus or minus 5 MHz
B. Plus or minus 600 kHz
C. Plus or minus 500 kHz
D. Plus or minus 1 MHz

Test time. Remember that some of these questions require you to think critically if you forgot or didn't memorize the answer. Let's say you forget the offset. By remembering the band range, you should eliminate some potential wrong answers. The 2m band is from 144 to 148 MHz. Only 4 MHz wide. So, having an offset in the MHz range would be unrealistic. Therefore, the likely answer would be in kilohertz (kHz).

The same applies to the UHF range (70 cm), which covers the band from 420 to 450 MHz. The wide 30 MHz range allows for a larger offset. An offset of kilohertz would be too small, so the answer is one of the MHz answers.

When you start to connect to repeaters, you will most often be given the repeater offset by looking them up in a directory, such as Repeater Book. Which can be found by a simple Google search for Repeater Book.

Listening in on a Repeater

When a station is listening on a repeater and looking for a contact, it typically announces its call sign, followed by the word "monitoring." For example, an operator might say, "This is K1ABC monitoring." This phrase indicates that K1ABC is actively listening and open to contacting any other station on the same repeater frequency. It's a way for operators to make themselves available for communication without initiating a specific call to another station.

For beginners, it's essential to recognize and use this standard practice to facilitate effective and courteous communication on repeaters. By announcing that you are "monitoring," you signal to other operators that you are ready and willing to engage in a conversation.

T2A09: Which of the following indicates that a station is listening on a repeater and looking for a contact?

A. "CQ CQ" followed by the repeater's call sign
B. The station's call sign followed by the word "monitoring"
C. The repeater call sign followed by the station's call sign
D. "QSY" followed by your call sign

Calling a Station on a Repeater

When you know the call sign of the station you want to contact on a repeater, the appropriate way to call them is first to say their call sign, followed by your own call sign. For example, if you want to call station K1ABC and your call sign is N1XYZ, you would say, "K1ABC, this is N1XYZ." This method clearly identifies both the calling and the receiving stations, facilitating smooth and efficient communication.

This practice helps ensure that the other station knows who is trying to contact them and can respond accordingly. It also maintains the standard communication protocol within the ham radio community, promoting orderly and respectful use of the repeater.

T2A04: What is an appropriate way to call another station on a repeater if you know the other station's call sign?

A. Say "break, break," then say the station's call sign
B. Say the station's call sign, then identify with your call sign
C. Say "CQ" three times, then the other station's call sign
D. Wait for the station to call CQ, then answer

Linked Repeaters

Linked repeaters are a network of repeaters connected to extend the range and coverage of radio communications. It is a system where multiple repeaters are interconnected, allowing signals received by one repeater to be transmitted by all the repeaters in the network.

By linking multiple repeaters, operators can communicate over much greater distances than possible with a single repeater. This is achieved through various methods, such as using the internet, dedicated radio links, or other communication technologies to interconnect repeaters.

Link Methods: Repeaters can be linked using different technologies. Some common methods include:

- **Internet Linking**: Systems like IRLP (Internet Radio Linking Project) or EchoLink use the internet to connect repeaters globally.

- **Dedicated Radio Links**: Microwave or UHF/VHF links directly connect repeaters without relying on the internet.

- **Satellite Links**: Satellites can be used to connect repeaters across vast distances.

Imagine you are a ham radio operator in New York City and want to communicate with a friend in Los Angeles. If both cities have repeaters that are part of a linked network, your transmission in New York can be relayed through the linked repeaters all the way to Los Angeles. This allows you to have a real-time conversation with your friend as if you were both on the same local repeater.

T2B03: Which of the following describes a linked repeater network?

A. A network of repeaters in which signals received by one repeater are transmitted by all the repeaters in the network
B. A single repeater with more than one receiver
C. Multiple repeaters with the same control operator
D. A system of repeaters linked by APRS

Simplex Channels & Repeaters in VHF/UHF

Simplex channels are designated in the VHF/UHF band plans to <u>facilitate direct communication between stations within range of each other without relying on repeaters.</u> This practice helps ensure that repeaters, which are valuable community resources, remain available for situations where they are truly needed, such as extending the communication range over greater distances or through obstacles. Operators can communicate directly by using simplex channels (remember: where you transmit and receive on the same frequency), reducing the load on repeaters and preventing unnecessary congestion.

T2B09: Why are simplex channels designated in the VHF/UHF band plans?
A. So stations within range of each other can communicate without tying up a repeater
B. For contest operation
C. For working DX only
D. So stations with simple transmitters can access the repeater without automated offset

Using simplex channels is especially beneficial during events, emergencies, or casual conversations where direct line-of-sight communication is sufficient. It promotes efficient use of the radio spectrum by allowing multiple simultaneous conversations in different areas without overburdening the repeater infrastructure.

Chapter 23

Space (Satellites and Stations)

SATELLITE COMMUNICATION REPRESENTS ONE of the most thrilling aspects of amateur radio, offering a unique blend of technical challenges and the sheer joy of making contacts that travel through space. In amateur radio, the control operator is responsible for ensuring that all transmissions follow the rules. When communicating through an amateur satellite or space station, the important thing to remember is that anyone with a valid amateur radio license who can transmit on the satellite's uplink frequency can be the control operator.

So yes, as a Technician licensee, you can engage with satellites specifically designed for amateur radio. Essentially, these satellites act as repeaters in space; they receive signals transmitted from the Earth and then rebroadcast them back down to other global locations. This ability to communicate via satellite opens up intercontinental QSOs (contacts) even with the relatively limited power and antenna systems permissible under a Technician license.

T1E02: Who may be the control operator of a station communicating through an amateur satellite or space station?

A. Only an Amateur Extra Class operator
B. A General class or higher licensee with a satellite operator certification
C. Only an Amateur Extra Class operator who is also an AMSAT member
D. Any amateur allowed to transmit on the satellite uplink frequency

---

Embarking on satellite communications requires specific equipment, but getting started can be practical and affordable. The most basic setup includes:

- A dual-band handheld transceiver (VHF / UHF).

- A simple handheld directional antenna.

- A means to track satellite passes.

Software applications, many of which are free, can predict when and where satellites will be visible over your location. These apps provide essential data like azimuth and elevation of each pass, helping you point your antenna accurately. A popular choice among beginners for your antenna is a handheld Yagi, which is relatively inexpensive and provides sufficient gain to make successful satellite contacts.

Preparation is vital when you're ready to make your first satellite QSO. Begin by selecting a suitable satellite. Choose a pass high enough over the horizon (usually more than 30 degrees) to ensure clear line-of-sight communication. Before the pass begins, set up your station—this includes adjusting your transceiver to the appropriate uplink frequency (for transmitting) and downlink frequency (for receiving) and preparing your antenna. As the satellite approaches, start listening to the downlink frequency. You might hear other operators making contacts; listen to these exchanges to get a sense of the rhythm and pace of satellite communication.

Initiating contact involves transmitting your call sign and listening for responses during the satellite pass. Remember, satellite passes typically last only about 10 minutes, so operations need to be efficient. Adjust your antenna position regularly to maximize signal strength, a process that can be challenging but becomes easier with practice. Making your first contact through a satellite is an exhilarating experience, often accompanied by a sense of achievement that bolsters your enthusiasm for further explorations in this high-tech area of amateur radio.

Websites like AMSAT (The Radio Amateur Satellite Corporation) offer information, including tutorials, operational news, and details of current and upcoming satellite missions. These resources are invaluable for both novices and experienced satellite operators. Engaging with these communities can provide practical advice, troubleshooting tips, and the camaraderie of shared interest.

Space Stations

According to the FCC Part 97 regulations, a space station is an amateur radio station located more than 50 kilometers (about 31 miles) above the Earth's surface. Remembering this definition is straightforward if you consider a space station as any equipment operating in the "space" above our atmosphere.

These stations facilitate global communications, allowing amateur radio operators to communicate over vast distances by relaying signals through these space-based stations.

SPACE STATION

T1A07: What is the FCC Part 97 definition of a space station?

A. Any satellite orbiting Earth
B. A manned satellite orbiting Earth
C. An amateur station located more than 50 km above Earth's surface
D. An amateur station using amateur radio satellites for the relay of signals

---

One of the most exciting and unique aspects of amateur radio is the ability to communicate with the International Space Station (ISS). This capability opens up a new frontier for ham radio enthusiasts, allowing unique opportunities to interact directly with astronauts aboard the ISS, bringing a thrilling dimension to the hobby.

Contacting the International Space Station (ISS)

Any amateur radio operator with a Technician class or higher license can contact the International Space Station (ISS) on VHF bands – no NASA approval needed! The ISS operates an amateur radio station that can be accessed by licensed amateurs worldwide, providing a unique and thrilling opportunity to communicate with astronauts in space. Typically, these contacts are made on the 2-meter band (144-148 MHz), a frequency range that Technician class licensees are authorized to use.

T1B02: Which amateurs may contact the International Space Station (ISS) on VHF bands?

A. Any amateur holding a General class or higher license
B. Any amateur holding a Technician class or higher license
C. Any amateur holding a General class or higher license who has applied for and received approval from NASA
D. Any amateur holding a Technician class or higher license who has applied for and received approval from NASA

Contacting the ISS involves using a VHF radio and a suitable antenna, often with a clear line of sight to the sky. Operators need to know the ISS's orbit and timing as it passes overhead quickly. This is where your software comes into play. This capability is one of the many privileges of holding a Technician class or higher license.

Why VHF Bands are Specifically Called Out for ISS Contacts

VHF bands are specifically called out to contact the International Space Station (ISS) because the ISS's amateur radio equipment primarily operates on the 2-meter band, which falls within the VHF frequency range (144-148 MHz). The VHF frequencies are well-suited for this purpose due to their ability to support clear, long-distance communication without being significantly affected by atmospheric conditions. Additionally, the VHF band balances range and equipment size, making it accessible for many amateur radio operators.

The 2-meter band is also widely available to amateur radio operators holding a Technician class license, making it an ideal choice for ISS contacts. Using VHF frequencies allows for reliable communication with the ISS as it orbits the Earth, enabling operators to take advantage of the relatively low power requirements and ease of antenna construction compared to higher frequency bands.

Receiving Telemetry from a Space Station

Telemetry from a space station, such as data on its health, status, and scientific measurements, can be received by anyone with the appropriate equipment. You do not need a specific license to receive and interpret these signals. Telemetry data is often broadcasted openly to provide valuable information to scientists, engineers, and amateur radio enthusiasts interested in space exploration and satellite technology.

T8B11: Who may receive telemetry from a space station?

A. Anyone
B. A licensed radio amateur with a transmitter equipped for interrogating the satellite
C. A licensed radio amateur who has been certified by the protocol developer
D. A licensed radio amateur who has registered for an access code from AMSAT

For amateur radio operators and space enthusiasts, receiving telemetry offers a fascinating glimpse into the operations of space missions and the functioning of satellites. With a suitable receiver and software, anyone can decode these signals, including information on the satellite's power levels, temperature, and other fundamental parameters. This accessibility encourages widespread interest and participation in space science, making it an engaging and educational aspect of amateur radio.

Satellite Communications

A Low-Earth Orbit (LEO) satellite orbits the Earth at relatively low altitudes, typically between 160 and 2,000 kilometers (100 to 1,240 miles) above the Earth's surface. These satellites move quickly, completing an orbit in about 90 to 120 minutes. LEO satellites are commonly used for various applications, including communications, Earth observation, and scientific research. Their proximity to Earth allows for lower communication latency and better imaging resolution.

T8B10: What is a LEO satellite?

A. A sun-synchronous satellite
B. A highly elliptical orbit satellite
C. A satellite in low energy operation mode
D. A satellite in low earth orbit

A satellite beacon is a transmission from a satellite containing status information about the satellite's health, operational status, and other data. These beacons are essential for tracking the satellite and ensuring it functions correctly. They provide valuable information (telemetry) to amateur radio operators and satellite controllers, including battery levels, temperature, and signal strength (the health and status of the satellite). By decoding these beacon signals, operators can monitor the satellite's condition and performance, making satellite beacons an important element in satellite communication and management.

T8B05: What is a satellite beacon?

A. The primary transmit antenna on the satellite
B. An indicator light that shows where to point your antenna
C. A reflective surface on the satellite
D. A transmission from a satellite that contains status information

---

T8B01: What telemetry information is typically transmitted by satellite beacons?

A. The signal strength of received signals
B. Time of day accurate to plus or minus 1/10 second
C. Health and status of the satellite
D. All these choices are correct

Common Modes of Transmission Used by Amateur Radio Satellites

Amateur radio satellites commonly use several transmission modes, including Single-Sideband (SSB), Frequency Modulation (FM), and Continuous Wave (CW) or data modes. Each serves a different purpose and offers unique advantages for satellite communication.

Single Sideband (SSB) is widely used for voice communications. It provides efficient bandwidth use, making it suitable for long-distance communication with minimal power. Frequency Modulation (FM) is popular for its simplicity and clear audio quality, making it ideal for voice contacts, especially for beginners. Continuous Wave (CW) involves Morse code, and various digital data modes are used for more robust communication, especially in weak signal conditions. These modes are also efficient regarding bandwidth and power usage, making them ideal for telemetry and control signals.

T8B04: What mode of transmission is commonly used by amateur radio satellites?

A. SSB
B. FM
C. CW/data
D. All these choices are correct

Satellite Signals

Spin Fading

Satellite signal spin fading is caused by the rotation of the satellite and its antennas as it orbits the Earth. As the satellite spins, the orientation of its antennas relative to the receiving station on Earth changes continuously. This variation in orientation affects the polarization and signal strength, leading to fluctuations in the received signal. These fluctuations are perceived as fading, where the signal strength periodically increases and decreases.

T8B09: What causes spin fading of satellite signals?

A. Circular polarized noise interference radiated from the sun
B. rotation of the satellite and its antennas
C. Doppler shift of the received signal
D. Interfering signals within the satellite uplink band

Understanding spin fading is vital for amateur radio operators to optimize satellite communication. It highlights the need for circularly polarized antennas or signal diversity techniques to mitigate the effects of spin fading. By knowing this phenomenon, operators can better prepare for and manage the challenges of maintaining a stable communication link with orbiting satellites.

Doppler Shift

Doppler shift refers to the observed change in signal frequency caused by the <u>relative motion between the satellite and the Earth station</u>. As a satellite moves toward the observer, its signal frequency increases; conversely, as it moves away, the frequency decreases. This phenomenon is similar to the change in pitch of a passing siren.

T8B07: What is Doppler shift in reference to satellite communications?

A. A change in the satellite orbit
B. A mode where the satellite receives signals on one band and transmits on another
C. An observed change in signal frequency caused by relative motion between the satellite and Earth station
D. A special digital communications mode for some satellites

Understanding and compensating for the Doppler shift will aid in maintaining clear satellite communication. The frequency changes can be significant, especially for higher frequency bands like VHF and UHF, and operators need to adjust their transmit and receive frequencies accordingly to keep the signal clear. Many modern transceivers and satellite tracking software can automatically adjust for Doppler shifts, simplifying the process.

Satellite Tracking

Satellite tracking programs are essential for radio operators engaged in satellite communications. These programs provide valuable real-time information that helps operators predict and track satellite positions. One of the key features of satellite tracking programs is the ability to <u>display maps showing the satellite's real-time position</u> and track over the Earth. This visual representation allows operators to see where the satellite is and where it will be, facilitating better planning for communication attempts.

Additionally, satellite tracking programs provide detailed <u>information on a satellite pass's time, azimuth, and elevation</u>. This includes the start, maximum altitude, and end of a pass, enabling operators to align their antennas accurately for optimal signal reception. Furthermore, these programs account for the Doppler shift, showing the <u>apparent frequency of the satellite transmission</u> as it changes due to the relative motion between the satellite and the Earth station.

T8B03: Which of the following are provided by satellite tracking programs?

A. Maps showing the real-time position of the satellite track over Earth
B. The time, azimuth, and elevation of the start, maximum altitude, and end of a pass
C. The apparent frequency of the satellite transmission, including effects of Doppler shift
D. All these choices are correct

Understanding Azimuth in Satellite Tracking

Azimuth is a term used in satellite tracking and navigation to describe an object's horizontal angle or direction, such as a satellite, from a specific observation point. It is measured in degrees, with 0 degrees representing true north. The angle increases clockwise, with 90 degrees corresponding to the east, 180 degrees to the south, and 270 degrees to the west. Essentially, azimuth helps to pinpoint the direction along the horizon from which the satellite will appear or disappear, allowing operators to align their antennas correctly.

For amateur radio operators, azimuth is a key parameter when tracking satellites. By knowing the azimuth angle, operators can adjust their antennas to accurately follow the satellite's path as it moves across the sky. This ensures optimal signal reception and transmission, making establishing and maintaining satellite communication easier. Understanding azimuth and how to use it in conjunction with elevation and other tracking data is essential for successful satellite operations.

Keplerian Elements

To function effectively, <u>satellite tracking programs require specific inputs known as Keplerian elements</u>. Keplerian elements are sets of parameters that describe the orbits of satellites. They include information such as the satellite's inclination, eccentricity, and mean anomaly, which are used to calculate its position and velocity at any given time. By regularly updating the Keplerian elements, tracking programs can provide accurate predictions and real-time tracking of satellite movements. Understanding how to input and utilize these elements ensures amateur radio operators can maximize their satellite communication endeavors, maintaining precise monitoring and effective communication with satellites in orbit.

T8B06: Which of the following are inputs to a satellite tracking program?

A. The satellite transmitted power
B. The Keplerian elements
C. The last observed time of zero Doppler shift
D. All these choices are correct

Satellite Uplink and Downlink Operations

In satellite communications, you need to ensure your uplink power is neither too low nor too high. One way to determine whether your satellite uplink power is appropriate is to <u>compare your signal strength on the downlink to the satellite's beacon signal</u>. Your signal strength should be about the same as the beacon. This balance ensures that your transmission is strong enough to be received clearly without overpowering other signals, maintaining optimal operation for all satellite users.

T8B12: Which of the following is a way to determine whether your satellite uplink power is neither too low nor too high?

A. Check your signal strength report in the telemetry data
B. Listen for distortion on your downlink signal
C. Your signal strength on the downlink should be about the same as the beacon
D. All these choices are correct

Using excessive effective radiated power (ERP) – remember, ERP is the measure of the power output of the radio transmitter - on a satellite uplink can have a significant negative impact. When one user's signal is too strong, it can block access for other users, preventing them from successfully transmitting and receiving signals through the satellite. This behavior, known as "hogging" the satellite, disrupts the shared nature of amateur satellite operations and diminishes the experience for the broader amateur radio community. By keeping your uplink power at an appropriate level, you help ensure that the satellite remains accessible to all operators.

T8B02: What is the impact of using excessive effective radiated power on a satellite uplink?

A. Possibility of commanding the satellite to an improper mode
B. Blocking access by other users
C. Overloading the satellite batteries
D. Possibility of rebooting the satellite control computer

Understanding satellite operating modes is also essential. When a satellite is described as operating in U/V mode, the uplink frequency is in the 70-centimeter band (UHF), and the downlink frequency is in the 2-meter band (VHF). This mode designation helps operators know which bands to transmit to and receive from the satellite. Familiarity with these modes is essential for effective satellite communication, as it ensures that operators use the correct frequencies and maximize the utility of the satellite for all users.

T8B08: What is meant by the statement that a satellite is operating in U/V mode?

A. The satellite uplink is in the 15-meter band, and the downlink is in the 10-meter band
B. The satellite uplink is in the 70-centimeter band, and the downlink is in the 2-meter band
C. The satellite operates using ultraviolet frequencies
D. The satellite frequencies are usually variable

Satellite and space station communication offers a new frontier for those with a Technician license, providing an engaging mix of technical challenges and the exciting opportunity to communicate through space.

As you delve deeper into this hobby, keep in mind that each new skill you learn and every boundary you push enriches your personal experience and contributes to the vibrant global community of amateur radio operators, even reaching the final frontier of space!

Chapter 24

Exploring Other Modes

Computers & the Internet

COMBINING COMPUTERS WITH RADIO equipment has opened many possibilities for amateur radio operators. Digital mode operations involve using computer software to encode and decode digital signals, which enables more efficient and varied forms of communication.

<u>In digital mode operation, specific signals are used in a computer-radio interface: receive audio, transmit audio and transmitter keying.</u> Receive audio is the signal from the radio that the computer needs to process. Transmit audio is the signal generated by the computer that needs to be sent to the radio for transmission. Transmitter keying is used to switch the transceiver between receive and transmit modes. Understanding these signals is essential for setting up a functional digital mode station.

T4A06: What signals are used in a computer-radio interface for digital mode operation?

A. Receive and transmit mode, status, and location
B. Antenna and RF power
C. Receive audio, transmit audio, and transmitter keying
D. NMEA GPS location and DC power

When connecting a computer to a transceiver for digital operations, one essential connection is <u>between the computer's "line in" port and the transceiver's speaker connector</u>. This connection allows the computer to receive audio signals from the transceiver, which can then be processed by digital software. This setup ensures the signals are correctly received and transmitted, enabling smooth digital communication.

T4A07: Which of the following connections is made between a computer and a transceiver to use computer software when operating digital modes?

A. Computer "line out" to transceiver push-to-talk
B. Computer "line in" to transceiver push-to-talk
C. Computer "line in" to transceiver speaker connector
D. Computer "line out" to transceiver speaker connector

You need to read carefully here. Did you see that answer 'D' was thrown in there to potentially catch you off guard?

Additionally, digital hot spots have become a popular tool for ham operators. A digital mode <u>hotspot is a device that connects to the internet</u> and allows for communication using digital voice or data systems. By linking a transceiver to a digital hot spot, operators can extend their communication reach via the internet, accessing a broader network of digital systems. This capability enhances the flexibility and utility of amateur radio, making it easier to connect with other operators worldwide. This is the same thing as using your phone as a hotspot!

T4A10: What function is performed with a transceiver and a digital mode hot spot?

A. communication using digital voice or data systems via the internet
B. FT8 digital communications via AFSK
C. RTTY encoding and decoding without a computer
D. High-speed digital communications for meteor scatter

Digital

Understanding Digital Communications Modes

Digital communications modes transmit data using digital signals rather than analog ones. Examples of digital modes include packet radio, IEEE 802.11 (Wi-Fi), and FT8. Packet radio sends data in packets over radio frequencies, commonly used for amateur radio networking and message forwarding. IEEE 802.11, more widely known as Wi-Fi, is a standard for wireless local area networking, enabling devices to communicate wirelessly over short distances. FT8 is a digital mode used in amateur radio for weak signal communication. It is known for its efficiency in low-bandwidth conditions and ability to decode signals barely above the noise level.

These digital modes offer various advantages, such as improved signal clarity, efficient bandwidth use, and the ability to transmit data reliably over long distances or through challenging conditions. These digital communication modes expand the possibilities for experimentation and practical communication in the field. Recognizing the differences and applications of these modes will help you in your exam preparation and practical amateur radio activities.

T8D01: Which of the following is a digital communications mode?

A. Packet radio
B. IEEE 802.11
C. FT8
D. All these choices are correct

To get started with digital modes like FT8 and DMR, you'll need a basic setup:

- A transceiver capable of the mode.

- A computer with suitable software.

- An interface to connect your transceiver to your computer.

There are numerous advantages to using digital modes. They allow communication across vast distances with minimal equipment, provide clear audio unaffected by the static and noise typically associated with analog modes, and offer new ways to interact and exchange information. Additionally, digital modes like FT8 are particularly useful in crowded band conditions, as their signals are much narrower than typical voice transmissions, allowing more communications to occur within the same frequency spectrum.

Numerous resources are available for those looking to dive deeper into digital modes. Online tutorials, many of which are free, can offer step-by-step guides on setting up and operating digital modes. Websites like the "ARRL" (American Radio Relay League) provide extensive documentation and forums where experienced amateurs share their knowledge and advice.

FT8

FT8, named after its creators Franke and Taylor and the "8" denoting mode's 8-frequency keying format, operates <u>under weak or low signal-to-noise operating</u> conditions and primarily makes quick, long-distance contacts. Its ability to work under challenging conditions where other modes might fail makes it particularly appealing to amateurs who enjoy making international contacts without needing elaborate setups.

T8D13: What is FT8?

A. A wideband FM voice mode
B. A digital mode capable of low signal-to-noise operation
C. An eight channel multiplex mode for FM repeaters
D. A digital slow-scan TV mode with forward error correction and automatic color compensation

---

Starting with FT8, the first step is installing <u>WSJT-X software</u>, explicitly designed for weak-signal communication by radio amateurs. Once installed, you'll configure the software to recognize your radio's settings, which involves selecting the correct port, <u>audio input, and output settings</u>. The true magic begins when you tune your radio to an FT8 frequency (commonly 14.074 MHz for the 20m band) and start decoding signals. Observing the software automatically decode transmissions and even auto-reply based on your settings is practical and thrilling.

T4A04: How are the transceiver audio input and output connected in a station configured to operate using FT8?

A. To a computer running a terminal program and connected to a terminal node controller unit
B. To the audio input and output of a computer running WSJT-X software
C. To an FT8 conversion unit, a keyboard, and a computer monitor
D. To a computer connected to the FT8converter.com website

Operating Activities Supported by WSJT-X Software

The WSJT-X software is renowned for supporting various specialized digital modes that facilitate unique operating activities in amateur radio. These activities include Earth-Moon-Earth (EME) communication, weak signal propagation beacons, and meteor scatter communication.

<u>Earth-Moon-Earth (EME) communication</u>, also known as moonbounce, involves transmitting a signal to the moon and receiving the reflected signal back on Earth. This mode allows for long-distance communication by leveraging the moon as a passive reflector. <u>Weak signal propagation</u> beacons are used to monitor and study propagation conditions. These beacons transmit low-power signals that can be received over great distances under the right conditions, providing valuable data on the behavior of radio waves. <u>Meteor scatter</u> communication takes advantage of ionized trails left by meteors entering the Earth's atmosphere. These trails reflect radio signals, allowing brief communication windows as the meteor trails dissipate.

These activities and how WSJT-X facilitates them help to illustrate amateur radio's advanced capabilities and innovative techniques.

T8D10: Which of the following operating activities is supported by digital mode software in the WSJT-X software suite?

A. Earth-Moon-Earth
B. Weak signal propagation beacons
C. Meteor scatter
D. All these choices are correct

Digital Mobile Radio (DMR)

On the other hand, DMR is a digital voice mode that offers clear audio communication and the ability to share data alongside voice messages. It uses time-division multiple access (TDMA) technology, allowing multiple users to share the same frequency channel, which increases efficiency and reduces interference.

DMR requires a different approach. First, you'll need a DMR-capable transceiver and a registration for a Radio ID from the DMR network. Setting up involves programming your radio with talkgroups and repeaters, which can be a complex process for beginners. Software tools like Tytera MD-380 are commonly used for programming DMR radios. These tools allow you to input different channels and talkgroups corresponding to your local DMR repeater settings. Once set up, using DMR can be as simple as selecting a channel and talking, with the technology handling the complexities of digital communication in the background.

A DMR (Digital Mobile Radio) "code plug" is a file that contains all the access information needed to operate a DMR radio. It is the programming file for your DMR radio, setting up frequencies, color codes, time slots, and talkgroups to enable seamless communication. Without a properly configured code plug, your DMR radio won't know how to connect to the network or communicate effectively.

Think of a code plug as the instruction manual for your DMR radio, telling it exactly how to access various repeaters and participate in different talkgroups. This setup ensures you can easily switch between different communication channels and networks.

T4B07: What does a DMR "code plug" contain?

A. Your call sign in CW for automatic identification
B. Access information for repeaters and talkgroups
C. The codec for digitizing audio
D. The DMR software version

DMR Repeaters

The DMR repeater system enhances communication by allowing multiple users to share the same frequency channel through time-division multiple access (TDMA). This effectively doubles the capacity of a single repeater, enabling more efficient use of available frequencies.

Key Features of the DMR Repeater System

- **Time-Division Multiple Access (TDMA)**: DMR uses TDMA to split a single frequency into two-time slots, allowing two separate conversations to co-occur on the same frequency. This increases the repeater system's efficiency and capacity.

- **Enhanced Audio Quality**: DMR provides superior audio quality to analog systems, especially in noisy environments. Digital signal processing helps reduce background noise and interference, resulting in clearer communications.

- **Advanced Features**: DMR supports a range of advanced features, including text messaging, GPS location services, and data transmission, making it a versatile tool for both amateur and professional users.

- **Network Connectivity**: DMR repeaters can be connected via the internet to form wide-area networks, allowing users to communicate across great distances beyond the reach of a single repeater. This connectivity enables global communication and linking multiple repeaters into a seamless network.

Relevance to Amateur Radio

DMR has become increasingly popular in amateur radio due to its efficiency and advanced capabilities. Operators use DMR repeaters to extend their communication range, improve audio quality, and access digital features.

Time-Division Multiple Access (TDMA)

DMR allows two separate voice signals to be transmitted on a single <u>12.5 kHz repeater channel through a technique called time-multiplexing</u>. This method divides the channel into two timeslots, enabling two conversations to occur simultaneously without interference. Each time slot alternates between the two voice signals, effectively doubling the repeater channel's capacity.

By using time-multiplexing, DMR allows more users to communicate over the same channel.

T8D07: Which of the following describes DMR?

A. A technique for time-multiplexing two digital voice signals on a single 12.5 kHz repeater channel
B. An automatic position tracking mode for FM mobiles communicating through repeaters
C. An automatic computer logging technique for hands-off logging when communicating while operating a vehicle
D. A digital technique for transmitting on two repeater inputs simultaneously for automatic error correction

---

Color Codes in DMR Repeater Systems

In DMR (Digital Mobile Radio) repeater systems, the color code functions similarly to CTCSS tones or DCS codes in analog systems. <u>The primary purpose of the color code is to ensure that only authorized users can access the repeater.</u> For a radio to communicate through a DMR repeater, its color code must match the repeater's color code. This helps to prevent interference from other nearby repeaters operating on the same frequency and maintains organized and secure communication channels.

T2B12: What is the purpose of the color code used on DMR repeater systems?

A. Must match the repeater color code for access
B. Defines the frequency pair to use
C. Identifies the codec used
D. Defines the minimum signal level required for access

For beginners, it's important to remember that the color code acts as a digital key. The repeater will only accept your transmission with the correct color code, even if you are on the right frequency and time slot.

Wait a minute. I still don't understand "color codes." For example, is this repeater "blue?" The term "color code" refers to a digital identifier rather than an actual color. These codes differentiate between different repeaters and networks operating on the same frequency. Each DMR repeater is assigned a color code, which is a number from 0 to 15. This code must be programmed into the user's radio to allow access to the repeater.

For example, if you want to use a DMR repeater with the frequency 445.500 MHz and it has a color code of 1 (the code is provided by the repeater owner or in the directory), you would set your radio to 445.500 MHz and program the color code to 1. Only then will your radio be able to communicate through that repeater.

Talkgroups

Talkgroups in Digital Mobile Radio (DMR) systems are a way to organize and manage communication channels. They function like virtual groups within the DMR network, allowing groups of users to communicate without interference from other groups. Each talkgroup is assigned a unique identifier, and users can switch between talkgroups to join different conversations or activities.

T8D02: What is a "talkgroup" on a DMR repeater?

A. A group of operators sharing common interests
B. A way for groups of users to share a channel at different times without hearing other users on the channel
C. A protocol that increases the signal-to-noise ratio when multiple repeaters are linked together
D. A net that meets at a specified time

Key Features of Talkgroups

1. **Organized Communication**: Talkgroups help organize communications by categorizing users into specific groups based on their interests, activities, or operational needs. For instance, there can be talkgroups for local chat, regional communication, emergency response, or specific clubs and events.

2. **Selective Calling**: By selecting a talkgroup, users ensure that their transmissions are only heard by other users in the same talkgroup. This prevents the network from becoming overloaded with unrelated conversations and keeps communications relevant to each group's purpose.

3. **Efficient Use of Resources**: Talkgroups use the available frequency spectrum and network infrastructure efficiently. Multiple talkgroups can share the same frequency without interfering with each other, as they are digitally separated within the DMR system.

How to Use Talkgroups

To use a talkgroup, you must program your DMR radio with the correct talkgroup ID. When you switch to a talkgroup, your radio will transmit and receive only within that group. For example, if you want to participate in a regional discussion, you would switch to the corresponding regional talkgroup ID.

Example Scenario

Imagine you are part of a ham radio club that uses a DMR repeater. The club might have a general talkgroup for everyday conversation, an emergency talkgroup for urgent communications, and a special events talkgroup for coordinating activities during club events. By selecting the appropriate talkgroup, you can communicate with fellow club members without disturbing other groups using the same repeater.

T2B07: How can you join a digital repeater's "talkgroup"?

A. Register your radio with the local FCC office
B. Join the repeater owner's club
C. Program your radio with the group's ID or code
D. Sign your call after the courtesy tone

Digital Voice Transceiver

A digital voice transceiver selects a specific group of stations by <u>entering the group's identification code</u>. This talkgroup ID allows the transceiver to filter and communicate with only the stations that are part of that designated group.

T4B09: How is a specific group of stations selected on a digital voice transceiver?

A. By retrieving the frequencies from transceiver memory
B. By enabling the group's CTCSS tone
C. By entering the group's identification code
D. By activating automatic identification

Using these identification codes, digital voice transceivers can ensure that communications are directed to the intended recipients without interference from other users on the same frequency.

D-Star

D-STAR, which stands for Digital Smart Technologies for Amateur Radio, is a digital voice and data protocol designed for amateur radio by the Japan Amateur Radio League (JARL). It provides advanced features and capabilities compared to traditional analog radio, including improved voice clarity, efficient use of the radio spectrum, and the ability to transmit data such as text messages, images, and GPS information.

Key Features of D-STAR:

- **Digital Voice and Data Communication:** D-STAR allows for clear digital voice communications and can also transmit data, which is beneficial for sending text messages, GPS coordinates, and other types of digital information.

- **Call Sign Routing:** One of D-STAR's unique aspects is its ability to route communications based on call signs. This means you can communicate with other amateur radio operators worldwide by simply entering their call sign, and the D-STAR network will handle the signal routing.

- **Internet Connectivity:** D-STAR repeaters can be linked via the internet, enabling worldwide communication. This is particularly useful for amateur radio operators who wish to connect with others beyond the range of local repeaters.

- **Automatic Repeater Linking:** D-STAR repeaters can automatically link to other repeaters, creating a wide-area communications network. This allows for greater coverage and the ability to participate in larger communication networks.

- **Data Transmission:** In addition to voice, D-STAR supports digital data transmission, which can include everything from short text messages to more complex data packets.

How D-STAR Works:

D-STAR uses digital modulation techniques, specifically GMSK (Gaussian Minimum Shift Keying), to transmit information. When you speak into a D-STAR radio, your voice is digitized and compressed, then transmitted as a digital signal. This signal can include your voice and additional data, such as your call sign and location.

Repeaters and gateways in the D-STAR network play a role in extending the range of communications. D-STAR repeaters receive digital signals and retransmit them, while gateways connect these repeaters to the internet, facilitating global communication.

Before transmitting with a D-STAR digital transceiver, it is essential to program your call sign into the device. D-STAR (Digital Smart Technologies for Amateur Radio) is a digital voice and data protocol explicitly developed for amateur radio. One of its key features is using call signs to identify and route digital communications. By entering your call sign, the transceiver can correctly identify you to other stations and ensure your transmissions are routed correctly through repeaters and digital networks.

> **T4B11: Which of the following must be programmed into a D-STAR digital transceiver before transmitting?**
>
> **A. Your call sign**
> B. Your output power
> C. The codec type being used
> D. All these choices are correct

Programming your call sign into the D-STAR transceiver is a straightforward process typically done through the radio's menu system. This step is not only for compliance with FCC regulations, which require identification of the transmitting station but also for the proper functioning of the D-STAR network.

Packet Radio

Packet radio is a digital communication mode that transmits data over radio frequencies. It allows the exchange of text messages, telemetry data, and even internet traffic by breaking the data into small packets. Each packet contains not only the data itself but also addressing and error-checking information to ensure the accurate delivery of the message.

How Packet Radio Works

- **Data Packaging**: Information is divided into packets, which include headers with addressing and control information and a payload with the actual data.

- **Transmission**: Packets are transmitted over the radio frequency using protocols such as AX.25, specifically designed for amateur radio use.

- **Receiving**: The receiving station collects the packets, checks for errors, and reassembles them into the original message. Error-checking information in the packets ensures that only accurate data is processed.

Benefits of Packet Radio

- **Error Correction**: The built-in error-checking mechanisms help ensure the accuracy of transmitted data, making packet radio highly reliable.

- **Efficient Use of Bandwidth**: By transmitting data in packets, the system can efficiently use the available bandwidth, allowing multiple users to share the same frequency.

- **Automatic Repeat Request (ARQ)**: If a packet is received with errors, the receiving station can request a retransmission, ensuring data integrity.

Practical Applications

Packet radio is used for various purposes in the amateur radio community:

- **Messaging**: Sending text messages between operators, similar to email.

- **Telemetry**: Transmits data from remote sensors or stations, which is helpful for weather stations, balloon experiments, and satellite communications.

- **Emergency Communications**: Providing a reliable and robust method for data transmission during emergencies when other communication systems may be down.

Components of Packet Radio Transmissions

Packet radio transmissions include several components that ensure the accuracy and reliability of data communication. One essential element is the checksum, which permits error detection. The checksum is a calculated value based on the data in the packet, allowing the receiving station to verify the integrity of the received information. If the data doesn't match the checksum, it indicates an error occurred during transmission.

Additionally, packet radio transmissions feature a header containing important control information, such as the station's call sign to which the data is being sent. This header ensures that the packet is correctly routed to its intended recipient. Another component is the automatic repeat request (ARQ). In case of an error detected by the checksum, the ARQ mechanism enables the receiving station to request a retransmission of the faulty packet, ensuring the data is received correctly.

Remembering the roles of the checksum, header, and ARQ will help you understand how packet radio maintains data integrity and accurate communication.

T8D08: Which of the following is included in packet radio transmissions?

A. A check sum that permits error detection
B. A header that contains the call sign of the station to which the information is being sent
C. Automatic repeat request in case of error
D. All these choices are correct

ARQ Transmission Systems

An ARQ (Automatic Repeat reQuest) transmission system is an error correction method used in digital communications. In an ARQ system, the receiving station continuously checks the incoming data for errors. If an error is detected, the receiving station requests the transmitting station to resend the erroneous data. This process ensures that the data received is accurate and error-free, providing reliable communication even over noisy or unstable channels.

T8D11: What is an ARQ transmission system?

A. A special transmission format limited to video signals
B. A system used to encrypt command signals to an amateur radio satellite
C. An error correction method in which the receiving station detects errors and sends a request for retransmission
D. A method of compressing data using autonomous reiterative Q codes prior to final encoding

---

ARQ in Amateur Radio

In amateur radio, ARQ transmission systems benefit digital modes where accurate data transmission is critical. For example, modes like PACTOR and some implementations of RTTY (Radio Teletype) use ARQ to maintain data integrity. PACTOR is commonly used for email, file transfers, and text communication, particularly in situations where other forms of communication might be unreliable or unavailable.

By automatically correcting errors, ARQ systems enable hams to communicate more effectively, reducing the need for manual intervention and retransmission requests. This error correction method ensures messages are received as intended, making ARQ a valuable tool for amateur radio operators engaged in digital communication.

Automatic Packet Reporting System (APRS)

APRS, or Automatic Packet Reporting System, is a digital communication protocol used in amateur radio to provide real-time tactical data and location information. One of the primary applications of APRS is to facilitate digital communications combined with a mapping system that shows the locations of participating stations. This feature is particularly useful in emergencies, public service events, and other scenarios where real-time tracking and information sharing is important.

T8D05: Which of the following is an application of APRS?

A. Providing real-time tactical digital communications in conjunction with a map showing the locations of stations
B. Showing automatically the number of packets transmitted via PACTOR during a specific time interval
C. Providing voice over internet connection between repeaters
D. Providing information on the number of stations signed into a repeater

---

How APRS Enhances Communication

By transmitting data packets that include GPS coordinates and other relevant information, APRS allows operators to visualize the positions and movements of stations on a map. This capability enhances situational awareness and coordination among operators, making it an invaluable tool for search and rescue operations, disaster response, and large-scale event management.

Types of Data Transmitted by APRS

APRS can transmit various data types to enhance real-time communication and situational awareness. The primary data transmitted by APRS includes GPS position data, which allows operators to track the locations of stations and moving objects on a map. This feature is handy for emergency response, search and rescue operations, and public service events where precise location tracking is critical.

In addition to GPS data, APRS can transmit <u>text messages</u>, enabling operators to exchange short, real-time messages. This capability supports efficient coordination and information sharing during events or operations. APRS also transmits weather data, providing <u>real-time weather conditions</u> from stations equipped with weather sensors. This information is valuable for monitoring and responding to changing weather conditions during outdoor and emergency activities.

T8D03: What kind of data can be transmitted by APRS?

A. GPS position data
B. Text messages
C. Weather data
D. All these choices are correct

Phase Shift Keying (PSK)

The abbreviation "PSK" stands for <u>Phase Shift Keying</u>, a digital modulation technique that transmits data over radio waves. In PSK, the carrier signal's phase is varied according to the digital data being sent. This phase variation represents the binary data, making it an efficient method for transmitting information, especially in noisy environments. There are different types of PSK, including Binary Phase Shift Keying (BPSK) and Quadrature Phase Shift Keying (QPSK), each differing in how the phase changes to encode data.

T8D06: What does the abbreviation "PSK" mean?

A. Pulse Shift Keying
B. Phase Shift Keying
C. Packet Short Keying
D. Phased Slide Keying

PSK in Amateur Radio

In amateur radio, PSK is commonly used in digital communication modes, particularly PSK31. PSK31 is a popular mode because it uses narrow bandwidth, allowing for effective communication even with low power and under poor conditions. It is especially favored for keyboard-to-keyboard communications, where operators can type messages to each other in real-time.

Voice Over Internet Protocol (VoIP)

Voice Over Internet Protocol (VoIP) is a technology that allows you to make voice calls using a broadband internet connection instead of a regular phone line. VoIP converts your voice into digital signals and transmits them over the internet. This method uses digital techniques to deliver clear and efficient voice communications, making it a popular choice for personal and business use.

Remembering VoIP is easy if you consider it the technology behind internet-based phone calls, like those made with services such as Skype, Zoom, or WhatsApp. VoIP leverages the internet instead of relying on traditional telephone infrastructure, offering advantages like lower costs, flexibility, and integration with other Internet services. <u>Knowing that VoIP delivers voice communications over the internet</u> using digital techniques will help you quickly recall this concept for your exam and understand its growing importance in modern communication.

T8C07: What is Voice Over Internet Protocol (VoIP)?

A. A set of rules specifying how to identify your station when linked over the internet to another station
B. A technique employed to "spot" DX stations via the internet
C. A technique for measuring the modulation quality of a transmitter using remote sites monitored via the internet
D. A method of delivering voice communications over the internet using digital techniques

A gateway in amateur radio is a station that connects other amateur radio stations to the internet. This allows radio signals to be transmitted over the internet, extending the reach of communication far beyond the traditional range of radio waves. Gateways bridge the radio frequency spectrum and the internet, enabling seamless communication between operators in different parts of the world.

Think of it as a station that provides the "gate" to the internet. This setup enhances amateur radio's capabilities by allowing operators to connect globally, regardless of geographic limitations.

T8C11: What is an amateur radio station that connects other amateur stations to the internet?

A. A gateway
B. A repeater
C. A digipeater
D. A beacon

Mesh Network

An amateur radio mesh network is a data network designed for amateur radio operators. It utilizes commercial Wi-Fi equipment modified with specialized firmware to enable its use within amateur radio bands. This adaptation allows amateur radio operators to create a flexible and decentralized network where data can be transmitted over long distances without relying on traditional internet infrastructure. The mesh network structure means that each node can communicate with multiple other nodes, creating a robust and resilient communication system that can be particularly useful in emergencies or for extending internet connectivity in remote areas.

T8D12: Which of the following best describes an amateur radio mesh network?

A. An amateur-radio based data network using commercial Wi-Fi equipment with modified firmware
B. A wide-bandwidth digital voice mode employing DMR protocols
C. A satellite communications network using modified commercial satellite TV hardware
D. An internet linking protocol used to network repeaters

Understanding how to set up and operate an amateur radio mesh network can significantly enhance an operator's ability to participate in and contribute to various amateur radio activities.

Internet Radio Linking Project (IRLP)

The Internet Radio Linking Project (IRLP) is a technique that connects amateur radio systems, like repeaters, via the internet using Voice Over Internet Protocol (VoIP). This allows amateur radio operators to communicate with each other over long distances, far beyond the normal range of their radio equipment, by linking their local repeaters to other repeaters worldwide through the internet.

Think of IRLP as a bridge between traditional radio communication and the internet. By using VoIP, IRLP enables clear and reliable voice communication between radio operators in different regions, enhancing the versatility and reach of amateur radio networks. Knowing that IRLP connects repeaters via the internet using VoIP will help you recall this concept for your exam, highlighting its role in expanding the capabilities of ham radio communication.

T8C08: What is the Internet Radio Linking Project (IRLP)?

A. A technique to connect amateur radio systems, such as repeaters, via the internet using Voice Over Internet Protocol (VoIP)
B. A system for providing access to websites via amateur radio
C. A system for informing amateurs in real time of the frequency of active DX stations
D. A technique for measuring signal strength of an amateur transmitter via the internet

Directly accessing IRLP (Internet Radio Linking Project) nodes over the air is accomplished by using Dual-Tone Multi-Frequency (DTMF) signals. DTMF signals are the audio tones you hear when pressing keys on a telephone keypad (digital tones). In the context of IRLP, these tones send commands from your radio to the IRLP node, instructing it to connect to other nodes or perform specific functions.

DTMF are signals that "dial" into the IRLP system using your radio. You can control the IRLP node and link with other repeaters worldwide by pressing the appropriate keys on your radio's keypad.

T8C06: How is over the air access to IRLP nodes accomplished?

A. By obtaining a password that is sent via voice to the node
B. By using DTMF signals
C. By entering the proper internet password
D. By using CTCSS tone codes

And...

T2B06: What type of signaling uses pairs of audio tones?

A. DTMF
B. CTCSS
C. GPRS
D. D-STAR

EchoLink

EchoLink is a protocol that allows an amateur station to transmit through a repeater without needing a traditional radio to initiate the transmission. Instead, EchoLink uses Voice Over Internet Protocol (VoIP) to connect amateur radio operators over the internet. This means you can use your computer or smartphone to access and transmit through repeaters anywhere worldwide, as long as they are connected to the EchoLink network. How cool is that!

EchoLink is like a modern bridge between internet technology and traditional ham radio. It enables greater flexibility and convenience, allowing operators to participate in amateur radio communication from virtually any location with internet access. Knowing that EchoLink allows transmission through repeaters using VoIP without a physical radio will help you quickly recall this concept for your exam, emphasizing its role in expanding the accessibility and reach of ham radio operations.

T8C09: Which of the following protocols enables an amateur station to transmit through a repeater without using a radio to initiate the transmission?

A. IRLP
B. D-STAR
C. DMR
D. EchoLink

Before using the EchoLink system, you must register your call sign and provide proof of your amateur radio license. This process ensures that only licensed radio operators can access the system, maintaining the integrity and security of the EchoLink network. Registration typically involves submitting your call sign and a copy of your license to the EchoLink administrators, who will verify your credentials.

Think of this registration step as a way to authenticate and validate your identity within the EchoLink community. Once registered, you can use your computer or smartphone to connect to repeaters and other stations worldwide.

T8C10: What is required before using the EchoLink system?

A. Complete the required EchoLink training
B. Purchase a license to use the EchoLink software
C. Register your call sign and provide proof of license
D. All these choices are correct

National Television System Committee (NTSC)

The term "NTSC" stands for National Television System Committee and refers to a type of analog fast-scan color TV signal. NTSC is a standard for analog television used predominantly in North America and parts of South America before the transition to digital broadcasting. This system encodes color video signals and synchronizes them for display on television screens, providing the capability to transmit moving images and sound over radio frequencies.

NTSC in Amateur Radio

In the context of amateur radio, NTSC transmissions can be used for Amateur Television (ATV) operations, where enthusiasts transmit live video signals. These transmissions allow for sharing video content such as public service announcements, live events, or technical demonstrations.

T8D04: What type of transmission is indicated by the term "NTSC?"

A. A Normal Transmission mode in Static Circuit
B. A special mode for satellite uplink
C. An analog fast-scan color TV signal
D. A frame compression scheme for TV signals

Now, that was a lot to take in and remember. But don't let it overwhelm you. Like any hobby, it's best to start with the basics and build your skills with the fundamentals. As you grow and gain experience, you'll naturally start to branch out into these other operating modes, adding depth and versatility to your radio operation.

Chapter 25

Troubleshooting

TROUBLESHOOTING ISSUES IS ESSENTIAL for any ham radio enthusiast. We'll cover common troubleshooting techniques to help you identify and resolve issues that may arise during operation. Mastering these skills will ensure smooth and reliable communication as a beginner, enhancing your overall ham radio experience. Let's dive into the essentials of radio operation and problem-solving to keep your equipment running optimally.

On the Air Issues

Radio Frequency Interference

Radiofrequency interference (RFI) can significantly impact the quality of communications in amateur radio. It can be caused by various factors, including fundamental overload, harmonics, and spurious emissions. Understanding each of these sources of interference and knowing how to address them is essential for maintaining clear and reliable transmissions.

T7B03: Which of the following can cause radio frequency interference?

A. Fundamental overload
B. Harmonics
C. Spurious emissions
D. All these choices are correct

Fundamental Overload: This occurs when a strong signal overwhelms the receiver's front end, causing it to malfunction. Fundamental overload can be mitigated by improving the receiver's selectivity or using an attenuator to reduce the strength of the incoming signal. Ensuring your equipment is calibrated correctly and not operating excessively high power levels can also help prevent this issue.

Harmonics: Harmonics are unwanted frequencies that are multiples of the fundamental frequency. They can cause interference on other bands and frequencies. To reduce harmonics, use low-pass filters on transmitters and ensure all equipment is properly shielded and grounded. Regularly checking your equipment for proper operation and maintenance can also help minimize harmonic emissions.

Spurious Emissions are unintended signals that are not harmonic but can still cause interference. They can result from faulty equipment or improper adjustments. Band-pass filters and maintaining your equipment in good working order can help eliminate these emissions. Additionally, ensuring your transmitter is tuned correctly and not operating beyond its designed parameters can reduce spurious emissions.

By understanding these causes of RFI and taking appropriate measures to mitigate them, amateur radio operators can ensure their transmissions are clear and free from interference.

Radio Frequency Interference with Consumer Electronics

Amateur radio transmissions can sometimes unintentionally interfere with consumer electronics, such as AM/FM radios and television sets. This interference is often due to the consumer device's inability to reject strong signals outside its intended frequency band. Knowing the causes and solutions to these issues is key to maintaining harmony between amateur radio operations and other electronic devices.

Unintentional Reception by AM/FM Radios

When a broadcast AM or FM radio unintentionally receives an amateur radio transmission, it is usually because the receiver cannot reject strong signals outside its intended band. This problem can be widespread in areas with strong amateur transmissions. Improving the receiver's ability to filter out unwanted signals through better internal filtering or external band-pass filters can help mitigate this issue.

> **T7B02: What would cause a broadcast AM or FM radio to receive an amateur radio transmission unintentionally?**
>
> **A. The receiver is unable to reject strong signals outside the AM or FM band**
> B. The microphone gain of the transmitter is turned up too high
> C. The audio amplifier of the transmitter is overloaded
> D. The deviation of an FM transmitter is set too low

Reducing VHF Transceiver Overload

VHF transceivers can experience overload from nearby commercial FM stations. This type of interference can be reduced by installing a band-reject filter, also known as a notch filter, which attenuates the specific frequency range of the commercial FM signals, thereby preventing them from overwhelming the transceiver.

> **T7B07: Which of the following can reduce overload of a VHF transceiver by a nearby commercial FM station?**
>
> A. Installing an RF preamplifier
> B. Using double-shielded coaxial cable
> C. Installing bypass capacitors on the microphone cable
> **D. Installing a band-reject filter**

Fundamental Overload in Non-Amateur Devices

Strong amateur signals might cause fundamental overload in non-amateur radios or TV receivers. This problem can often be resolved by installing a filter at the antenna input of the affected device. The filter blocks the amateur signal before it can enter the receiver, eliminating the interference.

T7B05: How can fundamental overload of a non-amateur radio or TV receiver by an amateur signal be reduced or eliminated?

A. Block the amateur signal with a filter at the antenna input of the affected receiver
B. Block the interfering signal with a filter on the amateur transmitter
C. Switch the transmitter from FM to SSB
D. Switch the transmitter to a narrow-band mode

TV Interference Solutions

One common issue is interference with non-fiber optic cable TV systems caused by amateur transmissions. The first step in resolving this type of interference is to <u>ensure that all TV feed line coaxial connectors are correctly installed</u>. Poorly installed connectors can act as entry points for unwanted signals, causing interference. Adequately secured and shielded connectors help maintain signal integrity and reduce the chances of interference.

T7B09: What should be the first step to resolve non-fiber optic cable TV interference caused by your amateur radio transmission?

A. Add a low-pass filter to the TV antenna input
B. Add a high-pass filter to the TV antenna input
C. Add a preamplifier to the TV antenna input
D. Be sure all TV feed line coaxial connectors are installed properly

By understanding these types of interference and implementing the appropriate solutions, amateur radio operators can minimize the impact of their transmissions on nearby consumer electronics. This ensures better operation of their equipment and helps maintain good relations with neighbors and the broader community.

Addressing Interference Issues with Neighbors

Interference between amateur radio stations and nearby consumer electronics can be a common issue. Knowing how to handle these situations responsibly and diplomatically is essential for maintaining good relations with your neighbors and ensuring your station operates within legal and technical standards.

When Your Station Causes Interference

If a neighbor informs you that your station's transmissions are interfering with their radio or TV reception, <u>the first step is to ensure that your station is functioning properly</u>. Check your equipment to verify that it does not cause interference to your own radio or television when tuned to the same channel. This self-check helps you confirm that your transmissions are clean and within acceptable parameters. Additionally, consider installing low-pass filters, ferrite chokes, or other interference mitigation devices on your equipment to reduce the possibility of causing unintended interference.

T7B06: Which of the following actions should you take if a neighbor tells you that your station's transmissions are interfering with their radio or TV reception?

A. Make sure that your station is functioning properly and that it does not cause interference to your own radio or television when it is tuned to the same channel
B. Immediately turn off your transmitter and contact the nearest FCC office for assistance
C. Install a harmonic doubler on the output of your transmitter and tune it until the interference is eliminated
D. All these choices are correct

When a Neighbor's Device Causes Interference

If you experience harmful interference to your amateur station from something in a neighbor's home, it is important to approach the situation politely and cooperate. First, work with your neighbor to identify the offending device. Once identified, explain that FCC rules prohibit the use of devices that cause harmful interference to licensed radio services. Offer practical solutions, such as using better-quality cables or adding ferrite chokes to the interfering device. At the same time, ensure that your station meets the standards of good amateur practice by maintaining proper grounding, using well-shielded equipment, and operating within legal power limits.

T7B08: What should you do if something in a neighbor's home is causing harmful interference to your amateur station?

A. Work with your neighbor to identify the offending device
B. Politely inform your neighbor that FCC rules prohibit the use of devices that cause interference
C. Make sure your station meets the standards of good amateur practice
D. All these choices are correct

You are committed to resolving interference issues amicably and responsibly by taking these steps. This helps maintain a good relationship with your neighbors and ensures that your amateur radio station operates smoothly and within FCC regulations.

Understanding RF Feedback

RF feedback, or radio frequency feedback, can significantly impact the quality of voice transmissions. A common symptom of RF feedback in a transmitter or transceiver is receiving reports of garbled, distorted, or unintelligible voice transmissions from other operators. This issue occurs when transmitted RF energy re-enters the transmitter's audio circuits, causing unwanted feedback that distorts the transmitted audio signal.

T7B11: What is a symptom of RF feedback in a transmitter or transceiver?

A. Excessive SWR at the antenna connection
B. The transmitter will not stay on the desired frequency
C. Reports of garbled, distorted, or unintelligible voice transmissions
D. Frequent blowing of power supply fuses

To mitigate RF feedback, ensure your station is grounded correctly, and all connections are secure. Using ferrite chokes on microphone cables and other audio lines can help prevent RF energy from coupling into the audio path. Additionally, keeping antenna feed lines away from the radio equipment and audio cables can reduce the likelihood of RF feedback.

Distorted or Unintelligible Audio Signals

Suppose you receive a report that your audio signal through an FM repeater is distorted or unintelligible. In that case, several factors could be causing the problem. One common issue is that your transmitter might be slightly off frequency. Even a slight deviation can result in poor signal quality, making your audio sound distorted or unclear. Ensuring your transmitter is calibrated correctly and operating on the correct frequency is needed for clear communication.

Another potential problem could be that your batteries are running low. Low battery power can lead to reduced transmitter output and poor audio quality. Always check your battery levels before operating, especially in portable or mobile setups, and replace or recharge them as necessary. Additionally, being in a poor location, such as an area with heavy obstructions or poor signal coverage, can cause audio issues. Moving to a higher or more open area often improves signal strength and clarity. Addressing these factors ensures your audio signal remains clear and intelligible when using FM repeaters.

T7B10: What might be a problem if you receive a report that your audio signal through an FM repeater is distorted or unintelligible?

A. Your transmitter is slightly off frequency
B. Your batteries are running low
C. You are in a bad location
D. All these choices are correct

Distorted FM Transmission Audio

If your FM transmission audio is distorted on voice peaks, one likely cause is that you are talking too loudly into the microphone. Speaking too loudly can overdrive the microphone input, causing the audio signal to become clipped and distorted. This distortion occurs because the microphone and transmitter circuits are designed to handle a specific range of audio levels, and excessive volume exceeds this range, resulting in poor sound quality.

To avoid this issue, speaking at a normal volume and maintaining an appropriate distance from the microphone is essential. Additionally, adjusting the microphone gain setting on your transceiver can help ensure that your audio levels remain within the optimal range for clear transmission.

T2B05: What would cause your FM transmission audio to be distorted on voice peaks?

A. Your repeater offset is inverted
B. You need to talk louder
C. You are talking too loudly
D. Your transmit power is too high

FM, or Frequency Modulation, is mentioned explicitly in the question because the nature of FM transmission makes it particularly sensitive to audio input levels. In FM, the frequency of the carrier wave is varied by the amplitude of the input audio signal. When you speak too loudly, the microphone captures a high-amplitude audio signal, which causes significant variations in the carrier frequency. This can lead to over-modulation, resulting in distortion, especially on voice peaks where the audio signal is strongest.

In other modulation methods like AM (Amplitude Modulation), the amplitude of the carrier wave is varied, and it has different characteristics and tolerances for audio input levels. FM's inherent sensitivity to input volume makes it essential for operators to manage their speaking levels carefully to avoid distortion. By highlighting FM, the question emphasizes the importance of maintaining proper audio levels, specifically in FM transmissions, to ensure clear and distortion-free communication.

Over-Deviating

Over-deviation in FM transceivers occurs when the transmitted signal's modulation exceeds the specified frequency deviation limits. Frequency deviation refers to the extent to which the carrier frequency varies in response to the transmitted audio signal. FM transceivers operate within certain deviation limits to ensure clear communication and avoid interference with adjacent channels.

This results in a too-wide signal for the designated channel, causing distortion and interference with adjacent channels. Talking farther from the microphone is a simple and effective solution if you are told your transceiver is over-deviating.

Increasing the distance between your mouth and the microphone reduces the audio input level, decreasing the modulation depth. This adjustment helps ensure your signal stays within the proper deviation limits, leading to clearer and more reliable communication.

T7B01: What can you do if you are told your FM handheld or mobile transceiver is over-deviating?

A. Talk louder into the microphone
B. Let the transceiver cool off
C. Change to a higher power level
D. Talk farther away from the microphone

Distorted Audio Caused by RF Current

Suppose you experience distorted audio due to RF current on the shield of your microphone cable. In that case, one effective solution is to use a ferrite choke. A ferrite choke, also known as a ferrite bead, is a passive electronic component that suppresses high-frequency noise and interference. When placed on the microphone cable, the ferrite choke helps to block or absorb the unwanted RF currents that can cause distortion in your audio signal.

T7B04: Which of the following could you use to cure distorted audio caused by RF current on the shield of a microphone cable?

A. Band-pass filter
B. Low-pass filter
C. Preamplifier
D. Ferrite choke

The ferrite choke impedes the RF current, reducing the amount of RF energy that travels along the cable shield. This simple addition can significantly improve audio quality by eliminating interference and ensuring clear and intelligible communication.

Single Side Band

Excessive microphone gain on SSB (Single Sideband) transmissions can result in distorted transmitted audio. When the microphone gain is set too high, the audio signal becomes too strong, causing the transmitter to over-modulate. This over-modulation leads to audio distortion, making the transmission sound unclear and challenging to understand. Additionally, excessive gain can cause splatter, where the signal spreads into adjacent frequencies, potentially causing interference to other stations. Setting the microphone gain at an appropriate level is essential to maintain clear and intelligible audio quality.

T4B01: What is the effect of excessive microphone gain on SSB transmissions?

A. Frequency instability
B. Distorted transmitted audio
C. Increased SWR
D. All these choices are correct

Suppose the voice pitch of a single-sideband signal returning to your CQ call seems too high or too low. You can use the RIT (Receiver Incremental Tuning) or Clarifier control in that case. The RIT or Clarifier allows you to fine-tune the receiver frequency without changing the transmitter frequency. This adjustment helps correct any slight frequency mismatches between your transmitter and the responding station, ensuring the audio pitch is accurate. Proper use of the RIT or Clarifier enhances communication clarity, making it easier to understand the incoming signal.

T4B06: Which of the following controls could be used if the voice pitch of a single-sideband signal returning to your CQ call seems too high or low?

A. The AGC or limiter
B. The bandwidth selection
C. The tone squelch
D. The RIT or Clarifier

Repeater Issues

Unable to Access, But can Hear

If you can hear a repeater's output but cannot access it, several factors could be causing the issue. One common reason is an improper transceiver offset. Repeaters receive signals on one frequency and transmit on another, typically with a standard offset. If your transceiver is not set to the correct offset, your signal won't reach the repeater's input frequency, preventing access.

Another possible reason is using the wrong CTCSS tone or DCS code. Repeaters often require a specific sub-audible tone (CTCSS) or digital code (DCS) to open the squelch and allow your signal to be retransmitted. If your transceiver is not programmed with the correct CTCSS tone or DCS code, the repeater will not recognize your signal, even though you can hear its output. Ensuring that your transceiver settings match the repeater's requirements is essential for successful communication.

T2B04: Which of the following could be the reason you are unable to access a repeater whose output you can hear?

A. Improper transceiver offset
B. You are using the wrong CTCSS tone
C. You are using the wrong DCS code
D. All these choices are correct

Communicating with a Distant Repeater Using a Directional Antenna

When buildings or other obstructions block the direct line of sight to a distant repeater, a directional antenna can help overcome this challenge by finding a path that reflects signals to the repeater. Instead of aiming your antenna directly at the repeater, you can point it towards a nearby surface, such as a building, hill, or other large structure, that can reflect the radio waves. This reflection can create an indirect path for the signals to reach the repeater, bypassing the obstruction.

T3A05: When using a directional antenna, how might your station be able to communicate with a distant repeater if buildings or obstructions are blocking the direct line of sight path?

A. Change from vertical to horizontal polarization
B. Try to find a path that reflects signals to the repeater
C. Try the long path
D. Increase the antenna SWR

This approach requires critical thinking and experimentation. As a ham operator, you should consider the environment and identify potential reflective surfaces. Adjust the direction of your antenna and monitor the signal strength and quality. By methodically testing different angles and positions, you can find the optimal reflection path that allows clear communication with the repeater. This troubleshooting technique enhances problem-solving skills and ensures reliable communication even in challenging environments.

As with any skill or hobby, real-life troubleshooting will be your best teacher. Don't hesitate to lean on mentors and peers for guidance and support as you navigate challenges and expand your knowledge.

Chapter 26

Privacy and Ethics

IN AMATEUR RADIO, WHERE the airwaves are open and accessible to anyone with the right equipment, privacy and ethics have a unique significance. While ham radio operates on a foundation of openness and community, specific legal and ethical standards are designed to protect the privacy of communications and ensure all operators engage with honesty and respect. You will need to understand these nuances in order to maintain the integrity and camaraderie that define the amateur radio community.

Privacy Laws

Unlike traditional forms of communication like telephony or emailing, where privacy can often be tightly controlled, the nature of amateur radio means that transmissions are inherently public. However, legal boundaries still govern what is considered acceptable regarding listening to and sharing the content of transmissions. It's important to note that the FCC prohibits using amateur radio frequencies for personal or commercial gain, and there are specific rules against broadcasting music, obscene content, or transmissions intended to facilitate a crime.

Eavesdropping, or listening to private transmissions without the consent of those involved, falls into a legal gray area in amateur radio. Since transmissions are technically public, listening in is not illegal unless the information is used for criminal or harmful purposes. However, the ethical implications of eavesdropping can be significant. It's considered poor practice and contrary to the spirit of amateur radio to use information gleaned from private conversations inappropriately. Moreover, the deliberate interception and sharing of encrypted transmissions are strictly illegal, emphasizing the balance between open communication and privacy.

Ethical Communication

Ethics in amateur radio extend beyond mere legal compliance; they touch upon how operators interact. Ethical communication is founded on honesty and integrity. This means identifying oneself accurately with a call sign, engaging in truthful and respectful exchanges, and providing help or advice in good faith. The amateur radio community prides itself on a spirit of friendship and support, often called the "ham spirit," which encourages seasoned operators to assist newcomers and promotes mutual respect among all participants, regardless of their level of expertise or geographical location.

In practical terms, ethical communication involves more than adhering to regulations; it's about fostering a positive, supportive environment. For instance, if you notice a fellow operator making a mistake in frequency use or protocol, the ethical approach would be to gently and privately correct them rather than publicly criticizing or shaming them. This helps maintain the individual's dignity and supports the broader community ethos of learning and improvement.

Handling Interference

Intentional or accidental interference can be one of the most challenging aspects of ham radio operation. Ethically handling interference involves first determining whether the interference is malicious or unintentional. If it's the latter, reaching out to the interfering station with a polite request to adjust their practices, such as changing frequency or modifying transmission power, can often resolve the issue. Keeping a calm and cooperative demeanor during these interactions is necessary, as it reflects respect and understanding, fostering a more amicable resolution.

Avoiding direct confrontation is advisable if interference is suspected to be intentional, which is rare but can occur. Instead, documenting the instances of interference and reporting them to the appropriate authorities or amateur radio organizations can ensure that the issue is addressed formally and fairly without escalating the conflict or disrupting the broader community harmony.

Managing Interference on the Same Frequency

When two stations transmit on the same frequency and interfere with each other, the recommended course of action is for the stations to negotiate continued use of the frequency. This involves operators communicating respectfully and collaboratively to decide how to share or alternate frequency use. Such negotiations help prevent ongoing interference and ensure that both stations can operate effectively without disrupting each other.

T2B08: Which of the following applies when two stations transmitting on the same frequency interfere with each other?

A. The stations should negotiate continued use of the frequency
B. Both stations should choose another frequency to avoid conflict
C. Interference is inevitable, so no action is required
D. Use subaudible tones so both stations can share the frequency

Public Service and Privacy

Amateur radio operators often play a fundamental role during emergencies, providing critical communication links when other systems fail. Balancing this commitment to public service with respect for individual privacy is essential. Handling all transmitted information sensitively during public service operations, such as emergency or disaster communications, is vital. Operators should avoid disclosing sensitive personal details over the airwaves. They should be cautious about sharing information about the situation that could lead to panic or misinformation.

Moreover, when handling emergency communications, maintaining operational security by restricting sensitive operational details to those who need to know them is a fundamental practice. This protects the privacy of those affected by the emergency and ensures the effectiveness and integrity of the response efforts.

Navigating the intricate balance between open communication, privacy, and ethics in amateur radio is about following rules and embracing a culture of respect, support, and responsibility. Adhering to these principles helps maintain the amateur radio community as a welcoming and valuable space for all who join. Whether communicating across towns or continents, your approach to privacy and ethics shapes your experience and the broader perception and effectiveness of the amateur radio community worldwide.

Rules & Regulations

NAVIGATING RADIO IS AKIN to learning the rules of the road before taking the wheel. Just as drivers must understand and adhere to traffic laws to ensure safety and order on the streets, amateur radio operators must familiarize themselves with specific regulations and operating procedures.

This chapter aims to demystify the regulatory landscape governed by the Federal Communications Commission (FCC), clarify the process of obtaining your Technician license, and discuss the importance of ethical operation, all of which form the backbone of responsible and enjoyable use of the amateur radio spectrum.

Chapter 27

Understanding the FCC's Role

Regulatory Overview

THE FCC PLAYS A core role in regulating amateur radio in the United States, maintaining an organized and fair use of the radio spectrum. As an independent U.S. government agency, the FCC oversees all domestic and international communications by radio, television, wire, satellite, and cable. In amateur radio, the FCC's responsibilities include licensing operators, allocating frequency bands, and ensuring that all communications adhere to established rules designed to prevent interference and ensure public safety.

T1A02: Which agency regulates and enforces the rules for the Amateur Radio Service in the United States?

A. FEMA
B. Homeland Security
C. The FCC
D. All these choices are correct

Critical rules that you must be aware of as an aspiring or practicing ham include regulations on frequency usage, power limits, and emission types. For example, the FCC dictates which frequency bands are available for amateur use and what power is acceptable for transmission in these bands.

The rationale behind these rules is multifaceted: They prevent interference with other communication services and users, ensure that amateur radio practices do not pose a hazard to public safety, and preserve the integrity and utility of the radio spectrum for all users.

Understanding these rules isn't just about compliance; it's about contributing to a harmonious, interference-free environment where the amateur radio community can thrive. The FCC's regulations are laid out in Part 97 of the Commission's rules, and familiarizing yourself with this document is an excellent start to understanding your obligations and rights as a radio operator.

And why does this all exist?

T1A01: Which of the following is part of the Basis and Purpose of the Amateur Radio Service?

A. Providing personal radio communications for as many citizens as possible
B. Providing communications for international non-profit organizations
C. Advancing skills in the technical and communication phases of the radio art
D. All these choices are correct

The FCC understands the benefits of radio, and a legion of ham operators ensures that the art of radio and the skills involved in its communication continue to advance and be maintained - an indispensable skill during emergencies.

The FCC provides definitions that we will review in this section. Those definitions often appear in questions on the exam, like this one...

T1D10: How does the FCC define broadcasting for the Amateur Radio Service?

A. Two-way transmissions by amateur stations
B. Any transmission made by the licensed station
C. Transmission of messages directed only to amateur operators
D. Transmissions intended for reception by the general public

When operating our radios, we transmit across open radio frequencies. Therefore, anyone may receive our transmissions, implying they are intended for the general public.

Speaking of emergencies:

T1A10: What is the Radio Amateur Civil Emergency Service (RACES)?

A. A radio service using amateur frequencies for emergency management or civil defense communications
B. A radio service using amateur stations for emergency management or civil defense communications
C. An emergency service using amateur operators certified by a civil defense organization as being enrolled in that organization
D. All these choices are correct.

RACES, or the Radio Amateur Civil Emergency Service, is a key component of amateur radio operations in the United States. The FCC established it to provide emergency communication support during civil emergencies. Thus, answer 'A' is correct; it's a radio service using amateur frequencies for emergency communications. Those communications happen on amateur stations/equipment, so answer 'B' is correct.

Managed by local, county, and state emergency agencies, RACES volunteers, who are licensed amateur radio operators, work closely with civil authorities to ensure reliable communication when traditional systems fail. Working closely with civil authorities, we are certified; you have to have a relationship with them ahead of time and enroll with them; you can't just show up and wave your hands around saying, "Pick Me." So, the answer 'C' is correct.

RACES operators are activated only during declared emergencies or drills. They participate in regular training to stay prepared. This service not only exemplifies the public service spirit of ham radio but also enhances community preparedness, provides communication redundancy, and bolsters public trust in amateur radio operators as essential resources during crises.

If answers A, B, and C are all correct by deduction, then 'D' is correct.

Restrictions for U.S. Amateurs in Secondary Segments

U.S. amateurs must be mindful of their operating practices in segments of the radio bands where the Amateur Radio Service is designated as secondary. This designation means primary users, including government or commercial stations, have priority access to those frequencies. As a result, amateur radio operators <u>must ensure they do not interfere</u> with these primary users. If interference occurs, the amateur station must either cease transmission or make necessary adjustments to eliminate the interference.

T1B08: How are U.S. amateurs restricted in segments of bands where the Amateur Radio Service is secondary?

A. U.S. amateurs may find non-amateur stations in those segments and must avoid interfering with them
B. U.S. amateurs must give foreign amateur stations priority in those segments
C. International communications are not permitted in those segments
D. Digital transmissions are not permitted in those segments

Understanding this restriction is essential for responsible amateur radio operations. It emphasizes the need for amateurs to be aware of the band plans and usage designations within the frequency spectrum they are utilizing. By adhering to these guidelines, amateurs help maintain harmonious and efficient use of the radio frequencies, ensuring that all services can coexist without disruption. This knowledge highlights the importance of respecting and cooperating with other spectrum users to preserve the integrity and functionality of the amateur radio service.

Licensing Requirements

To legally operate on amateur radio frequencies in the U.S., you must obtain a license from the FCC. That's why you are here!

T1C01: For which license classes are new licenses currently available from the FCC?

A. Novice, Technician, General, Amateur Extra
B. Technician, Technician Plus, General, Amateur Extra
C. Novice, Technician Plus, General, Advanced
D. Technician, General, Amateur Extra

The important thing to remember here is that other types of licenses were historically issued. But <u>today, we have the Technician, General, and Amateur Extra.</u>

T1A04: How many operator/primary station license grants may be held by any one person?

A. One
B. No more than two
C. One for each band on which the person plans to operate
D. One for each permanent station location from which the person plans to operate

Technically, your amateur radio license has two parts: an operator and a station license. The operator license is the personal authorization granted to you as an individual. This license signifies that the holder has demonstrated the knowledge and skills to safely and effectively use radio frequencies allocated for amateur radio use.

The primary station license is tied to the operator license. It authorizes the licensee to set up and operate an amateur radio station. <u>For us hams, the two are combined into one.</u>

T1A05: What proves that the FCC has issued an operator/primary license grant?

A. A printed copy of the certificate of successful completion of examination
B. An email notification from the NCVEC granting the license
C. The license appears in the FCC ULS database
D. All these choices are correct

As simple as it sounds, how do you prove you have a license? <u>You show up in their database.</u>

And showing up in their database is essential. Because...

T1C10: How soon after passing the examination for your first amateur radio license may you transmit on the amateur radio bands?

A. Immediately on receiving your Certificate of Successful Completion of Examination (CSCE)
B. As soon as your operator/station license grant appears on the ARRL website
C. As soon as your operator/station license grant appears in the FCC's license database
D. As soon as you receive your license in the mail from the FCC

That's right. You don't actually have a license – even if you passed the exam - until you show up in their database. Until then, no radio operations!

Being in their database means you must provide an active email address, which is very important.

T1C04: What may happen if the FCC is unable to reach you by email?

A. Fine and suspension of operator license
B. Revocation of the station license or suspension of the operator license
C. Revocation of access to the license record in the FCC system
D. Nothing; there is no such requirement

Yes! If you miss an email from the FCC, they can revoke your license. So, watch your email! Make sure the FCC isn't going to your spam or junk folder. In fact, they are so concerned about you maintaining your email address with them that they ask two questions about it!

T1C07: Which of the following can result in revocation of the station license or suspension of the operator license?

A. Failure to inform the FCC of any changes in the amateur station following performance of an R.F. safety environmental evaluation
B. Failure to provide and maintain a correct email address with the FCC
C. Failure to obtain FCC type acceptance prior to using a home-built transmitter
D. Failure to have a copy of your license available at your station

At this point, you have taken the exam, passed it, and ensured your email is correct in the FCC database and your record in their database is live. You are now officially licensed!

So, how long is your license good for?

T1C08: What is the normal term for an FCC-issued amateur radio license?

A. Five years
B. Life
C. Ten years
D. Eight years

You need to renew your license <u>every 10 years</u>. You will only have to retake the test if your license expires. But...

T1C09: What is the grace period for renewal if an amateur license expires?

A. Two years
B. Three years
C. Five years
D. Ten years

If your license expires, you have <u>two years to renew your license</u> before you have to retake the exam. However, during those two years...

T1C11: If your license has expired and is still within the allowable grace period, may you continue to transmit on the amateur radio bands?

A. Yes, for up to two years
B. Yes, as soon as you apply for renewal
C. Yes, for up to one year
D. No, you must wait until the license has been renewed

You must only continue transmitting on licensed amateur bands with an active license. Please don't do it! Go renew your license.

A final word about licenses: amateur radio clubs can also obtain a license from the FCC.

A club license allows amateur radio operators to collectively operate under a single call sign, facilitating group activities and promoting collaboration among members. <u>To apply for a club license, the club must have at least four members,</u> including a trustee with a current amateur radio license. The trustee is responsible for the club's station and compliance with FCC regulations.

T1F11: Which of the following is a requirement for the issuance of a club station license grant?

A. The trustee must have an Amateur Extra Class operator license grant
B. The club must have at least four members
C. The club must be registered with the American Radio Relay League
D. All these choices are correct

Obtaining a club license involves submitting an application (FCC Form 605) and paying any required fees. Once granted, the club call sign is used by all members when operating the club station, enabling the club to participate in events, contests, and public service activities under a unified identity. Club licenses help foster a sense of community and shared purpose among amateur radio enthusiasts, making organizing and coordinating activities that promote the hobby and its benefits easier.

Transmission Restrictions and Rules

One key aspect of the Federal Communications Commission's (FCC) oversight involves setting restrictions on transmissions to minimize interference and promote orderly communication. In this section, we will explore the various constraints imposed by the FCC on ham radio transmissions, understanding their importance in maintaining the integrity of amateur radio bands and ensuring safe and effective communication for all operators.

Willful Interference

Let's start with one of the most fundamental ones: willful interference with other amateur radio stations is never permitted. The FCC regulations strictly prohibit any actions that intentionally disrupt or interfere with the communications of other amateur radio operators. This rule is fundamental to maintaining the integrity and functionality of the amateur radio bands, ensuring that all operators can communicate effectively and without unnecessary disruption.

T1A11: When is willful interference to other amateur radio stations permitted?

A. To stop another amateur station that is breaking the FCC rules
B. At no time
C. When making short test transmissions
D. At any time, stations in the Amateur Radio Service are not protected from willful interference

You need to understand and abide by this prohibition. Intentional interference violates FCC regulations and undermines the spirit of cooperation and mutual respect central to the amateur radio community. By adhering to this rule, operators help create a respectful and efficient communication environment, fostering positive interactions and reliable transmissions across the amateur radio spectrum. Compliance with this regulation is essential for preserving the quality and reliability of amateur radio operations for everyone involved.

Broadcasting

In amateur radio, one-way transmissions are generally prohibited under the circumstance of broadcasting. Broadcasting refers to transmitting messages intended for reception by the general public. The FCC regulations strictly limit amateur radio operators from broadcasting to prevent using amateur frequencies for purposes other than personal, non-commercial communications. This rule helps maintain the amateur bands for their intended purpose, facilitating individual and emergency communications, technical training, and the advancement of radio art.

T1D02: Under which of the following circumstances are one-way transmissions by an amateur station prohibited?

A. In all circumstances
B. Broadcasting
C. International Morse Code Practice
D. Telecommand or transmissions of telemetry

There are exceptions, however, where amateur stations may transmit information typically associated with broadcasting. According to FCC rules, amateur radio operators can transmit information to support broadcasting, program production, or news gathering if no other means of communication is available and the situation involves the immediate safety of human life or property protection. These exceptions are rare and typically apply in emergency scenarios where alternative communication methods are not viable. Understanding these regulations ensures that amateur radio operations remain compliant with FCC rules and that amateur bands are used appropriately.

T1D09: When may amateur stations transmit information in support of broadcasting, program production, or news gathering, assuming no other means is available?

A. When such communications are directly related to the immediate safety of human life or protection of property
B. When broadcasting communications to or from the space shuttle
C. Where non-commercial programming is gathered and supplied exclusively to the National Public Radio network
D. Never

Indecent or Obscene

The FCC strictly prohibits the transmission of any language that may be considered indecent or obscene on amateur radio frequencies. This rule is in place to ensure that amateur radio remains a respectful and appropriate medium for communication, accessible to operators of all ages and backgrounds. The use of indecent or obscene language is not only against FCC regulations but also against the ethos of amateur radio, which promotes courteous and professional conduct among operators.

T1D06: What, if any, are the restrictions concerning transmission of language that may be considered indecent or obscene?

A. The FCC maintains a list of words that are not permitted to be used on amateur frequencies
B. Any such language is prohibited
C. The ITU maintains a list of words that are not permitted to be used on amateur frequencies
D. There is no such prohibition

Amateur radio is a community-driven activity that relies on mutual respect and adherence to established guidelines to function effectively. Violating these rules by transmitting indecent or obscene language can lead to severe penalties, including fines, suspension, or revocation of the operator's license.

Prohibited Communications

FCC-licensed amateur radio stations are prohibited from exchanging communications with any country that has notified the International Telecommunication Union (ITU) of its objection to such communications. This regulation ensures that amateur radio operators respect international agreements and the sovereignty of other nations. The ITU maintains a list of countries that have objected to amateur radio communications, and operators must stay informed about these restrictions to remain compliant with FCC rules.

T1D01: With which countries are FCC-licensed amateur radio stations prohibited from exchanging communications?

A. Any country whose administration has notified the International Telecommunication Union (ITU) that it objects to such communications
B. Any country whose administration has notified the American Radio Relay League (ARRL) that it objects to such communications
C. Any country banned from such communications by the International Amateur Radio Union (IARU)
D. Any country banned from making such communications by the American Radio Relay League (ARRL)

Understanding this regulation is important as it emphasizes the respect of international protocols and maintaining good relations within the global amateur radio community. By adhering to these restrictions, operators help ensure that amateur radio remains a cooperative and internationally respected hobby, free from political and diplomatic conflicts.

Encoded Transmissions

In amateur radio, messages encoded to obscure their meaning are generally prohibited. This rule ensures transparency and openness in amateur radio communications, fostering an environment where all operators can understand and participate in exchanges. However, this rule has specific exceptions, particularly when transmitting control commands to space stations or radio control craft. In these cases, encoding messages is permissible because it is necessary to maintain the integrity and security of the control functions, preventing unauthorized access or interference.

T1D03: When is it permissible to transmit messages encoded to obscure their meaning?

A. Only during contests
B. Only when transmitting certain approved digital codes
C. Only when transmitting control commands to space stations or radio control craft
D. Never

Operators must be aware that any attempt to encode messages for general communication purposes is prohibited and could lead to penalties. However, recognizing the exceptions for space stations and radio control craft allows operators to engage in these specialized activities without violating the rules. This balance helps maintain the transparency of amateur radio while accommodating the unique needs of certain advanced operations.

Special Conditions and Exceptions in Amateur Radio Operations

In amateur radio, certain transmissions that are typically prohibited can be authorized under specific conditions. For example, transmitting music using a phone emission is generally not allowed. However, there is an exception when the music is incidental to an authorized retransmission of manned spacecraft communications. This means that if music is part of the audio feed from a manned space mission being retransmitted, transmitting that music over amateur radio frequencies is permissible. This exception helps facilitate educational and public outreach efforts related to space missions.

T1D04: Under what conditions is an amateur station authorized to transmit music using a phone emission?

A. When incidental to an authorized retransmission of manned spacecraft communications
B. When the music produces no spurious emissions
C. When transmissions are limited to less than three minutes per hour
D. When the music is transmitted above 1280 MHz

Another exception involves using amateur radio stations to sell or trade equipment. Amateur radio operators can notify other amateurs of equipment availability for sale or trade, provided these activities are not conducted regularly and are limited to amateur radio equipment. This rule ensures that the amateur bands are not used for commercial purposes, maintaining their primary focus on communication, experimentation, and emergency preparedness.

T1D05: When may amateur radio operators use their stations to notify other amateurs of the availability of equipment for sale or trade?

A. Never
B. When the equipment is not the personal property of either the station licensee, or the control operator, or their close relatives
C. When no profit is made on the sale
D. When selling amateur radio equipment and not on a regular basis

Furthermore, amateur radio operators can receive compensation for operating their stations under specific circumstances. One such situation is when the communication is incidental to classroom instruction at an educational institution. Suppose an amateur radio station is part of a class or educational program. In that case, the control operator can be compensated for their involvement. This exception supports the educational use of amateur radio, helping to teach students about radio communications and technology in a hands-on learning environment.

T1D08: In which of the following circumstances may the control operator of an amateur station receive compensation for operating that station?

A. When the communication is related to the sale of amateur equipment by the control operator's employer
B. When the communication is incidental to classroom instruction at an educational institution
C. When the communication is made to obtain emergency information for a local broadcast station
D. All these choices are correct

By understanding these exceptions, amateur radio operators can more effectively navigate the rules and regulations, ensuring compliance while taking advantage of the unique opportunities that amateur radio offers.

International Communications

FCC-licensed amateur radio stations can engage in international communications under specific conditions. These communications must be incidental to the purposes of the Amateur Radio Service, which primarily include technical experimentation, emergency communication, and fostering international goodwill. Additionally, these communications should be personal, meaning they should not be conducted for commercial purposes, broadcasting, or any form of financial interest.

T1C03: What types of international communications are an FCC-licensed amateur radio station permitted to make?

A. Communications incidental to the purposes of the Amateur Radio Service and remarks of a personal character
B. Communications incidental to conducting business or remarks of a personal nature
C. Only communications incidental to contest exchanges; all other communications are prohibited
D. Any communications that would be permitted by an international broadcast station

The emphasis on personal and incidental communications helps ensure that amateur radio remains a hobby focused on experimentation, learning, and public service. For example, an amateur radio operator might converse with a fellow operator in another country about their equipment setups or discuss the technical aspects of radio wave propagation.

Transmission Locations

An FCC-licensed amateur radio station can transmit from any vessel or craft in international waters, provided the vessel is documented or registered in the United States. This is interesting, but it means that amateur radio operators can operate their stations while on ships, boats, or other maritime vessels as long as these vessels are legally recognized as U.S. entities. This allowance supports maritime communication needs, including safety, emergency communication, and continuing amateur radio activities at sea.

T1C06: From which of the following locations may an FCC-licensed amateur station transmit?

A. From within any country that belongs to the International Telecommunication Union
B. From within any country that is a member of the United Nations
C. From anywhere within International Telecommunication Union (ITU) Regions 2 and 3
D. From any vessel or craft located in international waters and documented or registered in the United States

---

It is important to remember this if you, as an amateur radio operator, find yourself traveling or working on vessels in international waters. This provision ensures amateur radio operators can maintain communication capabilities and comply with international and U.S. regulations. It allows for the continuous operation of amateur radio equipment, promoting safety and spreading technical knowledge even when operators are far from the land.

Restrictions on Third-Party Communications with Foreign Stations

Specific restrictions apply when a non-licensed person can speak to a foreign station using a station under the control of a licensed amateur operator. The fundamental limitation is that the foreign station must be in a country with which the United States has a third-party agreement. A third-party agreement is an international understanding that allows individuals who are not licensed amateur radio operators to communicate with foreign stations under the supervision of a licensed operator.

T1F07: Which of the following restrictions apply when a non-licensed person is allowed to speak to a foreign station using a station under the control of a licensed amateur operator?

A. The person must be a U.S. citizen
B. The foreign station must be in a country with which the U.S. has a third-party agreement
C. The licensed control operator must do the station identification
D. All these choices are correct

This regulation ensures that communications are conducted legally and within international agreements. Licensed operators must know these agreements to avoid unauthorized communications that could lead to regulatory issues. By adhering to this restriction, amateur radio operators help maintain the integrity of the Amateur Radio Service and foster positive international relations. Understanding these guidelines is essential for operators who wish to allow non-licensed individuals to experience and enjoy amateur radio communications legally and responsibly.

Definition of Third-Party Communications

Third-party communications in amateur radio refer to the transmission of messages by a licensed control operator to another amateur station control operator on behalf of a third person who is not a licensed amateur radio operator. It involves relaying messages between two stations where the content of the message originates from someone other than the licensed operators themselves. This practice is regulated to ensure that all parties involved adhere to legal and operational standards set by the FCC and international agreements.

T1F08: What is the definition of third party communications?

A. A message from a control operator to another amateur station control operator on behalf of another person
B. Amateur radio communications where three stations are in communications with one another
C. Operation when the transmitting equipment is licensed to a person other than the control operator
D. Temporary authorization for an unlicensed person to transmit on the amateur bands for technical experiments

Understanding third-party communications is important, especially involving non-licensed individuals. The control operator must ensure that the transmission complies with all relevant regulations, including verifying that the receiving station is in a country with a third-party agreement with the United States. This ensures that the communication is lawful and respects international protocols, maintaining the integrity and proper functioning of the amateur radio service. Familiarity with the concept and regulations surrounding third-party communications helps operators engage in this practice responsibly and within the legal framework.

Call Signs

Once you pass your FCC exam and obtain your Technician Class license, you will automatically be assigned a call sign based on your geographic location and the sequential issuing system.

Maverick, Goose, and Iceman are classic call signs. Unfortunately, ham radio call signs are a little different.

Understanding the structure and use of call signs in amateur radio is akin to learning how to properly address letters in the postal system. Each part of a call sign holds a specific meaning, and using them correctly ensures clear, efficient, and respectful communication across the airwaves. A call sign is your unique identifier as an amateur radio operator, and it serves multiple purposes:

- Legal identification

- A signal of your geographic location

- A reflection of your license class

Call signs in the United States and globally follow a structure set by international agreements and national regulations. Typically, a call sign consists of one or two letters (prefix), a numeral (indicating the geographic area), and one to three additional letters (suffix) assigned sequentially.

The most common type of Technician class call sign is 2 x 3. U.S. call signs start with a letter (K, N, or W) followed by another. Then, a number indicating your region. They end with three letters, which are assigned sequentially (meaning ABC will be assigned before ABD).

For instance, in the U.S., a call sign like KG9XYZ (just an example call sign) indicates that the operator is licensed in the U.S. and in the region represented by the numeral 9, which includes Wisconsin, Illinois, and Indiana.

Call Sign Number Indicating Region

T1C05: Which of the following is a valid Technician class call sign format?

A. KF1XXX
B. KA1X
C. W1XX
D. All these choices are correct

Many operators opt for vanity call signs—these are custom-selected by the operator and can include combinations that are easier to remember or might spell out initials or words relevant to the operator. Obtaining a vanity call sign involves applying through the FCC, and a fee is involved. Still, many find that the personalization is worth the effort. Vanity call signs are available to anyone who has already obtained their ham radio license.

Special event call signs are temporary and used to mark significant events or anniversaries in the amateur radio community. They are often sought after for their unique or commemorative value and help promote the event or celebration within and beyond the amateur radio community. For example, a special event station might use a call sign like W100NASA to mark NASA's 100th anniversary.

T1C02: Who may select a desired call sign under the vanity call sign rules?

A. Only a licensed amateur with a General or Amateur Extra Class license
B. Only a licensed amateur with an Amateur Extra Class license
C. Only a licensed amateur who has been licensed continuously for more than 10 years
D. Any licensed amateur

Using your call sign correctly during communications is not just a legal requirement—it fosters clear and effective exchanges. Whenever you initiate a contact or QSO, your call sign should be clearly stated at the beginning of the communication. This helps identify you with other operators. Additionally, <u>regulations require that you broadcast your call sign at least every ten minutes during and at the end of your communications</u>. This protocol ensures listeners can easily identify who is transmitting, enhancing transparency and accountability in the airwaves.

However, legally, you don't need to state your call sign at the beginning of a conversation. That's just part of being courteous and introducing yourself.

T1F03: When are you required to transmit your assigned call sign?

A. At the beginning of each contact and every 10 minutes thereafter
B. At least once during each transmission
C. At least every 15 minutes during and at the end of a communication
D. At least every 10 minutes during and at the end of a communication

Now, the FCC throws another almost identical question but phrases it using a specific scenario. Just remember, the answer doesn't change!

T1F02: How often must you identify with your FCC-assigned call sign when using tactical call signs such as "ace Headquarters."

A. Never, the tactical call is sufficient
B. Once during every hour
C. At the end of each communication and every ten minutes during a communication
D. At the end of every transmission

Your call sign is more than just a series of letters and numbers—it is your signature across the airwaves, representing your identity and ethics as an operator. Your reputation will precede you!

FCC and Amateur Self-Management

In the amateur radio community, voluntary management plays a central role in maintaining efficient and orderly use of the radio spectrum. This voluntary management cooperates with the Federal Communications Commission (FCC), the official regulatory body overseeing amateur radio operations in the United States. While the FCC establishes the legal framework and regulations for amateur radio, including frequency allocations, licensing requirements, and technical standards, volunteer organizations and individuals within the amateur radio community often manage and coordinate frequency use.

One key aspect of this voluntary management is the role of Frequency Coordinators. Local or regional amateur radio operators typically recognize these coordinators. They are responsible for recommending specific frequency allocations and operational parameters for repeaters, auxiliary stations, and other uses within the amateur bands. Although Frequency Coordinators are not formally appointed by the FCC, their recommendations are generally respected and followed by the amateur radio community to minimize interference and ensure effective spectrum use.

The cooperation between voluntary management and the FCC is based on mutual respect and the goal of maintaining an organized and functional radio environment. The FCC relies on the expertise and local knowledge of Frequency Coordinators and other volunteer organizations to manage the complexities of frequency use and to address issues such as interference and congestion. In turn, the amateur radio community adheres to the guidelines and regulations set forth by the FCC, ensuring that their operations are legal and compliant.

This collaborative approach allows the amateur radio community to self-manage many aspects of its operations while still adhering to the overarching regulatory framework established by the FCC. It leverages amateur radio operators' collective knowledge and experience to maintain efficient and effective spectrum use, ensuring the hobby is enjoyable and accessible for all participants.

Frequency Coordinator

A Frequency Coordinator is an individual or organization responsible for managing and assigning frequencies within amateur radio bands to ensure efficient and orderly spectrum use. This role is vital for coordinating repeater frequencies to prevent interference and ensure repeaters operate effectively within a given area. Frequency Coordinators work closely with local amateur radio clubs and operators to assign frequencies for repeaters, simplex channels, and other communication needs.

The primary goal of a Frequency Coordinator is to minimize conflicts and interference between different users of the amateur radio spectrum. They maintain databases of assigned frequencies and work to resolve any disputes or issues that arise. By coordinating frequencies, these individuals help maintain a harmonious and functional radio environment, allowing operators to enjoy clear and reliable communications. Understanding the role of a Frequency Coordinator is essential for amateur radio operators, especially those involved in setting up or operating repeaters, as it ensures that their equipment operates within the guidelines established to protect all spectrum users.

Local amateur radio operators recognize Volunteer Frequency Coordinators to recommend transmit/receive channels and other parameters for auxiliary and repeater stations. These coordinators analyze the needs of the local amateur community and suggest appropriate frequency allocations and operational parameters for repeaters and auxiliary stations to minimize conflicts and maximize the effective use of available frequencies. By coordinating these recommendations, Volunteer Frequency Coordinators help maintain orderly and reliable communication networks within the amateur radio bands. Their work is essential for preventing frequency conflicts that could disrupt communication and ensuring that repeaters and auxiliary stations operate smoothly.

T1A08: Which of the following entities recommends transmit/receive channels and other parameters for auxiliary and repeater stations?

A. Frequency Spectrum Manager appointed by the FCC
B. Volunteer Frequency Coordinator recognized by local amateurs
C. FCC Regional Field Office
D. International Telecommunication Union

Amateur operators select Frequency Coordinators in a local or regional area whose stations are eligible to be a repeater or auxiliary stations. These operators choose a coordinator based on the individual or organization's ability to manage and assign frequencies within the amateur radio bands. The selection process is often based on the coordinator's experience, knowledge of the local radio environment, and ability to work collaboratively with the amateur radio community. This collaborative selection process ensures that the coordinator has the support and trust of the local amateur radio operators, leading to more effective and harmonious frequency management.

T1A09: Who selects a Frequency Coordinator?

A. The FCC Office of Spectrum Management and Coordination Policy
B. The local chapter of the Office of National Council of Independent Frequency Coordinators
C. Amateur operators in a local or regional area whose stations are eligible to be repeater or auxiliary stations
D. FCC Regional Field Office

Understanding the selection and role of Frequency Coordinators highlights the importance of community involvement in maintaining an organized and functional amateur radio environment. Their work ensures efficient spectrum use, minimizes interference, and promotes reliable communication across the amateur radio community.

Chapter 28

What's Legal for a Technician?

It's important to understand that the Technician license has certain limitations regarding the radio spectrum and power output. Let's review those restrictions and what you will have access to.

Bands

This section will provide an overview of the privileges and opportunities available as a newly licensed Technician, helping you make the most of your amateur radio experience.

Understanding Band Plans

A band plan is a <u>voluntary guideline the amateur radio community uses to organize different modes and activities within an amateur band</u>. While the FCC establishes the legal frequency privileges for various license classes, band plans provide additional structure to ensure that the spectrum is used efficiently and harmoniously. Amateur radio organizations create these plans and reflect the collective agreement of operators on how to allocate the best frequencies for various modes of operation, such as voice, digital, CW (Morse code), and satellite communications.

T2A10: What is a band plan, beyond the privileges established by the FCC?

A. A voluntary guideline for using different modes or activities within an amateur band
B. A list of operating schedules
C. A list of available net frequencies
D. A plan devised by a club to indicate frequency band usage

Band plans help minimize interference and promote a more orderly use of the radio spectrum by designating specific segments for different activities. For example, one part of a band might be recommended for SSB phone operations, another for digital modes, and another for CW. By following these guidelines, amateur radio operators can enjoy better communication experiences and contribute to a more organized and cooperative use of amateur bands.

Band Plan Adherence

Adherence to designated band plans and frequency allocations is not merely a legal requirement but a best practice that all amateur operators are expected to follow. Band plans are developed by the amateur radio community and coordinated through the ARRL (American Radio Relay League). These plans outline which segments of the amateur bands are to be used for different types of communications, such as voice, Morse code, or digital modes.

Following these plans helps prevent interference between users operating different modes of communication within the same band. For instance, if digital modes are relegated to specific frequencies within a band, sticking to these guidelines ensures that your digital signals do not interfere with ongoing voice communications elsewhere in the band. Such structure facilitates more efficient spectrum use and enhances the overall experience by reducing conflicts and maximizing the clarity and reach of communications.

Band Segments

Technicians generally have access to all amateur radio bands above 30 MHz, which are ideal for local communications and some limited long-distance communications under certain atmospheric conditions.

This includes the popular 2-meter (VHF) and 70-centimeter bands (UHF), ideal for local and regional communication. Technicians also have limited privileges on specific HF bands (10-meter).

Here is a cheat sheet or quick reference guide to the band segments you can access with a technician's license: **www.MorseCodePublishing.com/TechExam**

VHF/UHF Band Segments Limited to CW Only

Technician class operators can access various VHF and UHF band segments, some designated exclusively for Continuous Wave (CW) operation.

The specific segments limited to CW only are 50.0 MHz to 50.1 MHz on the 6-meter band and 144.0 MHz to 144.1 MHz on the 2-meter band. These segments are reserved for Morse code communications.

T1B07: Which of the following VHF/UHF band segments are limited to CW only?

A. 50.0 MHz to 50.1 MHz and 144.0 MHz to 144.1 MHz
B. 219 MHz to 220 MHz and 420.0 MHz to 420.1 MHz
C. 902.0 MHz to 902.1 MHz
D. All these choices are correct

Memorization Trick for CW-Only Segments

To help you remember the specific segments limited to CW (Continuous Wave) only, consider these as small ranges at the start of their respective bands.

Here's a mnemonic to assist: "Start Small with CW" on both the 6-meter and 2-meter bands:

- 6-meter band: think of "50 to 50.1" as a tiny segment at the beginning.

- 2-meter band: "144 to 144.1" is the small starting segment.

Just remember, CW starts right at the beginning of these bands!

The allocation of these CW-only segments helps ensure that Morse code enthusiasts have dedicated frequencies where they can practice and communicate without interference from other modes. This setup promotes the preservation and continuation of CW as a skill and mode of communication.

Phone (Voice)

What is meant by a phone sub-band in amateur radio? No, you don't need a phone like you think. It refers to specific portions of the radio frequency spectrum <u>designated for voice communications</u>, also known as "phone" operations. These sub-bands are allocated by regulatory bodies such as the Federal Communications Commission (FCC). They are part of the broader amateur radio bands. Within these phone sub-bands, amateur radio operators can use various voice modulation methods, such as amplitude modulation (AM), frequency modulation (FM), and single sideband (SSB) modulation, to communicate.

The allocation of phone sub-bands ensures that voice communications do not interfere with other transmissions, such as digital modes or Morse code (CW). Each amateur radio band typically has specific sub-bands for different operations, allowing for organized and efficient spectrum use.

Technician class operators have phone (voice) privileges on only one HF band: the <u>10-meter band.</u> Specifically, Technician licensees can operate phone modes from <u>28.300 MHz to 28.500 MHz</u> within this band. This allocation provides an excellent opportunity for new operators to experience HF voice communication and make contacts over long distances, particularly during favorable propagation conditions.

Operating on the 10-meter band allows Technician class operators to enjoy the benefits of HF communication, including the potential for international contacts and participation in various amateur radio activities and contests.

T1B06: On which HF bands does a Technician class operator have phone privileges?

A. None
B. 10 meter band only
C. 80 meter, 40 meter, 15 meter, and 10 meter bands
D. 30 meter band only

And within the 10-meter band...

T1B01: Which of the following frequency ranges are available for phone operation by Technician licensees?

A. 28.050 MHz to 28.150 MHz
B. 28.100 MHz to 28.300 MHz
C. 28.300 MHz to 28.500 MHz
D. 28.500 MHz to 28.600 MHz

This one is tricky. If you glance, all the answers are very close and almost identical. You must memorize this one; remember, it's the 300-500 number you want. Good luck!

Here is what I do. I see this question, and each answer is 28 MHz, so I know we are talking about the '28 club.' Within the '28 club,' only 3 to 5 members are allowed in at once. So, the answer is 28.3 to 28.5! That's what I do.

Identifying Your Station on Phone Sub-Bands

When operating in a phone sub-band, it's essential to correctly identify your station to ensure clear and legal communication. The language used for identification must be <u>English</u>. This requirement helps maintain consistency and clarity across international communications, making it easier for operators worldwide to understand and identify each other accurately.

T1F04: What language may you use for identification when operating in a phone sub-band?

A. Any language recognized by the United Nations
B. Any language recognized by the ITU
C. English
D. English, French, or Spanish

The method of call sign identification required for a station transmitting phone signals is either <u>sending the call sign using a CW (Morse code) or a phone emission (voice).</u> This dual option allows flexibility while ensuring the station identification is clear and meets regulatory standards. Proper identification complies with FCC regulations and promotes good operating practices by ensuring that all operators can easily recognize each other on the airwaves.

T1F05: What method of call sign identification is required for a station transmitting phone signals?

A. Send the call sign followed by the indicator RPT
B. Send the call sign using a CW or phone emission
C. Send the call sign followed by the indicator R
D. Send the call sign using only a phone emission

Additionally, operators may include self-assigned indicators to provide additional information about their location or operating conditions when using phone transmissions. Acceptable self-assigned indicators include formats like "KL7CC <u>stroke</u> W3," "KL7CC <u>slant</u> W3," or "KL7CC <u>slash</u> W3." The words stroke, slant, or slash are just different regional ways to pronounce the symbol shown here: / as in KL7CC/W3

T1F06: Which of the following self-assigned indicators are acceptable when using a phone transmission?

A. KL7CC stroke W3
B. KL7CC slant W3
C. KL7CC slash W3
D. All these choices are correct

SSB Phone (Voice)

Single-sideband (SSB) phone operation is a widely used voice communication mode in amateur radio. It is valued for its efficient use of bandwidth and power. For technician-class operators, SSB phones can be used <u>in at least some segment of all amateur bands above 50 MHz.</u> This includes popular VHF and UHF bands such as the 6-meter band (50-54 MHz), the 2-meter band (144-148 MHz), and the 70-centimeter band (420-450 MHz).

T1B10: Where may SSB phone be used in amateur bands above 50 MHz?

A. Only in sub-bands allocated to General class or higher licensees
B. Only on repeaters
C. In at least some segment of all these bands
D. On any band, if the power is limited to 25 watts

The availability of SSB phone segments in these higher-frequency bands allows for versatile communication options, enabling long-distance contacts under favorable conditions.

Selecting the correct receiver filter bandwidth for Single-Sideband (SSB) reception is essential to achieving the best signal-to-noise ratio. A bandwidth of 2400 Hz (2.4 kHz) is ideal for SSB. This bandwidth is wide enough to pass the entire SSB signal, which typically ranges from about 300 Hz to 3000 Hz, while also narrow enough to exclude most of the noise and interference from adjacent channels.

T4B10: Which of the following receiver filter bandwidths provides the best signal-to-noise ratio for SSB reception?

A. 500 Hz
B. 1000 Hz
C. 2400 Hz
D. 5000 Hz

Using a 2400 Hz filter helps enhance the received signal's clarity and intelligibility. It ensures that the voice signals are not distorted or clipped, providing a clear and crisp audio output. This balance between signal fidelity and noise reduction is pivotal for effective communication, especially in crowded band conditions or weak signals.

Digital

The 219 to 220 MHz segment of the 1.25-meter band is explicitly allocated for fixed digital message forwarding systems only. This means amateur radio operators can use this frequency range exclusively for digital communications involving automated message forwarding between fixed stations. This allocation supports activities such as packet radio networks and other digital communication systems that rely on efficiently transferring data over long distances.

T1B05: How may amateurs use the 219 to 220 MHz segment of 1.25 meter band?

A. Spread spectrum only
B. Fast-scan television only
C. Emergency traffic only
D. Fixed digital message forwarding systems only

By restricting the 219 to 220 MHz segment to fixed digital message forwarding, the FCC ensures that this portion of the spectrum is used effectively for high-speed, reliable digital communications. This helps reduce congestion in other parts of the amateur bands and promotes the development and use of advanced digital communication technologies.

Power Output

As a Technician class licensee, you must know the specific power output restrictions that apply to your operating privileges. While Technician license holders enjoy access to various frequency bands and modes, the FCC has set limits on the maximum power you can use. These restrictions are in place to ensure safe and efficient radio spectrum use and minimize interference with other users.

Power Output for Frequencies Above 30 MHz

Technician class operators are allowed a maximum peak envelope power (PEP) output of 1500 watts on frequencies above 30 MHz. This high power limit allows for robust communications over longer distances, particularly on popular VHF and UHF bands like the 2-meter and 70-centimeter bands. However, it's essential to remember that while 1500 watts is the upper limit, effective communication often requires much less power. Using the minimum power necessary to maintain a reliable connection is a good practice to reduce interference and conserve energy.

T1B12: Except for some specific restrictions, what is the maximum peak envelope power output for Technician class operators using frequencies above 30 MHz?

A. 50 watts
B. 100 watts
C. 500 watts
D. 1500 watts

Power Output for HF Band Segments

For the HF band segments that Technician licensees are permitted to use, the maximum PEP output is <u>limited to 200 watts</u>. The 200-watt limit is designed to provide sufficient power for making contacts over longer distances typical of HF propagation while reducing the risk of interference with other HF users. As with VHF and UHF operations, using the minimum power necessary to make contact effectively is essential.

T1B11: What is the maximum peak envelope power output for Technician class operators in their HF band segments?

A. 200 watts
B. 100 watts
C. 50 watts
D. 10 watts

Understanding the Difference in Power Limits for Technician Licensees

The significant difference in power limits between HF and frequencies above 30 MHz for Technician licensees is primarily due to the different propagation characteristics and potential for interference associated with these frequency bands.

Power Limits for Frequencies Above 30 MHz

Frequencies above 30 MHz, including the VHF and UHF bands, have more predictable and generally shorter-range propagation characteristics. These frequencies are primarily used for local and regional communications, where higher power can help overcome obstacles, terrain, and other local interference to ensure reliable contacts. The 1500-watt power limit allows Technician operators to effectively communicate over greater distances within these bands, especially in challenging environments or during adverse conditions.

Power Limits for HF Band Segments

In contrast, HF bands (below 30 MHz) have different propagation characteristics, often allowing for long-distance (DX) communication due to ionospheric reflection. Signals in the HF bands can travel thousands of miles, bouncing off the ionosphere, especially during favorable propagation conditions. Because HF signals can cover much larger areas, there is a higher potential for interference with other users. The 200-watt power limit for Technician licensees in their HF segments helps minimize this risk. Lower power levels are typically sufficient for making long-distance contacts in the HF bands, and the 200-watt limit strikes a balance between effective communication and reducing the likelihood of causing interference.

Remember to grab the quick reference guide. Again, it's one of those things you print off and keep at your radio. Over time, it will become memorized. For now, refer to the guide!

Chapter 29

Public Service in Ham Radio

PUBLIC SERVICE IS A cornerstone of amateur radio, offering operators a unique opportunity to support their communities during emergencies, public events, and disasters. Ham radio operators provide reliable communication when traditional systems fail or are overloaded. Whether it's coordinating relief efforts during a natural disaster, supporting communication at a marathon, or participating in community events, amateur radio operators are there to assist.

Understanding the rules and regulations that govern public service activities in ham radio is essential. These rules ensure that operators use their skills and equipment responsibly and effectively. In this section, we will explore the various aspects of public service in amateur radio, including the types of events where ham operators can contribute, the specific regulations that apply, and best practices for ensuring efficient and effective communication.

RACES and ARES

RACES (Radio Amateur Civil Emergency Service) and ARES (Amateur Radio Emergency Service) are vital organizations within the amateur radio community that provide vital communication services during emergencies.

RACES (Radio Amateur Civil Emergency Service)

RACES is a service regulated by the Federal Communications Commission (FCC) - a unique amateur radio service regulated under Part 97 of the FCC rules - and operates under the auspices of local, county, and state emergency management agencies. It is designed to provide emergency communication services during civil emergencies, such as natural disasters, terrorist attacks, or other significant incidents. RACES activation is usually coordinated by government emergency management agencies, and its members are trained and prepared to operate under the direction of these agencies during emergencies.

T2C04: What is RACES?

A. An emergency organization combining amateur radio and citizens band operators and frequencies
B. An international radio experimentation society
C. A radio contest held in a short period, sometimes called a "sprint."
D. An FCC part 97 amateur radio service for civil defense communications during national emergencies

---

ARES (Amateur Radio Emergency Service)

ARES is a volunteer-based organization organized by the American Radio Relay League (ARRL). Unlike RACES, ARES is more flexible and can be activated for a broader range of emergency situations, including non-governmental and community events. ARES members provide communication support during disasters, public service events, and other community activities. Membership in ARES is voluntary and open to any licensed amateur radio operator, and training focuses on developing the skills needed to provide effective emergency communications.

T2C06: What is the Amateur Radio Emergency Service (ARES)?

A. A group of licensed amateurs who have voluntarily registered their qualifications and equipment for communications duty in the public service
B. A group of licensed amateurs who are members of the military and who voluntarily agreed to provide message handling services in the case of an emergency
C. A training program that provides licensing courses for those interested in obtaining an amateur license to use during emergencies
D. A training program that certifies amateur operators for membership in the Radio Amateur Civil Emergency Service

Key Differences and Collaboration

- **Regulation and Activation**: RACES operates under government regulation and is activated by governmental agencies. ARES is more community-oriented and can be activated to meet various public service needs.

- **Flexibility**: ARES can be used in a broader range of situations than RACES, which is specifically for civil emergencies.

- **Membership and Training**: Both organizations require training, but RACES is typically more formal and structured due to its government affiliation.

Despite these differences, RACES and ARES often collaborate to provide comprehensive emergency communication services. Understanding the roles and functions of both organizations is key for amateur radio operators who wish to contribute effectively to public service and emergency response efforts.

Here's a simplified way to remember the difference between the two groups: When you see 'ARES,' think of the first 'A' for "Amateur," which suggests a less formal, volunteer-based organization. For 'RACES,' the 'C' stands for "Civil," which reminds you of "Civil service" and indicates that it's a government-run organization. This isn't a perfect rule, but it can be a helpful way to distinguish between the two.

American Radio Relay League (ARRL)

The ARRL (American Radio Relay League) is the national association for amateur radio in the United States, founded in 1914. It advocates for hams' interests, provides educational resources, supports emergency communication through ARES, coordinates licensing and testing, fosters community through events, and advances technical standards. The ARRL plays a role in promoting and supporting amateur radio, ensuring the hobby's growth and vitality.

Their website has many great, and often very technical, radio readings and resources.

Nets in Ham Radio

A "net" in ham radio refers to a scheduled on-air meeting of amateur radio operators, organized for various purposes such as sharing information, coordinating activities, providing emergency communications, or simply socializing. Nets are typically run by a net control station (NCS), which facilitates the orderly flow of communication, ensures that all participants have a chance to speak, and maintains the structure and purpose of the net.

Key Features and Types of Nets

1. **Regularly Scheduled**: Nets usually occur at specific times and frequencies, allowing operators to plan their participation. Schedules for nets can often be found on club websites, newsletters, or through on-air announcements.

2. **Net Control Station (NCS)**: The NCS plays a role in managing the net, <u>calling the net to order, directing and coordinating the order of speakers,</u> and ensuring the net runs smoothly and efficiently. The NCS helps to prevent chaos and overlapping transmissions by controlling who speaks and when.

T2C02: Which of the following are typical duties of a Net Control Station?

A. Choose the regular net meeting time and frequency
B. Ensure that all stations checking into the net are properly licensed for operation on the net frequency
C. Call the net to order and direct communications between stations checking in
D. All these choices are correct

Types of Nets:

- **Traffic Nets**: These nets handle formal message traffic, ensuring that important messages are relayed accurately and efficiently across distances.

- **Emergency Nets**: Activated during emergencies or disasters, these nets provide communication links for coordinating relief efforts, reporting conditions, and supporting emergency services.

- **Social Nets**: These are informal gatherings where operators check in, share news, discuss topics of interest, and enjoy fellowship.

- **Special Interest Nets**: Focused on specific topics, such as technical discussions, DXing (long-distance communications), contesting, or modes like digital or CW (Morse code).

Participating in a Net

When participating in a net, <u>it is standard practice to transmit only when directed by the net control station (NCS) unless you report an emergency</u>. The NCS manages the flow of communication, ensuring that all participants can speak in an orderly manner without causing interference or confusion. By waiting for the NCS to call on you, you help maintain the net's structure and efficiency, allowing for clear and effective communication.

T2C07: Which of the following is standard practice when you participate in a net?

A. When first responding to the net control station, transmit your call sign, name, and address as in the FCC database
B. Record the time of each of your transmissions
C. Unless you are reporting an emergency, transmit only when directed by the net control station
D. All these choices are correct

This practice enables the smooth operation of the net, as it prevents overlapping transmissions and ensures that important messages are conveyed accurately. You should immediately report the situation in an emergency, as emergency traffic always takes priority. Understanding and following this protocol is essential for all amateur radio operators, as it demonstrates respect for the net's organization and enhances the overall communication experience.

Net Operation

In the context of net operation, <u>"traffic" refers to the messages exchanged by net stations</u>. These messages can be routine communications, formal messages, or emergency information that needs to be relayed to other stations. Managing traffic efficiently is essential for maintaining clear and organized communication, especially during emergency operations or coordinated events.

Traffic in net operations is typically categorized and prioritized to ensure that urgent and important messages are handled promptly. Formal traffic handling follows a structured format to ensure accuracy and clarity, making it easier for operators to pass messages accurately without misunderstandings. Understanding the concept of traffic and how it is managed in net operations is essential for all amateur radio operators, as it ensures effective communication and coordination within the network. This knowledge is also part of exam preparation, highlighting the organized nature of amateur radio communications.

T2C05: What does the term "traffic" refer to in net operation?

A. Messages exchanged by net stations
B. The number of stations checking in and out of a net
C. Operation by mobile or portable stations
D. Requests to activate the net by a served agency

Good Traffic Handling

Good traffic handling in amateur radio involves passing messages precisely as received, without alteration or interpretation. This precision is beneficial because even minor changes in the message content can lead to misunderstandings, incorrect actions, or failed communications, especially during emergencies. The integrity of the message must be maintained to ensure that the recipient receives the information as the sender intended.

T2C08: Which of the following is a characteristic of good traffic handling?

A. Passing messages exactly as received
B. Making decisions as to whether messages are worthy of relay or delivery
C. Ensuring that any newsworthy messages are relayed to the news media
D. All these choices are correct

Traffic handling refers to relaying messages between stations in a network, often within organized nets. This can include routine communications, formal messages, and emergency information. Operators must follow standardized procedures and use specific formats to ensure clarity and accuracy. This involves using the standard phonetic alphabet for clarity (spelling out the words), confirming the receipt of each message, and maintaining detailed logs of all traffic handled. By adhering to these practices, amateur radio operators ensure reliable and effective communication, which is key in coordinated operations, especially during emergencies.

T2C03: What technique is used to ensure that voice messages containing unusual words are received correctly?

A. Send the words by voice and Morse code
B. Speak very loudly into the microphone
C. Spell the words using a standard phonetic alphabet
D. All these choices are correct

The Parts of the Message

Understanding the Preamble in Formal Traffic Messages

The preamble of a formal traffic message contains essential information needed to track the message as it moves through various stations. This message section typically includes details such as the message number, the originating station's call sign, the destination, the date and time of origination, and the message's priority level. This information ensures the message can be accurately tracked, identified, and managed as it is relayed from one operator to another.

Including a detailed preamble is vital for effective traffic handling. It allows operators to verify the authenticity and urgency of the message, ensure it reaches the correct destination, and maintain an accurate log of all communications. This structured approach helps prevent errors, provides accountability, and enhances the reliability of the communication process.

T2C10: What information is contained in the preamble of a formal traffic message?

A. The email address of the originating station
B. The address of the intended recipient
C. The telephone number of the addressee
D. Information needed to track the message

Radiogram

A radiogram is a standardized format for sending formal messages in amateur radio communications, particularly within traffic nets. Developed by the National Traffic System (NTS), radiograms facilitate clear and precise message handling. Each radiogram includes a preamble (containing the check, the message number, origin, destination, and priority), the recipient's address, the text of the message, and the sender's signature. This format ensures that messages are transmitted consistently, making them easier to relay accurately across multiple stations. Understanding the components of a radiogram, including the "check," is essential for effective participation in traffic handling and for preparing for the amateur radio exam.

In a radiogram header, "check" <u>refers to the number of words or word equivalents</u> in the message's text portion. This count helps ensure that the message has been received accurately and completely. When the receiving operator verifies the check, they count the words in the message text and compare them to the number in the header. If the numbers match, it confirms that the entire message has been transmitted without omissions or errors.

T2C11: What is meant by "check" in a radiogram header?

A. The number of words or word equivalents in the text portion of the message
B. The call sign of the originating station
C. A list of stations that have relayed the message
D. A box on the message form that indicates that the message was received and/or relayed

---

Rules in Emergencies and Public Service

Effective communication is vital during emergencies and in group settings, and amateur radio operators play a vital role in these scenarios. Whether coordinating disaster relief efforts, supporting public events, or participating in organized nets, following established rules and protocols ensures that communication remains clear, efficient, and effective.

This section will delve into the specific rules that apply during emergencies and when operating within groups. By adhering to these guidelines, amateur radio operators can provide invaluable support during critical times, enhancing safety and operational success.

FCC Rules and Emergency Operations in Amateur Radio

<u>FCC rules always apply to the operation of an amateur station, even during emergencies</u>. This ensures that all communication remains orderly, efficient, and within legal guidelines. Adherence to FCC rules becomes even more important during emergencies as it helps manage the limited communication resources and prevents interference with key operations.

T2C01: When do FCC rules NOT apply to the operation of an amateur station?

A. When operating a RACES station
B. When operating under special FEMA rules
C. When operating under special ARES rules
D. FCC rules always apply

While the primary focus in emergencies is facilitating effective communication, operators must still follow established protocols such as identifying their stations, using appropriate frequencies, and maintaining proper power levels. These regulations ensure that all stations can operate harmoniously and that emergency communications are prioritized without interference.

Operating Outside Frequency Privileges in Emergencies

Under normal circumstances, amateur station control operators must adhere strictly to the frequency privileges of their license class. However, there is an exception to this rule in situations involving the immediate safety of human life or property protection. In such emergencies, operators can operate outside their assigned frequency privileges to facilitate urgent communications necessary to address the crisis.

This exception ensures amateur radio operators can provide communication support when most needed, regardless of their license class restrictions. Despite this allowance, operators must still follow all other FCC rules and regulations, such as identifying their stations and using appropriate communication protocols. Understanding this exception is crucial for both practical emergency response and exam preparation. It highlights the flexibility within the regulatory framework to prioritize human safety and property protection in emergency situations, even though FCC rules always apply.

T2C09: Are amateur station control operators ever permitted to operate outside the frequency privileges of their license class?

A. No
B. Yes, but only when part of a FEMA emergency plan
C. Yes, but only when part of a RACES emergency plan
D. Yes, but only in situations involving the immediate safety of human life or the protection of property

Not confusing at all... is it!

Public service is a cornerstone of amateur radio, allowing operators to support their communities during emergencies, public events, and disasters. Understanding the rules and regulations governing public service activities is essential for responsible and effective communication.

Chapter 30

The Amateur Station and Control

IN THIS CHAPTER, WE will delve into the essentials of station control and the operational aspects of your ham radio. Understanding these fundamentals is necessary for the effective and compliant operation within the amateur radio service. From learning the responsibilities of the control operator to mastering the procedures for safe and efficient station management, this section will provide you with the knowledge needed to control and operate your ham radio station confidently.

Control Operator

A control operator is an individual responsible for properly operating an amateur radio station. This person holds a valid amateur radio license and ensures all transmissions comply with FCC rules and regulations. The control operator must be present and control the station whenever it is transmitting. Their duties include managing the station's frequency usage, adhering to power limits, and ensuring all communications are conducted legally and ethically.

The control operator's role is central to maintaining the integrity of the amateur radio service. By overseeing the station's operation, the control operator helps prevent interference with other communications, ensures the safety of radio operations, and upholds the standards the FCC sets. This responsibility highlights the importance of knowledge and adherence to regulations in the amateur radio community, making the control operator a key figure in any station's operation.

Presumed Control Operator of an Amateur Station

The FCC presumes the station licensee to be the control operator of an amateur station unless there is documentation in the station records indicating otherwise. This means that, by default, the person or entity that holds the license for the amateur radio station is responsible for its operation and ensuring compliance with all FCC regulations. The station licensee is assumed to be in control unless it is clearly documented that another qualified operator has been designated as the control operator for specific periods or operations.

T1E11: Who does the FCC presume to be the control operator of an amateur station unless documentation to the contrary is in the station records?

A. The station custodian
B. The third party participant
C. The person operating the station equipment
D. The station licensee

Designation of the Station Control Operator

The station licensee must designate the station control operator. The station licensee is the individual or entity that holds the license for the amateur radio station and is ultimately responsible for its operations. By designating a control operator, the licensee ensures that a qualified individual can manage the station's transmissions and adhere to FCC regulations.

T1E03: Who must designate the station control operator?

A. The station licensee
B. The FCC
C. The frequency coordinator
D. Any licensed operator

This designation process helps ensure that all operations are conducted legally and ethically, with a responsible person overseeing the use of the radio equipment. It emphasizes the importance of accountability and proper management within the amateur radio community, reinforcing the need for all stations to operate under the supervision of a knowledgeable and licensed control operator.

Control Operator Privileges

Under normal circumstances, a Technician class licensee may never be the control operator of a station operating in an Amateur Extra Class band segment. The FCC regulations strictly limit the operating privileges of amateur radio licensees based on their license class, and each class grants access to specific frequency bands. Technician class licensees are restricted to certain VHF and UHF bands and do not have the authority to operate in the frequency segments reserved for Amateur Extra Class licensees.

T1E06: When, under normal circumstances, may a Technician class licensee be the control operator of a station operating in an Amateur Extra Class band segment?

A. At no time
B. When designated as the control operator by an Amateur Extra Class licensee
C. As part of a multi-operator contest team
D. When using a club station whose trustee holds an Amateur Extra Class license

This restriction ensures that only those who have passed the more advanced licensing exams and demonstrate a deeper understanding of radio theory, regulations, and operating practices can access the broader and more powerful segments of the amateur bands.

Responsibility for Proper Operation

When the control operator is not the station licensee, both the control operator AND the station licensee share the responsibility for the proper station operation. This means that while the control operator is actively managing the station's transmissions, ensuring compliance with FCC regulations and operational standards, the station licensee retains overall responsibility for the station's activities and must ensure that any designated control operators are qualified and compliant with all relevant rules.

T1E07: When the control operator is not the station licensee, who is responsible for the proper operation of the station?

A. All licensed amateurs who are present at the operation
B. Only the station licensee
C. Only the control operator
D. The control operator and the station licensee

This dual responsibility ensures that checks and balances are in place, promoting accountability and adherence to regulations. The control operator must operate within the privileges of their license class and follow all operational guidelines. At the same time, the station licensee must ensure that their station is used appropriately and legally.

Repeaters

When a repeater inadvertently retransmits communications that violate FCC rules, the control operator of the originating station is held accountable. This means the individual who controls the station from which the violating communication originated is responsible for ensuring compliance with FCC regulations. Even though the repeater may automatically retransmit the communication, it is the responsibility of the originating station's control operator to ensure that their transmissions adhere to the rules.

T1F10: Who is accountable if a repeater inadvertently retransmits communications that violate the FCC rules?

A. The control operator of the originating station
B. The control operator of the repeater
C. The owner of the repeater
D. Both the originating station and the repeater owner

This accountability emphasizes the importance of operators being vigilant and always responsible for their transmissions. Operators must monitor their content and avoid prohibited communications, such as obscene language, deliberate interference, or unlicensed operations.

Identifying Your Station During On-The-Air Test Transmissions

When making on-the-air test transmissions, it is essential to identify the transmitting station. This requirement is part of the FCC regulations to ensure that all transmissions can be traced back to a licensed operator, maintaining accountability and order on the amateur radio bands. Proper identification involves stating your call sign at regular intervals during the test transmission, typically every ten minutes and at the end of the transmission.

T2A06: Which of the following is required when making on-the-air test transmissions?

A. Identify the transmitting station
B. Conduct tests only between 10 p.m. and 6 a.m. local time
C. Notify the FCC of the transmissions
D. All these choices are correct

Identifying your station during test transmissions complies with regulatory requirements and helps other operators know who is conducting the tests. This transparency prevents confusion and interference, allowing the amateur radio community to function smoothly.

When Identification is Not Required for Transmissions

Under FCC regulations, an amateur station may transmit without identifying on the air only in specific circumstances, such as when transmitting signals to <u>control model craft</u>. This exemption is designed to facilitate the operation of remote-controlled devices, like model airplanes or boats, which often require rapid and frequent transmission of control signals. In these cases, continuously identifying the station could interfere with the effective control of the model craft.

T1D11: When may an amateur station transmit without identifying on the air?

A. When the transmissions are of a brief nature to make station adjustments
B. When the transmissions are unmodulated
C. When the transmitted power level is below 1 watt
D. When transmitting signals to control model craft

This narrow and specific exception emphasizes the importance of proper station identification in most amateur radio operations, allowing for accountability and ensuring that transmissions comply with FCC rules.

Amateur Station

Control Operator Requirement for Amateur Stations

An amateur station is <u>NOT permitted</u> to transmit without a control operator. According to FCC regulations, a control operator must be present and in charge of the station's operations whenever it is transmitting. The control operator ensures that all transmissions comply with FCC rules and regulations, including frequency use, power limits, and communication content.

T1E01: When may an amateur station transmit without a control operator?

A. When using automatic control, such as in the case of a repeater
B. When the station licensee is away and another licensed amateur is using the station
C. When the transmitting station is an auxiliary station
D. Never

Understanding this requirement is needed for all operators, as it emphasizes the responsibility of operating a station. Transmissions could easily violate regulations without a control operator, leading to potential interference, legal issues, and disruption of the amateur radio community.

Frequency Privileges

The transmitting frequency privileges of an amateur station <u>are determined by the operator license class</u> held by the control operator. The FCC issues different classes of amateur radio licenses, each granting varying levels of access to the radio frequency spectrum.

T1E04: What determines the transmitting frequency privileges of an amateur station?

A. The frequency authorized by the frequency coordinator
B. The frequencies printed on the license grant
C. The highest class of operator license held by anyone on the premises
D. The class of operator license held by the control operator

The control operator's license class dictates which frequencies can be used, ensuring that operators transmit only on frequencies they are authorized to use.

Station and Records for FCC Inspection

The FCC requires that an amateur radio station and its associated records be available for inspection at any time upon request by an FCC representative. This rule ensures that amateur radio operators comply with FCC regulations and promptly demonstrate their adherence to proper operational standards. The records typically include the station log, licenses, and documentation of control operators, which collectively provide evidence of lawful operation and proper station management.

T1F01: When must the station and its records be available for FCC inspection?

A. At any time ten days after notification by the FCC of such an inspection
B. At any time upon request by an FCC representative
C. At any time after written notification by the FCC of such inspection
D. Only when presented with a valid warrant by an FCC official or government agent

Preparing for an FCC inspection means keeping all records accurate, up-to-date, and easily accessible. This readiness demonstrates a commitment to compliance and helps prevent any potential legal issues arising from improper station management.

Amateur Station's Control Point

An amateur station's control point is where the control operator performs their function. This is where the operator can manage and control the station's transmissions, ensuring compliance with FCC regulations. The control point is critical because it is where the operator can monitor and adjust the station's operations in real-time, maintaining proper frequency use, power levels, and adherence to communication protocols.

T1E05: What is an amateur station's control point?

A. The location of the station's transmitting antenna
B. The location of the station's transmitting apparatus
C. The location at which the control operator function is performed
D. The mailing address of the station licensee

Knowing the concept of a control point is essential for amateur radio operators because it defines the operational center of their station. Whether the control point is a physical location, like a designated radio room, or a remote setup managed via computer software, it is the hub where the operator exercises authority and responsibility. This setup ensures that every transmission is supervised and that the station operates within the legal and technical parameters established by the FCC.

Automatic Retransmission of Signals

Certain amateur stations, specifically repeater, auxiliary, and space stations, are permitted to retransmit the signals of other amateur stations automatically. Repeaters are commonly used to extend the range of communications, receiving a signal on one frequency and retransmitting it on another, typically with higher power and from a better location. This allows operators to communicate over greater distances than possible with direct point-to-point communication.

T1D07: What types of amateur stations can automatically retransmit the signals of other amateur stations?

A. Auxiliary, beacon, or Earth stations
B. Earth, repeater, or space stations
C. Beacon, repeater, or space stations
D. Repeater, auxiliary, or space stations

The key word in this question is automatically.

Auxiliary stations are similar but are used to support the primary operations of a station, often facilitating control or linking between repeaters. Space stations like those on satellites can also automatically retransmit signals, allowing for global communication coverage. Understanding these capabilities is key as it helps utilize the full potential of the amateur radio infrastructure, enhancing communication possibilities and operational flexibility. These systems play a vital role in expanding the reach and effectiveness of amateur radio networks, making it possible to maintain reliable communication in various scenarios.

Automatic Control

Automatic control in amateur radio refers to the operation of a station without the need for continuous real-time human intervention. A prime example of this is repeater operation. Repeaters are designed to receive signals on one frequency and automatically retransmit them on another, allowing for extended communication ranges. This process is entirely automated, meaning the repeater can operate independently, continually receiving and retransmitting signals without a control operator always being present.

T1E08: Which of the following is an example of automatic control?

A. Repeater operation
B. Controlling a station over the internet
C. Using a computer or other device to send CW automatically
D. Using a computer or other device to identify automatically

Remote Control Operation

Remote control operation in amateur radio involves managing a station from a location other than where the radio equipment is physically situated. Specific requirements must be met for this type of operation to ensure compliance with FCC regulations. First, a control operator must be at the control point where the station is remotely operated. This ensures that a qualified individual constantly oversees the station's transmissions and adheres to all relevant rules and guidelines.

Additionally, a control operator is always required during remote control operations. This constant oversight is there to maintain the integrity and proper functioning of the amateur radio station. The control operator must indirectly manipulate the station's controls, typically software or other remote-control technologies, to manage frequency changes, power levels, and other operational parameters. Understanding these requirements helps amateur radio operators ensure that their remote control operations are legal, safe, and effective, maintaining the high standards of the amateur radio community.

T1E09: Which of the following are required for remote control operation?

A. The control operator must be at the control point
B. A control operator is required at all times
C. The control operator must indirectly manipulate the controls
D. All these choices are correct

Example of Remote Control in Amateur Radio

As defined in Part 97 of the FCC regulations, remote control in amateur radio involves operating a station from a distance without being physically present at the station's location. <u>A typical example of this is operating the station over the internet.</u> This setup allows a licensed amateur radio operator to control their equipment from virtually anywhere, using computer software and internet connectivity to adjust frequencies, power levels, and other operational settings.

T1E10: Which of the following is an example of remote control as defined in Part 97?

A. Repeater operation
B. Operating the station over the internet
C. Controlling a model aircraft, boat, or car by amateur radio
D. All these choices are correct

This is beneficial for operators who need to manage their stations while away from home or for those who want to remotely take advantage of better propagation conditions or more favorable locations. Remote control helps operators utilize modern technology to enhance their radio operations, ensuring they remain compliant with FCC rules while enjoying the flexibility and convenience that remote control offers. It also underscores the importance of maintaining proper control and supervision, as the control operator must still ensure all transmissions adhere to regulatory standards.

Beacons

According to FCC Part 97, <u>a beacon is defined as an amateur station transmitting communications for the purpose of observing propagation or conducting related experimental activities.</u> Beacons are tools in the amateur radio community, as they provide valuable information about the conditions of the ionosphere and other propagation mediums. By continuously transmitting a signal, beacons allow operators to monitor the strength and stability of signals over various frequencies and distances, helping to predict the best times and frequencies for effective communication.

T1A06: What is the FCC Part 97 definition of a beacon?

A. A government transmitter marking the amateur radio band edges
B. A bulletin sent by the FCC to announce a national emergency
C. A continuous transmission of weather information authorized in the amateur bands by the National Weather Service
D. An amateur station transmitting communications for the purposes of observing propagation or related experimental activities

---

Beacons serve an essential function in propagation studies and experimentation. By analyzing beacon signals, operators can gain insights into the current state of the radio bands and make informed decisions about when and how to conduct their transmissions.

Expanding Your Ham Experience

EMBARKING ON THE PATH of amateur radio is akin to setting out on a vast sea with endless horizons to explore. Each new skill you acquire and every piece of knowledge you gain is like catching a favorable wind, pushing you further along this exciting voyage, looking for more adventures.

After you get licensed, many opportunities open up to you. Let's examine a few of them.

Chapter 31

Hams Having Fun

Contests

HAM RADIO CONTESTS, OFTEN known as "radiosport," provide an exhilarating opportunity for operators of all skill levels to test their abilities in a structured competitive environment. For you as a newly licensed Technician, participating in these contests can significantly enhance your operating skills, broaden your understanding of radio propagation, and deepen your engagement with the regional amateur radio community.

Contests vary widely, ranging from local VHF competitions to worldwide DX contests. Each contest has its own set of rules, objectives, and modes of operation, which can include making as many contacts as possible within a certain period or contacting as many geographic areas as possible.

T8C03: What operating activity involves contacting as many stations as possible during a specified period?

A. Simulated emergency exercises
B. Net operations
C. Public service events
D. Contesting

To begin your adventure in radiosport, the first step is finding contests that align with your license privileges and interests. Many contests are designed to include newcomers and may have categories or power restrictions ensuring a level playing field.

The American Radio Relay League (ARRL) website is a tremendous resource, listing various upcoming contests with detailed descriptions of their rules and eligibility criteria. Platforms like ContestCalendar.com aggregate information about ham radio contests worldwide, providing a comprehensive schedule to help you plan your participation months in advance.

When choosing a contest, consider the required bands, modes, and duration to ensure it matches your available equipment and personal schedule. Some contests, like those dedicated to digital modes, also need specific software for logging contacts or participation.

Understanding and adhering to contest etiquette ensures fair play and fosters participant camaraderie. One fundamental aspect of contest etiquette is precise and efficient communication. Unlike regular QSOs, contest exchanges are typically brief, focusing on the essential information required by the contest rules, such as signal report, call sign, and location. Maintaining a brisk but polite pace helps keep the contest moving and allows more participants to make contacts.

T8C04: Which of the following is good procedure when contacting another station in a contest?

A. Sign only the last two letters of your call if there are many other stations calling
B. Contact the station twice to be sure that you are in his log
C. Send only the minimum information needed for proper identification and the contest exchange
D. All these choices are correct

Respecting the frequency rights of others is also essential; if you find a frequency in use, it's courteous to move to another one rather than causing interference. Additionally, self-spotting, or announcing your frequency on spotting networks to attract contacts, is generally frowned upon unless explicitly allowed by the contest rules.

Logging during contests is more than just a formality—it's a core part of the strategy and integrity of radiosport. Each contact made during the contest must be meticulously logged, with details including the time of contact, the call sign of the other station, and the exchange information. Accuracy is paramount, as errors can lead to penalties or disqualification from the contest results. Many operators use specialized logging software that helps record contacts and check for duplicate contacts, which are usually not allowed. These programs can also provide real-time scores and projections to help you adjust your strategy.

Grid Locators

A grid locator is a letter-number designator assigned to a specific geographic location, used primarily in amateur radio to identify the exact position of a station. This system, known as the Maidenhead Locator System, divides the world into a grid of squares, each represented by a unique combination of letters and numbers. For example, a grid locator might look like "FN31pr," where "FN" identifies a larger geographic area, "31" pinpoints a more specific region within that area, and "pr" provides an even more precise location.

Grid locators are essential for several reasons. They allow amateur radio operators to accurately log their contacts while participating in contests.

Understanding that a grid locator is a geographic designator consisting of letters and numbers helps operators easily communicate their location, making it a crucial tool in the ham radio community.

T8C05: What is a grid locator?

A. A letter-number designator assigned to a geographic location
B. A letter-number designator assigned to an azimuth and elevation
C. An instrument for neutralizing a final amplifier
D. An instrument for radio direction finding

After the contest, submitting your log is the final step in your participation. Each contest has specific rules about how and when logs should be submitted, often through a web upload interface or email. Timely submission is essential, as most contests have strict deadlines for log entries, typically within a few days to a week after the event concludes. Once submitted, your log will be checked against the logs of other participants, a process that helps ensure the accuracy and fairness of the scoring. After all logs are evaluated, results are published, detailing scores and standings. These results can provide valuable insights into your performance and areas for improvement.

Partaking in contests and striving for awards brings excitement and recognition and significantly contributes to your growth as an amateur radio operator. Each contest challenges you to apply your skills, adapt to changing conditions, and interact with a diverse array of operators, each of which enriches your experience and builds your proficiency in the hobby. As you continue to participate, you'll likely find that the thrill of competition, the joy of achievement, and the camaraderie of the community become enduring motivators in your ongoing journey in amateur radio.

Hidden Transmitter Hunts: Fox Hunts

A fox hunt, also known as a hidden transmitter hunt, is a popular activity in the amateur radio community. Participants use radio direction-finding techniques to locate a hidden transmitter called the "fox." Equipped with directional antennas and radio equipment, hunters search for the fox by following the strongest signal directions. The goal is to find the exact location of the hidden transmitter as quickly as possible. Fox hunts are fun and competitive, help improve radio direction-finding skills, and promote fellowship among radio enthusiasts. This engaging activity demonstrates the practical applications of radio technology and provides valuable experience in signal tracking and problem-solving.

A directional antenna is an indispensable tool for fox hunting. This type of antenna focuses on the reception of signals in a specific direction, making it easier to determine the direction from which a signal is coming. By rotating the directional antenna and observing where the signal strength is strongest, hunters can pinpoint the location of the hidden transmitter.

T8C02: Which of these items would be useful for a hidden transmitter hunt?

A. Calibrated SWR meter
B. A directional antenna
C. A calibrated noise bridge
D. All these choices are correct

Using a directional antenna simplifies the process of locating the hidden transmitter by providing a clear indication of direction, which is much more efficient than using a non-directional antenna that receives signals equally from all directions.

This technique is essential for hidden transmitter hunts, allowing participants to quickly and accurately find the transmitter. In addition, because a directional antenna helps identify the direction of signals, it can help us locate sources of interference or jamming.

Radio direction finding is a method for locating noise interference or jamming sources by determining the direction from which a radio signal originates. This technique uses directional antennas that can focus on signals from specific directions. By rotating the antenna and noting the direction where the signal is strongest, operators can triangulate and pinpoint the source of the interference.

T8C01: Which of the following methods is used to locate sources of noise interference or jamming?

A. Echolocation
B. Doppler radar
C. Radio direction finding
D. Phase locking

Welcome to the excitement known as ham contests and fox hunts, where you can test your skills, challenge yourself, and enjoy friendly competition. Whether you're chasing signals or hunting hidden transmitters, these activities offer a fun and engaging way to hone your radio expertise and connect with fellow enthusiasts.

Chapter 32

Upgrading Your License: From Technician to General

As you begin to feel more comfortable with your Technician License, you may look towards broader seas, yearning to delve deeper and reach further. This is where upgrading your license from Technician to General comes into play, opening up a world of enhanced frequencies and international communication opportunities.

Benefits of Upgrading

Upgrading to a General class license is not just a step up—it's a gateway to a new realm of possibilities in the amateur radio community. This upgrade allows you to access additional frequency bands below 30 MHz, primarily for high-frequency (HF) communications. These bands are prized for their ability to communicate internationally, harnessing the ionosphere to bounce radio waves across oceans and continents. This capability expands your operational reach and immerses you in a global network of amateur radio operators. Engaging with this community can enrich your experience with diverse cultural exchanges and deepen your understanding of international communication dynamics from the comfort of your radio station.

Moreover, the General class license opens up more bandwidth in the HF bands, allowing you to experiment with a broader range of modes and activities, including contesting and emergency communications on a global scale. These activities enhance your technical skills and contribute to your personal growth within the hobby, offering new challenges and the satisfaction of overcoming them.

Keeping the Hobby Alive

NOW THAT YOU HAVE everything you need to pass your Technician Class License exam and dive into the exciting world of ham radio, it's time to share your newfound knowledge and help others on their journey.

Would you help someone you've never met, even if you never got credit for it?

That person might be a lot like you were when you first picked up this book—curious about ham radio, eager to learn, and looking for guidance. Your honest review can be the light that guides them to success.

At Morse Code Publishing, we aim to make ham radio accessible to everyone. But to reach more people and make a bigger impact, we need your help.

Most people judge a book by its cover—and its reviews. You can make a difference by taking just 60 seconds to leave a review. Your review could help:

- one more beginner discover amateur radio

- one more student pass their Technician exam

Your review won't cost you anything but could change someone's life.

Thank you for your support. Ham radio thrives when we pass on our knowledge—and with your help, we can keep the passion alive and growing.

Welcome to the club! Your biggest fan,

Morse Code Publishing

PS – When you help someone else, you become part of their success story. If you believe this book will help others, consider sharing it with them. Let's keep the goodwill going!

Conclusion

As we turn the final page of this journey together, take a moment to reflect on the incredible path you've navigated—from a curious beginner to a proud holder of the Technician Class license (or soon-to-be license holder) poised at the threshold of the vast and vibrant hobby of ham radio. Throughout this book, we have traversed the essentials of ham radio operations, tackled the intricacies of FCC regulations, and mastered the knowledge necessary to pass the Technician Class exam. But more importantly, you've taken your first steps into a global community rich with continuous learning, friendship, and service opportunities.

Remember, the license you've worked so hard to acquire is much more than a license - it's your ticket to endless exploration and innovation in radio communications. The real adventure begins now, with each frequency you tune into and every connection you make expanding your horizons.

Connect with local clubs, participate in online forums, and attend hamfests. These platforms are not just avenues for deepening your technical knowledge—they are gateways to forging lasting friendships and contributing to a hobby that thrives on active participation and mutual support.

Ham radio offers unparalleled opportunities for personal growth and public service. Whether it's experimenting with digital modes, communicating through satellites, or providing emergency comms, you have the power to make a significant impact.

Now, I call on you to share your journey. Inspire others by recounting your experience obtaining your Technician Class license. Whether through social media, a blog, or casual conversations within your community, share your story of successfully obtaining your license.

Looking ahead, consider the next challenge. The General and Extra Class licenses await, opening new frequencies and possibilities.

Thank you for allowing us to participate in your entry into the ham radio world. It's been a privilege to guide and accompany you on this initial leg of what I hope will be a long and fulfilling adventure. Continue to push the boundaries of what you can achieve, and remember that each frequency tuned and each call made is a note in the ongoing symphony of global communication.

Finally, view your Technician license as the beginning of a continuous learning path in amateur radio. Continuous learning will expand your operational capabilities and deepen your enjoyment and engagement with the hobby. Participate in forums, join clubs, attend ham fests, and perhaps most importantly, stay curious. Amateur radio is vast and varied, and there is always something new to discover.

Welcome to the community—we can't wait to hear your call sign on the air.

- 73

Jared Johnson, KF0RTU @ Morse Code Publishing

Exam Question Index

Glossary

- **Absorbed:** When a radio wave loses energy as it passes through a medium, such as a building or the atmosphere, reducing signal strength.

- **ALARA:** An acronym for "As Low As Reasonably Achievable," a safety principle in ham radio to minimize exposure to radio frequency energy, especially in antenna placement and power levels.

- **Alternating Current (AC):** Electrical current that flows first in one direction in a wire and then in the other. The applied voltage is also changing polarity. This direction reversal continues at a rate that depends on the frequency of the AC.

- **Amateur Operator:** A person holding a written authorization to be the control operator of an amateur station.

- **Amateur Service:** A radiocommunication service for the purpose of self-training, intercommunication, and technical investigations carried out by amateurs, that is, duly authorized persons interested in radio technique solely with a personal aim and without pecuniary interest.

- **Amateur Station:** A station licensed in the amateur service, including necessary equipment, used for amateur communication.

- **Ammeter:** A test instrument that measures current.

- **Ampere (A):** The basic unit of electrical current. Current is a measure of the electron flow through a circuit.

- **Amplitude Modulation (AM):** A modulation technique that encodes information by varying the amplitude of the radio carrier signal. Used in AM broadcast radio, it is more susceptible to noise than FM.

- **Analog:** Representation of signals as continuous waveforms rather than discrete digital data. Analog voice transmission is common in handheld radios.

- **Antenna:** A device that picks up or sends out radio frequency energy.

- **Antenna Gain:** The measure of how well an antenna directs or concentrates radio frequency energy in a specific direction compared to an isotropic radiator. Higher gain indicates a more focused signal, often measured in dBi (decibels relative to an isotropic antenna).

- **Antenna Switch:** A switch used to connect one transmitter, receiver, or transceiver to several different antennas.

- **Antenna Tuner:** A device used to match the impedance of an antenna to the transmitter or receiver, ensuring efficient power transfer and reducing signal reflection. It allows the use of antennas that might not be naturally resonant at the operating frequency.

- **APRS (Automatic Packet Reporting System):** A digital communication protocol used in amateur radio to transmit real-time data such as location, weather information, and messages.

- **Attenuation:** The decrease in signal strength as a radio wave travels through space or materials. It can be caused by distance, physical obstructions, and atmospheric conditions.

- **Autopatch:** A device that allows repeater users to make telephone calls through a repeater.

- **Automatic Gain Control (AGC):** A feature in receivers that automatically adjusts the gain to maintain a constant output level, even when the received signal strength varies.

- **Balun:** Contraction for balanced to unbalanced. A device to couple a balanced load to an unbalanced source, or vice versa.

- **Band Plans:** Agreed-upon allocations of frequency ranges within the amateur radio bands, designating specific uses like voice, CW, or digital modes to minimize interference.

- **Band Spread:** A receiver quality used to describe how far apart stations on different nearby frequencies will seem to be. Band spread determines how easily signals can be tuned.

- **Bandwidth:** The range of frequencies a radio can receive or transmit on. Wider bandwidth allows receiving multiple signals simultaneously.

- **Base Station:** A fixed radio installation, typically at a home or command center, used for effective long-range communication. It often includes a more powerful radio unit and larger antenna systems than portable units.

- **Battery:** A device that converts chemical energy into electrical energy.

- **Beacon:** A continuous or periodic signal transmitted by a station to indicate its presence and location, often used in propagation studies to assess the conditions of a particular band.

- **Beam Antenna:** A directional antenna. A beam antenna must be rotated to provide coverage in different directions.

- **Bluetooth:** A wireless technology standard for exchanging data over short distances, used in radios for audio accessories, data transfer, and programming.

- **Break:** A term used during voice communication to indicate a pause or request to interrupt ongoing communication, often to pass an urgent message.

- **Call Sign:** A unique identifier assigned to a radio operator by their country's telecommunications authority. It's used to legally identify the operator or station during transmissions.

- **Carrier Squelch:** A radio receiver function that mutes audio unless a signal is detected, preventing static noise. It activates with an incoming transmission's carrier wave, filtering out background noise.

- **Channel:** A labeled radio memory location containing preset receive and transmit frequencies for convenient recall, allowing quick switching between saved frequencies.

- **Chirp:** A slight shift in transmitter frequency each time you key the transmitter.

- **Closed Repeater:** A repeater that restricts access to those who know a special code.

- **Coaxial Cable:** A type of feed line with one conductor inside the other.

- **CQ:** A general call to all stations, indicating that the operator is available for communication. For example, "CQ CQ CQ, this is K1ABC calling CQ."

- **CTCSS (Continuous Tone-Coded Squelch System):** A subaudible tone transmitted along with voice signals to prevent interference from other users on the same frequency.

- **CW (Continuous Wave):** Morse code telegraphy.

- **DCS (Digital Code Squelch):** Digitally encoded subaudible codes transmitted with your audio to block unwanted receptions unless the code matches.

- **Decibel (dB):** A logarithmic unit used to measure the relative difference in power, often used in describing gains, losses, and signal strength in radio systems.

- **Delta Loop Antenna:** A variation of the cubical quad with triangular elements.

- **Digital:** Representation of signals as discrete binary data rather than continuous waveforms. Digital modes convert voice and data to binary for transmission.

- **Digipeater:** A packet-radio station used to retransmit signals that are specifically addressed to be retransmitted by that station.

- **Dipole Antenna:** A simple and widely used antenna consisting of two equal-length conductive elements that radiate radio waves.

- **Directional Antenna:** An antenna that concentrates the signal in one direction.

- **Duplex:** Using two distinct frequencies for transmitting and receiving, allowing simultaneous sending and receiving.

- **Duplexer:** A device that allows simultaneous transmission and reception on different frequencies using the same antenna. It's commonly used in repeaters.

- **Dummy Load:** A station accessory that allows you to test or adjust transmitting equipment without sending a signal out over the air. Also called dummy antenna.

- **Dual-Band Radio:** A radio that can operate on two distinct frequency bands, like VHF and UHF, allowing increased versatility.

- **Dual Watch:** Monitoring two different frequencies or channels simultaneously by rapidly switching between them.

- **DXing:** The hobby of receiving or sending long-distance communications. "DX" is shorthand for "distance" or "distant."

- **Earth Ground:** A circuit connection to a ground rod driven into the Earth or to a cold-water pipe made of copper that goes into the ground.

- **Earth-Moon-Earth (EME):** A mode of communication that involves bouncing radio signals off the moon's surface to reach distant stations. Also known as "moonbounce."

- **Electromagnetic Interference (EMI):** Unwanted noise or signals that interfere with the operation of electronic devices, often caused by other electrical equipment or power lines.

- **Elmer:** An informal term in the ham radio community for an experienced ham who mentors or guides newcomers.

- **Emergency:** A situation where there is a danger to lives or property.

- **Emergency Traffic:** Messages with life and death urgency or requests for medical help and supplies that leave an area shortly after an emergency.

- **Emission:** The transmitted signal from an amateur station.

- **Feedline:** The cable or transmission line that connects the radio to the antenna, crucial for efficiently transferring power. Types include coaxial cable and ladder line.

- **Ferrite Beads:** Small components made of ferrite material, used to suppress high-frequency noise in electronic circuits by absorbing unwanted signals.

- **Final:** The last transmission in a contact before signing off.

- **FM Repeater:** A device that receives an FM signal on one frequency, amplifies it, and retransmits it on another frequency, typically with an offset. Repeaters extend the communication range of handheld and mobile radios.

- **Fox Hunt:** An amateur radio activity where participants use radio direction-finding techniques to locate a transmitter at an unknown location.

- **Frequency:** The number of cycles a radio wave completes per second, measured in Hertz (Hz).

- **Frequency Modulation (FM):** A modulation technique that encodes information by varying the frequency of the radio carrier signal. Used in two-way handheld radios for voice communication.

- **Gain:** The increase in signal strength or amplification provided by an antenna or amplifier. Gain is usually measured in decibels (dB) and can enhance the range and clarity of communication.

- **Ground Loop:** An unwanted electrical current that flows in a looped grounding system, often causing noise and interference in radio equipment.

- **Ground Wave:** A type of radio wave that travels along the Earth's surface, typically used for short to medium-range communication, especially at lower frequencies.

- **Grounding:** The practice of connecting equipment to the Earth or a common ground point to reduce electrical noise, enhance safety, and improve signal quality.

- **Ham Radio (Amateur Radio):** A hobby involving radio frequencies for non-commercial communication, experimentation, and emergency services. It requires licensing.

- **High Gain Antenna:** An antenna that amplifies signal strength in a specific direction, enhancing long-distance communication.

- **Hertz (Hz):** The unit of frequency, equal to one cycle per second.

- **HT (Handheld Transceiver):** A portable, handheld radio used for two-way communication. HTs are popular among new ham radio operators for their portability and ease of use.

- **Impedance:** The resistance to the flow of alternating current (AC) in a circuit, combining resistance and reactance.

- **Intermediate Frequency (IF):** A lower frequency to which a received signal is shifted to simplify processing and amplification in a receiver.

- **IP Rating (Ingress Protection):** A standard that defines the levels of protection against intrusion from foreign bodies and moisture in electrical enclosures.

- **IP67 Rating:** Indicates that a device is entirely dust-tight and can be submerged in water up to 1 meter deep for 30 minutes without damage.

- **Ionosphere:** A region of electrically charged (ionized) gases high in the atmosphere that bends radio waves as they travel through it, returning them to Earth.

- **Isotropic Radiator:** A theoretical antenna that radiates equally in all directions, used as a reference point for measuring antenna gain.

- **ITU (International Telecommunication Union):** A specialized agency of the United Nations that coordinates global telecommunications standards, including frequency allocations for amateur radio.

- **J-Pole Antenna:** An omnidirectional antenna, typically for VHF/UHF bands, characterized by a distinctive 'J' shape. It's popular for its simplicity and low SWR.

- **Li-ion (Lithium-ion):** A battery chemistry commonly used in modern handheld radios. Offers good energy density and requires protection circuitry for safe charging.

- **Linking (Repeater Linking):** Connecting multiple repeaters via the internet or RF to create a larger coverage area.

- **Mobile Radio:** A radio designed for vehicle use, providing better performance than handhelds but less than base stations.

- **Mode:** The type of signal modulation used in radio communication, such as AM, FM, SSB (Single Sideband), CW (Continuous Wave), or digital modes like PSK31 or RTTY.

- **MR (Memory Recall):** A mode that allows users to access and operate on frequencies previously programmed and stored in the radio's memory channels.

- **Net Control:** The designated station in an amateur radio net that manages communications and directs net operations.

- **NiMH (Nickel-Metal Hydride):** A battery chemistry often used in older handheld radios, more susceptible to performance degradation over time than Li-ion.

- **NVIS (Near Vertical Incidence Skywave):** A propagation method where radio signals are directed nearly vertically to reflect off the ionosphere and return to Earth over short to medium distances.

- **Offset:** The difference between the receive and transmit frequencies required for repeaters. It allows simultaneous reception and transmission.

- **Oscillator:** An electronic circuit that produces a continuous, oscillating electrical signal, often used as a frequency reference in transmitters and receivers.

- **Packet Radio:** A digital mode of communication that transmits data in packets using amateur radio frequencies.

- **Parity:** A method of adding a simple checksum to transmitted data bits to detect errors in received data packets, used in digital transmission schemes.

- **PL Tone (Private Line):** A pilot tone squelch system using subaudible tones similar to CTCSS for selectively receiving transmissions encoded with a matching tone.

- **PL-259 Connector:** A type of coaxial connector commonly used in ham radio for connecting coaxial cable to radios, antennas, and other equipment.

- **Power Supply:** A device that provides electrical power to a radio, typically converting AC from a wall outlet to the DC required by the radio.

- **Propagation:** The way radio waves travel through the atmosphere and environment between transmitting and receiving antennas, influenced by factors like frequency, distance, terrain, and atmospheric conditions.

- **PSK31 (Phase Shift Keying, 31 Baud):** A popular digital mode used in ham radio for low-power communications, known for its narrow bandwidth and efficiency in poor conditions.

- **PTT (Push To Talk):** A button that initiates voice transmissions in half-duplex radio systems. The transceiver switches to transmit mode when the operator presses PTT.

- **Q Codes:** A standardized collection of three-letter message encodings, used in Morse code and voice communication to convey common phrases efficiently.

- **QRM:** A Q code used to describe interference from other stations. For example, "I am receiving QRM" means there is interference affecting your reception.

- **QRN:** A Q code used to describe natural interference, such as static from thunderstorms.

- **QRP Operation:** Transmitting with low power, often less than 5 watts, in ham radio. It's a challenge to make distant contacts with minimal power output.

- **QSL Card:** A written confirmation of two-way radio communication between two amateur radio stations or a one-way reception of a signal.

- **QSO:** A radio communication or conversation between two ham radio operators.

- **Rag Chew:** An informal, often lengthy conversation via radio on non-technical subjects.

- **Radiosonde:** A balloon-borne device that measures various atmospheric parameters and transmits the data back to a ground receiver via radio.

- **RDIS (Radio Data Information Service):** Digital data streams encoded alongside analog FM voice transmissions to convey identifying and messaging information.

- **Receiver:** A device that picks up radio signals from an antenna, processes them to extract the desired information, and converts them to a form that can be heard or read.

- **Reflected:** Describes a radio wave that bounces off a surface, such as the ground or an object, affecting signal strength and quality.

- **Refracted:** Describes a radio wave that bends as it passes through different layers of the atmosphere or materials, affecting the signal path.

- **Repeater:** A device that receives radio signals on one frequency and retransmits them at higher power on another frequency to extend range.

- **Repeater Directory:** A listing or database of repeater frequencies and information, often used by ham radio operators to find local repeaters.

- **Repeater Offset:** The difference between the repeater's receive frequency (input) and transmit frequency (output). Allows simultaneous reception and retransmission without interference between the frequencies.

- **RF (Radio Frequency):** The range of electromagnetic waves used in wireless communication, oscillating between 20 kHz and 300 GHz, enabling wireless communication.

- **RF Gain:** Adjusts the radio's receiver sensitivity to incoming signals, allowing for better signal reception.

- **RFA (Radio Frequency Adapter):** A device that converts between different radio bands or frequency ranges, allowing interoperability between diverse radio types.

- **RF Interference (RFI):** Unwanted reception of RF signals that disrupt normal radio operations, a common issue in densely populated areas.

- **RSSI (Received Signal Strength Indicator):** A measurement of the strength or power level of received radio signals, displayed by the radio as a signal meter.

- **S-Meter:** A meter on a receiver that measures the strength of the received signal, usually calibrated in S-units.

- **Silent Key (SK):** A term used to refer to a deceased ham radio operator.

- **Simplex:** Communication where signals are transmitted and received on the same frequency channel but not simultaneously.

- **Skip:** A phenomenon where radio waves reflect off the ionosphere, allowing long-distance communication beyond the normal line-of-sight range.

- **Sky Wave:** A radio wave that is refracted or reflected back to Earth by the ionosphere, allowing long-distance communication beyond the horizon.

- **SMA (SubMiniature version A):** A type of RF coaxial connector used in various applications, including antennas, radios, and telecommunications equipment.

- **Squelch:** Mutes the radio speaker when no transmission is received to reduce background noise. Adjusting the squelch threshold filters out weak or irrelevant signals.

- **Spurious Emissions:** Unwanted signals generated by a transmitter that are outside the intended transmission band.

- **Standing Wave:** A stationary wave pattern formed on a transmission line when there is a mismatch between the load and the transmission line, often indicated by a high SWR.

- **Stealth Antenna:** A type of antenna designed to be unobtrusive or hidden, often used by amateur radio operators in areas with restrictions on visible antennas.

- **SSB (Single Sideband):** A mode of amplitude modulation that eliminates the carrier and one of the sidebands, making it more efficient for voice communication, especially on HF bands.

- **SWR (Standing Wave Ratio):** A measure of the radio antenna system's efficiency in transmitting power. A lower SWR indicates better efficiency.

- **SWR Meter:** A device used to measure the Standing Wave Ratio (SWR) of an antenna system. It helps ensure the antenna is properly tuned and the radio is operating efficiently.

- **Tactical Call Signs:** Temporary call signs used for simplicity and clarity during emergency operations or events.

- **Time-Out Timer (TOT):** A safety feature that limits the maximum duration of radio transmissions to prevent overheating or hogging the channel.

- **TPU (Thermoplastic Polyurethane) Case:** A case that blends rubber's flexibility with plastic's strength, offering protection against impacts and scratches for electronic devices.

- **Traffic:** Messages passed via radio, especially within nets, for relaying information such as health and welfare updates during emergencies.

- **Transceiver:** A single device that functions as both a transmitter and a receiver, commonly used in ham radios for two-way communication.

- **Transmitter:** A device that generates and sends radio frequency signals through an antenna to communicate with other radios.

- **Tropospheric Ducting:** A type of VHF propagation that can occur when warm air overruns cold air, creating a duct in which radio signals can travel over longer distances.

- **UHF (Ultra High Frequency):** Radio frequency ranges from 300 MHz to 3 GHz, used for two-way radio, Wi-Fi, Bluetooth, and other applications.

- **VE (Volunteer Examiner):** A licensed amateur radio operator who is certified to administer licensing exams to other amateur radio candidates.

- **VFO (Variable Frequency Oscillator):** Allows a radio receiver or transmitter to be tuned to different frequencies manually.

- **VHF (Very High Frequency):** Radio frequency ranges from 30–300 MHz, commonly used for two-way radio, FM, and television broadcasting.

- **VOX (Voice-Operated Transmission):** A feature that allows hands-free automatic transmission based on detecting your speech input via a microphone.

- **Wavelength:** The physical length of one cycle of a radio wave, inversely related to frequency.

- **Yagi Antenna:** A directional antenna with multiple elements, including a driven element, reflector, and directors, used for focusing radio signals in a specific direction.

- **Zero Beat:** The point at which two frequencies match exactly, often used when tuning in a signal in CW or SSB modes to ensure accurate frequency alignment.

References

- Aberle, S. (2017, June 21). HAM radio in emergency operations. Domestic Preparedness. https://domesticprepar edness.com/articles/ham-radio-in-emergency-operations

- American Radio Relay League (ARRL). (n.d.). Amateur radio communication: Amateur radio emergency service. https://www.arrl.org/amateur-radio-emergency-communication

- American Radio Relay League (ARRL). (n.d.). Building simple antennas. https://www.arrl.org/building-simple -antennas

- American Radio Relay League (ARRL). (n.d.). NTS standard net procedures. https://www.arrl.org/chapter-five -nts-standard-net-procedures

- American Radio Relay League (ARRL). (n.d.). What is ham radio? https://www.arrl.org/what-is-ham-radio

- Britannica. (n.d.). Amateur radio. https://www.britannica.com/technology/amateur-radio

- Federal Communications Commission (FCC). (n.d.). Amateur radio service. https://www.fcc.gov/wireless/burea u-divisions/mobility-division/amateur-radio-service

- Federal Communications Commission (FCC). (n.d.). 47 CFR Part 97 -- Amateur radio service. https://www.ecfr .gov/current/title-47/chapter-I/subchapter-D/part-97

- Ham Radio for Nontechies. (2024). A beginner's guide to HAM radio. https://hamradiofornontechies.com/the -definitive-guide-to-ham-radio/

- Ham Radio Prep. (2024). Ham radio emergency communications guide. https://hamradioprep.com/ham-radio-i n-emergencies/

- Institute For Building Technology and Safety. (2024). Tips for HAM radio operators: Natural disaster communica- tions best practices. https://ibtsonhand.org/resource/tips-for-ham-radio-operators-natural-disaster-communicati ons-best-practices/

- Racoma, A. J. (2020, October 6). The prepper's guide to emergency communications. https://n2rac.com/the-pre ppers-guide-to-emergency-communications-37f438db250

- Radio Design Group, Inc. (2023, October 9). The vital role of HAM radio in disaster preparedness. https://www .radiodesigngroup.com/blog/the-vital-role-of-ham-radio-in-disaster-preparedness

- Stathis, J. (2020, September 16). There's no better time to be an amateur radio geek. Wired. https://www.wired.c om/story/amateur-radio-disaster-preparedness/

- Stryker Radios. (2023, June 22). Ham radio emergency frequencies and common uses. https://strykerradios.com /ham-radios/ham-radio-emergency-frequencies-common-uses/

Book 2: Ham Radio General Class License Study Guide

Upgrade Your License! Move Beyond the Basics of Amateur Radio
Ace the FCC Exam and Expand Your On-Air Reach

Morse Code Publishing

Introduction

Welcome to the next chapter in your amateur radio journey!

I'm thrilled to be a part of it. Not long ago, I was right where you are—excited to go beyond my Technician Class License and eager to earn my General.

This upgrade opened up a new world of long-distance communications that I was excited to try out. I wrote this book to help you experience that excitement based on what I found helped me pass my exam.

This book is your study guide to that next chapter – the General Class License.

With *"Ham Radio General Class License Study Guide: Upgrade Your License! Move Beyond the Basics of Amateur Radio, Ace the FCC Exam, and Expand Your On-Air Reach,"* you're getting a roadmap designed to help you ace the FCC exam and gain access to new frequencies, modes, and connections.

This guide goes beyond simply passing the test; it is about building a richer understanding of amateur radio. I've included all the FCC exam questions, correct answers, practice exams, and practical strategies to help make the material stick. My goal is to simplify your study process, keep you engaged, and give you the confidence to tackle this exam.

Consider this book as your reliable companion on your General Class license path. It's here to make even the trickiest parts of the exam feel straightforward, preparing you not only to pass but to truly appreciate the expanded privileges that come with your new license.

So, if you're a current ham operator ready to take that next big step or someone passionate about deepening your understanding of amateur radio, this book is for you.

The book is organized into four main sections: Rules & Regulations, Operating Your Radio, The Physical Radio Station, and The Science of Radio. Each section aligns with the sub-elements of the General Class License exam. This structure ensures comprehensive coverage of all topics, making it easy to navigate and focus on specific areas.

One of the unique points of this book is its practical approach. It combines theoretical knowledge with real-world applications, making the learning process relevant and engaging. Memorable mnemonics, clear explanations, and practical examples help reinforce one's understanding of key concepts.

Now, it is your turn! As you finish this book, you will be equipped and ready to upgrade your license. The path to becoming a General Class License holder is within your reach. This book is your guide, mentor, and companion on this exciting journey.

The airwaves are waiting for you to expand your reach. Welcome to the General Class!

Jared Johnson (KF0RTU/AG) @ Morse Code Publishing

The General Class License Exam

THE GENERAL CLASS LICENSE is your gateway to expanded privileges. With this upgrade, you'll gain access to a wider range of frequencies and operating modes, opening the door to global communication and advanced technical skills.

Earning this license demonstrates your commitment to learning and going beyond the basics of radio. The exam covers a variety of topics, ensuring you're prepared for the challenges and opportunities of being a General Class operator. By passing this exam, you'll not only enhance your skills but also become a more versatile and capable member of the ham radio community.

Chapter 1

Design of this Book

THE AIM OF THIS book is to help you achieve three goals:

1. **Move beyond the basics and expand your knowledge of amateur radio.**

2. **Prepare for and pass the FCC exam to obtain your General license.**

3. **Equip you to expand your reach and connect globally over the airwaves!**

This book is divided into four core sections: The Science of Radio, The Physical Radio Station, Operating Your Radio, and Rules & Regulations. We conclude with a final section on further expanding your skills and knowledge.

Each part of this book forms a critical element in our teaching approach. Together, they work to strengthen your understanding and knowledge of ham radio, leading you toward success.

Rules & Regulations	Science of Radio
The Radio Station	Operating the Radio

The Science of Radio

This section covers the technical side of radio communication. We'll break down essential topics like wave propagation, modulation, and frequency, along with the types of signals and how they move through the atmosphere. These concepts are the building blocks for understanding how radio waves work and how you can use them to communicate.

The Physical Radio Station

Here, we explore the physical setup of your station. From antennas and feedlines to transceivers and power supplies, you'll learn about the key components of a working radio station. We'll also touch on how to design your station and troubleshoot problems so you're always ready to connect with others.

Operating Your Radio

This section is all about hands-on experience. You'll learn to make contacts, use repeaters, and operate in CW, SSB, and digital modes. We'll also cover the basics of clear communication, handling interference, and safety. By the end, you'll be confident in running your station efficiently and responsibly.

Rules & Regulations

This area covers the legal side of amateur radio. We'll review FCC regulations, band plans, licensing, and operator responsibilities. Knowing these rules is critical to ensuring safe and lawful use of the radio spectrum while avoiding interference with other users.

Our Teaching Framework

Our approach to teaching integrates all four sections—Rules & Regulations, The Radio Science, The Physical Radio Station, and Operating Your Radio—so that each part builds on the others. For example, understanding wave propagation will help you improve your station setup and operating skills while knowing the rules ensures you follow proper procedures.

This interconnected method will give you a well-rounded and deeper understanding of amateur radio, ensuring you're ready for the General exam. At the end of the book, you'll find a complete index to each exam question. This will enable you to quickly and easily reference key materials to review as needed.

A quick word of encouragement: expanding your skills in amateur radio can seem like a lot at first, but it's all about focusing on the core ideas. You don't have to memorize everything—just focus on the fundamental principles that drive the field. Once you understand these, you'll see that most other details are variations or applications of these basics.

This book is designed to be straightforward and practical, providing just the information you need without getting bogged down in unnecessary detail. By mastering the core principles and seeing how they fit together, you'll find that what once seemed complicated is now clear and manageable.

Keeping Information Accurate

We've worked hard to make sure the material in this book is accurate and up to date. However, the world of amateur radio continues to evolve, and sometimes mistakes can slip through. If you spot any errors or outdated information, please let us know! Your feedback helps us keep this book current and useful for future readers.

You can reach me at **Jared@MorseCodePublishing.com**.

Thank you for helping us keep this resource valuable for everyone!

Chapter 2

Why Upgrade to General?

UPGRADING FROM THE TECHNICIAN to the General license brings several exciting and practical benefits, significantly expanding an amateur radio operator's capabilities.

One of the most substantial upgrades is access to a broader range of high-frequency (HF) bands, which allows operators to communicate over long distances, including global communication, rather than being mainly limited to local and regional contacts on VHF and UHF.

You'll gain access to portions of popular bands like 80, 40, 20, and 15 meters, which are beneficial for long-distance communication (DXing). But that's not all—you'll also have operating privileges on portions of less commonly used bands such as 160, 60, 30, 17, 12, and 10 meters. This broader spectrum opens up a wide range of frequencies and modes, enhancing your ability to connect with stations across the globe, often under various propagation conditions.

With these additional bands, General license holders can explore broader access with modes such as SSB (Single Sideband), CW (Morse Code), and various digital modes like FT8 and PSK31, which are popular for DX communication.

One of the most compelling reasons to upgrade is the increased power limit. As a General Class licensee, you can legally transmit using up to 1500 watts of PEP (Peak Envelope Power) on most HF bands. This allows you to boost your signal strength, which can be invaluable during challenging propagation conditions or when trying to reach distant stations.

This upgrade also enhances the opportunity to experiment with and fine-tune operating techniques and radio equipment, making it ideal for operators who enjoy the technical aspects of radio. Additionally, access to expanded HF privileges opens doors to participating in more contests, emergency communications, and social networks with amateur operators worldwide, fostering a richer and more versatile ham radio experience.

By upgrading, you'll expand your capabilities in terms of spectrum and power and the variety of ways you can communicate. The General Class license is a gateway to a broader and more exciting amateur radio experience, whether chasing DX, experimenting with digital modes, or enjoying more flexible operating privileges.

Overall, the General license offers both practical tools and a greater sense of freedom on the air.

Onward and upward to General!

Chapter 3

The Exam

IMAGINE THE SENSE OF accomplishment as you key your first transmission using your new General call sign. To reach this milestone, a well-structured and carefully laid out study plan is needed.

Upgrading from a Technician to a General license involves steps designed to ensure you are well-prepared for the additional responsibilities and privileges. The first step is to commit to studying for the General exam, and as you are already reading this part of the book, I think we are good here!

This exam is more challenging than the Technician exam, requiring a deeper understanding of various topics. Consistent study sessions are important and should be scheduled. Allocate specific times each week dedicated to studying different sections of the exam.

Preparing for the Exam

In the first book in this series, *Ham Radio Technician Class License Study Guide: From Beginner to Licensed! Master the Fundamentals of Amateur Radio, Ace the FCC Exam, and Get on the Air with Confidence,* we provided an entire chapter on how to study and feel confident in your study sessions. We will not repeat that here but will provide you with a free downloadable copy of that chapter to use if needed.

How to Study

As you work through this book, treat the questions and answers as part of the regular content—they're integrated in a way to help reinforce your learning. Each question appears alongside four possible answers, with the correct one **bolded** for easy identification. Additionally, important details in the surrounding text will be *underlined*, signaling key information directly related to the exam question.

For instance, take a look at this exam question.

> **What is a geomagnetic storm? (G3A06)**
>
> A. A sudden drop in the solar flux index
> B. A thunderstorm that affects radio propagation
> C. Ripples in the geomagnetic force
> **D. A temporary disturbance in Earth's geomagnetic field**

Read the question and answers as part of the regular text. They are designed to be read as you find them. Now, focus on the bolded answer and examine the incorrect distractors. Do you understand why they are incorrect? This teaching method will emphasize both the correct answers and the insights provided by the distractors. It's not just about memorizing this one question but also about grasping the core knowledge surrounding it.

Practice Exams

Taking practice exams as part of your study plan is much like rehearsing for a big performance—it offers a critical snapshot of how well you've retained the material and highlights areas that may need more focus. Regularly taking practice exams helps gauge your understanding and prepares you for the exam's structure and timing.

We've included 16 practice exams — yes, 16! Why in the world are there so many? It's important to experience every possible exam question at least once in the format you'll encounter on test day. Section G8C has sixteen potential questions, and since only one will appear on the actual exam, we constructed 16 practice tests to ensure you're ready for whichever one comes your way.

And so be confident. If you can pass these 16 practice exams, the actual exam will be a breeze.

Practice Exams

Scheduling Your Practice Exams

As your exam day draws nearer, try to increase the frequency of your practice exams. In the final weeks before the test, aim to recreate exam conditions as closely as possible. Find a quiet place where you won't be disturbed and time yourself just as you would in the exam setting. This method not only aids in solidifying your knowledge but also helps reduce any nerves by familiarizing you with the pacing and structure of the test. It's also a great way to practice time management, ensuring you can confidently complete all sections within the given timeframe.

The Exam Format

The exam is designed to assess your knowledge across various topics. While it includes similar subject matter to the Technician exam, the General Class questions dive deeper into the technical details, ensuring you're well-prepared for the expanded privileges.

The General Class exam includes 35 multiple-choice questions, and to pass, you'll need to correctly answer at least 26, which is a score of 74%.

The Sub-Elements of the Exam

Understanding how the exam content is weighted helps prioritize your study efforts. The exam sub-elements are weighted differently, which means that some sub-elements will have more questions on the exam than others. This insight directs your focus to areas needing a more robust review.

For instance, you will see two times as many questions about FCC rules (5 questions) as you will for Safety (2). While having well-rounded knowledge is important, emphasizing these heavier-weighted areas can increase your chances of a successful outcome. See below for each sub-element and the number of questions for each sub-element on the exam.

Sub-elements on the General exam:

1. Commission's Rules: (5 questions from this sub-element will be on the exam)

2. Operating Procedures: (5 questions from this sub-element will be on the exam)

3. Radio Wave Propagation (3 questions)

4. Amateur Radio Practices (5 questions)

5. Electrical Principles (3)

6. Circuit Components (2)

7. Practical Circuits (3)

8. Signals and Emissions (3)

9. Antennas and Feed Lines (4)

10. Safety (2)

For a total of 35 exam questions.

Taking Your Exam

We are going to make an assumption at this point. The assumption is that you took and passed your Technician's exam. Therefore, you know about FRN (FCC Registration Number) and how to navigate the FCC Website.

You also should know how to schedule your exam since you've already done it for the Technician's. All this to say, we will list some high-level resources below but won't go into as much detail as we did in the *Technician's Study Guide*. If you do have questions, please reach out, and we would be happy to fill in any details for you, just in case it's been a while since your Technician's exam: **Jared@MorseCodePublishing.com**

Schedule Your Exam

Visit the ARRL website and navigate to the search for exam session page. Register for the exam that works for you.

- https://www.arrl.org/find-an-amateur-radio-license-exam-session

  - Or type into Google, "Find an Amateur Radio License Exam in Your Area," and click on the ARRL site that comes up

Exam Day

Exam day represents the culmination of all your hard work and preparation as you take a significant step toward upgrading your amateur radio license. Focusing on mental and physical readiness is important to perform at your best.

Bring a valid photo ID, such as a driver's license or passport. Minors may need a school ID and possibly a guardian's ID. You'll also need any previous Amateur Radio licenses or Certificates of Successful Completion (CSCE), two pencils, a pen, a calculator with cleared memory, and the exam fee. Cell phones and other electronic devices with calculator functions are prohibited.

This should all be a reminder.

After successfully passing, you'll need to complete and submit the necessary paperwork and fees with assistance from the Volunteer Examiners (VEs). Once everything is processed, your new General Class license will be issued, granting you broader operating privileges.

With that overview out of the way, let's begin studying for your General Class License!

The Science of Radio

THE SCIENCE OF RADIO forms the foundation of everything we do as amateur radio operators. Understanding the principles behind how radio waves travel, interact with the environment, and carry information is essential to becoming a skilled and effective operator. This section dives into the technical concepts that make radio communication possible.

By exploring these core topics, you'll gain a deeper appreciation for the physics and engineering behind the signals you transmit and receive. Whether you're building antennas, optimizing your equipment, or troubleshooting issues, the knowledge in this section will empower you to tackle challenges with confidence and curiosity. Mastering the science of radio is not only a key step in passing the General Class License exam but also in unlocking the full potential of amateur radio.

Chapter 4

Propagation

UNDERSTANDING PROPAGATION IS A cornerstone of effective radio communication. Propagation refers to the way radio waves travel from a transmitter to a receiver, and it's influenced by a variety of factors, including frequency, atmospheric conditions, and the environment.

For General Class license holders, knowing how propagation works is necessary to maximize the reach and reliability of their transmissions. Whether it's bouncing signals off the ionosphere for long-distance communication or using ground waves for local coverage, mastering propagation concepts will empower you to adapt to different operating conditions and make the most of your equipment.

This chapter explores key propagation topics, giving you the knowledge needed to communicate effectively across the bands and pass your exam.

Propagation Types

Line-of-Sight

Line-of-sight propagation is the simplest form, applicable primarily to VHF and UHF communications. These signals travel in a straight line from the transmitter to the receiver, making them ideal for short-range communications. However, obstacles like buildings or hills can block these signals, limiting their range. This mode is commonly used in urban environments and local communications, where clear paths between antennas are maintained.

The Technician's exam largely covered aspects of line-of-sight propagation.

Ground

Ground wave propagation is another particularly relevant mode for HF communications over short distances. These waves follow the contour of the Earth, bending slightly to maintain contact with the ground. Ground waves are most effective at lower frequencies, covering distances up to several hundred miles. This mode is less affected by obstacles than line-of-sight, making it reliable for regional communications. However, ground wave signals gradually lose strength as they travel, and the Earth's curvature and terrain limit their range.

Skywave

Skywave propagation, also known as ionospheric reflection or skip propagation, is the key to long-distance HF communications. In this mode, radio waves are reflected or refracted ("skipped") by the ionosphere, a layer of the Earth's atmosphere filled with charged particles. This natural phenomenon allows signals to travel beyond the horizon, bouncing between the ionosphere and the Earth's surface to cover thousands of miles. Skywave propagation (skip propagation) is highly effective for global communication, enabling contact with stations on different continents.

The distance a signal can "skip" depends on factors like the frequency, the ionosphere layer it bounces off, and the current solar activity. Higher ionospheric layers, like the F2 region, typically enable longer skip distances.

Skip Zone

A transmitting station's "skip zone" is an area surrounding the transmitter where radio signals sent via skywave propagation cannot be received. This happens because the ionosphere refracted the radio waves at distances farther away from the station, skipping over the region closest to the transmitter. The signal is reflected to Earth farther out, leaving a gap between the point where ground wave signals weaken and the area where skywave signals return.

To visualize it, imagine the radio signal bouncing off the ionosphere like a ball. The signal travels upward, skipping over nearby areas and returning to Earth farther away. The skip zone is the area in between—where neither ground waves nor skywave signals reach effectively. Think of this as a kind of no man's land. This can be an issue for stations trying to communicate with nearby locations using higher-frequency signals that rely on ionospheric reflection.

Scatter

However, with all things related to radio waves, it is not quite that simple. Scatter propagation is a type of propagation that allows signals to be heard in the transmitting station's skip zone. Instead of the signal being fully absorbed or reflected to Earth by the ionosphere, a portion of the signal is scattered in different directions by irregularities in the ionosphere or the Earth's atmosphere. This scattering can cause weak signals to reach areas within the skip zone, where normal ground-wave or skywave signals would not be heard.

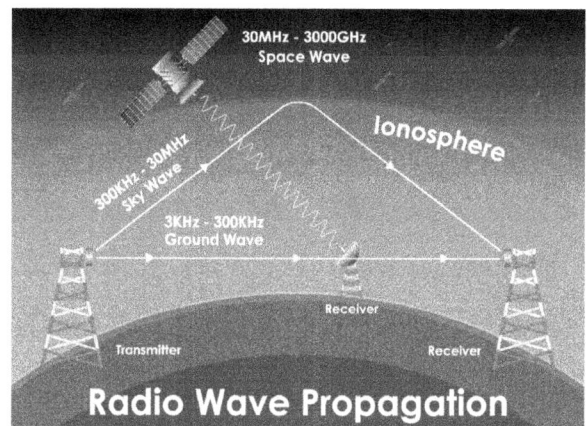
Radio Wave Propagation

Think of scatter propagation as "filling in the gap." It lets signals get through to areas typically skipped over by direct ionospheric reflection, making communication possible even in the zones close to the transmitter that are usually silent.

What type of propagation allows signals to be heard in the transmitting station's skip zone? (G3C09)

A. Faraday rotation
B. Scatter
C. Chordal hop
D. Short-path

HF Scatter

HF scatter signals often have a characteristic fluttering or wavering sound. This occurs because the signal is scattered in multiple directions by irregularities in the ionosphere, causing parts of the signal to take different paths and arrive at slightly different times. The result is a fluctuating, unstable signal that creates a noticeable "flutter" when received.

What is a characteristic of HF scatter? (G3C06)

A. Phone signals have high intelligibility
B. Signals have a fluttering sound
C. There are very large, sudden swings in signal strength
D. Scatter propagation occurs only at night

HF scatter signals sound distorted because the radio wave's energy is scattered into the skip zone through several paths. As the signal encounters irregularities in the ionosphere, parts of it are reflected or refracted in multiple directions. These scattered signal portions take slightly different routes, meaning they arrive at the receiver at different times. This causes the signal to blend together unevenly, resulting in a distorted or fluttering sound.

Scatter propagation creates a "multi-path effect," where the signal is split and reassembled imperfectly at the receiver. This scattering process produces a signal that sounds less clear and distorted than a direct or single-path signal.

What makes HF scatter signals often sound distorted? (G3C07)

A. The ionospheric region involved is unstable
B. Ground waves are absorbing much of the signal
C. The E region is not present
D. Energy is scattered into the skip zone through several different paths

---

And finally, HF scatter signals in the skip zone are usually weak because only a small portion of the signal's energy is scattered into this area. Instead of the entire signal being reflected to Earth, most of the signal follows the regular skywave propagation path, skipping over the zone. The small amount of energy that does get scattered by irregularities in the ionosphere is spread out in various directions, meaning only a tiny fraction reaches the skip zone, resulting in a much weaker signal.

Why are HF scatter signals in the skip zone usually weak? (G3C08)

A. Only a small part of the signal energy is scattered into the skip zone
B. Signals are scattered from the magnetosphere, which is not a good reflector
C. Propagation is via ground waves, which absorb most of the signal energy
D. Propagation is via ducts in the F region, which absorb most of the energy

---

Short & Long Path

Short-path and long-path propagation refer to two different routes a radio signal can take when traveling using skywave propagation.

1. **Short-path propagation** is the signal's most direct route between two locations, typically following the shortest distance over the Earth's surface. The signal bounces off the ionosphere and returns to Earth after covering the shortest distance between the transmitter and receiver. For example, suppose you're communicating with someone on the opposite side of the continent. In that case, the signal will likely travel in a short arc across the globe.

2. **Long-path propagation** is when the signal travels the long way around the Earth, taking the opposite direction. Instead of following the shortest route, the signal bounces off the ionosphere. It travels around the other side of the globe before reaching the receiver. This long-path route often takes several times the distance of the short path, which causes the signal to arrive slightly later.

Both types of propagation can happen simultaneously, especially on higher HF bands. When a skywave signal reaches your location through both short-path and long-path propagation, you might hear a slightly delayed echo.

This happens because the signal takes two different routes to get you: the short path, the most direct route, and the long path, which travels around the Earth in the opposite direction. Since the long-path signal has a greater distance to travel, it arrives a fraction of a second later, creating the echo effect.

What is a characteristic of skywave signals arriving at your location by both short-path and long-path propagation? (G3B01)

A. Periodic fading approximately every 10 seconds
B. Signal strength increased by 3 dB
C. The signal might be cancelled, causing severe attenuation
D. A slightly delayed echo might be heard

An easy way to remember this for the exam is to think of the signal as taking "two routes" to get to you. The short path is the quick way, while the long path takes the scenic route around the globe. The result is that slight delay, which you hear as an echo, making this a distinctive characteristic of skywave signals.

Ionospheric Propagation

Picture yourself standing at the edge of a vast ocean. You know there are islands in the distance, but you cannot see them. Instead, you rely on the waves to carry messages across the expanse. In the world of amateur radio, the ionosphere serves a similar purpose. It is an invisible but powerful medium that allows your signals to travel vast distances, bouncing off its layers like stones skipping across water. As you upgrade to a General Class license, understanding ionospheric propagation becomes more essential for optimizing your communication reach, given your expanded HF privileges.

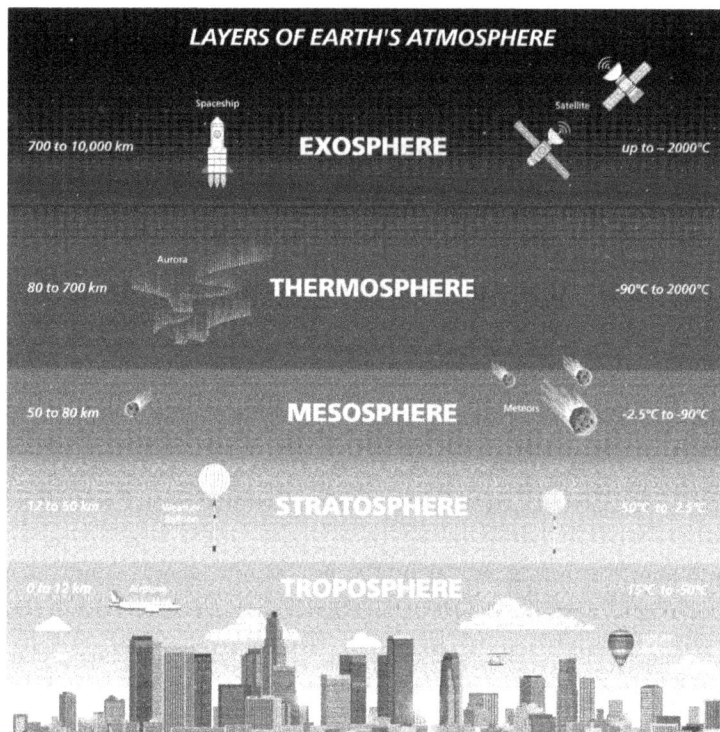

The ionosphere is a region of the Earth's upper atmosphere that overlaps with parts of the mesosphere, thermosphere and exosphere. It is crucial in long-distance radio communication. It is composed of several layers filled with ions and free electrons created by the interaction of solar radiation with atmospheric particles. These layers are not rigid but dynamic, fluctuating in response to solar activity and atmospheric conditions. The ionosphere is typically divided into three primary layers: the D layer, the E layer, and the F layer. Each of these layers has unique characteristics that influence radio wave propagation.

When radio waves have frequencies below the Maximum Usable Frequency (MUF) and above the Lowest Usable Frequency (LUF), the ionosphere refracts or bends them back to Earth. More on MUF and LUF coming up. This allows long-distance communication by "skipping" the signal between the Earth and the ionosphere. The ionosphere acts like a mirror for these frequencies, reflecting them rather than letting them escape into space.

How does the ionosphere affect radio waves with frequencies below the MUF and above the LUF? (G3B05)

A. They are refracted back to Earth
B. They pass through the ionosphere
C. They are amplified by interaction with the ionosphere
D. They are refracted and trapped in the ionosphere to circle Earth

D Layer

The D layer, situated between 50 and 90 kilometers (30-90 miles) above the Earth, is the lowest of the ionospheric layers and closest to the Earth. It forms during the daytime when high-energy X-rays and ultraviolet (UV) light from the Sun ionize atmospheric particles. The D layer is known for its absorbent properties, particularly affecting low-frequency signals. As radio waves pass through this layer, they can experience significant attenuation, reducing their strength.

Which ionospheric region is closest to the surface of Earth? (G3C01)

A. The D region
B. The E region
C. The F1 region
D. The F2 region

During the day, it plays an key role in absorbing low-frequency radio waves, which can weaken or block those signals from traveling far. The D region of the ionosphere is the most absorbent of signals below 10 MHz during daylight hours.

Which ionospheric region is the most absorbent of signals below 10 MHz during daylight hours? (G3C11)

A. The F2 region
B. The F1 region
C. The E region
D. The D region

As a result, long-distance communication during the day on the 40-, 60-, 80-, and 160-meter bands is more difficult because the D region of the ionosphere absorbs signals at these frequencies. The Sun's radiation increases the ionization in the D region, which causes it to absorb lower-frequency/longer-wavelength signals, making it harder for them to travel long distances.

Why is long-distance communication on the 40-, 60-, 80-, and 160-meter bands more difficult during the day? (G3C05)

A. The F region absorbs signals at these frequencies during daylight hours
B. The F region is unstable during daylight hours
C. The D region absorbs signals at these frequencies during daylight hours
D. The E region is unstable during daylight hours

However, the D region almost disappears at night, allowing these signals to travel greater distances.

An easy way to remember this is to think of "D" for "Daytime Absorption." The D region is most active and absorbent when the Sun is out. At night, the D region loses much of its ionization, allowing those same signals to travel farther with less interference. So, if you're trying to communicate on these lower frequencies, nighttime will be your friend!

E Layer

Above the D layer lies the E layer, which extends from about 90 to 120 kilometers (60-70 miles) above the Earth. This layer is less dense than the D layer but still significantly influences radio wave propagation.

When a radio signal bounces off the E region of the ionosphere, it typically covers a distance of up to 1,200 miles (about 1,930 kilometers) in one "hop" before it returns to Earth. The E region can reflect signals in the 3 to 30 MHz range (HF Range). This makes it useful for medium-distance communication, allowing operators to reach stations within a 1,200-mile radius without relying on ground-wave propagation.

> **What is the approximate maximum distance along the Earth's surface normally covered in one hop using the E region? (G3B10)**
>
> A. 180 miles
> **B. 1,200 miles**
> C. 2,500 miles
> D. 12,000 miles

One notable phenomenon associated with the E layer is Sporadic E propagation. Sporadic E occurs when dense patches of ionization form within the E layer, reflecting VHF signals over long distances. This can enable unexpected and fascinating communication opportunities, particularly during the summer months when Sporadic E activity peaks. However, the E layer weakens at night, reducing its impact on signal propagation.

If you aim to cover even greater distances, multiple hops or higher layers like the F region come into play. But for one hop, the E region is perfect for those medium-range contacts!

F Layer

The F layer, located between 240 and 500 kilometers above the Earth (150-300 miles), is the highest and most critical layer for long-distance HF communication. This greater altitude (highest) allows radio signals to travel much farther before bouncing back to Earth, resulting in longer skip distances.

> **Why is skip propagation via the F2 region longer than that via the other ionospheric regions? (G3C03)**
>
> A. Because it is the densest
> B. Because of the Doppler effect
> **C. Because it is the highest**
> D. Because of temperature inversions

During the daytime, the F layer splits into two sub-layers: F1 and F2.

The F1 layer resides at lower altitudes and has moderate ionization levels, while the F2 layer, higher up, has the highest concentration of ions and free electrons.

The F2 region is the highest ionospheric layer. It remains ionized at night, making it ideal for long-distance communication, especially on the higher HF bands (3 to 30 MHz). Signals can reflect off the F2 layer, allowing them to travel thousands of kilometers and enabling global communication. In fact, in one hop, signals can typically cover a distance of up to 2,500 miles (about 4,000 kilometers) along the Earth's surface. Basically, double the distance of the E layer.

What is the approximate maximum distance along the Earth's surface normally covered in one hop using the F2 region? (G3B09)

A. 180 miles
B. 1,200 miles
C. 2,500 miles
D. 12,000 miles

This is why ham radio operators can reach contacts halfway around the world with the help of the F2 layer, particularly during periods of good solar activity. To remember this, think of the "F2" layer as the "Far-reaching" layer, enabling signals to travel great distances, up to 2,500 miles in just one bounce.

Critical Frequency

The term "critical frequency" refers to the highest frequency that can be refracted or bent back to Earth by the ionosphere at a given incidence angle (the angle at which the signal is transmitted). Suppose a signal is transmitted at a frequency higher than the critical frequency. In that case, it will pass through the ionosphere and escape into space instead of being reflected back to Earth.

What is meant by the term "critical frequency" at a given incidence angle? (G3C02)

A. The highest frequency which is refracted back to Earth
B. The lowest frequency which is refracted back to Earth
C. The frequency at which the signal-to-noise ratio approaches unity
D. The frequency at which the signal-to-noise ratio is 6 dB

Think of the "critical frequency" as the upper limit for successful skywave communication at a specific angle. When your transmission frequency is below the critical frequency, the ionosphere will reflect the signal back to Earth. If you go above it, the signal won't return, making it necessary to know this limit for effective long-distance communication.

Critical Angle

The "critical angle" in radio wave propagation refers to the highest angle at which a radio signal can be transmitted and still return to Earth by the ionosphere. Suppose the signal is sent at a higher angle than the critical angle. In that case, it will pass through the ionosphere and not be reflected, effectively disappearing into space. The critical angle is determined by specific ionospheric conditions, including the density and height of the ionized layers, which can vary based on factors like solar activity and time of day.

What does the term "critical angle" mean, as applied to radio wave propagation? (G3C04)

A. The long path azimuth of a distant station
B. The short path azimuth of a distant station
C. The lowest takeoff angle that will return a radio wave to Earth under specific ionospheric conditions
D. The highest takeoff angle that will return a radio wave to Earth under specific ionospheric conditions

---

To determine the critical angle for radio wave propagation, you need to consider two main factors: the frequency of the signal you're transmitting and the ionospheric conditions.

Here's a simplified explanation of how it works:

1. **Ionospheric Density**: The ionosphere's ability to reflect a signal depends on its level of ionization, which varies by time of day, solar activity, and atmospheric conditions. The more ionized the layer, the higher the frequency of signals it can reflect.

2. **Frequency**: The higher the frequency of your signal, the smaller the critical angle. If the signal is transmitted at an angle that is too high, it will escape into space instead of being reflected back to Earth.

Calculating the critical angle requires complex formulas, but in practice, it is often determined through real-time propagation prediction tools or trial and error by adjusting your antenna's takeoff angle and experimenting with different frequencies.

An easy way to remember this is to consider the critical angle as the "ceiling" for successful signal reflection. If you go above that ceiling, your signal escapes the ionosphere. Still, if you stay below it, your signal will bounce back to Earth, allowing you to communicate over long distances.

Maximum & Lowest Usable Frequency

In radio communication, MUF (Maximum Usable Frequency) and LUF (Lowest Usable Frequency) are terms used to describe the limits within which a radio frequency can be successfully transmitted over long distances through the ionosphere.

- **MUF (Maximum Usable Frequency):** <u>This is the highest/ maximum frequency at which a radio wave can be transmitted (used) between two points by refracting off the ionosphere under specific conditions.</u> Signals above the MUF will typically pass through the ionosphere into space rather than being refracted back to Earth. The MUF changes with factors like time of day, season, and solar activity, generally increasing with stronger solar activity.

- **LUF (Lowest Usable Frequency):** <u>This is the lowest frequency that can propagate between two points without significant signal degradation due to ionospheric absorption or other losses.</u> Frequencies below the LUF will often be absorbed in the D-layer of the ionosphere, especially during the daytime. The LUF is affected by atmospheric conditions and can vary based on the same factors as the MUF. However, it tends to be more susceptible to increased ionization and lower at night.

What does MUF stand for? (G3B08)

A. The Minimum Usable Frequency for communications between two points
B. The Maximum Usable Frequency for communications between two points
C. The Minimum Usable Frequency during a 24-hour period
D. The Maximum Usable Frequency during a 24-hour period

What does LUF stand for? (G3B07)

A. The Lowest Usable Frequency for communications between two specific points
B. Lowest Usable Frequency for communications to any point outside a 100-mile radius
C. The Lowest Usable Frequency during a 24-hour period
D. Lowest Usable Frequency during the past 60 minutes

LUF marks the minimum frequency needed to avoid excessive absorption by the ionosphere, ensuring your signal can travel the distance. Staying above the LUF helps to maintain a strong, stable connection.

Radio waves with frequencies below the Lowest Usable Frequency (LUF) are usually attenuated, or weakened before they can reach their intended destination. This happens because lower frequencies are more susceptible to absorption by the ionosphere, particularly in the D layer during the daytime. As a result, these signals lose strength and can fade out before they have the chance to refract back to Earth and complete the long-distance communication path.

What usually happens to radio waves with frequencies below the LUF? (G3B06)

A. They are refracted back to Earth
B. They pass through the ionosphere
C. They are attenuated before reaching the destination
D. They are refracted and trapped in the ionosphere to circle Earth

Frequencies below this threshold get absorbed rather than reflected, which means they won't be able to travel far enough to connect with distant stations. Staying above the LUF ensures that your signal is strong enough to overcome ionospheric absorption and reach its destination.

MUF (Maximum Usable Frequency) and LUF (Lowest Usable Frequency) guide operators in selecting the ideal frequency range for long-distance HF communication. Staying within this "sweet spot" ensures signals are refracted back to Earth rather than escaping into space (above the MUF) or being absorbed by the atmosphere (below the LUF). This balance is key to reliable, effective communication.

Factors Impacting MUF

The MUF is influenced by several key factors: path distance and location, time of day and season, and solar radiation and ionospheric disturbances.

- **Path distance and location** affect the MUF because longer paths generally require higher frequencies to maintain a connection, as the ionosphere needs to refract the signal over a greater distance.

- **Time of day and season** are also important—during daylight hours and summer months, increased solar radiation causes more ionization in the ionosphere, which typically raises the MUF. Conversely, ionization decreases at night and in winter, often lowering the MUF.

- **Solar radiation and ionospheric disturbances** play a significant role, too. High solar activity, such as sunspots and solar flares, increases ionization and can raise the MUF, while disturbances like geomagnetic storms can disrupt the ionosphere and reduce the MUF.

What factors affect the MUF? (G3B02)

A. Path distance and location
B. Time of day and season
C. Solar radiation and ionospheric disturbances
D. All these choices are correct

Remember MUF as being shaped by the "when" (time and season), the "where" (path and location), and the "what" (solar activity) of your signal path. Together, these elements determine the highest frequency that will reliably refract back to Earth for effective communication.

How to Determine MUF & LUF

For long-distance skip propagation, a frequency just below the Maximum Usable Frequency (MUF) will experience the least attenuation. The MUF represents the highest frequency that can be reliably refracted back to Earth by the ionosphere. Using a frequency just below the MUF ensures that the signal will be strong enough to avoid excessive absorption and reach distant stations with minimal loss of strength.

Which frequency will have the least attenuation for long-distance skip propagation? (G3B03)

A. Just below the MUF
B. Just above the LUF
C. Just below the critical frequency
D. Just above the critical frequency

By operating just below this boundary, you're optimizing your signal for minimal attenuation, ensuring it can travel long distances with reduced fading and signal loss. This choice of frequency helps maintain a clear, stable connection over vast areas.

To figure out the Maximum Usable Frequency (MUF) and Lowest Usable Frequency (LUF), operators rely on a combination of propagation tools, real-time data, and practical experience. Here's a simplified process:

Determining MUF

1. **Ionospheric Data**: Operators use online resources like ionospheric maps, space weather updates, or tools such as the VOACAP (Voice of America Coverage Analysis Program) to check real-time ionospheric conditions.

2. **Calculation or Software**: MUF is typically calculated based on the critical frequency (the highest frequency that will reflect at vertical incidence) and the transmission angle. Many propagation prediction tools automatically perform this calculation.

3. **Trial and Error**: Operators can test various frequencies within their band privileges to find the highest frequency that supports reliable communication over the intended path.

Determining LUF

1. **Atmospheric Noise and Absorption**: The LUF is influenced by atmospheric conditions, signal strength, and noise levels. Operators look at propagation tools and band reports to assess how lower frequencies are performing.

2. **Trial and Error**: Testing the lower frequencies and monitoring signal quality can help determine the point below which communication becomes unreliable.

Practical Tools

- **Software Solutions**: Tools like VOACAP, HamCAP, or even smartphone apps provide predictions for MUF and LUF based on input parameters like time, date, location, and solar activity.

- **Online Resources**: Websites and live propagation maps offer up-to-date MUF and LUF values, as well as solar indices like the K-index or solar flux, which affect propagation.

By combining these tools with experience, operators can effectively select the optimal frequencies for communication under varying propagation conditions.

LUF exceeds MUF

<u>When the Lowest Usable Frequency (LUF) exceeds the Maximum Usable Frequency (MUF), ordinary skywave communication over that path becomes impossible.</u> This situation occurs because the ionosphere can no longer support the reflection of any frequencies within the usable range. The ionosphere absorbs signals below the LUF, and signals above the MUF pass through it without being refracted back to Earth, leaving no viable frequencies for long-distance communication.

What happens to HF propagation when the LUF exceeds the MUF? (G3B11)

A. Propagation via ordinary skywave communications is not possible over that path
B. HF communications over the path are enhanced
C. Double-hop propagation along the path is more common
D. Propagation over the path on all HF frequencies is enhanced

---

To remember this for the exam, think of the LUF and MUF as the boundaries of a usable "window" for HF propagation. If the LUF rises above the MUF, this "window" closes, preventing effective skywave communication on that path. In such cases, operators might need to wait until conditions change and the LUF drops below the MUF to restore reliable HF propagation.

When could this happen?

The situation where the Lowest Usable Frequency (LUF) exceeds the Maximum Usable Frequency (MUF) typically occurs during severe ionospheric disturbances or under extreme conditions that disrupt normal skywave propagation. Here are some specific scenarios when this might happen:

1. Severe Solar Storms

- During a major solar flare or coronal mass ejection (CME), the D-layer of the ionosphere becomes highly ionized, drastically increasing absorption of lower frequencies. This causes the LUF to rise significantly.

- Simultaneously, the upper ionospheric layers (E and F layers) may become unstable or depleted, reducing the MUF as they lose their ability to refract higher frequencies.

2. Geomagnetic Storms

- Intense geomagnetic activity can disrupt the ionosphere's structure, creating turbulent and irregular ionization patterns. This can cause both higher absorption (raising the LUF) and a loss of higher-frequency reflectivity (lowering the MUF).

3. Nighttime on Low-Solar-Activity Cycles

- During periods of very low solar activity (solar minimum), combined with nighttime conditions, the ionosphere becomes weakly ionized. The MUF drops significantly as the F-layer loses its ability to reflect higher frequencies, while increased D-layer absorption raises the LUF.

4. Polar Regions During Auroral Events

- In polar regions, auroral events caused by charged particles from the sun can severely disrupt the ionosphere. This leads to extreme absorption at lower frequencies (high LUF) while destabilizing higher-frequency propagation (low MUF).

Practical Example:

Imagine you're trying to use HF communication during a solar storm. The usual bands, like 20 meters, which might normally support propagation for long-distance communication, fail because the LUF has risen to, say, 18 MHz, while the MUF has dropped to 12 MHz. There is no frequency band in the HF range that can propagate via skywave, effectively blocking communication on that path.

Understanding these conditions allows operators to switch to alternative methods, such as VHF/UHF communication or digital modes that rely on internet-linked systems during such challenging times.

Near Vertical Incidence Skywave

Near Vertical Incidence Skywave (NVIS) propagation is a method of using <u>medium-frequency (MF) or high-frequency (HF) radio waves to communicate over short distances by transmitting signals at high elevation angles</u>—almost straight up into the sky. These signals bounce off the ionosphere and are reflected back to Earth, covering areas within a range of about 200 to 400 miles (300 to 600 kilometers). This technique is instrumental in hilly or mountainous regions where direct line-of-sight communication isn't possible.

NVIS propagation is closely related to the concept of critical angle. In NVIS, radio signals are transmitted to the ionosphere at a very high angle—almost straight up. For the signal to be refracted back to Earth (rather than passing through the ionosphere into space), the frequency must be below the critical frequency, and the angle of transmission must be lower than the critical angle.

What is near vertical incidence skywave (NVIS) propagation? (G3C10)

A. Propagation near the MUF
B. Short distance MF or HF propagation at high elevation angles
C. Long path HF propagation at sunrise and sunset
D. Double hop propagation near the LUF

The critical angle determines the maximum angle at which a signal can still be refracted by the ionosphere and returned to Earth. For NVIS to work, the signal is transmitted at a near-vertical angle below the critical angle. This ensures the signal is bent back down over a short distance, enabling effective local communication. So, NVIS relies on using frequencies low enough (below the critical frequency) to allow this high-angle reflection.

Solar Activity

The Sun and its activities play an interesting role in radio communication, particularly for operators using HF bands. The Sun constantly emits energy in the form of electromagnetic radiation and charged particles, which interact with Earth's atmosphere and magnetic field. This solar energy, especially during periods of high activity like solar flares and increased sunspot numbers, enhances the ionosphere's ability to refract radio waves to Earth, making long-distance communication possible.

However, solar activity can be a double-edged sword. While it can improve signal propagation, intense solar events like geomagnetic storms can also disrupt communications.

Solar Cycle

The solar cycle spans approximately 11 years and alternates between solar maximum and minimum periods. During solar maximum, increased solar activity boosts ionization, while during solar minimum, activity is reduced and propagation conditions are weaker.

The cycle begins at a solar minimum when the Sun has few or no sunspots and low solar activity. It gradually builds up to a solar maximum, where sunspot numbers and solar activity peak. After the maximum, the activity declines again, leading to the following solar minimum.

The solar cycle is crucial for radio operators because solar activity directly affects ionospheric conditions, which influences HF radio wave propagation. During high solar activity (solar maximum) periods, more sunspots increase ionization in the ionosphere, improving HF propagation, especially on higher frequencies. Conversely, lower solar activity can lead to poorer HF propagation during solar minimums, particularly at higher frequencies.

As of January 2025, we are in Solar Cycle 25, which began in December 2019. Solar cycles typically last about 11 years, with activity peaking around the midpoint. Predictions indicate that Solar Cycle 25 will reach its maximum in July 2025.

Increased solar activity during this period can enhance ionospheric propagation, benefiting long-distance radio communications. However, it also raises the likelihood of solar flares and geomagnetic storms, which can disrupt communications and affect satellite operations. Staying informed about solar activity is crucial for radio operators to optimize performance and mitigate potential disruptions.

Exam question time.

The 20-meter band (14 MHz) typically supports worldwide propagation during daylight hours at any point in the solar cycle. The 20-meter band is well-suited for long-distance communication, regardless of whether solar activity is high or low. While higher solar activity (during solar maximum) enhances propagation on many bands, the 20-meter band remains reliable for worldwide communication even during solar minimum, making it a popular choice for amateur operators seeking consistent long-distance contacts.

At what point in the solar cycle does the 20-meter band usually support worldwide propagation during daylight hours? (G3A07)

A. At the summer solstice
B. Only at the maximum point
C. Only at the minimum point
D. At any point

Remember, the solar cycle is a natural rhythm of the Sun's activity that peaks and wanes every 11 years, and this cycle heavily impacts radio communication conditions.

28 Day Cycle

HF propagation conditions vary in a 26- to 28-day cycle due to the Sun's rotation around its axis. The Sun rotates once approximately every 27 days, causing different features on its surface, such as sunspots and coronal holes, to periodically face Earth. As these solar features come into view, they affect Earth's ionosphere by altering how radio waves are refracted, resulting in changes to HF propagation. For instance, when an active region with sunspots or a coronal hole rotates into view, it can either enhance or disrupt HF communication, depending on the solar activity.

What causes HF propagation conditions to vary periodically in a 26- to 28-day cycle? (G3A10)

A. Long term oscillations in the upper atmosphere
B. Cyclic variation in Earth's radiation belts
C. Rotation of the Sun's surface layers around its axis
D. The position of the Moon in its orbit

Think of the Sun as a rotating "influencer" that regularly changes the conditions we experience on Earth. As the Sun's surface turns, different areas emit solar wind and radiation that impact HF propagation, leading to a repeating cycle of conditions roughly every 27 days.

Understanding the impact of solar activity helps radio operators predict and adapt to changing conditions for optimal performance. Also, if you follow any Ham Radio groups on Facebook, there is always lots of chatter and excitement around the Sun's activities.

Solar Flares

Solar flares are a sudden burst of intense radiation. These flares can cause sudden ionospheric disturbances (SIDs), leading to short-term radio fade-outs or enhanced absorption of electromagnetic waves.

The increased ultraviolet (UV) and X-ray radiation from a solar flare affects radio propagation on Earth in approximately 8 minutes. This is because it takes about 8 minutes for the radiation from the Sun to travel 93 million miles to Earth, moving at the speed of light. Once this radiation reaches our planet, it can immediately cause disturbances in the ionosphere, particularly on the sunlit side of Earth, leading to short-term radio signal blackouts or fading.

> **Approximately how long does it take the increased ultraviolet and X-ray radiation from a solar flare to affect radio propagation on Earth? (G3A03)**
>
> A. 28 days
> B. 1 to 2 hours
> **C. 8 minutes**
> D. 20 to 40 hours

Solar flares send bursts of intense radiation that disrupt radio communications quickly after they happen, giving operators little time to prepare once the flare occurs.

Impact from Sudden Ionospheric Disturbances

A sudden ionospheric disturbance (SID) disrupts daytime ionospheric propagation, especially affecting lower-frequency signals. When a SID occurs—often triggered by a solar flare—it floods the ionosphere with intense ultraviolet and X-ray radiation. This extra energy causes increased ionization, particularly in the D layer, which absorbs lower-frequency HF signals and weakens or blocks them from being refracted back to Earth.

> **What effect does a sudden ionospheric disturbance have on the daytime ionospheric propagation? (G3A02)**
>
> A. It enhances propagation on all HF frequencies
> **B. It disrupts signals on lower frequencies more than those on higher frequencies**
> C. It disrupts communications via satellite more than direct communications
> D. None, because only areas on the night side of the Earth are affected

While higher frequencies may still get through, lower frequencies are much more likely to be absorbed, disrupting long-distance communication on those bands during the day.

Sunspots

A sunspot is a temporary dark spot that appears on the surface of the Sun due to lower temperatures compared to the surrounding areas. These spots are caused by intense magnetic activity, which inhibits the transfer of heat to the surface, making them cooler and darker than the surrounding solar material. While they may look small in images, sunspots can be massive, often much larger than Earth.

Sunspots are important to radio operators because they are closely linked to solar activity, which can enhance ionospheric conditions for radio wave propagation. <u>A higher sunspot number generally improves HF propagation, particularly at higher frequencies.</u> Sunspots are indicators of increased solar activity, which boosts the amount of solar radiation reaching Earth's ionosphere. This extra energy causes more ionization in the ionospheric layers, allowing higher-frequency radio waves (such as those on the 10, 12, and 15-meter bands) to be reflected to Earth more efficiently. As a result, long-distance communication on these higher frequencies becomes more reliable and effective.

How does a higher sunspot number affect HF propagation? (G3A01)

A. Higher sunspot numbers generally indicate a greater probability of good propagation at higher frequencies
B. Lower sunspot numbers generally indicate greater probability of sporadic E propagation
C. A zero sunspot number indicates that radio propagation is not possible on any band
D. A zero sunspot number indicates undisturbed conditions

Conversely, low sunspot numbers result in reduced ionization, negatively affecting long-distance communication. So, while sunspots may look dark and calm, they are actually dynamic regions driving much of the sun's activity that directly impacts amateur radio propagation.

Coronal Mass Ejection

A Coronal Mass Ejection (CME) is a large burst of plasma and magnetic field from the Sun's corona, the outermost layer of the Sun's atmosphere. During a CME, billions of tons of charged particles are ejected into space, traveling at high speeds. When these charged particles interact with Earth's magnetic field, they can cause geomagnetic storms, which can impact radio communications, satellite operations, and power grids.

For radio operators, CMEs are significant because they can disrupt HF radio signals by disturbing the ionosphere, leading to degraded signal quality or complete signal loss. Conversely, CMEs can also produce auroras, enhancing VHF signal propagation for short periods by providing an additional reflective surface for signals.

<u>A CME takes much longer than a solar flare to affect radio propagation on Earth, usually between 15 hours and several days.</u> CMEs are massive clouds of plasma and magnetic fields ejected from the Sun, and they travel much slower than the radiation from a solar flare. When a CME reaches Earth, it can cause significant disturbances in the geomagnetic field.

How long does it take a coronal mass ejection to affect radio propagation on Earth? (G3A11)

A. 28 days
B. 14 days
C. 4 to 8 minutes
D. 15 hours to several days

A CME is a "slow but powerful storm" compared to the almost immediate impact of a solar flare. While solar flare effects are felt in minutes, CMEs take much longer to travel from the Sun to Earth, but their impacts can last longer and be more widespread once they arrive.

Solar Flare vs Coronal Mass Ejection

A coronal mass ejection (CME) is different from a solar flare. However, they both originate from the Sun and can affect radio communications on Earth.

- **Solar Flare**: A solar flare is a sudden, intense burst of energy and radiation from the Sun's surface, mainly in the form of electromagnetic waves like X-rays and ultraviolet light. Their impact is felt within 8 minutes. Solar flares can affect radio propagation almost immediately, especially on Earth's daylit side, causing disturbances in the ionosphere. This can lead to short-term radio signal blackouts on HF bands, known as radio blackouts or fade-outs.

- **Coronal Mass Ejection (CME)**: Conversely, a CME involves the ejection of large amounts of plasma and magnetic fields into space. Unlike solar flares, CMEs take longer (15 hours to several days) to reach Earth, but they can cause geomagnetic storms when they do. These storms affect radio communications and satellite operations and can also cause auroras, which sometimes improve VHF propagation.

In summary, solar flares impact radio signals quickly and briefly. CMEs have a delayed but often longer-lasting effect due to their impact on Earth's magnetic field. Both disrupt HF communications but in different ways.

Coronal Holes

A solar coronal hole is a region on the Sun's corona (the outer atmosphere) where the Sun's magnetic field is open, allowing solar wind to escape more readily into space. When viewed in ultraviolet or X-ray light, these regions appear darker than the surrounding areas because they are cooler and less dense. Coronal holes are sources of high-speed solar wind streams, which can travel faster than the solar wind from other parts of the Sun.

When charged particles from solar coronal holes reach Earth, they can disturb HF communication. The high-speed solar wind released from these coronal holes interacts with Earth's magnetic field, causing geomagnetic disturbances. These disturbances disrupt the ionosphere. As a result, long-distance HF communication can become unreliable, with signals fading or wholly blocked.

How is long distance radio communication usually affected by the charged particles that reach Earth from solar coronal holes? (G3A14)

A. HF communication is improved
B. HF communication is disturbed
C. VHF/UHF ducting is improved
D. VHF/UHF ducting is disturbed

The charged particles they send toward Earth create geomagnetic disturbances that make it harder for radio signals to travel long distances, leading to poor HF propagation until the disturbance passes. Unlike solar flares or coronal mass ejections, the effects of coronal holes are usually less intense but can last longer.

Geomagnetic Storms

What is a geomagnetic storm? (G3A06)

A. A sudden drop in the solar flux index
B. A thunderstorm that affects radio propagation
C. Ripples in the geomagnetic force
D. A temporary disturbance in Earth's geomagnetic field

A geomagnetic storm is a temporary disturbance in Earth's geomagnetic field caused by solar activity like solar flares or coronal mass ejections (CMEs). When these bursts of solar energy reach Earth, they interact with our planet's magnetic field, disrupting it.

Just like a thunderstorm can disrupt your day, a geomagnetic storm disrupts radio signals by causing disturbances in the ionosphere, making communication more challenging until the storm subsides.

A geomagnetic storm can significantly degrade HF propagation, especially in high-latitude regions. When solar activity disturbs Earth's geomagnetic field, it disrupts the ionosphere, which is essential for reflecting HF radio signals back to Earth. This disturbance can cause HF signals to fade, become unreliable, or even be blocked entirely, particularly in areas closer to the poles with the strongest geomagnetic effects.

High-latitude operators may experience poor or no signal reception during these storms because the ionosphere becomes too unstable to effectively reflect HF signals to Earth.

How can a geomagnetic storm affect HF propagation? (G3A08)

A. Improve high-latitude HF propagation
B. Degrade ground wave propagation
C. Improve ground wave propagation
D. Degrade high-latitude HF propagation

On the other hand, high geomagnetic activity can benefit radio communications by creating auroras, which can reflect VHF signals. During intense solar activity, such as a geomagnetic storm, charged particles from the Sun interact with Earth's magnetic field and atmosphere, creating beautiful auroras (northern and southern lights). These auroras can act like mirrors in the sky, reflecting VHF signals and allowing communication over longer distances than usual. Hopefully you remember this piece from the tech exam. Or maybe you even tried it!

While geomagnetic storms often disrupt HF communications, they can enhance VHF communications by temporarily allowing signals to reflect off the ionized particles in the auroras, making long-distance VHF contacts possible.

How can high geomagnetic activity benefit radio communications? (G3A09)

A. Creates auroras that can reflect VHF signals
B. Increases signal strength for HF signals passing through the polar regions
C. Improve HF long path propagation
D. Reduce long delayed echoes

Measuring Earth's Geomagnetic Field

The A-index and K-index are two key measurements that help radio operators understand the stability of Earth's geomagnetic field and how it affects radio wave propagation. While both indices measure geomagnetic activity, they do so over different time frames.

The K-index focuses on short-term changes, providing a snapshot of geomagnetic disturbances over a 3-hour period, which helps track immediate conditions that might disrupt HF communications.

It is a measurement of the short-term stability of Earth's geomagnetic field. It ranges from 0 to 9 and indicates how much Earth's magnetic field is being disturbed by solar activity. A low K-index (0-1) means the geomagnetic field is calm and stable. At the same time, a higher K-index (5 or above) indicates significant disturbances, such as those caused by solar storms. High K-index values can lead to disruptions in HF communications and can also affect satellite systems.

What does the K-index measure? (G3A12)

A. The relative position of sunspots on the surface of the Sun
B. The short-term stability of Earth's geomagnetic field
C. The short-term stability of the Sun's magnetic field
D. The solar radio flux at Boulder, Colorado

In contrast, the A-index offers a longer-term view, averaging geomagnetic activity over an entire 24-hour period to show the overall stability of the Earth's magnetic field.

The A-index measures the long-term stability of Earth's geomagnetic field. Unlike the K-index, which tracks short-term fluctuations, the A-index provides an average value over 24 hours to show the overall level of geomagnetic activity. It is expressed on a scale from 0 to 400, with lower numbers indicating stable conditions and higher numbers reflecting more significant disturbances. A low A-index means conditions are calm, which is ideal for radio propagation. At the same time, a high A-index suggests ongoing geomagnetic instability, which can degrade HF communication.

What does the A-index measure? (G3A13)

A. The relative position of sunspots on the surface of the Sun
B. The amount of polarization of the Sun's electric field
C. The long-term stability of Earth's geomagnetic field
D. The solar radio flux at Boulder, Colorado

These indices give radio operators a comprehensive picture of geomagnetic conditions, helping them anticipate short-term and long-term effects on signal propagation.

Low Solar Activity

The 15-meter, 12-meter, and 10-meter bands are the least reliable for long-distance communications during periods of low solar activity. These higher-frequency HF bands depend heavily on strong ionization in the ionosphere to reflect signals to Earth for global communication. During low solar activity, like at the solar minimum, less solar radiation reaches the ionosphere, reducing its ability to reflect signals at these higher frequencies. As a result, signals on these bands are more likely to pass through the ionosphere instead of being refracted back to Earth, making them unreliable for long-distance contact.

Which of the following are the least reliable bands for long-distance communications during periods of low solar activity? (G3A04)

A. 80 meters and 160 meters
B. 60 meters and 40 meters
C. 30 meters and 20 meters
D. 15 meters, 12 meters, and 10 meters

Think of these bands as "high-frequency fair-weather friends." They perform best when solar activity is high, and the ionosphere is energized. Still, they struggle to support long-distance propagation during quiet periods, making lower-frequency bands like 20 meters more dependable in those conditions.

Solar Flux Index (SFI)

The Solar Flux Index (SFI) is a measure of solar radiation at a wavelength of 10.7 centimeters (about 2800 MHz), which is emitted by the Sun's outer layers. This index is used as an indicator of solar activity, as higher values generally correlate with increased ionization in the ionosphere. Higher SFI readings often mean better HF propagation, mainly on higher frequencies, since the increased solar radiation boosts the ionosphere's ability to reflect radio waves.

What is the solar flux index? (G3A05)

A. A measure of the highest frequency that is useful for ionospheric propagation between two points on Earth
B. A count of sunspots that is adjusted for solar emissions
C. Another name for the American sunspot number
D. A measure of solar radiation with a wavelength of 10.7 centimeters

Think of the Solar Flux Index as a "solar energy gauge." When the SFI is high, it signals that the Sun is more active, which typically enhances HF communication conditions by energizing the ionosphere. Monitoring the SFI helps radio operators predict when conditions will be favorable for long-distance communication.

Why did the 10.7 cm get called out? The 10.7-centimeter wavelength was specifically chosen for the Solar Flux Index because it is a reliable and consistent indicator of solar activity that correlates well with the levels of ionizing radiation impacting the ionosphere. The Sun's outer layers emit this particular wavelength and is relatively unaffected by the Earth's atmosphere, making it ideal for monitoring.

Scientists found that measuring solar emissions at 10.7 centimeters provided a stable and accurate representation of solar activity, especially sunspot activity. It directly reflects changes in the Sun's magnetic activity, making it a valuable tool for predicting HF radio propagation conditions. The 10.7-centimeter measurement has been used since the 1940s. It is a standard in solar monitoring, helping radio operators and scientists gauge solar conditions that impact communication.

Seasonal Impacts

Seasonal variations also affect propagation conditions. Higher temperatures and increased atmospheric activity can enhance VHF and UHF propagation through phenomena like tropospheric ducting in the summer.

However, during the summer, HF frequencies often experience high atmospheric noise or static levels. This noise is mainly caused by natural events like thunderstorms, which are more frequent in summer, particularly in tropical and subtropical regions. The lightning discharges create static that gets picked up by radio receivers, making it harder to hear signals clearly on these lower frequencies.

Which of the following is typical of the lower HF frequencies during the summer? (G3B12)

A. Poor propagation at any time of day
B. World-wide propagation during daylight hours
C. Heavy distortion on signals due to photon absorption
D. High levels of atmospheric noise or static

The more storms and atmospheric disturbances, the more static you'll likely encounter. This makes communication on lower HF bands noisier during the summer months compared to other times.

With their lower temperatures and reduced atmospheric turbulence, winter months can improve HF propagation, making long-distance contacts more reliable. Understanding these seasonal patterns helps operators plan their activities and maximize communication opportunities.

Determining Propagation Conditions

Propagation conditions frequently change based on atmospheric factors like solar activity and ionospheric conditions. Knowing how signals will behave on a specific band helps operators choose the best frequency for their transmissions. There are several ways to determine current propagation conditions, but one of the most reliable ways <u>to determine current propagation conditions is by using a network of online automated receiving stations.</u>

These networks, such as the Reverse Beacon Network (RBN) or PSKReporter, automatically listen for signals from stations like yours and provide real-time feedback on where your transmissions are being received around the world. By sending a short transmission, you can quickly determine how well your signal propagates and where it's reaching. This method gives you an immediate and reliable indication of propagation conditions on the desired band, making it easier to decide whether to continue using that band or switch to another for better coverage.

Which of the following is a way to determine current propagation on a desired band from your station? (G3B04)

A. Use a network of automated receiving stations on the internet to see where your transmissions are being received
B. Check the A-index
C. Send a series of dots and listen for echoes
D. All these choices are correct

These tools allow you to optimize your radio operations based on current conditions, ensuring your transmissions reach their intended destinations.

Signal Reports

Exchanging signal reports is standard practice at the start of an HF (high frequency) contact in amateur radio. These reports provide a quick assessment of signal quality between stations. This exchange helps operators adjust their equipment, such as transmitter power or antenna orientation, to improve communication-based on current conditions. The reports usually follow the RST system, which evaluates Readability, Signal strength, and Tone, allowing both operators to optimize their setup during the contact.

Why are signal reports typically exchanged at the beginning of an HF contact? (G2D11)

A. To allow each station to operate according to conditions
B. To be sure the contact will count for award programs
C. To follow standard radiogram structure
D. To allow each station to calibrate their frequency display

---

Signal Reports in HF Contacts

<u>Signal reports are typically exchanged at the beginning of an HF contact to ensure that both stations receive each other clearly and adapt to the prevailing propagation conditions.</u> By sharing a report using the RST (Readability, Signal Strength, and Tone) system, operators can assess how well their signals are being received. This helps them decide whether to adjust transmitter power, reposition their antennas, or modify other technical settings. For example, if the signal strength is weak, an operator might increase power or adjust their antenna to improve reception. This simple exchange helps ensure that the rest of the communication proceeds smoothly.

Chapter Summary

Understanding radio wave propagation is necessary for maximizing your ham radio experience as a General license holder. Propagation affects everything from local contacts to long-distance (DX) communication. It can vary with the time of day, season, solar activity, and the frequency band you use. By learning how factors like the ionosphere, tropospheric effects, and even ground reflections influence your signal, you gain the skills to select optimal frequencies and operating modes.

With this knowledge, you can maximize your reach, anticipate propagation conditions, and make informed choices for effective communication across the bands. Armed with these insights, your ability to adapt and experiment will grow, deepening your understanding and enjoyment of radio.

Chapter 5

Modulation and Bandwidth

MODULATION IS THE PROCESS of encoding information, such as voice or data, onto a carrier wave, allowing it to travel through the atmosphere. Different modulation types have unique characteristics and applications, like AM, FM, and digital methods.

Bandwidth, the range of frequencies a signal occupies, is closely tied to modulation, as each type requires different amounts of bandwidth to transmit clearly and efficiently.

Together, modulation and bandwidth form the backbone of signal quality, efficiency, and reach, making them critical concepts for every radio operator. This chapter explores how these elements work and interact.

Modulation

Modulation is the process of varying a carrier signal to transmit information such as voice, data, or video. At its core, modulation involves altering one or more properties of a carrier wave—typically its amplitude, frequency, or phase—to encode the information you wish to send. The primary purpose of modulation is to make the signal compatible with the transmission medium, allowing it to travel over long distances and through various environments. Without modulation, the raw signal would be too weak or unstable to reach its destination reliably.

You already learned a lot about basic modulation to obtain your Technician's license. For the General test, we will assume you have a baseline foundation of modulation and dive right into the deep end of the modulation pool.

Amplitude Modulation

Amplitude Modulation (AM) is one of the earliest and most straightforward methods of modulating a radio frequency (RF) signal. In AM, the carrier wave's amplitude (or strength) is varied in proportion to the amplitude of the input signal, such as voice or music. The carrier's frequency remains constant, but its instantaneous power level fluctuates based on the information being transmitted. AM has been widely used in broadcasting, particularly for AM radio stations (I'm shocked!) and in some types of voice communication systems. While it is more susceptible to noise and interference than other modulation types, it is still important in radio communication.

Generating an AM signal begins with the modulator circuit, which combines the audio signal with the carrier wave. This modulated signal then passes through an amplifier to boost its strength before being sent to the antenna and transmitted. On the receiving end, the AM receiver uses a demodulator circuit to extract the audio signal from the modulated carrier. This involves rectifying the signal to remove the negative half of the waveform, followed by filtering to remove the high-frequency carrier, leaving only the original audio.

WAVE SHAPES

Analog data signal

Digital data signal

Amplitude modulation (AM)

Digital AM

Frequency modulation (FM)

Carrier signal

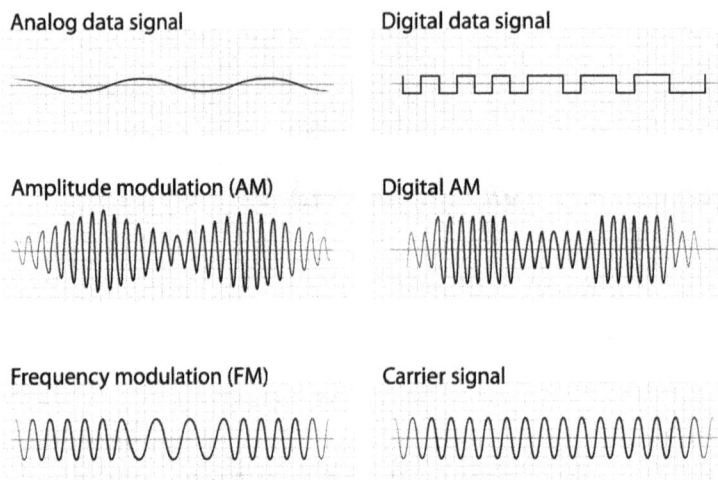

What type of modulation varies the instantaneous power level of the RF signal? (G8A05)

A. Power modulation
B. Phase modulation
C. Frequency modulation
D. Amplitude modulation

Remember amplitude modulation as controlling the "power level" or strength of the signal. As the signal's power fluctuates based on the input, the information is embedded in the varying amplitude of the carrier wave. This contrasts with frequency modulation, where the frequency of the carrier is varied instead of its amplitude.

Modulation Envelopes

In AM, a modulation envelope is a key concept to understanding how information is transmitted. The envelope represents the overall shape of the modulated signal, showing how the amplitude of the carrier wave changes in response to the input signal. By visualizing the envelope, you can see the peaks and troughs that correspond to the variations in the signal's amplitude, which carry the information being transmitted. This helps operators monitor signal quality and ensure proper modulation levels.

The modulation envelope of an AM signal is the waveform created by connecting the peak values of the modulated signal. Basically, the signal's outline or "shape" shows how the carrier wave's amplitude varies over time as it carries the information. The carrier itself oscillates rapidly at its own frequency. Still, the envelope gives a clear picture of how the signal's strength changes based on the input being transmitted.

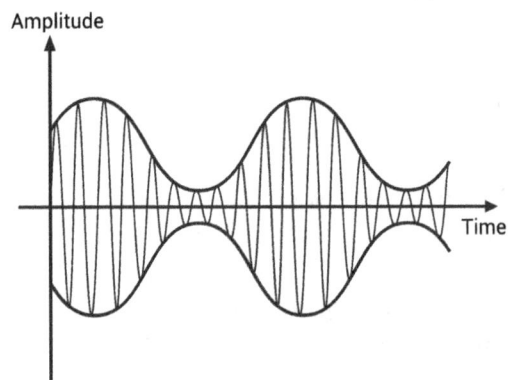

What is the modulation envelope of an AM signal? (G8A11)

A. The waveform created by connecting the peak values of the modulated signal
B. The carrier frequency that contains the signal
C. Spurious signals that envelop nearby frequencies
D. The bandwidth of the modulated signal

The envelope is a visual representation that helps you see how well the signal is being modulated. A clean, undistorted envelope indicates proper modulation, while a distorted or clipped envelope may indicate over-modulation or flat-topping.

Signal Distortion in AM

With AM, the goal is to transmit a clear signal by varying the amplitude of the carrier wave in direct relation to the input. However, when the input levels (such as the drive or speech levels) are too high, it can lead to signal distortion. Distortion not only reduces the quality of the signal but can also cause interference with adjacent frequencies, leading to poor communication.

Flat-topping occurs when an amplitude-modulated phone/ voice signal becomes distorted due to excessive drive or speech levels. When the signal is overdriven, the transmitter cannot accurately follow the peaks of the modulation, causing the signal to "clip" or flatten at the top rather than smoothly following the waveform. This results in a distorted signal, which not only reduces the clarity of the transmitted voice or data but can also cause interference to other stations.

What is meant by the term "flat-topping," when referring to an amplitude-modulated phone signal? (G8A10)

A. Signal distortion caused by insufficient collector current
B. The transmitter's automatic level control (ALC) is properly adjusted
C. Signal distortion caused by excessive drive or speech levels
D. The transmitter's carrier is properly suppressed

Flat-topping "clips" the peaks of the signal due to overdriving the transmitter. The solution is to reduce the drive or speech levels to ensure the transmitter can handle the full range of the signal without distortion.

To prevent overdriving, carefully manage input levels to ensure they stay within the transmitter's linear range. Properly adjusting the audio or modulating input levels helps keep signal integrity and avoid distortion. Using a modulation monitor allows operators to confirm that the transmitter is not exceeding 100% modulation. Additionally, ensuring that microphone and preamp gain settings are correctly adjusted will further reduce the risk of overdriving. By following these practices, operators can maintain clean and compliant transmissions, free from distortion or interference.

Frequency Modulation

Frequency Modulation (FM) works by varying the frequency of the carrier signal according to the information signal. Unlike AM, where the amplitude changes, FM keeps the amplitude constant while altering the carrier's frequency. The frequency deviation corresponds to the audio signal's instantaneous amplitude. An FM signal consists of a carrier and frequency deviations that encode the information. One of FM's primary advantages is its resilience to amplitude variations, making it less susceptible to noise and interference. This results in clearer and more reliable communication, particularly in environments with high levels of electromagnetic noise. However, FM requires more complex circuitry and a wider bandwidth than AM.

Generating an FM signal involves using a Voltage-Controlled Oscillator (VCO) in the transmitter. The VCO's audio signal modulates the VCO's frequency, resulting in a frequency-modulated carrier. The modulated signal then passes through an amplifier to increase its strength before transmission. On the receiving end, FM receivers use demodulator circuits to extract the audio signal from the frequency-modulated carrier, which locks onto the carrier's frequency and produces a voltage corresponding to the frequency deviations. This voltage is then converted back into the original audio signal.

FM is the process that changes the instantaneous frequency of a radio frequency wave to convey information. In FM, the frequency of the carrier wave is varied in response to the signal being transmitted. For example, when transmitting voice or music, the frequency of the carrier wave changes in sync with the audio signal's amplitude. This allows the signal to be encoded and then decoded by the receiver to reproduce the original sound or data.

> **What is the name of the process that changes the instantaneous frequency of an RF wave to convey information? (G8A03)**
>
> A. Frequency convolution
> B. Frequency transformation
> C. Frequency conversion
> **D. Frequency modulation**

Both AM and FM have unique advantages and drawbacks for ham radio. Understanding the differences and principles enhances your ability to choose the appropriate modulation technique for your communication needs.

Phase Modulation

Phase Modulation (PM) is a type of modulation used in radio communication where the phase angle of a carrier signal is changed to encode information. In simple terms, the phase of the carrier wave is shifted in response to the amplitude of the input signal, which could be voice or data. This modulation technique is advantageous in certain types of digital communication and is closely related to frequency modulation. However, the two operate on different principles.

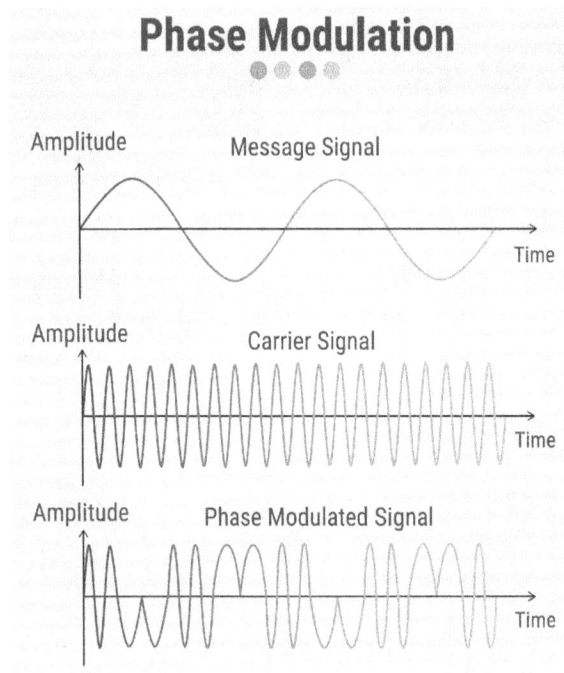

Phase Modulation

Phase modulation (PM) is the process of changing the phase angle of a radio frequency signal to convey information. The phase angle refers to the position of the waveform relative to time. This angle is adjusted in phase modulation according to the input signal. By shifting the phase of the carrier signal in proportion to the input, the information is encoded and can be transmitted over the air.

What is the name of the process that changes the phase angle of an RF signal to convey information? (G8A02)

A. Phase convolution
B. Phase modulation
C. Phase transformation
D. Phase inversion

Phase modulation is the "shifting of the angle" of the carrier wave to carry data. This differs from amplitude modulation, where the signal's amplitude changes, and from frequency modulation, where the frequency varies. PM is all about how the signal's phase shifts to encode information.

Reactance Modulators

A reactance modulator is a specific device used with a transmitter's RF amplifier stage to produce phase modulation. Reactance modulators work by varying the reactance (which can be inductive or capacitive) in a circuit, thereby shifting the phase of the carrier signal.

When a reactance modulator is connected to a transmitter's RF amplifier stage, it produces phase modulation. The reactance modulator adjusts the carrier wave's phase by changing the circuit's reactive components (inductance or capacitance). This results in the RF signal's varied phase following the input signal, which conveys the transmitted information. Phase modulation alters the timing of the wave, unlike amplitude modulation, which changes the strength of the wave.

What emission is produced by a reactance modulator connected to a transmitter RF amplifier stage? (G8A04)

A. Multiplex modulation
B. Phase modulation
C. Amplitude modulation
D. Pulse modulation

To remember this for the exam, associate a reactance modulator with phase shifts—it's a way of encoding information by adjusting the phase of the signal.

Digital Modulation

Digital modulation has transformed ham radio by offering significant advantages over traditional analog techniques. Instead of varying continuous signals as analog modulation does, digital modulation encodes information into discrete signals. This approach enhances the signal-to-noise ratio, resulting in clearer and more reliable transmissions, even in challenging conditions. It also maximizes bandwidth efficiency, enabling more data to be transmitted over narrower frequency ranges compared to the broader bandwidth requirements of analog modulation, which is more prone to noise and interference.

While the variety of digital modes may initially seem daunting, understanding their underlying modulation techniques—such as frequency shift, phase shift, or amplitude shift—helps operators choose the best mode for their specific needs. Each digital mode has a distinct purpose, whether it's long-distance communication under weak-signal conditions, efficient data transmission in noisy environments, or voice clarity over local repeaters.

By organizing digital modes by modulation type and use case, this section simplifies the learning process and prepares operators to adapt to diverse operating scenarios. Whether you're working DX with FT8, engaging in low-bandwidth text exchanges with PSK31, or using DMR for local digital voice communication, you'll gain the insights needed to match the right mode to the right purpose.

Types of Digital Modulation

Here's an organized list of digital modulation techniques, their modes, and when a ham radio operator would typically use each type:

Amplitude Shift Keying (ASK)

- **On-Off Keying (OOK)**

  - **Use Case:** Rarely used in modern ham radio but foundational for understanding Morse code. OOK was one of the earliest methods for transmitting data via simple on/off carrier pulses.

  - **When to Use:** Historical experiments or educational demonstrations of early communication methods.

Frequency Shift Keying (FSK)

- **RTTY (Radio Teletype)**

  - **Use Case:** Reliable text communication over long distances, often on HF bands.

  - **When to Use:** When you need a robust text-based mode with moderate bandwidth, use in contests or routine communication.

- **AFSK (Audio Frequency Shift Keying)**

  - **Use Case:** Used in APRS (Automatic Packet Reporting System) and some packet radio for real-time data transmission.

  - **When to Use:** For tracking, weather reporting, and sharing text messages or small files over VHF/UHF.

- **FT8 and FT4**

  - **Use Case:** Ultra-efficient weak-signal communication.

  - **When to Use:** When conditions are poor (low sunspot activity, high noise) or when working DX stations with minimal power.

- **Packet Radio (AX.25)**

  - **Use Case:** Reliable data transfer and file sharing, often on VHF/UHF.

  - **When to Use:** For digital communication in local ham networks, message relays, and emergency communications.

MODULATION AND BANDWIDTH   303

- **JT65/JT9**

  - **Use Case:** Ultra-weak signal communication with precise timing and narrow bandwidth.

  - **When to Use:** For very long-distance communication, often on HF or when atmospheric noise levels are high.

- **WSPR (Weak Signal Propagation Reporter)**

  - **Use Case:** Propagation analysis to map signal paths globally with very low power.

  - **When to Use:** To test antennas, propagation conditions, or experimental setups.

Phase Shift Keying (PSK)

- **BPSK (Binary Phase Shift Keying)**

  - **Use Case:** Simple phase modulation for low-power and robust communications.

  - **When to Use:** When working with very narrow bandwidth modes or basic digital communication.

- **QPSK (Quadrature Phase Shift Keying)**

  - **Use Case:** Efficient high-speed data transmission.

  - **When to Use:** For advanced digital modes where higher data rates are needed, such as satellite communication.

- **PSK31**

  - **Use Case:** Keyboard-to-keyboard text communication with narrow bandwidth.

  - **When to Use:** For casual ragchews and when you want to operate at low power but still achieve reliable text exchange on HF.

- **MFSK (Multiple Frequency Shift Keying)**

  - **Use Case:** Combines FSK with multiple tones for robustness against noise and interference.

  - **When to Use:** When operating in noisy HF conditions or requiring error-resistant text-based communication.

Quadrature Amplitude Modulation (QAM)

- **QAM (General Use)**

  - **Use Case:** Common in high-data-rate systems like digital TV or satellite links.

  - **When to Use:** Rarely directly used in amateur radio but foundational for SDR and modern communication systems.

- **DVB-S2 (Digital Video Broadcasting Satellite)**

  - **Use Case:** For high-bandwidth applications like amateur TV (ATV).

  - **When to Use:** When transmitting or receiving video or multimedia via satellite or terrestrial amateur radio TV systems.

Hybrid or Other

- **Olivia**

  ○ **Use Case:** Highly resilient text-based mode that works well in weak and noisy conditions.

  ○ **When to Use:** When other modes fail due to noise or weak signals, especially for reliable HF communication.

- **Pactor**

  ○ **Use Case:** Fast and efficient data transfer over HF.

  ○ **When to Use:** For email or file transfers, often in Winlink systems during emergencies or maritime operations.

- **Winlink**

  ○ **Use Case:** A global email network that uses various digital modes like Pactor.

  ○ **When to Use:** When sending or receiving emails from remote locations without internet access.

- **Digital Voice Modes (D-STAR, DMR, Fusion)**

  ○ **Use Case:** Clear digital voice communication with additional data capabilities.

  ○ **When to Use:** For local or regional voice communication with linked repeaters or during events and emergencies.

- **FSQ (Fast Simple QSO)**

  ○ **Use Case:** Fast text messaging between stations, including selective calling.

  ○ **When to Use:** When efficiency and rapid communication are required, such as during nets or emergencies.

Frequency Shift Keying

Frequency Shift Keying (FSK) is a digital modulation technique where the frequency of the carrier signal is changed to represent digital information, such as binary data.

In binary FSK (BFSK), two distinct frequencies are used to represent binary "1" and "0." This method is widely used in data communication systems because it is simple to implement and provides reliable communication, even in noisy environments. FSK's ability to shift frequencies to convey data makes it a robust and efficient method for transmitting digital signals.

The two distinct frequencies are identified as mark and space in an FSK signal. The mark frequency represents a binary "1" and the space frequency represents a binary "0." The FSK signal can transmit digital data by shifting between these two frequencies. The concept of "mark" and "space" comes from the early days of telegraphy, where the terms were used to describe the presence and absence of a signal.

How are the two separate frequencies of a Frequency Shift Keyed (FSK) signal identified? (G8C11)

A. Dot and dash
B. On and off
C. High and low
D. Mark and space

Direct binary FSK modulation is generated by directly changing an oscillator's frequency with a digital control signal. In this method, a digital signal (representing 1s and 0s) controls the frequency of the oscillator. One frequency represents a binary "1," and another describes a binary "0." By switching between these two frequencies, the oscillator encodes the binary data onto the carrier wave, allowing digital information to be transmitted over radio waves.

How is direct binary FSK modulation generated? (G8A01)

A. By keying an FM transmitter with a sub-audible tone
B. By changing an oscillator's frequency directly with a digital control signal
C. By using a transceiver's computer data interface protocol to change frequencies
D. By reconfiguring the CW keying input to act as a tone generator

---

To remember this for the exam, think of binary FSK as "direct control" of an oscillator's frequency. Each binary state (1 or 0) causes the oscillator to shift to a corresponding frequency, making it a straightforward yet effective method of modulation for digital communication systems.

Baudot Code

Baudot code is an early form of character encoding used for transmitting text in telecommunication systems, particularly in radio teletype (RTTY) communication. It was one of the first digital telegraph and early digital communication codes. Baudot is still in use today in some ham radio applications, especially for RTTY, where it allows text to be sent efficiently over radio waves. The Baudot code is relatively simple and is known for its use of 5 bits to represent each character, allowing for a streamlined method of text communication.

The Baudot code is a 5-bit code that includes additional start and stop bits to ensure the receiving system can synchronize with the incoming characters. Since 5 bits can only represent a limited number of characters, Baudot also uses a system of shift codes to switch between letters and figures, allowing more symbols to be represented in the same 5-bit structure. The start and stop bits help the receiver detect the beginning and end of each character, making the system more reliable for long-distance text communication.

Which of the following describes Baudot code? (G8C04)

A. A 7-bit code with start, stop, and parity bits
B. A code using error detection and correction
C. A 5-bit code with additional start and stop bits
D. A code using SELCAL and LISTEN

Radio Teletype (RTTY)

RTTY (Radio Teletype) is a method of sending text over radio waves using a system of digital signals. It was one of the earliest digital modes in ham radio and telecommunications. In RTTY, characters are encoded into a digital format (often using Baudot code) and then transmitted by shifting between two tones, commonly referred to as frequency-shift keying (FSK).

RTTY signals are characterized by a specific baud rate, commonly 45.45 baud, and distinct mark and space frequencies representing the binary states. RTTY is widely used for text-based communication, allowing operators to exchange messages over long distances. Its robustness makes it a preferred mode for emergency messaging, where reliable and quick communication is vital. RTTY's simplicity and effectiveness have ensured its continued popularity.

FT8

FT8, developed for weak-signal communication, showcases the cutting-edge capabilities of digital modulation. FT8 operates using very short transmission cycles of 15 seconds, allowing for rapid exchanges of information. This mode is designed to work under extremely poor propagation conditions, making it possible to achieve contact even when signals are barely above the noise floor. FT8's narrow bandwidth further enhances its effectiveness, concentrating the signal's power and reducing the likelihood of interference. Practical applications of FT8 include DXing, where operators aim to contact distant stations, and low-power communication, making it an excellent choice for operators with limited transmission power.

Here is a fun, irrelevant fact about FT8: the man who developed FT8 is a Nobel Prize-winning physicist, Joe Taylor. But don't worry, that's not on the exam.

FT8 uses 8-tone frequency shift keying (8-FSK) as its modulation method. In FT8, there are eight possible frequencies (or tones), each representing a different symbol, allowing the transmission of small data packets quickly and efficiently. This method makes FT8 highly effective at transmitting under weak-signal conditions, as the use of multiple tones enables reliable decoding even when signals are faint or compromised by noise.

What type of modulation is used by FT8? (G8A09)

A. 8-tone frequency shift keying
B. Vestigial sideband
C. Amplitude compressed AM
D. 8-bit direct sequence spread spectrum

FT8 is capable of receiving signals with very low signal-to-noise ratios (SNR), making it one of the best narrowband digital modes for weak-signal communication. The mode can decode signals with SNRs as low as -24 dB, meaning it can pull a usable signal out of background noise where many other modes would fail. This makes FT8 especially valuable in crowded or noisy bands or during poor propagation conditions, where other modes might not perform as well.

Which of the following narrowband digital modes can receive signals with very low signal to noise ratios? (G8C07)

A. MSK144
B. FT8
C. AMTOR
D. MFSK32

To remember this for the exam, think of FT8 as a mode that "hears whispers" in the noise. It excels at pulling out weak signals, making it the go-to digital mode when signal strength is very low, but communication is still needed.

FT8 Signal Reports

In FT8, signal reports are expressed in decibels (dB) relative to the noise floor. This signal-to-noise ratio (SNR) indicates how well the signal stands out from the background noise. FT8 is designed to decode signals with very low SNRs, making it a powerful tool for long-distance communication, even under poor conditions.

An FT8 signal report of +3 dB means that the signal received is 3 dB above the noise floor in a 2.5 kHz bandwidth. In simpler terms, the signal is stronger than the background noise by a factor of +3 dB, making it relatively easy to decode. Positive signal reports like +3 indicate a strong signal relative to the noise, while negative values would indicate weaker signals.

What does an FT8 signal report of +3 mean? (G8C15)

A. The signal is 3 times the noise level of an equivalent SSB signal
B. The signal is S3 (weak signals)
C. The signal-to-noise ratio is equivalent to +3dB in a 2.5 kHz bandwidth
D. The signal is 3 dB over S9

And, as a reminder, the 2.5 kHz bandwidth is specified in this question because it represents the typical bandwidth used for receiving digital modes like FT8. Signal-to-noise ratio (SNR) is often measured relative to the bandwidth being used, and in FT8, this is standardized at 2.5 kHz.

Packet Radio

Packet radio is a digital communication mode used to send and receive data over radio waves, similar to how data is transmitted over the internet. In packet radio, data is broken into small units called packets, which are sent individually, allowing for error-checking and reliable communication over long distances. This system enables the transfer of text, files, and even email-like messages between operators, often using VHF or UHF frequencies.

Each packet of data includes a header (with routing & handling information) and the actual message or data. Packet radio networks often use a terminal node controller (TNC) to handle the packetizing of data, error detection, and retransmission of lost packets. The system is popular in ham radio because it can send data even in poor conditions. It is used for messaging, telemetry, and connecting to packet bulletin board systems (BBS) or the internet.

What part of a packet radio frame contains the routing and handling information? (G8C03)

A. Directory
B. Preamble
C. Header
D. Trailer

In essence, packet radio turns radio signals into a form of digital communication, making it possible for hams to exchange data much like emails or text messages, but over the airwaves.

Weak Signal Propagation Reporter

WSPR (Weak Signal Propagation Reporter) is a digital mode for low-power transmission. It is widely used by operators to assess HF propagation. WSPR is part of a system that allows operators to send and receive low-power signals to study how radio waves propagate through the atmosphere. The primary goal of WSPR is to gather data on how signals travel over long distances and through different atmospheric conditions, making it an excellent tool for understanding propagation at various times of the day and in different seasons.

WSPR is a digital mode that acts as a low-power beacon to assess HF propagation. Operators use WSPR to send weak signals that can be decoded even under poor conditions. These signals are transmitted at very low power, often just a few watts or less, and are received by other WSPR stations around the world. By tracking which signals are received and where, amateur radio operators can assess how well HF signals are propagating under current conditions. This makes WSPR an invaluable tool for studying propagation without needing high-power equipment.

Which digital mode is used as a low-power beacon for assessing HF propagation? (G8C02)

A. WSPR
B. MFSK16
C. PSK31
D. SSB-SC

WSPR is a "whisper" in the world of radio. This low-power signal travels far and helps assess how well radio waves move through the atmosphere. It's ideal for propagation testing because it works even with weak signals and challenging conditions.

Phase Shift Keying

Phase Shift Keying 31

PSK31, or Phase Shift Keying 31 baud, exemplifies the advantages of digital modulation. It works by varying the phase of the carrier signal to encode data, providing a robust and efficient communication method. PSK31's signal characteristics include a very narrow bandwidth of approximately 31.25 Hz, which allows multiple signals to coexist within the same frequency range without significant interference. The mode supports low-power communication, making it ideal for operators looking to achieve long-distance contact with minimal power. PSK31 also incorporates error correction features, enhancing the reliability of data transmission. This mode is prevalent in digital contests, where efficient use of spectrum and low power requirements make for a challenging contest.

Varicode

One of the key elements of PSK31 is its use of Varicode, a character encoding system specifically designed for PSK31. Unlike fixed-length codes like ASCII, Varicode uses shorter codes for more frequently used characters and longer codes for less common ones, making data transmission faster and more efficient.

Varicode is a code used in PSK31 to send characters. It is a variable-length encoding system, meaning that more ordinary characters, like vowels or spaces, are represented by shorter sequences of bits. In comparison, less common characters are defined by longer sequences. This makes Varicode very efficient, especially in a mode like PSK31, where bandwidth is limited, and efficiency is key. Using Varicode, PSK31 ensures faster data transmission while maintaining clear communication with minimal bandwidth.

Which type of code is used for sending characters in a PSK31 signal? (G8C12)

A. Varicode
B. Viterbi
C. Volumetric
D. Binary

Varicode is a more innovative way to encode text and makes PSK31 fast and efficient for real-time keyboard-to-keyboard digital communication.

In PSK31, using uppercase letters will slow the transmission because they require longer Varicode bit sequences than lowercase letters or other common characters. This difference in encoding makes uppercase transmissions take longer, which can slightly reduce communication efficiency in this mode. The encoding system is designed this way to optimize for speed in typical usage, where lowercase letters dominate.

Which of the following statements is true about PSK31? (G8C08)

A. Upper case letters are sent with more power
B. Upper case letters use longer Varicode bit sequences and thus slow down transmission
C. Error correction is used to ensure accurate message reception
D. Higher power is needed as compared to RTTY for similar error rates

Quadrature Phase Shift Keying

QPSK (Quadrature Phase Shift Keying) is a digital modulation technique used in radio communication. In QPSK, the phase of the carrier signal is changed to represent data, but unlike simpler phase-shift keying methods (such as binary PSK, which only shifts the phase by 180 degrees), QPSK uses four distinct phase shifts: 0°, 90°, 180°, and 270°. This allows QPSK to transmit two bits of data per symbol, effectively doubling the data rate compared to binary PSK.

Said again, QPSK modulation transmits digital data by shifting the phase of the carrier signal in one of four possible angles: 0°, 90°, 180°, or 270°. Each phase shift represents a pair of bits, also known as a "dibit." For example, 0° might represent "00," 90° might represent "01," and so on. Using four distinct phase shifts, QPSK efficiently encodes two bits per symbol, allowing faster data transmission without requiring more bandwidth.

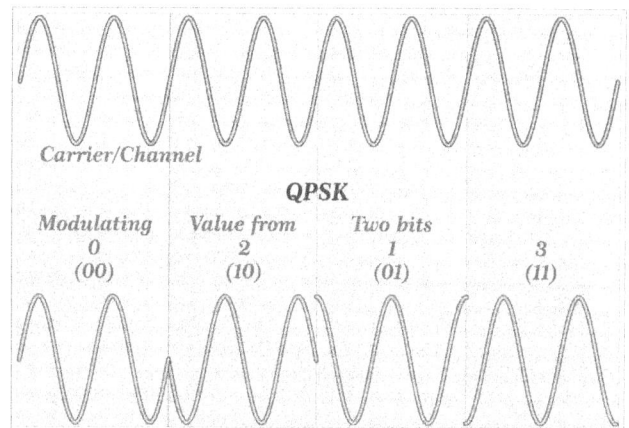

What is QPSK modulation? (G8A12)

A. Modulation using quasi-parallel to serial conversion to reduce bandwidth
B. Modulation using quadra-pole sideband keying to generate spread spectrum signals
C. Modulation using Fast Fourier Transforms to generate frequencies at the first, second, third, and fourth harmonics of the carrier frequency to improve noise immunity
D. Modulation in which digital data is transmitted using 0-, 90-, 180- and 270-degrees phase shift to represent pairs of bits

QPSK transmits two bits of data with each phase shift, making it more efficient than simpler schemes like BPSK, which uses only two-phase shifts. This efficiency makes QPSK ideal for high-speed systems that require fast data transmission.

By encoding two bits per symbol, QPSK improves bandwidth efficiency while maintaining robustness against noise. It is commonly used in satellite communications, wireless networks, and digital TV for its ability to handle larger data loads.

QPSK31

QPSK31 is a digital mode used in ham radio communication that combines Quadrature Phase Shift Keying with a baud rate of 31.25 (hence the "31" in the name). QPSK31 is similar to PSK31, but it offers greater robustness due to its error correction capabilities. Like PSK31, it is narrowband, making it very efficient in terms of bandwidth use. QPSK31's encoding method allows for reliable communication, even under poor signal conditions, and it is particularly suitable for low-power, long-distance communication, commonly known as "weak signal" work.

QPSK31 characteristics:

- Sideband Sensitive: QPSK31 is sensitive to the sideband (upper or lower) in which it is transmitted. If the sideband is incorrect, the signal might not decode properly, making it crucial to use the correct sideband for transmission.

- Error Correction: QPSK31 incorporates error correction coding, which makes it more resilient to interference and signal degradation. This means that the system can automatically detect and correct some transmission errors, improving the reliability of the communication.

- Narrow Bandwidth: QPSK31 has a very narrow bandwidth, approximately the same as BPSK31 (Binary Phase Shift Keying). Both modes use around 31.25 Hz of bandwidth, making them ideal for communicating over long distances using low power.

Which of the following is characteristic of QPSK31? (G8A06)

A. It is sideband sensitive
B. Its encoding provides error correction
C. Its bandwidth is approximately the same as BPSK31
D. All these choices are correct

Focus on the fact that QPSK31 is sideband sensitive, uses error correction to improve signal reliability, and operates within the same narrow bandwidth as BPSK31, which helps with efficient, long-distance communication.

But wait... what's BPSK31?

Binary Phase Shift Keying

BPSK31 stands for Binary Phase Shift Keying 31.25 baud, and it is a popular digital mode used in ham radio for text-based communication. It operates by modulating the phase of a carrier signal between two distinct states (0° and 180°), which represents binary data (1s and 0s). The "31" in BPSK31 refers to the baud rate, which is 31.25 baud, meaning it transmits 31.25 symbols per second, making it an efficient mode for weak signal and low-power communication.

BPSK31 is well known for its narrow bandwidth of approximately 31 Hz. Since it uses minimal bandwidth, it can work effectively even when signal strength is low, and it is often used for keyboard-to-keyboard digital communication. Operators appreciate BPSK31's simplicity and ability to maintain communication when voice modes fail due to poor signal conditions.

Overmodulation

Overmodulation occurs when the amplitude of the modulating signal exceeds the limits that the transmitter or system is designed to handle. In simpler terms, it's when the input signal is too strong for the transmitter, leading to distortion and interference in the transmitted signal. This often causes the signal to "spill over" into adjacent frequencies, creating unwanted side effects like distortion and interference with other communications.

In amplitude modulation (AM), for example, overmodulation can cause the peaks of the modulated signal to be "clipped," which leads to signal distortion. In frequency modulation (FM) or digital modes, overmodulation results in a signal that occupies more bandwidth than intended, causing interference and degraded transmission quality.

Preventing overmodulation is crucial to ensure a clear, high-quality signal and to avoid interfering with nearby frequencies or other operators. It can often be managed by reducing the input signal level to the transmitter.

AM Overmodulation

In amplitude modulation (AM) and other modulation methods, maintaining a proper balance between the input signal and the carrier wave is essential for clear transmission. Suppose the modulation goes beyond the allowable level. In that case, it can lead to signal distortion, poor transmission quality, and unintended interference with other signals. Overmodulation is a common issue that can result in unwanted side effects, such as excessive bandwidth, splatter, or interference with adjacent frequencies.

One of the main effects of overmodulation is the production of excessive bandwidth. When a signal is overmodulated, it causes the transmitter to generate unwanted harmonics and sidebands that spread beyond the intended frequency range. This results in the signal occupying more bandwidth than necessary, potentially causing interference with adjacent channels and other signals.

Which of the following is an effect of overmodulation? (G8A08)

A. Insufficient audio
B. Insufficient bandwidth
C. Frequency drift
D. Excessive bandwidth

This not only degrades the quality of the transmission but also causes interference, which is detrimental to both the operator and others sharing the frequency spectrum.

Digital Overmodulation

Waterfall Displays

A waterfall display visualizes signals received over time in a particular frequency range, commonly used in digital modes and radio teletype (RTTY). On this display, time flows downward while frequency is shown across the horizontal axis, and the intensity or strength of the signal is displayed using color gradients. This tool helps operators visualize multiple signals, their strength, and any interference in real-time. Understanding how to interpret the patterns on a waterfall display is key for identifying issues like overmodulation and adjusting your transmission settings.

On a waterfall display, overmodulation appears as one or more vertical lines on either side of the data mode or RTTY signal. Overmodulation occurs when the input level to the transmitter is too high, causing distortion and unwanted side effects like these vertical lines. This can interfere with other signals and reduce the clarity of your own transmission. The vertical lines indicate that the signal has exceeded the intended modulation limits, spilling into adjacent frequencies, which can cause interference.

What is indicated on a waterfall display by one or more vertical lines on either side of a data mode or RTTY signal? (G8C13)

A. Long path propagation
B. Backscatter propagation
C. Insufficient modulation
D. Overmodulation

Frequency is represented along the horizontal axis, showing a range of frequencies, you are monitoring. Signal strength is indicated by the intensity or color brightness; stronger signals appear more intense or brighter. Time is represented vertically, meaning older signals are shown at the top, and more recent signals are displayed at the bottom as the display updates in real-time.

Which of the following describes a waterfall display? (G8C14)

A. Frequency is horizontal, signal strength is vertical, time is intensity
B. Frequency is vertical, signal strength is intensity, time is horizontal
C. Frequency is horizontal, signal strength is intensity, time is vertical
D. Frequency is vertical, signal strength is horizontal, time is intensity

---

Remember overmodulation as "spilling over" onto other frequencies, and the vertical lines on the waterfall display are a visual sign that the signal is too strong and needs to be adjusted to prevent interference.

Unmodulated Carrier

An unmodulated carrier is a continuous radio frequency signal transmitted without any additional information or modulation. In radio communication, the carrier wave serves as the base signal that can be modulated by varying its amplitude, frequency, or phase to carry voice, data, or other information. However, when the carrier is unmodulated, it remains constant in amplitude and frequency, and no information is transmitted.

An unmodulated carrier is often used as a reference signal or during testing and calibration of transmitters and receivers. It doesn't carry any meaningful communication, but it provides a foundation for modulation to be added later.

What is the ratio of PEP to average power for an unmodulated carrier? (G5B11)

A. 0.707
B. 1.00
C. 1.414
D. 2.00

The ratio of Peak Envelope Power (PEP) to average power for an unmodulated carrier is 1.00. This means that the PEP and the average power are the same for an unmodulated carrier. Since an unmodulated carrier is a steady, continuous signal with no variations in amplitude, there are no peaks in power beyond the average level, so the PEP is equal to the average power.

To remember this for the exam, think of an unmodulated carrier as a signal without fluctuations. Since PEP measures the maximum power in a signal's envelope, and an unmodulated carrier has no variations, the maximum power (PEP) and the average power are identical, giving a 1:1 ratio.

Try this one:

What is the output PEP of an unmodulated carrier if the average power is 1060 watts? (G5B13)

A. 530 watts
B. 1060 watts
C. 1500 watts
D. 2120 watts

An unmodulated carrier's output PEP equals its average power, a 1 to 1 ratio. In the case of an unmodulated carrier, there are no variations in the signal, so the average power and PEP are the same. Therefore, if the average power is 1060 watts, the PEP is also 1060 watts.

Remember that there are no peaks or dips in power for an unmodulated carrier, as it is a steady signal. Since PEP measures the maximum power of a signal's envelope, and an unmodulated carrier is constant, its PEP is the same as its average power.

Data Reliability and Networking

In the world of digital communication, transmitting data accurately and efficiently is as critical as sending the signal itself. While modulation techniques lay the foundation for encoding and transmitting information, maintaining data integrity and managing networks are the keys to seamless communication.

This section explores the tools and strategies that ensure reliable data exchange, even in challenging conditions. From Automatic Repeat reQuest (ARQ), which handles retransmissions to correct errors, to Forward Error Correction (FEC), which minimizes disruptions by proactively encoding redundancy, we'll dive into how these systems work together to enhance communication.

Additionally, we'll discuss mesh networks, an innovative approach to creating resilient and adaptive connections, especially in emergency or remote environments. By understanding these concepts, you'll gain a comprehensive view of how modern radio systems achieve reliability and adaptability, empowering you to operate more effectively in any scenario.

Automatic Repeat reQuest

ARQ (Automatic Repeat reQuest) is a communication protocol used in digital radio and data transmission systems to ensure the accurate delivery of data. In ARQ mode, the transmitter sends a packet of data and then waits for an acknowledgment from the receiver. If the receiver detects any errors in the data, it sends a request back to the transmitter to resend the data. This process continues until the data is correctly received and acknowledged, ensuring reliable communication even in noisy or weak signal conditions.

Key Features of ARQ

- Error detection: The receiver checks for errors in the transmitted data.

- Automatic retransmission: If errors are detected, the receiver requests a retransmission.

- Reliability: ARQ ensures that the data is accurately transmitted by repeating the process until the data is error-free.

ARQ is often used in systems where data accuracy is critical, such as in digital communication modes and internet communication protocols like TCP/IP. In amateur radio, it is used in modes like Pactor and other digital modes requiring error-free communication.

This mode is highly efficient for maintaining the integrity of transmitted data, particularly in situations where interference or weak signals may corrupt the information being sent.

Transmission Attempts

In ARQ mode, the system will drop the connection if there are excessive transmission attempts without successfully exchanging information. This happens because there is a limit to how many times the system will try to resend a packet after receiving a NAK (Negative Acknowledgment) due to errors. If too many attempts fail, the connection is terminated, as it becomes clear that the communication link is unreliable at that moment.

What action results from a failure to exchange information due to excessive transmission attempts when using an ARQ mode? (G8C06)

A. The checksum overflows
B. The connection is dropped
C. Packets will be routed incorrectly
D. Encoding reverts to the default character set

Consider ARQ, a system that knows when to "give up" after too many failed retries. If it can't send data correctly after a set number of tries, the connection is dropped to prevent wasting time and resources on an unstable link.

Negative Acknowledgment

In ARQ mode, a NAK (Negative Acknowledgment) response means that the receiver has detected an error in the transmitted packet and is requesting the transmitter to retransmit the packet. The NAK serves as a signal to the transmitter that the data was not correctly received. This process continues until the receiver gets the correct data and sends an acknowledgment (ACK). By using NAK and ACK responses, ARQ ensures accurate communication by automatically handling errors without manual intervention.

In an ARQ mode, what is meant by a NAK response to a transmitted packet? (G8C05)

A. Request retransmission of the packet
B. Packet was received without error
C. Receiving station connected and ready for transmissions
D. Entire file received correctly

To remember this for the exam, think of NAK as a "No, send it again" response. It's part of the system that ensures reliable data transfer by requesting retransmission when needed.

Forward Error Correction

Forward Error Correction (FEC) is a technique used in digital communication systems to enhance the reliability of data transmission by allowing the receiver to detect and correct errors without needing to request a retransmission. In FEC, redundant information is added to the original data before transmission. This redundancy allows the receiver to identify and fix errors that may occur due to noise, interference, or weak signals. FEC is commonly used in radio communication, satellite links, and digital modes to improve communication in challenging conditions.

FEC works by sending redundant information along with the actual data. For example, suppose a bit of data is lost or corrupted during transmission. In that case, the redundant information provides clues that help the receiver reconstruct the correct data. This method improves communication reliability, especially in environments where retransmissions are not practical, or signals are weak.

How does forward error correction (FEC) allow the receiver to correct data errors? (G8C10)

A. By controlling transmitter output power for optimum signal strength
B. By using the Varicode character set
C. By transmitting redundant information with the data
D. By using a parity bit with each character

FEC is a way of "building in safety nets" with extra data, so even if part of the message is lost or distorted, the receiver can still piece it together correctly without asking for a resend.

Mesh Networks

Mesh networks are a type of communication network where each node, or station, is connected to multiple other nodes, creating a robust and flexible network. These nodes communicate with each other to forward data. If one node fails, the network can dynamically reroute the data through alternate nodes. This makes mesh networks highly reliable, especially when individual nodes become unavailable or damaged. Mesh networks are commonly used in microwave communications. They are becoming more popular in amateur radio for their reliability and ease of setup.

<u>In a mesh network of microwave nodes, if one node fails or becomes unreachable, the network can reroute the data packet through another available node to ensure that the packet still reaches its target.</u> This is possible because each node in a mesh network is connected to multiple other nodes, allowing alternate paths to be used if the primary route is blocked. This redundancy makes mesh networks highly resilient to node failures and other issues that might disrupt communication in more traditional network setups.

Which is true of mesh network microwave nodes? (G8C09)

A. Having more nodes increases signal strengths
B. If one node fails, a packet may still reach its target station via an alternate node
C. Links between two nodes in a network may have different frequencies and bandwidths
D. More nodes reduce overall microwave out of band interference

---

Tools like ARQ and FEC enhance accuracy and minimize errors, while mesh networks offer flexibility and adaptability, especially in complex or challenging environments. Together, these technologies form the backbone of modern digital communication, empowering operators to maintain seamless connections under varying conditions. By mastering these principles, you'll be well-equipped to ensure reliable communication, whether you're exchanging data in everyday operations or tackling the challenges of emergency scenarios.

Bandwidth

Bandwidth refers to the range of frequencies a signal occupies. Different modulation methods use different amounts of bandwidth, which affects how efficiently the frequency spectrum is used and how much interference might occur. Phone emissions, such as AM, FM, and single sideband, vary in bandwidth requirements. For efficient communication, especially in crowded frequency bands, using narrower bandwidth signals is often preferred because it reduces the chance of interference and allows more signals to occupy the same frequency range.

Single Sideband

<u>SSB is a form of amplitude modulation that uses the narrowest bandwidth among phone emissions</u>. In SSB, only one of the two sidebands (either the upper sideband or the lower sideband) is transmitted, and the carrier is either significantly reduced or entirely suppressed. This results in a much narrower signal than traditional AM or FM, making SSB more efficient in spectrum use. SSB typically occupies about 2.5 to 3 kHz of bandwidth, while AM can use up to 6 kHz or more.

Which of the following phone emissions uses the narrowest bandwidth? (G8A07)

A. Single sideband
B. Vestigial sideband
C. Phase modulation
D. Frequency modulation

SSB is the "efficient" modulation mode, using the smallest bandwidth while allowing clear voice communication. This makes it a popular choice in amateur radio and long-distance communication, where minimizing bandwidth is crucial.

Frequency Modulation

In frequency modulation (FM), the total bandwidth of a signal depends on two key factors: the deviation and the modulating frequency. The deviation refers to how much the carrier frequency shifts from its center frequency in response to the input signal. In contrast, the modulating frequency is the highest in the signal being transmitted (like voice or music).

To calculate the total bandwidth of an FM transmission, you use Carson's Rule, which estimates the bandwidth as follows:

$$Total\ Bandwidth = 2 \times (Deviation + Modulating\ Frequency)$$

What is the total bandwidth of an FM phone transmission having 5 kHz deviation and 3 kHz modulating frequency? (G8B06)

A. 3 kHz
B. 5 kHz
C. 8 kHz
D. 16 kHz

Given an FM transmission with a 5 kHz deviation and a 3 kHz modulating frequency, Carson's Rule gives us the following:

$$Total\ Bandwidth = 2 \times (5\ kHz + 3\ kHz) = 2 \times 8\ kHz = 16\ kHz$$

Thus, the total bandwidth of the FM transmission is 16 kHz.

Carson's Rule helps you estimate the bandwidth of an FM signal by doubling the sum of the deviation and the modulating frequency. This formula gives you a quick way to figure out the approximate bandwidth an FM signal uses.

Symbol Rate and Bandwidth

Symbol rate refers to how many symbols are transmitted per second in a digital communication system. Each symbol represents a unit of data, which can include one or more bits. The bandwidth is the range of frequencies required to transmit the signal without distortion. Understanding the relationship between symbol rate and bandwidth is crucial in radio communication because the faster you send data (higher symbol rate), the more bandwidth is required to transmit it effectively.

A direct relationship exists between transmitted symbol rate and bandwidth: higher symbol rates require wider bandwidth. When you transmit more symbols per second, you are packing more information into the signal, increasing the frequency range needed to carry that information. The transmitted data will become distorted or lost if the bandwidth is too narrow for the symbol rate. The bandwidth must be wide enough to accommodate the higher symbol rate to ensure clear, efficient communication.

What is the relationship between transmitted symbol rate and bandwidth? (G8B10)

A. Symbol rate and bandwidth are not related
B. Higher symbol rates require wider bandwidth
C. Lower symbol rates require wider bandwidth
D. Bandwidth is half the symbol rate

Consider symbol rate as how fast you are "talking" in data terms. The faster you "talk" (higher symbol rate), the more room (wider bandwidth) you need for the signal to pass through clearly.

Chapter Summary

In summary, modulation and bandwidth are fundamental aspects of radio communication, influencing signal clarity and transmission range. Modulation enables signals to convey voice, data, or video by modifying a carrier wave with information. Bandwidth, on the other hand, determines the spectrum space that modulated signals occupy and affects the ability to differentiate multiple signals within the same frequency range.

Understanding how modulation types impact bandwidth helps radio operators select appropriate methods for specific applications, balancing clarity, efficiency, and signal range to optimize communication.

Chapter 6

Frequency

RADIO WAVE PROPAGATION AND frequency are closely related because the frequency of a radio wave determines how it interacts with the atmosphere and the ionosphere. Lower frequencies, like those in the HF range, can be refracted or bent back to Earth by the ionosphere, enabling long-distance communication.

However, this ability depends on factors like solar activity and the state of the ionosphere at the time. Higher frequencies, like VHF and UHF, typically pass through the ionosphere and are not reflected, which limits them to line-of-sight communication.

Low vs High HF Bands

Time for a definition review. You've probably seen this idea already in this book: low high-frequency and high high-frequency. So, what are they?

The terms lower high-frequency (HF) bands and higher high-frequency (HF) bands refer to specific segments within the HF spectrum, which spans from 3 MHz to 30 MHz. The HF spectrum is widely used in amateur radio for long-distance communication due to its ability to reflect off the ionosphere.

- **Lower HF Bands**: Typically refers to frequencies closer to the **3 MHz to 10 MHz** range. These include bands like **80 meters (3.5–4.0 MHz)** and **40 meters (7.0–7.3 MHz)**. These frequencies:

  - Are more reliable for nighttime communication due to better ionospheric conditions.

  - Are more prone to ground wave propagation, making them useful for shorter-range communication during the day.

  - Tend to have higher atmospheric noise levels, such as static from thunderstorms.

- **Higher HF Bands**: Typically refers to frequencies in the **10 MHz to 30 MHz** range, including bands like **20 meters (14.0–14.35 MHz)**, **15 meters (21.0–21.45 MHz)**, and **10 meters (28.0–29.7 MHz)**. These frequencies:

  - Are more effective for daytime ionospheric propagation, especially during high solar activity.

  - Are better suited for long-distance communication (DX) as they reflect well off the ionosphere.

  - Generally, experience lower atmospheric noise, leading to cleaner signals.

In Summary:

- **Lower HF bands**: Better at night, shorter-range, more noise.

- **Higher HF bands**: Better during the day, long-distance, less noise.

Understanding how these bands behave helps operators choose the right frequency for optimal communication based on time, conditions, and desired range.

Intermediate Frequency (IF)

Intermediate Frequency (IF) is a crucial concept in superheterodyne receivers, which are widely used in radio communication. A superheterodyne receiver aims to convert the incoming radio signal to a lower, fixed frequency—called the intermediate frequency—before the signal is processed further. This conversion makes it easier to filter, amplify, and demodulate the signal because working with a fixed frequency simplifies the design and improves the performance of the receiver.

Why Use an Intermediate Frequency?

When a radio receives signals from different frequencies, directly amplifying and processing those signals can be difficult, especially at very high frequencies. By mixing the incoming signal with a signal generated by the local oscillator, the receiver creates new frequencies: the sum and difference of the incoming signal and the oscillator's frequency. The difference frequency, which is lower than the original, becomes the intermediate frequency (IF).

Once the signal is converted to IF, the radio can use highly effective filters and amplifiers explicitly designed for that frequency. This makes the receiver more selective and sensitive, improving its ability to pick out weak signals and reject interference from nearby frequencies.

Example of How IF Works:

If a receiver is designed to work with an intermediate frequency of 455 kHz, and it is tuned to receive a signal at 10 MHz, the local oscillator might be set to 10.455 MHz. When the 10 MHz signal mixes with the 10.455 MHz signal from the local oscillator, it creates a difference frequency of 455 kHz, which is the intermediate frequency that the receiver will process.

Key Points to Remember:

- Intermediate frequency simplifies signal processing by converting high-frequency signals to a lower, fixed frequency.

- IF allows for better filtering, amplification, and tuning in a radio receiver, making the radio more effective at isolating and processing the desired signal.

- Superheterodyne receivers, which use IF, are the most common type of radio receivers due to their superior performance.

Signal Processing and Frequency Control

Mixers & Local Oscillators

In a radio receiver, particularly a superheterodyne receiver, the process of converting incoming radio signals to a lower, more manageable frequency called the intermediate frequency (IF) is necessary for efficient signal processing. This conversion happens through a mixer component, which combines the incoming radio signal and the local oscillator (LO) signal. By adjusting or tuning the local oscillator, the receiver can convert a wide range of incoming frequencies to the same intermediate frequency, making it easier to filter, amplify, and demodulate the desired signal.

Which mixer input is varied or tuned to convert signals of different frequencies to an intermediate frequency (IF)? (G8B01)

A. Image frequency
B. Local oscillator
C. RF input
D. Beat frequency oscillator

The local oscillator is the part of the mixer circuit that is varied or tuned to convert signals of different frequencies to a fixed intermediate frequency (IF). The local oscillator generates a signal that mixes with the incoming radio signal in the mixer. This mixing results in two new frequencies: the sum and difference of the local oscillator frequency and the incoming signal frequency. The receiver then selects the difference frequency, which is the intermediate frequency, for further processing.

Mixers and Frequency Conversion

A mixer is a crucial component in radio receivers, and it is used to shift the frequency of incoming radio signals to a lower intermediate frequency (IF). The mixer combines the incoming signal (RF input) with a signal generated by the Local Oscillator (LO). This process creates two new frequencies: one at the sum of the two input frequencies and another at the difference. These new frequencies are key because they allow the receiver to convert a wide range of signals to a fixed intermediate frequency for easier processing.

What combination of a mixer's Local Oscillator (LO) and RF input frequencies is found in the output? (G8B11)

A. The ratio
B. The average
C. The sum and difference
D. The arithmetic product

When an RF signal enters a mixer, it is combined with the frequency generated by the Local Oscillator (LO). The output of the mixer includes two new frequencies: the sum and the difference of the LO frequency and the RF input frequency. For example, if the RF input is 10 MHz and the LO is set to 11 MHz, the mixer output will include frequencies at 21 MHz (sum of 11 MHz + 10 MHz) and 1 MHz (difference of 11 MHz – 10 MHz).

The receiver typically selects the difference frequency (the intermediate frequency, or IF) for further processing, as it's easier to handle a lower frequency in terms of filtering and amplification.

Heterodyning

Let's use some more repetition here to help this concept stick. Heterodyning is when two signals of different frequencies are mixed to produce new frequencies called sum and difference frequencies. Said another way, heterodyning is the mixing of two RF signals. The result of this mixing can generate a signal that is easier to process or decode, especially in receivers. Heterodyning is the core principle behind how most radio receivers work.

What is another term for the mixing of two RF signals? (G8B03)

A. Heterodyning
B. Synthesizing
C. Frequency inversion
D. Phase inversion

In a superheterodyne receiver, the incoming radio frequency (RF) signal is mixed with a locally generated signal (called the local oscillator, or LO). This mixing creates two new frequencies: one at the sum of the two signals and one at the difference. The receiver then selects the difference frequency, called the intermediate frequency (IF), which is lower and easier to amplify and process than the original high-frequency signal.

Heterodyning allows radios to operate more efficiently by shifting high-frequency signals to lower, more manageable frequencies for further processing. This process is vital in many radio technologies, from ham radio receivers to commercial AM and FM radio, as it improves the ability to filter, amplify, and demodulate signals.

Frequency Multipliers

A frequency multiplier is a circuit or device that increases the frequency of an input signal by generating harmonics of that signal. These harmonics are multiples of the original frequency, such as twice, three times, or four times the input frequency. The multiplier circuit selects the desired harmonic and uses it to generate a higher-frequency output signal.

For example, if a signal at 10 MHz is fed into a frequency multiplier set to triple the frequency, the output would be 30 MHz. Multipliers are commonly used in radio transmitters, especially in VHF and UHF systems, to efficiently produce higher frequencies from lower-frequency oscillators, which are easier to control and stabilize.

Key Points to Remember:

- **Multiplies frequency**: Takes an input signal and produces a higher-frequency output (a harmonic multiple).

- **Common in radio transmitters**: Especially useful in VHF/UHF transmitters to generate higher operating frequencies.

VHF FM transmitters

In VHF FM transmitters, sometimes it's easier to generate a lower-frequency signal and then multiply that frequency to reach the desired operating frequency. This is where a frequency multiplier comes into play. A multiplier is a circuit that takes a lower-frequency signal and generates a harmonic (or multiple) of that signal. This method allows the transmitter to efficiently reach higher operating frequencies without directly generating those higher frequencies, which would require more complex and precise oscillators.

In a VHF FM transmitter, the stage that generates a harmonic of a lower frequency signal to reach the desired operating frequency is called a multiplier. For example, suppose the transmitter needs to operate at 150 MHz. In that case, it might start with a signal at 50 MHz and use a frequency multiplier to produce the third harmonic (3 × 50 MHz = 150 MHz). This allows the transmitter to generate high-frequency signals more easily by initially working with lower, more manageable frequencies.

What is the stage in a VHF FM transmitter that generates a harmonic of a lower frequency signal to reach the desired operating frequency? (G8B04)

A. Mixer
B. Reactance modulator
C. Balanced converter
D. Multiplier

A multiplier is a "frequency booster" in a VHF FM transmitter. It takes the original lower frequency and creates a harmonic, or multiple, to bring the signal to the higher operating frequency required for VHF communication.

Frequency Deviation

Frequency deviation in an FM transmitter refers to how much the frequency of the carrier wave shifts from its central frequency in response to the modulation signal. This deviation is directly related to the strength of the input signal (such as voice or data) and is an important parameter because it affects both the quality of the transmission, and the bandwidth used. In radio communication, controlling the deviation is necessary to ensure that the transmitted signal stays within the legal limits and provides clear communication without interference.

What is the frequency deviation for a 12.21 MHz reactance modulated oscillator in a 5 kHz deviation, 146.52 MHz FM phone transmitter? (G8B07)

A. 101.75 Hz
B. 416.7 Hz
C. 5 kHz
D. 60 kHz

First, some key terms in this question.

Key terms:

- **Frequency deviation:** The amount the carrier frequency shifts above or below its center frequency during modulation.

- **Reactance modulated oscillator:** A stage in the transmitter where frequency modulation begins by varying the reactance.

- **Frequency multiplication:** The process of using frequency multipliers in the transmitter to achieve the desired operating frequency (in this case, 146.52 MHz).

FM transmitters often use lower-frequency oscillators modulated with the audio signal, then multiply the frequency to reach the final transmission frequency. During multiplication, the frequency deviation also scales by the same factor as the frequency.

Key Data:

- **Reactance modulated oscillator frequency (initial frequency):** 12.21 MHz

- **Final carrier frequency:** 146.52 MHz

- **Final frequency deviation:** 5 kHz

We must determine the deviation at the **reactance modulated oscillator** before the multiplication occurs.

Step 1: Determine the multiplication factor

The multiplication factor tells us how many times the reactance modulated oscillator's frequency is multiplied to reach the final carrier frequency.

$$Multiplication\ Factor = \frac{Final\ Frequency}{Oscillator\ Frequency}$$

$$Multiplication\ Factor = \frac{146.52\ MHz}{12.21\ MHz} = 12$$

Step 2: Relating deviation to multiplication

The frequency deviation scales by the same factor as the frequency multiplication. Thus:

$$Deviation\ at\ Oscillator = \frac{Final\ Deviation}{Multiplication\ Factor}$$

$$Deviation\ at\ Oscillator = \frac{5\ kHz}{12} = 0.4167\ kHz\ or\ 416.7\ Hz$$

Final Answer

The frequency deviation at the 12.21 MHz reactance modulated oscillator is **416.7 Hz**.

By following this process, you can confidently solve similar questions and understand the interplay between frequency, deviation, and multiplication in FM transmitters.

Interference Types

Image Response

In radio receivers, particularly superheterodyne receivers, image response is an important concept to understand. This occurs when a signal at a different frequency than the desired one gets processed by the receiver, leading to interference. Specifically, image response happens when a signal at twice the intermediate frequency (IF) away from the desired signal frequency is mixed in the receiver. Understanding image response and how to minimize it is critical to ensure clear reception and avoid interference from unwanted signals, especially in crowded frequency environments.

What is the term for interference from a signal at twice the IF frequency from the desired signal? (G8B02)

A. Quadrature response
B. Image response
C. Mixer interference
D. Intermediate interference

An image response is an interference caused by a signal twice the intermediate frequency (IF) away from the desired signal. In a superheterodyne receiver, the local oscillator (LO) generates a frequency that mixes with the incoming radio signal to produce the IF. However, suppose there is a signal at a frequency offset by twice the IF from the desired frequency. In that case, it can also create the same IF, leading the receiver to mistakenly process this unwanted signal. This results in interference and distorted reception.

To prevent image response interference:

1. Use better RF filtering to reject signals at the image frequency before they reach the mixer.

2. Choose a higher IF, as this increases the separation between the desired and image frequencies, making filtering more effective.

By following these steps, you can systematically solve for image response interference and understand how to address it in receiver design.

Think of image response as a "mirror" problem—where a signal that is not intended to be received gets mixed in, creating interference. Image response can be reduced by using a well-designed receiver with adequate filtering to block these unwanted signals.

Intermodulation

Intermodulation occurs when <u>two or more signals are combined in a non-linear circuit, such as an amplifier or antenna system, and produce unwanted spurious outputs</u>, also called intermodulation products. These products are additional frequencies that are not part of the original signals, and they can cause interference with other nearby communications. Intermodulation distortion is common in crowded environments where multiple transmitters are operating, and it is a significant concern in ensuring clean, interference-free communication in ham radio.

What process combines two signals in a non-linear circuit to produce unwanted spurious outputs? (G8B12)

A. Intermodulation
B. Heterodyning
C. Detection
D. Rolloff

Let's sidestep here to discuss non-linear circuits.

Non-Linear Circuits

A non-linear circuit is one in which the relationship between the input and output signals does not follow a straight-line (linear) pattern, meaning the output is not directly proportional to the input. In other words, changes in the input don't produce proportional changes in the output. This behavior occurs because some components in the circuit, like diodes, transistors, and other active devices, have non-linear characteristics. These components do not consistently obey Ohm's law ($V = IR$) across different voltages and currents.

Non-linear circuits are essential in many applications, especially radio and signal processing. For instance, they are used in oscillators, modulators, amplifiers, and frequency mixers, where creating and processing complex waveforms is necessary. Non-linearities in these circuits allow them to perform tasks like signal amplification and frequency modulation, which are vital for communication systems. However, non-linear circuits can also introduce distortion and harmonics, sometimes requiring careful design considerations to minimize unwanted effects.

Intermodulation happens when two signals mix in a non-linear circuit, creating new frequencies that are combinations of the original signals.

These spurious outputs can appear as sums and differences of the original signals. For example:

$$f1 + f2 \quad or \quad f1 - f2$$

Intermodulation products are classified by their "order," which refers to the sum of the coefficients in the mixing process. For example:

$$2f1 - f2 \quad or \quad 3f1$$

These unwanted frequencies can interfere with other communications, causing distortion or making it difficult to receive the intended signal.

Intermodulation is a "mixing" problem—two signals interact in a non-linear way, producing additional, unwanted signals that can cause interference. Proper equipment design and isolation can help reduce intermodulation's effects and improve the clarity of radio communication.

How to Achieve It:

1. **Design for Linearity:** Use high-quality components designed to operate within their linear range. Amplifiers, for instance, should be chosen based on their power-handling capacity to avoid overloading and non-linear behavior.

2. **Implement Filtering:** Use bandpass filters to block out-of-band signals that might otherwise interact and create intermodulation products.

3. **Ensure Proper Isolation:** Incorporate shielding, grounding, and physical separation to reduce the coupling of strong external signals into the circuit.

4. **Reduce Signal Overlap:** Design the system to limit the mixing of multiple strong signals, such as by optimizing antenna placement and avoiding high-power transmitters near sensitive receivers.

By addressing intermodulation at its source through design and isolation, you can significantly improve the reliability and clarity of radio communication, especially in environments with many competing signals.

Odd-Order

Odd-order intermodulation products are the ones that appear closest to the original signal frequencies. For example, third-order products (like 2f1−f2) are typically much closer to the original signals than higher-order products. Odd-order products are more problematic for radio operators because they fall near the frequency you're trying to use, potentially causing interference that can be difficult to filter out.

Which intermodulation products are closest to the original signal frequencies? (G8B05)

A. Second harmonics
B. Even-order
C. Odd-order
D. Intercept point

To recall this for the exam, remember that odd-order intermodulation products—such as third-order products—are closer to your operating frequencies and more likely to cause interference. This is why managing intermodulation distortion is important, especially in environments with multiple strong signals.

This topic is still a bit fuzzy to me. Let's try to break it down and simplify it.

When two or more signals mix in a non-linear circuit, they can produce unwanted signals called intermodulation products. These extra signals aren't part of the original ones you want and can interfere with your signals and others nearby.

Odd-order intermodulation products, like third-order ones, are the most troublesome because they show up close to the frequencies you're using. This makes them harder to filter out or avoid. To simplify for the exam: Odd-order products are the ones to watch out for because they're nearest to your important signals and can easily cause interference.

Let's break it down into simpler parts:

Odd-Order Intermodulation

Intermodulation products come in different "orders" based on how they are created. An odd-order product of frequencies F1 and F2 means the new signal is created by an odd-number combination of the original signals. For example, a third-order intermodulation product might look like this:

$$2f1 - f2$$

Which of the following is an odd-order intermodulation product of frequencies F1 and F2? (G8B13)

A. 5F1-3F2
B. 3F1-F2
C. 2F1-F2
D. All these choices are correct

This combination of two signals (f1 and f2) is mixed in the non-linear circuit. The result (like 2f1 - f2) is called a third-order product because the numbers add up to 3.

But I still don't understand the 3.

The "order" of an intermodulation product is determined by how many times the original signals are used in a combination. Each signal contributes a number, and we add them up to get the order of the product.

For example, let's look at **2f1 - f2**:

- 2f1 means we're using the first signal (f1) twice.

- f2 means we're using the second signal (f2) once.

So, you add up how many times you're using the signals:

- 2 (from 2f1) + 1 (from f2) = 3 (an odd number so odd-order).

Since 2f1- f2 adds up to 3, this is a third-order product.

Let's look at the other possible answers.

- A. 5F1-3F2 = 5 + 3 = 8 (even number, so an even order)

- B. 3F1-F2 = 3 + 1 = 4 (even number, so an even order)

Chapter Summary

This chapter explored the fundamental role of frequency in amateur radio, explaining how different bands behave and why selecting the right frequency is essential for effective operation. With an overview of how frequencies interact with the ionosphere, detailing why HF bands enable long-distance communication while VHF and UHF bands primarily support line-of-sight transmission. It highlighted the distinctions between lower and higher HF bands, their advantages, and the best times for use. Key factors like time of day, atmospheric conditions, and solar activity are examined to help operators optimize their radio performance.

We then delved into intermediate frequency (IF) and its critical role in superheterodyne receivers, breaking down how mixers and local oscillators convert signals to more manageable frequencies for improved filtering and reception. Essential concepts such as heterodyning, frequency multipliers, and frequency deviation are introduced to explain how signals are processed and refined for clear transmission.

Finally, this chapter addressed common interference issues, including image response and intermodulation, and provides strategies for minimizing unwanted signals that can disrupt communication. By the end of this chapter, readers will have a solid understanding of frequency behavior, signal processing, and the techniques needed to optimize their radio setup for maximum performance.

Chapter 7

Electrical Components

IMAGINE YOURSELF AT YOUR workbench, surrounded by various components, each playing a vital role in bringing your ham radio projects to life. As you prepare to upgrade your license, advancing beyond the basics of electrical principles becomes increasingly important. This chapter goes beyond the fundamental elements of electrical circuits and prepares you for the advanced topics on the General exam.

Components

Understanding how electrical components like capacitors, inductors, and their properties interact is fundamental for optimizing your equipment's performance. These concepts—capacitor, inductance, reactance, and resonance—are profoundly connected and influence how radio circuits handle electrical signals. Together, they determine how your radio can tune into different frequencies and filter out unwanted signals.

Reactance

A capacitor stores and releases electrical energy, while capacitance measures its ability to do so. Capacitors play a crucial role in tuning and signal filtering by allowing alternating current to pass while blocking direct current. On the other hand, an inductor stores energy in a magnetic field and opposes changes in current. The property of inductance determines how much resistance an inductor creates to changes in current flow.

Both capacitors and inductors create reactance, which opposes the flow of AC. Still, they do so in opposite ways—capacitors decrease reactance with higher frequencies, and inductors increase it.

> **What is reactance? (G5A02)**
>
> A. Opposition to the flow of direct current caused by resistance
> **B. Opposition to the flow of alternating current caused by capacitance or inductance**
> C. Reinforcement of the flow of direct current caused by resistance
> D. Reinforcement of the flow of alternating current caused by capacitance or inductance

The opposition to the flow of alternating current in a capacitor is called reactance. Unlike resistance, which affects both direct current and AC, reactance only impacts AC signals. In a capacitor, reactance decreases as the frequency of the AC increases. This means that at lower frequencies, the capacitor provides more opposition to the current flow, while at higher frequencies, the opposition (or reactance) becomes smaller, allowing more current to pass through.

Which of the following is opposition to the flow of alternating current in a capacitor? (G5A04)

A. Conductance
B. Reluctance
C. Reactance
D. Admittance

The opposition to the flow of alternating current (AC) in an inductor is called reactance. Just like capacitors, inductors create reactance, but they do so in a different way. In an inductor, reactance increases as the frequency of the AC rises. This means that the inductor offers more opposition to the current flow at higher frequencies, making it harder for the signal to pass through. Reactance in an inductor is called inductive reactance, and it plays a key role in controlling the flow of AC in circuits, especially in tuning and filtering applications in radios.

Which of the following is opposition to the flow of alternating current in an inductor? (G5A03)

A. Conductance
B. Reluctance
C. Admittance
D. Reactance

The unit used to measure reactance is the ohm (Ω), the same unit used to measure resistance. While resistance applies to both direct current and alternating current, reactance explicitly describes the opposition to the flow of AC in components like capacitors and inductors – repetition that builds memory! Just like resistance limits the flow of current, reactance does the same, but its value changes depending on the frequency of the AC signal.

What unit is used to measure reactance? (G5A09)

A. Farad
B. Ohm
C. Ampere
D. Siemens

To remember this for the exam, keep in mind that although reactance behaves differently from resistance (since it varies with frequency), it is still measured in ohms. Whether it's a capacitor or inductor, their reactance in an AC circuit is quantified using the familiar unit of ohms.

What letter is used to represent reactance? (G5A11)

A. Z
B. X
C. B
D. Y

The letter used to represent reactance is X. And no, you don't need to Tweet about that – sorry, that's what they call a bad ham joke! In electrical formulas and circuit analysis, X is used to denote reactance, distinguishing it from resistance, which is defined by the letter R. Reactance can either be from a capacitor (capacitive reactance) or an inductor (inductive reactance). X_c represents capacitive reactance, and X_i represents inductive reactance. Both types of reactance oppose the flow of alternating current.

Inductor & Capacitor Reaction to AC

That was a lot to take in. Let's review two more questions and use that repetitive memory-building technique.

Inductor

An inductor opposes AC flow through inductive reactance. The key thing to remember is that <u>as the frequency of the AC increases, the inductor's reactance also increases</u>. This means that at higher frequencies, the inductor provides more opposition to the current, limiting how much of the signal can pass through. Inductors are often used in circuits where you want to block or filter out higher-frequency signals.

How does an inductor react to AC? (G5A05)

A. As the frequency of the applied AC increases, the reactance decreases
B. As the amplitude of the applied AC increases, the reactance increases
C. As the amplitude of the applied AC increases, the reactance decreases
D. As the frequency of the applied AC increases, the reactance increases

---

Capacitor

A capacitor, on the other hand, behaves in the opposite way. <u>As the frequency of the AC increases, the capacitive reactance decreases.</u> This means capacitors allow more current to pass through as the frequency rises. A capacitor offers more opposition at lower frequencies, but as the frequency increases, it becomes less of a barrier, letting higher-frequency signals pass more easily.

How does a capacitor react to AC? (G5A06)

A. As the frequency of the applied AC increases, the reactance decreases
B. As the frequency of the applied AC increases, the reactance increases
C. As the amplitude of the applied AC increases, the reactance increases
D. As the amplitude of the applied AC increases, the reactance decreases

---

To Sum it Up

Inductors and capacitors oppose AC's flow, but they do so in opposite ways. <u>As the frequency of the applied AC increases, the reactance of an inductor increases,</u> meaning it resists higher frequencies more.

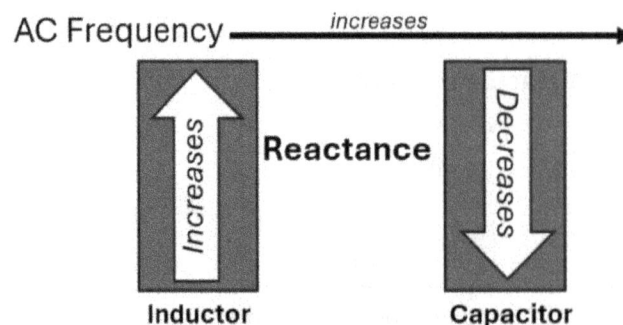

In contrast, <u>a capacitor's reactance decreases as the frequency increases,</u> allowing more current to flow at higher frequencies. This balance between inductors and capacitors is critical in radio circuits, particularly for tuning and filtering different frequencies. Understanding how they react to AC helps you manage which frequencies pass through a circuit and which are blocked.

Impedance

What is impedance? (G5A08)

A. The ratio of current to voltage
B. The product of current and voltage
C. The ratio of voltage to current
D. The product of current and reactance

Impedance is the ratio of voltage to current in an AC circuit. It's similar to resistance in a DC circuit, but impedance considers not only resistance but also the reactance caused by capacitors and inductors in an AC circuit. Impedance is measured in ohms (Ω) and determines how much the circuit resists the flow of alternating current.

Impedance is the overall "opposition" in an AC circuit that combines resistance and reactance. The higher the impedance, the more the circuit resists current flow for a given voltage. The lower the impedance, the easier it is for current to flow through the circuit. This concept is crucial when designing and tuning radio circuits to ensure the right balance between voltage and current for efficient signal transmission and reception.

Admittance

What is the term for the inverse of impedance? (G5A07)

A. Conductance
B. Susceptance
C. Reluctance
D. Admittance

The term for the inverse of impedance is admittance. While impedance measures how much a circuit opposes the flow of AC, admittance measures how easily AC flows through the circuit. In other words, if impedance is the "resistance" to current, admittance is the "permission" for current to flow. Admittance is measured in Siemens (S), the inverse unit of ohms.

Remember, impedance "blocks" current, while admittance "allows" current to pass. The higher the admittance, the easier it is for current to flow through a circuit. This relationship between impedance and admittance is important in understanding how circuits behave, especially in radio and communication systems, where efficient power transfer is needed.

Resonance

When resonance occurs, the reactance of the capacitor and inductor cancels out, allowing the circuit to operate efficiently at a specific frequency. It's the point where the energy exchange between a capacitor and an inductor in an LC circuit is perfectly balanced. When this happens, the circuit can either amplify or filter signals at a specific frequency with maximum efficiency.

Think of it this way: just like a swing moves higher with perfectly timed pushes, a circuit at resonance operates at its best when the alternating current (AC) matches the natural frequency of the LC components. At this resonant frequency, the opposition (reactance) from the capacitor and the inductor cancels each other out, allowing the circuit to pass or amplify the signal with minimal energy loss. This concept is fundamental in radios, where tuning to a particular frequency involves adjusting the circuit to resonate with the signal you want to receive.

What's an LC Circuit?

An LC circuit (also known as a tuned circuit or resonant circuit) is an electrical circuit consisting of an inductor (L) and a capacitor (C) connected together. The main feature of an LC circuit is that it can resonate at a specific frequency, known as the resonant frequency, which is determined by the values of the inductor and the capacitor.

When energy is applied to the circuit, the capacitor stores electrical energy, and the inductor stores magnetic energy. These energies oscillate back and forth between the two components, creating a resonant frequency where the circuit can either filter out or amplify signals. In radios, LC circuits are commonly used to select or tune into a specific frequency, allowing operators to pick up only the desired signals while filtering out unwanted frequencies.

What occurs in an LC circuit at resonance? (G5A12)

A. Current and voltage are equal
B. Resistance is cancelled
C. The circuit radiates all its energy in the form of radio waves
D. Inductive reactance and capacitive reactance cancel

In an LC circuit at resonance, the inductive reactance (opposition to AC from the inductor) and the capacitive reactance (opposition to AC from the capacitor) cancel each other out. This happens because, at the resonant frequency, the reactance from the inductor and the capacitor are equal in magnitude but opposite in direction. As a result, their effects balance out, and the circuit allows the AC signal to pass through with minimal opposition.

LC circuits are used in tuning, filtering, and signal processing in radio because they allow precise control over which frequencies are received or transmitted.

Electrical Circuits

In electrical circuits, how components like capacitors, inductors, and resistors are arranged—either in series or parallel—affects the overall behavior and performance of the circuit. Understanding these arrangements is necessary as they impact signal processing, filtering, and tuning.

In a series circuit, components are connected one after another, so the same current flows through each component, but the voltage is divided among them.

In contrast, in a parallel circuit, components are connected across the same two points, so the voltage is the same across each component, but the current is divided.

Whether components are in series or parallel affects the total reactance, impedance, and overall circuit performance, influencing how your radio equipment handles different frequencies. This section will explore the key differences and how these configurations are used in practical radio circuits.

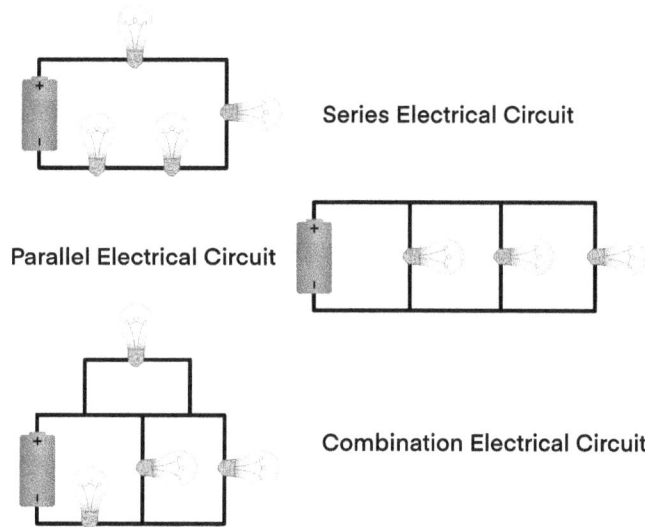

Series Electrical Circuit

Parallel Electrical Circuit

Combination Electrical Circuit

Series

Capacitance in Series

Capacitance refers to a capacitor's ability to store electrical energy in an electric field. Capacitors are widely used in radio circuits to filter signals, block DC currents, and tune circuits by controlling frequency response. When capacitors are connected in series or parallel, their total capacitance changes. Understanding how to calculate total capacitance in different configurations is needed to tune and adjust circuits.

When capacitors are connected in series, the total capacitance is always less than the smallest individual capacitor. The formula for calculating the total capacitance in series is the reciprocal method. This involves taking the reciprocal (1 divided by the value) of each capacitor's capacitance, adding those reciprocals together, and then taking the reciprocal of the total to find the overall capacitance. This approach ensures the proper calculation for capacitors working together in a series configuration.

Let's work through an exam question together.

What is the capacitance of three 100-microfarad capacitors connected in series? (G5C09)

A. 0.33 microfarads
B. 3.0 microfarads
C. 33.3 microfarads
D. 300 microfarads

First, we start with our formula:

$$\frac{1}{C_{total}} = \frac{1}{C_1} + \frac{1}{C_2} + \frac{1}{C_3}$$

Second, substitute values. For three 100-microfarad capacitors connected in series:

$$\frac{1}{C_{total}} = \frac{1}{100} + \frac{1}{100} + \frac{1}{100}$$

Third, solve the equation:

$$\frac{1}{C_{total}} = \frac{1}{100} + \frac{1}{100} + \frac{1}{100} = \frac{3}{100}$$

Now, take the reciprocal to find the total capacitance (flip the answer upside down):

$$\frac{1}{C_{total}} = \frac{100}{3} = 33.3 \; microfarads$$

So, the total capacitance of three 100-microfarad capacitors in series is 33.3 microfarads.

Keep in mind that when capacitors are connected in series, the total capacitance is lower than the smallest individual capacitor. In our example, the lowest capacitance was 100 microfarads, and our answer was 33.3.

Let's try another question.

What is the capacitance of a 20-microfarad capacitor connected in series with a 50-microfarad capacitor? (G5C12)

A. 0.07 microfarads
B. 14.3 microfarads
C. 70 microfarads
D. 1,000 microfarads

To calculate the total capacitance in a series configuration, you need to use the reciprocal formula:

$$\frac{1}{C_{total}} = \frac{1}{C_1} + \frac{1}{C_2}$$

For this question, you are given a 20-microfarad capacitor and a 50-microfarad capacitor. Using our formula, substitute in the values:

$$\frac{1}{C_{total}} = \frac{1}{20} + \frac{1}{50}$$

Solve by adding the fractions. Since they don't share a common denominator, it will be easier to add them as decimals.

$$\frac{1}{20} = 0.05 \; and \; \frac{1}{50} = 0.02$$

Which equals:

$$\frac{1}{C_{total}} = 0.05 + 0.02 = 0.07$$

Now, take the reciprocal to find the total capacitance:

$$\frac{1}{C_{total}} = \frac{1}{0.07} = 14.3 \; microfarads$$

So, the total capacitance of a 20-microfarad capacitor connected in series with a 50-microfarad capacitor is 14.3 microfarads. When capacitors are connected in series, the total capacitance (14.3 microfarads) is less than the smallest individual capacitor in the series (20 microfarads).

For the exam, remember that when capacitors are connected in series, their total capacitance is always less than the smallest capacitor in the series.

Inductance in Series

Inductance is the property of an inductor that allows it to store energy in a magnetic field when current flows through it. Inductors are components in radio circuits used for tuning and filtering signals. When designing or adjusting circuits, there are times when you may need to increase the total inductance. Depending on your goals, this can be achieved by modifying the configuration of inductors in the circuit, either in series or parallel.

Which of the following components should be added to an inductor to increase the inductance? (G5C14)

A. A capacitor in series
B. A capacitor in parallel
C. An inductor in parallel
D. An inductor in series

To increase the inductance in a circuit, you should <u>add another inductor in series</u> with the existing one. When inductors are connected in series, their inductances add together, increasing the total inductance. The formula is straightforward:

$$L_{total} = L_1 + L_2 + ...$$

For example, if you have a 10-millihenry inductor and add another 10-millihenry inductor in series, the total inductance becomes 20 millihenries.

To remember this, consider inductors in series as "stacking" their magnetic fields, increasing the total inductance. This is the opposite of capacitors in series, where the total capacitance decreases. Adding inductors in series is a simple and effective way to increase inductance in radio and electronic circuits.

Let's try a question.

What is the inductance of a circuit with a 20-millihenry inductor connected in series with a 50-millihenry inductor? (G5C11)

A. 7 millihenries
B. 14.3 millihenries
C. 70 millihenries
D. 1,000 millihenries

When inductors are connected in series, the total inductance is simply the sum of the individual inductances. This happens because, in a series circuit, the inductors share the same current, and their magnetic fields combine, leading to a straightforward addition of their inductances.

Step-by-Step Calculation:

To calculate the total inductance for a circuit with a 20-millihenry inductor connected in series with a 50-millihenry inductor, you use the following formula:

$$L_{total} = L_1 + L_2$$

Substituting the given values:

$$L_{total} = 20 \text{ mH} + 50 \text{ mH}$$

Solve the formula:

$$L_{total} = 20 \text{ mH} + 50 \text{ mH} = 70 \text{ millihenries}$$

<u>The total inductance of a 20-millihenry inductor connected in series with a 50-millihenry inductor is 70 millihenries.</u>

Inductors connected in series are added together, making it a straightforward calculation to increase inductance in a circuit.

Resonance in Series

Resonance is a critical concept in electronics, especially in radio circuits. In an LC circuit, which consists of an inductor (L) and a capacitor (C), resonance occurs when the inductive reactance (the opposition to current flow by the inductor) equals the capacitive reactance (the opposition by the capacitor). When these reactances are equal, the circuit operates at its resonant frequency, a frequency at which the circuit behaves in a highly efficient manner.

What happens when inductive and capacitive reactance are equal in a series LC circuit? (G5A01)

A. Resonance causes impedance to be very high
B. Impedance is equal to the geometric mean of the inductance and capacitance
C. Resonance causes impedance to be very low
D. Impedance is equal to the arithmetic mean of the inductance and capacitance

The circuit reaches resonance when inductive and capacitive reactance are equal in a series LC circuit. At resonance, the opposing effects of inductors and capacitors cancel each other out, <u>causing the circuit's impedance (total opposition to current flow) to be very low.</u> This means the circuit allows maximum current to flow through it at the resonant frequency with very little resistance.

At this point, the LC circuit can easily pass resonant frequency signals while blocking signals of other frequencies. This property of low impedance at resonance makes LC circuits highly effective for tuning radios to desired frequencies or filtering out unwanted signals.

Consider resonance the "sweet spot" where the inductive and capacitive reactance cancel each other, resulting in low impedance and efficient energy transfer.

Parallel

Capacitance in Parallel

To increase the total capacitance in a circuit, you should <u>add another capacitor in parallel</u> with the existing one. When capacitors are connected in parallel, their capacitances add together. This is because each capacitor provides an additional path for storing electrical charge, effectively increasing the overall capacity of the circuit to store energy. The formula for total capacitance in parallel is:

$$C_{total} = C_1 + C_2 + ...$$

So, by adding another capacitor in parallel, you increase the total capacitance, allowing the circuit to store more energy.

Which of the following components should be added to a capacitor to increase the capacitance? (G5C13)

A. An inductor in series
B. An inductor in parallel
C. A capacitor in parallel
D. A capacitor in series

To remember this for the exam, think of parallel capacitors as "working together" to increase capacitance. This is different from capacitors in series, where the overall capacitance decreases. Adding capacitors in parallel can be helpful when fine-tuning filters or other components that rely on specific capacitance values.

Exam Time

> **What is the equivalent capacitance of two 5.0-nanofarad capacitors and one 750-picofarad capacitor connected in parallel? (G5C08)**
>
> A. 576.9 nanofarads
> B. 1,733 picofarads
> C. 3,583 picofarads
> **D. 10.750 nanofarads**

When capacitors are connected in parallel, their capacitances add together to form the total capacitance. In this case, you have two 5.0-nanofarad capacitors and one 750-picofarad capacitor (remember the two(2) 5.0 nanofarad capacitors. I didn't register the written word two the first time I did this! Also, note the different units: nanofarad and picofarad. We need to convert!

First, convert the 750 picofarads to nanofarads to calculate the total capacitance because the units need to match. Since 1 nanofarad equals 1,000 picofarads, 750 picofarads is equivalent to 0.750 nanofarads.

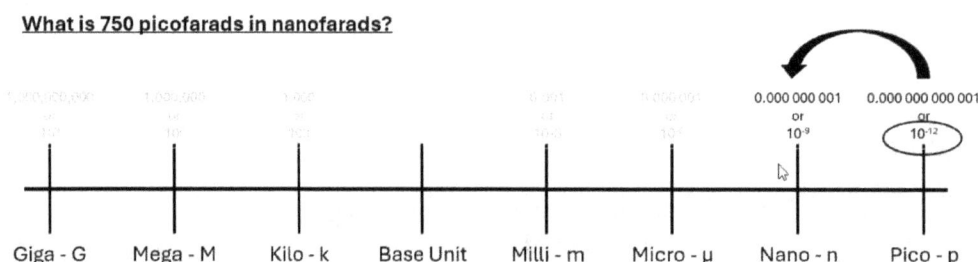

What is 750 picofarads in nanofarads?

$$750 \text{ pF} = 0.750 \text{ nF}$$

Moving to the Left the 'Big Number' Out Front Gets **SMALLER**

Now, you can add all the capacitances together:

$$C_{total} = 5.0 \text{ nF} + 5.0 \text{ nF} + 0.750 \text{ nF} = 10.750 \text{ nanofarads}$$

So, the equivalent capacitance of the three capacitors connected in parallel is 10.750 nanofarads.

Remember parallel capacitors as "adding up" their ability to store charge. This simple addition rule applies no matter how many capacitors you have in parallel, and it makes calculating the total capacitance straightforward.

Inductance in Parallel

Inductance is a fundamental property of electrical circuits, particularly those involving coils or inductors. It refers to the ability of an inductor to store energy in a magnetic field when current flows through it.

When inductors are connected in parallel, the total inductance is always less than the smallest individual inductor. The formula for calculating the total inductance in parallel is similar to how resistors are calculated in parallel:

$$\frac{1}{L_{total}} = \frac{1}{L_1} + \frac{1}{L_2} + \frac{1}{L_3}$$

What is the inductance of three 10-millihenry inductors connected in parallel? (G5C10)

A. 0.30 henries
B. 3.3 henries
C. 3.3 millihenries
D. 30 millihenries

By now, this math should be looking pretty straightforward!

For three 10-millihenry inductors connected in parallel:

$$\frac{1}{L_{total}} = \frac{1}{10} + \frac{1}{10} + \frac{1}{10} = \frac{3}{10}$$

Now, take the reciprocal to find the total inductance:

$$L_{total} = \frac{10}{3} = 3.3 \; millihenries$$

So, the total inductance of three 10-millihenry inductors in parallel is 3.3 millihenries.

To remember this for the exam, note that when inductors are connected in parallel, the total inductance is lower than the smallest inductor, and you use the reciprocal method to calculate the total value.

Resistors in Parallel

In a circuit with parallel resistors, the total current flowing through the circuit equals the sum of the currents flowing through each branch. This happens because, in a parallel configuration, each resistor provides an independent path for the current to flow. The more branches you have, the more paths there are for the current, so the total current increases as it splits among the different resistors.

The formula for total current (I_{total}) is:

$$I_{total} = I_1 + I_2 + ...$$

Each 'I' represents the current through an individual branch. Since the voltage across each resistor in parallel is the same, the current through each resistor depends on its resistance according to Ohm's Law.

How does the total current relate to the individual currents in a circuit of parallel resistors? (G5B02)

A. It equals the average of the branch currents
B. It decreases as more parallel branches are added to the circuit
C. It equals the sum of the currents through each branch
D. It is the sum of the reciprocal of each individual voltage drop

---

Think of parallel circuits as "splitting" the total current into separate branches. Each branch carries part of the total current, and when you add them together, you get the total current flowing into the circuit. This is different from series circuits, where the same current flows through all components.

Resistance in Parallel

When resistors are connected in parallel, the total resistance is always less than the smallest individual resistor. The formula for calculating total resistance (R_{total}) in a parallel circuit with two resistors is:

$$\frac{1}{R_{total}} = \frac{1}{R_1} + \frac{1}{R_2} + \frac{1}{R_3}$$

What is the approximate total resistance of a 100- and a 200-ohm resistor in parallel? (G5C04)

A. 300 ohms
B. 150 ohms
C. 75 ohms
D. 67 ohms

The two resistors have 100 ohms and 200 ohms values in this case.

Using the same type of math from the 'Capacitance in Series' section, we can solve the formula:

$$\frac{1}{R_{total}} = \frac{1}{100} + \frac{1}{200} = 0.01 + 0.005 = 0.015$$

Now, take the reciprocal to find the total resistance:

$$R_{total} = \frac{1}{0.015} \approx 66.7 \ ohms$$

So, the approximate total resistance of a 100-ohm and a 200-ohm resistor in parallel is 66.7 or 67 ohms.

To recall this for the exam, remember that for resistors in parallel, the total resistance is always lower than the smallest individual resistor, and the combined resistance can be calculated using the reciprocal formula.

Resistors & Resistance: How These Last Two Ideas Relate

Adding more resistors in parallel lowers the circuit's total resistance, allowing more total current to flow through it. As total resistance decreases, total current increases. Essentially, the more paths you give the current (by adding more resistors in parallel), the less resistance the circuit offers, and the more current will flow. So, these two concepts are linked: decreasing total resistance increases the total current in a parallel circuit.

Parallel Circuit

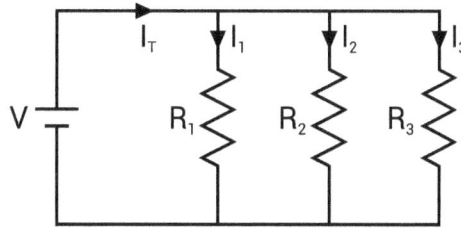

In a parallel circuit, the total current goes up when the total resistance goes down because the current has more paths to take through the circuit.

Let's answer another question.

What is the total resistance of a 10-, a 20-, and a 50-ohm resistor connected in parallel? (G5C03)

A. 5.9 ohms
B. 0.17 ohms
C. 17 ohms
D. 80 ohms

You use the reciprocal formula to find the total resistance of resistors connected in parallel. For three resistors with values of 10 ohms, 20 ohms, and 50 ohms, the formula is:

$$\frac{1}{R_{total}} = \frac{1}{R_1} + \frac{1}{R_2} + \frac{1}{R_3}$$

Substituting the resistor values:

$$\frac{1}{R_{total}} = \frac{1}{10} + \frac{1}{20} + \frac{1}{50}$$

Since the denominators aren't the same, let's switch to decimals.

$$\frac{1}{10} = 0.1 \ \ and \ \ \frac{1}{20} = 0.05 \ \ and \ \ \frac{1}{50} = 0.02$$

$$0.1 + 0.05 + 0.02 = 0.17$$

Now, take the reciprocal of 0.17 to find the total resistance:

$$R_{total} = \frac{1}{0.17} \approx 5.88 \ ohms$$

So, the total resistance of the 10-ohm, 20-ohm, and 50-ohm resistors connected in parallel is approximately 5.88, or rounded to 5.9 ohms, which is lower than the smallest resistor (10 ohms).

To remember this for the exam, note that the total resistance in a parallel circuit is always less than the smallest individual resistor. By using the reciprocal formula, you can calculate how the total resistance combines in parallel circuits.

Transformers

A transformer is an electrical device used to transfer electrical energy between two or more circuits through electromagnetic induction. It consists of two or more coils of wire, known as windings, wrapped around a shared core, typically made of iron. When alternating current flows through one coil (the primary winding), it creates a magnetic field that induces a voltage in the other coil (the secondary winding). This allows the transformer to increase or decrease voltage levels while maintaining the same frequency.

Transformers allow impedance matching, voltage regulation, and isolation between different circuit parts. For example, transformers help match the impedance between a transmitter and an antenna, ensuring that power is transferred efficiently without signal loss or reflections. By adjusting the voltage and current ratios, transformers play a role in optimizing the performance of radio equipment.

When an AC voltage source is connected to the primary winding of a transformer, it creates a changing magnetic field around the winding. This changing magnetic field induces a voltage in the secondary winding through mutual inductance. Mutual inductance occurs because the magnetic field generated by the current in the primary winding interacts with the secondary winding, causing a voltage to appear across it. This is the fundamental principle of how transformers work, allowing them to transfer electrical energy from one circuit to another without a direct electrical connection.

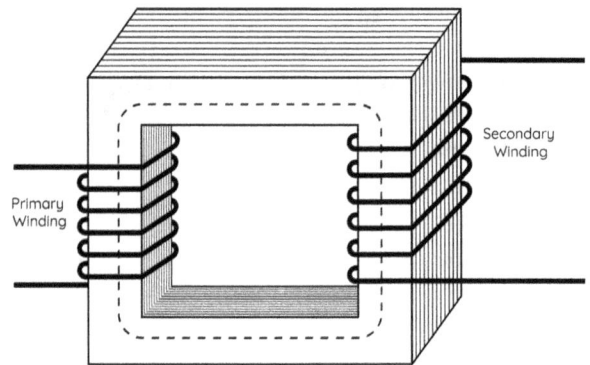

What causes a voltage to appear across the secondary winding of a transformer when an AC voltage source is connected across its primary winding? (G5C01)

A. Capacitive coupling
B. Displacement current coupling
C. Mutual inductance
D. Mutual capacitance

Mutual inductance is the "magnetic link" between the primary and secondary windings. The AC in the primary winding generates a magnetic field, which "induces" or transfers voltage into the secondary winding through this magnetic connection. This is why transformers can step up or down voltages in AC circuits, making them key in radio and electrical systems.

Transformer Design

Step-Up Transformer

In a voltage step-up transformer, the primary winding wire is usually larger than the secondary winding wire to handle the higher current flowing through it. When a transformer steps up the voltage, it simultaneously steps down the current. This means the primary winding, which operates at a lower voltage, must carry a higher current, while the secondary winding, operating at a higher voltage, carries a lower current. Since larger wire sizes are better suited to handle higher currents without overheating, the primary winding is made with thicker wire to manage the increased current safely.

Why is the primary winding wire of a voltage step-up transformer usually a larger size than that of the secondary winding? (G5C05)

A. To improve the coupling between the primary and secondary
B. To accommodate the higher current of the primary
C. To prevent parasitic oscillations due to resistive losses in the primary
D. To ensure that the volume of the primary winding is equal to the volume of the secondary winding

In a step-up transformer, the voltage goes up on the secondary side, but the current goes down. Therefore, the primary winding needs to handle more current, requiring a thicker wire, while the secondary winding can use a thinner wire since it carries less current. This balance ensures the transformer operates efficiently and safely.

Step-Down Transformer

A step-down transformer reduces voltage from a higher level to a lower level. It does this by having fewer turns (windings) on the secondary coil than on the primary coil. When alternating current is applied to the primary winding, the transformer decreases the voltage proportionally according to the turns ratio.

What is the output voltage if an input signal is applied to the secondary winding of a 4:1 voltage step-down transformer instead of the primary winding? (G5C02)

A. The input voltage is multiplied by 4
B. The input voltage is divided by 4
C. Additional resistance must be added in series with the primary to prevent overload
D. Additional resistance must be added in parallel with the secondary to prevent overload

Ok, so this is a tricky question because they are asking it "in reverse."

In a voltage step-down transformer with a 4:1 turns ratio, the transformer is designed to reduce the voltage when applied to the primary winding. However, suppose you reverse the setup and apply the input signal to the secondary winding instead of the primary. In that case, the transformer acts in reverse, becoming a step-up transformer. As a result, the input voltage is multiplied by the same ratio, which in this case is 4.

For example, apply 10 volts to the secondary winding of this 4:1 step-down transformer. The output voltage from the primary winding will be 40 volts. This happens because the transformer boosts the voltage by the same factor (4) it would normally step down when used in the conventional configuration. To recall this for the exam, remember that switching the input to the secondary winding reverses the transformer's function, multiplying the voltage by the turns ratio.

Transformer Math

When understanding the math associated with transformers, breaking things down step by step is important, especially for beginners. Transformers work based on the principle of proportionality between voltage, current, and the number of windings in the coils.

The formulas revolve around these relationships.

Voltage and Turns Ratio: The voltage in a transformer is directly proportional to the number of turns (or windings) in the coils. This means:

$$\frac{V_{primary}}{V_{secondary}} = \frac{N_{primary}}{N_{secondary}}$$

Where Vprimary is the voltage in the primary winding, Vsecondary is the voltage in the secondary winding, and Nprimary and Nsecondary are the number of turns in the primary and secondary windings, respectively. If the secondary coil has more turns than the primary, the voltage increases (step-up transformer). If it has fewer turns, the voltage decreases (step-down transformer).

Current and Turns Ratio: The current behaves inversely to the number of turns in the transformer. When the voltage is stepped up, the current is stepped down, and vice versa. The formula for current is:

$$\frac{I_{primary}}{I_{secondary}} = \frac{N_{secondary}}{N_{primary}}$$

Where Iprimary is the current in the primary winding, and Isecondary is the current in the secondary winding. So, the current decreases as the voltage increases in the secondary coil.

The takeaway for beginners is that voltage and current are related to the number of turns in the windings, and they trade off. If one goes up, the other goes down, and the math helps us calculate exactly how much! In the upcoming questions, I'll break down each concept so that even those new to this topic can easily follow along.

Let's try some questions.

What is the voltage output of a transformer with a 500-turn primary and a 1500-turn secondary when 120 VAC is applied to the primary? (G5C06)

A. 360 volts
B. 120 volts
C. 40 volts
D. 25.5 volts

The information from the previous section provides the basic understanding needed to answer this question. It uses the relationship between the number of turns in the transformer windings and the voltage, known as the voltage turns ratio.

Here's how to solve this:

$$\frac{V_{primary}}{V_{secondary}} = \frac{N_{primary}}{N_{secondary}}$$

This question is asking about voltage, so we'll use the transformer turns ratio for voltage.

Where:

- $V_{primary}$ is the voltage applied to the primary winding

- $V_{secondary}$ is the voltage output of the secondary winding

- $N_{primary}$ is the number of turns in the primary winding

- $N_{secondary}$ is the number of turns in the secondary winding

Step-by-step solution:

First, identify the values given:

- Primary turns ($N_{primary}$) = 500 turns

- Secondary turns ($N_{secondary}$) = 1500 turns

- Primary voltage ($V_{primary}$) = 120 VAC (volts alternating current)

Second, set up the formula:

$$\frac{120}{V_{secondary}} = \frac{500}{1,500}$$

Finally, cross multiply to eliminate the denominator: 120 * 1500 = 180,000, then divide by 500 = 360

Answer: The voltage output of the secondary winding is 360 VAC.

Since the secondary winding has three times more turns than the primary winding, the voltage in the secondary will be three times higher than the primary voltage. This transformer is a step-up transformer, increasing the voltage from 120 VAC to 360 VAC.

Transformers & Impedance

In radio systems, impedance matching is necessary to ensure efficient signal transfer between components, such as between an antenna and a transmission line. When the impedance of these components doesn't match, a portion of the signal is reflected back, causing power loss and reduced performance. This is where transformers come into play. By adjusting the ratio of turns in the primary and secondary windings, a transformer can match the impedance of two components, allowing maximum power transfer with minimal loss.

Transformers work by adjusting the voltage and current ratios, and since impedance is related to both, transformers can also change the impedance between circuits. For example, suppose an antenna has a 600-ohm impedance and needs to connect to a 50-ohm coaxial cable. In that case, a transformer can be used to "step down" the impedance. This is done by choosing a transformer with the correct turns ratio—in this case, about 3.46:1—to match the higher antenna impedance to the lower cable impedance, ensuring efficient signal transfer and system performance.

What transformer turns ratio matches an antenna's 600-ohm feed point impedance to a 50-ohm coaxial cable? (G5C07)

A. 3.5 to 1
B. 12 to 1
C. 24 to 1
D. 144 to 1

So, what's the math behind this, and how do we get 3.46 or rounded up to 3.5?

You need a transformer that adjusts the impedance between the two to match an antenna's 600-ohm feed point impedance to a 50-ohm coaxial cable. The turns ratio of a transformer that matches impedances is determined by the square root of the impedance ratio between the feed point and the coaxial cable. The formula is:

$$Turns\ Ratio = \sqrt{\frac{Z_{primary}}{Z_{secondary}}}$$

Where:

- Zprimary is the 600-ohm impedance of the antenna feed point.

- Zsecondary is the 50-ohm impedance of the coaxial cable.

Using the formula:

$$Turns\ Ratio = \sqrt{\frac{Z_{primary}}{Z_{secondary}}} = \sqrt{\frac{600}{50}} = \sqrt{12} \approx 3.46\ or\ 3.5\ to\ 1$$

This means the transformer needs a turns ratio of about 3.46, rounded to 3.5:1, to effectively match the 600-ohm feed point to the 50-ohm coaxial cable. This allows maximum power transfer between the antenna and the transmission line with minimal signal loss due to impedance mismatch.

Chapter Summary

In this chapter on Electrical Principles, we've explored some advanced concepts and math that form the backbone of radio and electronics. Beginning with electrical components and moving through configurations like series and parallel circuits, you have gained insight into how these elements combine to build functional circuits.

We then examined how reactance affects current flow in AC circuits, leading to a deeper understanding of impedance and admittance. These principles are crucial for managing circuit behavior, optimizing energy flow, and achieving resonance—a phenomenon at the core of frequency selection and tuning in radio applications. Finally, transformers illustrated a practical application of these principles, enabling energy transfer and signal modification across circuits. Together, these topics build a comprehensive foundation for understanding the electrical mechanics of radio communication.

Chapter 8

Power

POWER IS A FUNDAMENTAL concept in electronics and radio communication, directly influencing how efficiently your equipment operates and how far your signals can travel. In simple terms, power is the rate at which energy is used or transmitted. It's measured in watts (W) and calculated by multiplying voltage and current. Understanding power is critical for radio operators because it affects signal strength, transmission range, and compliance with legal transmission limits.

Power is especially important in radio systems because it determines how well you can send and receive signals. Whether tuning your transmitter or adjusting your antenna, knowing how to manage and calculate power ensures that your signal is strong enough to communicate clearly without exceeding legal limits or causing interference. This section will dive into the various types of power and explain how to calculate and manage power.

Power Formula

When we talk about consuming watts of electrical power, we're referring to the amount of energy a device or circuit uses over time. A watt is a unit of power that measures the rate at which energy is consumed or produced. In electrical terms, power (in watts) is calculated by multiplying the voltage (in volts) by the current (in amperes). The formula for power is:

Power	Current	Voltage
$\triangle$ P̃ / I E	$\triangle$ P / Ĩ E	$\triangle$ P / I Ẽ
P = E × I	I = P / E	E = P / I

$$P = I \times E$$

In a radio transmitter, the watts consumed determine how much energy is drawn from the power source to generate radio signals. The more watts a transmitter consumes, the more powerful the signal it can send, which often translates to greater range and clarity.

However, managing power consumption is also important, as too much power can lead to overheating, wear on equipment, or exceeding legal power limits. By understanding how watts relate to your equipment's performance, you can better balance power needs with efficiency and longevity in your radio setup.

Let's try some exam questions.

Here is a basic one to start with.

> **How many watts of electrical power are consumed by a 12 VDC light bulb that draws 0.2 amperes? (G5B04)**
>
> **A. 2.4 watts**
> B. 24 watts
> C. 6 watts
> D. 60 watts

Remembering back to our Technician's test, to calculate the number of watts consumed by a 12 VDC light bulb that draws 0.2 amperes, you can use the basic power formula because we want Power (P) and we have Voltage (V) and Current (I)

$$P = I \times E$$

Where:

- P is the power in watts,

- E is the voltage (12 VDC),

- I is the current in amperes (0.2 A).

Substitute the given values into the formula:

$$P = 12 \times 0.2 = 2.4 \; watts$$

So, the light bulb consumes 2.4 watts of electrical power.

Next, let's try a couple more advanced problems.

> **How many watts of electrical power are consumed if 400 VDC is supplied to an 800-ohm load? (G5B03)**
>
> A. 0.5 watts
> **B. 200 watts**
> C. 400 watts
> D. 3200 watts

First, VDC.

VDC stands for Volts Direct Current, which refers to the voltage supplied in a direct current (DC) circuit. In DC circuits, the current flows in only one direction, as opposed to alternating current (AC), where the direction of the current changes periodically.

VDC is used to specify the voltage level in DC systems. For example, a 12 VDC power supply provides a steady 12 volts in a single direction, which is typical for devices like batteries or power supplies for electronics.

Understanding VDC is important because it helps differentiate between direct current (constant, one-way flow of electrons) and alternating current (where the flow of electrons changes direction). In many radio systems, DC voltage powers the equipment, making VDC a critical concept.

But honestly, VDC vs DC still wasn't clear to me.

The terms **VDC** (Volts Direct Current) and **DC** (Direct Current) are closely related but have slightly different uses. **DC** refers to the type of current where electricity flows in a constant direction, while **VDC** specifically measures the voltage in a DC circuit. For example, a power source may supply DC power, but its voltage would be described as 12 VDC or 13.8 VDC. This distinction is important for precision, as many devices, including ham radios, require specific voltage levels for safe and efficient operation. Using "VDC" clarifies that the voltage is from a direct current source, avoiding confusion with AC voltage (VAC).

So now, to calculate how many watts of electrical power are consumed when 400 VDC is supplied to an 800-ohm load, you can use Ohm's Law and the power formula.

Voltage	Current	Resistance
E = I.R	I = E/R	R = E /I

Step 1: There are two key formulas you need:

Ohm's Law

Ohm's Law: $E = I \times R$

- E: Voltage in volts

- I: Current in amperes

- R: Resistance in ohms

Rearranging this formula can help us find current (I) if voltage (V) and resistance (R) are known:

$$I = \frac{E}{R}$$

Power Formula

Power Formula: $P = E \times I$

- P: Power in watts

- E: Voltage in volts

- I: Current in amperes

Step 2: Combine the Formulas

From **Ohm's Law**, we know that:

$$I = \frac{E}{R}$$

substitue this expression for I into the Power Formula

$$P = E \times \frac{E}{R}$$

Simplify:

$$P = \frac{E^2}{R}$$

This combined formula allows us to calculate power directly when we know the voltage (V) and resistance (R).

You are now left with two options: One, memorize another formula, or two, just remember Power and Ohms and how to combine them.

Step 3: Solve the Problem

$$P = \frac{E^2}{R}$$

Where:

- P is the power in watts,

- E is the voltage (400 VDC),

- R is the resistance (800 ohms).

Now, substitute the given values into the formula:

$$P = \frac{400^2}{800} = \frac{160,000}{800} = 200 \; watts$$

So, 200 watts of electrical power are consumed by the 800-ohm load when 400 VDC is supplied.

How many watts are consumed when a current of 7.0 milliamperes flows through a 1,250-ohm resistance? (G5B05)

A. Approximately 61 milliwatts
B. Approximately 61 watts
C. Approximately 11 milliwatts
D. Approximately 11 watts

Again, let's start with our two known formulas. **Ohm's Law** and the **formula for power**.

Step 1: These are two key formulas you need:

Ohm's Law:

Ohm's Law: $E = I \times R$

- E: Voltage in volts

- I: Current in amperes

- R: Resistance in ohms

Power Formula:

Power Formula: $P = E \times I$

- P: Power in watts

- E: Voltage in volts

- I: Current in amperes

Step 2: Combine the Formulas

From **Ohm's Law**, we know that:

$$E = I \times R$$

substitue this expression for E into the Power Formula

$$P = (I \times R) \times I$$

Simplify:

$$P = I^2 \times R$$

Why did we do it this way? Because we are being asked for Power and have been given current and resistance. So, we need a formula that only uses current and resistance.

This combined formula allows us to calculate power directly when we know the voltage (E) and resistance (R).

You are now left with two options. One, memorize another formula, or two, just remember Power and Ohms and how to combine them together.

Step 3: Solve the Problem

$$P = I^2 \times R$$

Where:

- P is the power in watts (?)
- I is the current (7.0 milliamps)—this is given as milliamps, but we want to use amps (the base unit) in our math. So 7.0 milliamps = 0.007 amps or 7.0×10^{-3} amps.
- R is the resistance (1,250 ohms)

Now, substitute the given values into the formula:

$$P = (7.0 \times 10^{-3})^2 \times 1,250$$

Do the math:

$$(7.0 \times 10^{-3})^2 = 49 \times 10^{-6} = 0.000049$$

$$P = 0.000049 \times 1,250 = 0.06125 \; watts$$

Final Answer

The power consumed is 0.0612 watts, or converted to a whole number, 61.25 milliwatts are consumed when a current of 7.0 milliamps flows through a 1,250 ohm resistance.

Batteries

Batteries with Low Internal Resistance

One of the advantages of batteries with low internal resistance is their ability to deliver high discharge current. Internal resistance is the inherent opposition within a battery to the flow of current. When a battery has low internal resistance, it can provide more current to power devices efficiently without losing too much energy as heat. This makes such batteries ideal for applications that require high power output, such as in portable ham radio setups, where reliable and strong energy flow is needed to keep communications running smoothly.

What is an advantage of batteries with low internal resistance? (G6A02)

A. Long life
B. High discharge current
C. High voltage
D. Rapid recharge

Batteries with low internal resistance ensure that the radio station operates smoothly, especially during high-demand activities like transmitting at higher power levels. When the internal resistance is low, the battery can maintain stable voltage levels under load, resulting in better performance and longer operating times for radio equipment.

Batteries with low internal resistance can provide higher discharge currents, making them better suited for high-power applications and ensuring consistent and reliable radio performance. Understanding this concept lets you choose the correct battery for your ham radio equipment.

Lead-Acid Battery Discharge and Lifespan

To maximize the lifespan of a standard 12-volt lead-acid battery, it's important to avoid discharging it too deeply. The minimum allowable discharge voltage for these batteries is 10.5 volts. This means that when the battery's voltage drops to 10.5 volts under load, it's time to stop using it and recharge it. Discharging below this point can cause permanent damage, reducing the battery's lifespan and ability to hold a charge in the future.

What is the minimum allowable discharge voltage for maximum life of a standard 12-volt lead-acid battery? (G6A01)

A. 6 volts
B. 8.5 volts
C. 10.5 volts
D. 12 volts

Understanding how to manage the discharge of a lead-acid battery is key for radio operators, especially during portable or emergency operations. Keeping the battery above 10.5 volts ensures optimal performance. It avoids damaging the battery, which would affect your ability to operate your radio reliably over time.

Decibels and Gain

Decibels (dB) are a fundamental unit of measurement used to quantify signal strength and gain. A decibel is a logarithmic unit that measures the ratio between two values, such as power levels or signal strengths. This logarithmic scale is handy because it can represent a wide range of values in a more manageable way. For instance, rather than dealing with large numbers, decibels allow you to express these quantities in simpler terms. In ham radio, decibels are used to compare signal levels, assess antenna gain, and evaluate amplifier performance. Understanding dB helps you make informed decisions about your equipment and operating conditions, ensuring optimal performance.

Gain is another critical concept in radio communication. It refers to the increase in signal strength achieved by an amplifier or antenna. Gain can be categorized into voltage, current, and power gain. Voltage gain refers to the increase in voltage, current gain to the rise in current, and power gain to the increase in power output. High gain is vital for improving signal reach and clarity, especially in weak signal conditions. It allows your transmissions to travel farther and be received more clearly, which is particularly important for DXing and other long-distance communications.

What dB change represents a factor of two increase or decrease in power? (G5B01)

A. Approximately 2 dB
B. Approximately 3 dB
C. Approximately 6 dB
D. Approximately 9 dB

A change of approximately 3 dB (decibels) represents either a doubling or halving of power – a factor of two increase). Specifically, a +3 dB change means the power has increased by a factor of two, while a -3 dB change means the power has decreased by half. This relationship is important because decibels are a logarithmic scale used to express ratios of power or voltage in a way that is easier to manage, especially when dealing with large ranges of values.

The 3 dB change in power comes from the logarithmic relationship between power ratios and decibels. The formula to calculate decibels when comparing two power levels is:

$$dB = 10 \times log_{10}\left(\frac{P_2}{P_1}\right)$$

Hopefully, this looks familiar from your Technician's exam!

Where:

- P2 is the final power level

- P1 is the initial power level

To understand why 3 dB represents a doubling or halving of power, let's consider the cases where the power ratio is 2 (doubling) or 0.5 (halving):

$$\text{Doubling Power: } dB = 10 \times log_{10}(2) \approx 10 \times 0.3010 = 3 \ dB$$

$$\text{Halving Power: } dB = 10 \times log_{10}(0.5) \approx 10 \times (-0.3010) = -3 \ dB$$

So, when the power increases by a factor of 2, the result is approximately +3 dB, and when the power decreases by half, the result is approximately -3 dB. This logarithmic relationship allows us to easily express large changes in power with relatively small dB values.

Another math question.

What percentage of power loss is equivalent to a loss of 1 dB? (G5B10)

A. 10.9 percent
B. 12.2 percent
C. 20.6 percent
D. 25.9 percent

A loss of 1 dB corresponds to a power loss of approximately 20.6 percent. This means that if you decrease the power of a signal by 1 dB, about 20.6% of the original power is lost, leaving you with roughly 79.4% of the original power. The decibel scale is logarithmic, which means that even small changes in dB represent significant changes in power.

To calculate the percentage of power loss corresponding to a 1 dB change, we can use the formula that relates decibels (dB) to the ratio of power levels:

$$dB = 10 \times log_{10}\left(\frac{P_2}{P_1}\right)$$

Where:

- P2 is the final power level.

- P1 is the initial power level.

Step-by-Step Calculation:

First, Set dB to -1 for a 1 dB loss:

$$-1 = 10 \times log_{10}\left(\frac{P_2}{P_1}\right)$$

Next, Divide by 10 to isolate the logarithm:

$$-0.1 = log_{10}\left(\frac{P_2}{P_1}\right)$$

Then, remove the logarithm by taking the inverse (antilog) of both sides:

$$\left(\frac{P_2}{P_1}\right) = 10^{-0.1} \approx 0.794$$

This means the final power (P2) is 79.4% of the initial power (P1) after a 1 dB loss.

Finally, calculate the percentage of power loss:

$$Power\ Loss = 100\% - 79.4\% = 20.6\%$$

So, a 1 dB loss corresponds to a 20.6% decrease in power.

Practical applications of dB and gain are numerous. Measuring antenna gain is one everyday use. Antenna gain is often expressed in dB, comparing the antenna's performance to a standard reference, such as an isotropic radiator (dBi) or a dipole antenna (dBd). For instance, an antenna with a gain of 6 dBd indicates it performs better than a standard dipole by 6 dB. Assessing amplifier performance is another application. By calculating the increase in signal strength provided by an amplifier, you can determine its effectiveness and suitability for your needs. Signal reports also frequently use dB to quantify improvements or degradations in signal strength. For example, an operator might report that your signal improved by 3 dB after you adjusted your antenna, indicating a noticeable enhancement in signal clarity.

Decibels and gain are concepts that allow you to measure and improve signal strength and quality, ensuring clear and effective transmissions. Whether comparing antennas, assessing amplifiers, or interpreting signal reports, a solid grasp of dB and gain empowers you to make informed decisions and enhance your overall operating experience.

Link Budget

A link budget is an essential calculation used to predict the performance of a communication link, especially over long distances. It accounts for all the factors that affect the signal from the transmitter to the receiver. By considering transmit power, antenna gains, and various system losses (such as cable losses or environmental factors), a link budget helps operators determine if the signal will be strong enough for reliable communication. Understanding link budgets is necessary to designing effective communication systems and ensuring signals are received clearly over the desired range.

A link budget is the calculation of the total power available for communication at the receiver's end. It starts with the transmit power, adds any gains provided by the transmitting and receiving antennas, and then subtracts the losses caused by factors like cable losses, free-space path loss, or atmospheric conditions. The result gives an estimate of how much signal power will be left when it reaches the receiver.

What is a link budget? (G8A13)

A. The financial costs associated with operating a radio link
B. The sum of antenna gains minus system losses
C. The sum of transmit power and antenna gains minus system losses as seen at the receiver
D. The difference between transmit power and receiver sensitivity

Measuring a link budget in real life involves testing and analyzing each component of the communication link. Start by measuring transmitter power with a power meter and accounting for cable losses. Verify antenna gain using manufacturer specs or field tests. Calculate or measure path loss using the free space formula or real-world tools like signal strength meters. Assess other losses, such as coaxial cable or environmental factors, using a network analyzer and online attenuation calculators. Finally, measure the received signal strength with an S-meter or spectrum analyzer and compare it to the receiver's sensitivity to ensure reliable communication. By combining theoretical calculations with real-world testing, you can optimize your station for consistent performance.

To remember this for the exam, think of a link budget as a balance sheet for your signal. You start with the signal's initial strength, account for any improvements (gains), and subtract the losses the signal encounters during transmission. This calculation helps ensure that your system has enough power to communicate effectively, especially over long distances.

Link Margin

Link margin helps ensure a reliable connection between a transmitter and receiver. It represents the "buffer" or extra power the received signal has above the minimum required level for proper reception. Environmental factors, interference, and equipment aging can reduce signal strength, and having a sufficient margin helps maintain the quality of the communication link even when conditions degrade. A higher link margin means more reliable communication, especially over long distances or in noisy environments.

Link margin is the difference between the received power level and the minimum required signal level needed for the receiver to function properly. In other words, the extra signal strength ensures reliable communication. Suppose the received power is only slightly above the required level. In that case, the link margin is small, and interference or other factors could easily disrupt the connection. A larger link margin provides more room for error, helping to maintain a stable connection even in challenging conditions.

What is link margin? (G8A14)

A. The opposite of fade margin
B. The difference between received power level and minimum required signal level at the input to the receiver
C. Transmit power minus receiver sensitivity
D. Receiver sensitivity plus 3 dB

To remember this for the exam, think of the link margin as a safety net—it's the extra power you have beyond what's minimally required to keep the communication link working reliably. This concept is key to designing systems that must function well even when faced with signal degradation or interference.

Chapter Summary

Understanding power principles is vital for managing and optimizing radio communication systems. Each concept ensures efficient signal transmission, from choosing batteries as power sources to the detailed calculations of decibels and gain. Radio operators can account for and compensate for losses by applying link budget and link margin principles, achieving clearer and more reliable connections. Mastery of these power fundamentals provides a solid foundation for advancing in the world of radio, where precise power management is critical for effective communication across distances and conditions.

The Physical Radio Station

YOUR RADIO STATION IS the hub of all amateur radio activities, where theory meets practice. Whether you're assembling your first setup or optimizing an existing one, understanding the physical components of your station is key to smooth and reliable operation. Each element, from antennas and feedlines to power sources and circuit design, plays a critical role in ensuring clear communication and efficient performance.

In this section, we'll discuss the technical standards, equipment essentials, and best practices for building, maintaining, and upgrading your radio station. This will empower you to create a setup that meets your needs and keeps you on the air with confidence, all while fully preparing for your General exam.

Chapter 9

Equipment

UPGRADING TO A GENERAL Class license unlocks exciting new privileges but demands a deeper understanding of the equipment that powers your station. This chapter dives into the critical components of a functional ham radio station to fully prepare for the exam and the challenges of expanded operation.

At the heart of your station is the transceiver, your primary tool for transmitting and receiving signals. Choosing a transceiver that covers the General Class frequency bands and offers features like Digital Signal Processing (DSP) ensures optimal performance and clarity. Equally important is a reliable antenna and feedline setup—your connection to the world. From dipoles to Yagi antennas, understanding these components and their installation is key to maximizing signal strength and communication success. Let's explore these topics to help you operate confidently and effectively while preparing for your exam.

Transceiver

Transceivers are the heart of any ham radio station, integrating transmitting and receiving functions into one device. Whether operating from a permanent base station or on the go, selecting the right transceiver is critical for effective communication. Base station transceivers offer high power and advanced features but are larger and pricier. Mobile and handheld transceivers provide flexibility and portability but may lack the robust capabilities of base units.

Three standout models combine performance and reliability for General-class operators. The Icom IC-7300 is highly regarded for its advanced features, including a high-resolution touchscreen, built-in DSP, and an automatic antenna tuner covering all HF bands. The Yaesu FT-991A offers versatility, covering HF, VHF, and UHF bands, with a compact design and easy-to-use interface suited for both base and mobile operations. The Kenwood TS-590SG delivers reliability and user-friendly operation for exceptional receiver performance and advanced filtering, making it a strong choice for serious operators.

By understanding the strengths of these transceiver options, you can confidently choose the best fit for your General Class privileges and communication goals.

Checklist: Selecting Your Transceiver

- **Type**: Decide between base station, mobile, and handheld transceivers based on your operating preferences and space.

- **Frequency Coverage**: Ensure the transceiver covers all HF bands available to General Class operators.

- **Power Output**: Look for adjustable power settings to match different operating conditions.

- **User Interface**: Choose a model with a clear display, intuitive menus, and easily accessible controls.

- **Additional Features**: Consider DSP, automatic antenna tuners, and memory channels for enhanced performance.

Regular calibration and maintenance are required to optimize performance. Proper alignment prevents interference and ensures reliable operation, while external filters and preamplifiers enhance signal quality. High-pass and low-pass filters block unwanted frequencies, and preamplifiers boost weak signals. Additionally, keeping connectors clean, inspecting cables, and applying firmware updates can improve performance and extend the life of your equipment.

Voice Operated Exchange

In amateur radio, there are two common ways to control a radio's transmission: Voice Operated Exchange (VOX) and Push-To-Talk (PTT). These methods allow operators to switch between listening and transmitting. PTT requires the operator to manually press a button when they wish to transmit. VOX automatically triggers the transmitter when it detects the operator's voice. Each method has advantages and depends on the situation and the operator's preferences.

The key difference between VOX operation and PTT operation is that VOX allows "hands-free" operation. With VOX enabled, the radio begins transmitting as soon as it detects the operator's voice without needing to press a button. This can be especially useful when the operator needs to use both hands for something else, like when using a computer, logging contacts, or adjusting equipment.

Which of the following statements is true of VOX operation versus PTT operation? (G2A10)

A. The received signal is more natural sounding
B. It allows "hands-free" operation
C. It occupies less bandwidth
D. It provides more power output

In contrast, PTT operation requires the operator to press and hold a button every time they want to transmit, which provides more manual control but requires active engagement. The convenience of VOX can be balanced by its sensitivity to background noise, which may accidentally trigger transmissions, so operators should adjust VOX settings accordingly.

Two-Tone Test

A two-tone test is a standard method for evaluating the linearity and performance of a transmitter's modulation and amplifier stages. This test injects two pure, non-harmonic audio signals/ tones of equal amplitude into the transmitter's input, typically through a microphone or audio interface. These tones are chosen so their frequencies fall within the transmitter's bandwidth but do not interfere with each other.

When the transmitter amplifies the signal, its output is analyzed using a spectrum analyzer or oscilloscope. In an ideal scenario, the output will show only the original tones and their sum and difference frequencies (the sidebands) without additional distortion or spurious signals. However, if the transmitter exhibits non-linearity or overdrive, unwanted harmonics or intermodulation products will appear, indicating distortion.

This test is critical for ensuring the transmitter delivers a clean signal, minimizing interference with other users on the band. It helps operators identify and correct issues like improper gain settings or amplifier inefficiencies, ensuring compliance with FCC standards and optimal on-air performance.

What signals are used to conduct a two-tone test? (G4B07)

A. Two audio signals of the same frequency shifted 90 degrees
B. Two non-harmonically related audio signals
C. Two swept frequency tones
D. Two audio frequency range square wave signals of equal amplitude

The critical transmitter performance parameter analyzed by the two-tone test is linearity. Linearity ensures that the transmitted signals remain clean and without distortion, especially in amplitude modulation and single-sideband (SSB) operations. If the transmitter is non-linear, the output will include intermodulation products—unwanted signals generated by the interaction of the two tones, which can cause interference and degrade signal quality.

What transmitter performance parameter does a two-tone test analyze? (G4B08)

A. Linearity
B. Percentage of suppression of the carrier and undesired sideband for SSB
C. Percentage of frequency modulation
D. Percentage of carrier phase shift

To perform a two-tone test, combine two pure audio tones (e.g., 700 Hz and 1900 Hz) of equal amplitude using a signal generator and inject them into the transmitter's audio input. Connect the transmitter to a dummy load and begin transmitting. A spectrum analyzer or oscilloscope is used to observe the output signal at the dummy load, checking for clean tones and sidebands without intermodulation distortion or harmonics. Adjust the transmitter's audio gain or drive levels and retest if unwanted signals are present. This test ensures the transmitter operates linearly, producing a clean signal while meeting FCC standards. Always use a dummy load and proper equipment for safe and accurate results.

The two-tone test helps amateur operators analyze the linearity of their transmitters by introducing two different audio tones. The output is checked to ensure that no distortion or unwanted frequencies are produced, as non-linear operation can affect communication quality. The goal is to ensure clean, reliable transmissions.

Power Amplifiers

Automatic Level Control

Automatic Level Control (ALC) is a feature used with RF power amplifiers to prevent excessive drive, which could damage the amplifier or cause signal distortion. The ALC system works by monitoring the amplifier's input power and, if the drive signal becomes too strong, reducing it to maintain safe operating levels. This ensures that the amplifier functions efficiently and avoids pushing the equipment beyond its designed limits. Without ALC, transmitting with excessive drive could lead to overheating, distortion of the transmitted signal, and interference with other users on the frequency.

Why is automatic level control (ALC) used with an RF power amplifier? (G4A05)

A. To balance the transmitter audio frequency response
B. To reduce harmonic radiation
C. To prevent excessive drive
D. To increase overall efficiency

Side note: Drive refers to the input signal provided to an RF power amplifier, typically originating from a transceiver or exciter. The drive level determines how much power the amplifier amplifies and outputs.

However, when transmitting AFSK (Audio Frequency Shift Keying) data signals, the ALC system should be disabled because its action can distort the signal. AFSK signals rely on precise frequency shifts to encode data. If the ALC kicks in to reduce drive, it can alter the intended signal, resulting in data errors or poor transmission quality. Avoiding ALC when handling such digital modes is important to ensure signal clarity and reliable data transmission.

Why should the ALC system be inactive when transmitting AFSK data signals? (G4A11)

A. ALC will invert the modulation of the AFSK mode
B. The ALC action distorts the signal
C. When using digital modes, too much ALC activity can cause the transmitter to overheat
D. All these choices are correct

While ALC helps keep RF power amplifiers safe from overdrive, it must be used carefully, especially in digital data modes like AFSK, where precision is needed.

Vacuum-Tube RF Power Amplifier

When operating a vacuum-tube RF power amplifier, understanding how to properly adjust the TUNE and LOAD controls is crucial for optimal performance and avoiding damage. The TUNE control helps match the amplifier's internal circuitry to the antenna system, optimizing power transfer. A pronounced dip in plate current identifies the correct setting of this control. This dip indicates that the amplifier is tuned correctly, allowing maximum power output with minimal stress on the components.

What is the effect on plate current of the correct setting of a vacuum-tube RF power amplifier's TUNE control? (G4A04)

A. A pronounced peak
B. A pronounced dip
C. No change will be observed
D. A slow, rhythmic oscillation

Similarly, the LOAD or COUPLING control is used to adjust the amplifier's power output without exceeding the maximum allowable plate current. This setting ensures that the amplifier operates efficiently while staying within safe limits. By carefully adjusting this control, you achieve the desired power output without exceeding the maximum allowable plate current.

What is the correct adjustment for the LOAD or COUPLING control of a vacuum tube RF power amplifier? (G4A08)

A. Minimum SWR on the antenna
B. Minimum plate current without exceeding maximum allowable grid current
C. Highest plate voltage while minimizing grid current
D. Desired power output without exceeding maximum allowable plate current

The TUNE and LOAD controls are critical for tuning vacuum-tube RF amplifiers. The TUNE control helps achieve the correct resonant point, seen as a dip in plate current. In contrast, the LOAD control adjusts the output power to prevent damage by keeping the plate current within safe limits.

Duty Cycle

The duty cycle in radio communication refers to the percentage of time a transmitter is actively transmitting versus being idle. Different modes, such as CW (Morse code), SSB (Single Sideband), or digital modes like FT8, have varying duty cycles. Modes with high-duty cycles require the transmitter to be on an output power more frequently, which can place significant stress on the equipment. Understanding the duty cycle of your mode helps avoid overheating and damaging your transmitter, especially during extended transmissions.

Knowing the mode's duty cycle is important because some modes have high-duty cycles that could push your transmitter beyond its average power rating. For example, digital modes like FT8 often have a nearly 100% duty cycle, meaning the transmitter constantly transmits without breaks. This continuous output can cause the transmitter to generate excessive heat and possibly exceed its designed power-handling capability, leading to equipment failure.

Why is it important to know the duty cycle of the mode you are using when transmitting? (G8B08)

A. To aid in tuning your transmitter
B. Some modes have high-duty cycles that could exceed the transmitter's average power rating
C. To allow time for the other station to break in during a transmission
D. To prevent overmodulation

To remember this for the exam, think of the duty cycle as the "workload" on your transmitter. Modes with high-duty cycles require careful monitoring of your equipment's power ratings and cooling system to avoid damage during prolonged use.

Receiver

Receiver performance metrics such as sensitivity, selectivity, and dynamic range significantly impact communication quality. Sensitivity determines a receiver's ability to detect weak signals, which is needed for long-distance communication in challenging conditions. Selectivity allows the receiver to isolate desired signals in crowded bands, rejecting adjacent-channel interference. Dynamic range ensures the receiver can handle both weak and strong signals without distortion, maintaining signal clarity.

A receiver's noise reduction control minimizes unwanted background noise, enhancing the clarity of weak signals. However, as the noise reduction control is increased, more aggressive algorithms are used to suppress noise. While this can help eliminate unwanted sound, it may also result in signal distortion. Reducing noise can inadvertently filter out parts of the desired signal, especially if the noise reduction is set too high.

What happens as a receiver's noise reduction control level is increased? (G4A07)

A. Received signals may become distorted
B. Received frequency may become unstable
C. CW signals may become severely attenuated
D. Received frequency may shift several kHz

Operators should carefully adjust the noise reduction control for optimal use, balancing between reducing background noise and maintaining the signal's integrity. Pushing the control too far may result in garbled or unintelligible audio, even though the background noise is reduced. Hence, finding the right level where signals remain clear without too much distortion is crucial.

What is the purpose of using a receive attenuator? (G4A13)

A. To prevent receiver overload from strong incoming signals
B. To reduce the transmitter power when driving a linear amplifier
C. To reduce power consumption when operating from batteries
D. To reduce excessive audio level on strong signals

A receive attenuator is a device that reduces the strength of incoming signals before they reach the receiver. The primary purpose of using an attenuator is to prevent receiver overload from powerful signals. When a signal is too strong, it can overwhelm the receiver, causing distortion, loss of sensitivity, and difficulty hearing weaker signals. By reducing the signal strength with an attenuator, you can ensure the receiver performs better, especially when strong local signals might interfere with reception.

For example, if you are trying to listen to weak signals from a distant station but a nearby station is transmitting a powerful signal, the receiver might become overloaded. In such cases, activating the attenuator helps manage this situation by reducing the power of the strong signal without significantly affecting weaker signals. This keeps your receiver from being overwhelmed and allows you to hear the signals more clearly.

Receiver Sensitivity and Bandwidth

Receiver sensitivity and bandwidth are crucial factors in radio system performance. Sensitivity determines the receiver's ability to detect weak signals, while bandwidth affects how much of the signal is passed through the receiver without being filtered out. Optimizing these parameters ensures clear reception with minimal noise and interference. Let's explore key aspects related to radio receivers' sensitivity, bandwidth, and filter performance.

Which parameter affects receiver sensitivity? (G7C08)

A. Input amplifier gain
B. Demodulator stage bandwidth
C. Input amplifier noise figure
D. All these choices are correct

Receiver sensitivity is determined by a few critical factors that dictate how well a receiver can pick up weak signals. The input amplifier gain boosts the strength of incoming signals, while the demodulator stage bandwidth must be set to allow the desired signal through without letting in too much noise. The input amplifier's noise figure also plays a big role—lower noise means better sensitivity, allowing the receiver to detect weaker signals without getting overwhelmed by interference.

The bandwidth of a band-pass filter is measured between what two frequencies? (G7C14)

A. Upper and lower half-power
B. Cutoff and rolloff
C. Pole and zero
D. Image and harmonic

The bandwidth of a band-pass filter is the range of frequencies it allows to pass through. It is measured between the upper and lower half-power points, which correspond to the frequencies where the signal is reduced by 3 dB from its maximum level. This ensures that only the desired signal frequencies are passed while filtering out those outside the target range, leading to better signal clarity and less interference.

What term specifies a filter's maximum ability to reject signals outside its passband? (G7C13)

A. Notch depth
B. Rolloff
C. Insertion loss
D. Ultimate rejection

Ultimate rejection refers to a filter's ability to block or attenuate signals outside its passband, ensuring that unwanted signals don't interfere with the desired transmission. A filter with a high ultimate rejection will effectively isolate the desired signal from outside noise and interference, providing clearer communication.

Understanding receiver sensitivity and bandwidth and how they are influenced by gain, noise, and filtering is key to designing and operating efficient radio systems. Properly balancing these factors ensures that weak signals can be received clearly while unwanted interference is minimized.

Filters and Attenuation

In radio communication systems, filters are critical for shaping and controlling signals by allowing certain frequencies to pass through while attenuating others. Attenuation refers to the reduction of signal strength as it passes through a filter or over a distance. Filters, particularly digital signal processing (DSP) filters, offer advanced flexibility, while key terms like insertion loss and cutoff frequency help quantify the performance of these filters. Let's explore some key questions about these concepts.

Which of the following is an advantage of a digital signal processing (DSP) filter compared to an analog filter? (G7C06)

A. A wide range of filter bandwidths and shapes can be created
B. Fewer digital components are required
C. Mixing products are greatly reduced
D. The DSP filter is much more effective at VHF frequencies

A digital signal processing (DSP) filter offers significant advantages over traditional analog filters, especially in terms of flexibility. DSP filters can be reprogrammed to create a wide range of bandwidths and shapes, allowing precise control over which frequencies are passed or attenuated. This versatility makes DSP filters more adaptable for different communication needs, such as reducing noise or improving signal clarity, without physical hardware changes.

What term specifies a filter's attenuation inside its passband? (G7C07)

A. Insertion loss
B. Return loss
C. Q
D. Ultimate rejection

Insertion loss refers to the attenuation of signal strength within a filter's passband, which is the range of frequencies that the filter is designed to pass. Ideally, a filter should allow all signals within its passband to pass through with minimal loss. However, in practice, some attenuation occurs even within this range, and this reduction is quantified as insertion loss. Low insertion loss is crucial for maintaining signal integrity in communication systems.

What is the frequency above which a low-pass filter's output power is less than half the input power? (G7C12)

A. Notch frequency
B. Neper frequency
C. Cutoff frequency
D. Rolloff frequency

The cutoff frequency of a low-pass filter is the point at which the filter begins to significantly attenuate higher frequencies. Beyond this frequency, the output power of the signal is reduced to less than half of the input power. The cutoff frequency is an essential characteristic of any filter because it defines the threshold at which unwanted high-frequency signals are attenuated, helping to maintain clear signal transmission within the desired frequency range.

Filters and attenuation are integral to maintaining signal clarity in radio communications. Digital signal processing (DSP) filters offer incredible flexibility in shaping signals, while important concepts like insertion loss and cutoff frequency help engineers design systems that minimize unwanted noise and maximize signal strength. Understanding these terms will help you optimize and analyze filters for better communication performance.

Noise Blanker

A noise blanker is a feature in many modern radio receivers designed to reduce a receiver's gain during sudden bursts of noise. These bursts of noise create short, high-intensity pulses that can overwhelm the radio signal, making it difficult to hear the desired transmission. The noise blanker works by detecting these pulses and temporarily reducing the receiver's gain, or sensitivity, during the duration of the noise pulse. This process minimizes the impact of the noise, allowing the desired signal to come through more clearly.

How does a noise blanker work? (G4A03)

A. By temporarily increasing received bandwidth
B. By redirecting noise pulses into a filter capacitor
C. By reducing receiver gain during a noise pulse
D. By clipping noise peaks

A noise blanker acts like a filter, automatically lowering the volume when a loud, unwanted noise is detected, ensuring that the signal you want to hear remains intelligible. This feature is especially useful for operators who work in environments with a lot of electronic interference, helping them maintain clear communication.

In summary, as a General, you must now understand the advanced intricacies of transmitters and receivers for effective communication. Ensuring your equipment is optimized for sensitivity, selectivity, and dynamic range will enhance your experience, enabling you to maximize your General license.

Signal Strength Meter

A Signal Strength Meter (S Meter) measures the strength of a received signal in decibels (dB) relative to a standard reference point. It is commonly found on most amateur radio transceivers. It gives an operator a quick visual indication of how strong an incoming signal is. The meter typically ranges from S1 to S9, with each number representing a stronger signal. Beyond S9, additional signal strength can be indicated in decibel increments, like 10 dB over S9. This helps operators gauge the quality of a communication link, making adjustments to ensure clear reception.

What does an S meter measure? (G4D04)

A. Carrier suppression
B. Impedance
C. Received signal strength
D. Transmitter power output

Let's look at some math questions about S Meter.

How does a signal that reads 20 dB over S9 compare to one that reads S9 on a receiver, assuming a properly calibrated S meter? (G4D05)

A. It is 10 times less powerful
B. It is 20 times less powerful
C. It is 20 times more powerful
D. It is 100 times more powerful

When you see a signal on your S meter that reads "20 dB over S9," it means that the signal is significantly stronger than one that simply reads S9. In technical terms, 20 dB over S9 indicates that the signal is 100 times more powerful than an S9 signal.

The relationship between decibels and signal strength is logarithmic, meaning each increase of 10 dB represents a tenfold increase in power. Therefore, 20 dB over S9 means two jumps of 10 dB – one jump from 0 to 10 and then the second jump from 11-20, which results in a signal that is 100 times more powerful ($10 \times 10 = 100$).

Understanding this relationship is important because it shows how much stronger one signal is compared to another. It also helps you adjust your receiver settings properly to enhance weaker signals or manage interference from stronger ones.

Another one.

How much change in signal strength is typically represented by one S unit? (G4D06)

A. 6 dB
B. 12 dB
C. 15 dB
D. 18 dB

One S unit on a signal strength meter (S-meter) typically represents a change of 6 decibels (dB) in signal strength. This means that for every increase of one S unit, the signal strength roughly doubles in power (or decreases by half for a decrease of one S unit). For example, a signal that measures S5 is approximately twice as strong as one at S4.

To answer this question effectively, remember:

- **Key Concept**: 1 S unit = 6 dB change in signal strength.

- **Application**: This applies to the calibration of most S-meters on HF transceivers, where a reading of S9 is the standard for a strong signal.

Knowing this relationship is essential for understanding signal reports and troubleshooting receiver performance.

And one final exam question.

How much must the power output of a transmitter be raised to change the S meter reading on a distant receiver from S8 to S9? (G4D07)

A. Approximately 1.5 times
B. Approximately 2 times
C. Approximately 4 times
D. Approximately 8 times

To change the S-meter reading on a distant receiver from S8 to S9, the transmitter's power output must be increased by a factor of 4.

One S-unit represents a change of 6 decibels (dB) in signal strength, and a 6 dB increase requires the signal power to be quadrupled.

Why Quadruple?

The decibel (dB) scale is logarithmic:

- Every **3 dB** increase doubles the power.

- A **6 dB** increase (one S-unit) is the same as **doubling the power twice:** $2 \times 2 = 4$ **times the power**.

Example:

Let's say your transmitter is currently outputting **25 watts**, and the receiving station's S-meter shows **S8**. To move the reading to **S9**:

- Multiply your current power by 4: **25 watts** $\times$ **4 = 100 watts**.

This relationship applies universally, so you can find the required increase by multiplying it by 4 for any starting power.

Transceiver

A transceiver is a device that combines both a transmitter and a receiver into a single unit, allowing for two-way communication. The name comes from combining the words "transmitter" and "receiver."

The transceiver allows the operator to communicate with other radio operators, both locally and globally, by transmitting their voice, data, or Morse code and receiving signals from others. Many modern transceivers are highly versatile, offering multiple modes of communication such as voice, digital, and even video over radio frequencies.

Notch Filter

Several features help improve communication clarity and ensure smooth operation with external equipment when operating an HF transceiver. One of these features is the notch filter, designed to reduce interference from carriers that may be present within the receiver's passband. A carrier is a continuous signal that can cause unwanted noise or interference, making it challenging to hear weaker signals. The notch filter helps by "notching out" or removing these unwanted carriers, allowing for clearer reception of the desired signal.

What is the purpose of the notch filter found on many HF transceivers? (G4A01)

A. To restrict the transmitter voice bandwidth
B. To reduce interference from carriers in the receiver passband
C. To eliminate receiver interference from impulse noise sources
D. To remove interfering splatter generated by signals on adjacent frequencies

---

Receiver Passband

A receiver passband refers to the range of frequencies that a radio receiver can process and pass through its filters to the output, allowing the operator to hear or decode the signals within that range. It is determined by the receiver's intermediate frequency (IF) filters or digital signal processing (DSP).

- **Purpose**: The passband defines the bandwidth of the receiver, ensuring it includes the desired signal while excluding adjacent signals and noise. For example, a single-sideband signal typically requires a passband of about 2.4–3 kHz, while a CW signal may only need 500 Hz or less.

- **Adjustment**: Many modern receivers allow operators to adjust the passband width and center frequency to optimize reception under different conditions, such as crowded bands or weak signals.

- **Relevance**: The passband directly affects the receiver's ability to isolate desired signals and reject interference, making it a critical factor in effective communication.

Delayed Output

Another feature is the delayed RF output when using an external amplifier. When you key up your transceiver, there needs to be a brief delay before RF energy is sent to the amplifier. This delay allows the amplifier to switch the antenna between the transceiver and its output. Without this delay, RF could be transmitted before the amplifier is ready, potentially causing signal distortion or damage to the equipment.

What is the purpose of delaying RF output after activating a transmitter's keying line to an external amplifier? (G4A09)

A. To prevent key clicks on CW
B. To prevent transient overmodulation
C. To allow time for the amplifier to switch the antenna between the transceiver and the amplifier output
D. To allow time for the amplifier power supply to reach operating level

---

Dual VFO

The dual-VFO (Variable Frequency Oscillator) feature is commonly used for transmitting on one frequency and listening on another. This is particularly useful in situations like split-frequency operation, where one operator transmits on a different frequency than they receive. This allows for more flexible and efficient communication, especially in DX (long-distance) operations.

Which of the following is a common use of the dual-VFO feature on a transceiver? (G4A12)

A. To allow transmitting on two frequencies at once
B. To permit full duplex operation -- that is, transmitting and receiving at the same time
C. To transmit on one frequency and listen on another
D. To improve frequency accuracy by allowing variable frequency output (VFO) operation

---

Speech Processor

A speech processor is a device used in transceivers to enhance voice clarity and loudness by manipulating the audio signal before it's transmitted. It works by compressing the dynamic range of the voice signal, which means it amplifies quieter parts while limiting louder parts, resulting in a more consistent and louder signal. This is particularly useful in weak signal conditions or noisy environments, helping to ensure that the operator's voice stands out clearly over long distances.

What is the purpose of a speech processor in a transceiver? (G4D01)

A. Increase the apparent loudness of transmitted voice signals
B. Increase transmitter bass response for more natural-sounding SSB signals
C. Prevent distortion of voice signals
D. Decrease high-frequency voice output to prevent out-of-band operation

---

A speech processor in a transceiver is a device designed to enhance the clarity and effectiveness of voice communication by increasing the apparent loudness of transmitted voice signals. This is especially useful in environments where signals may be weak, or there is interference. The speech processor works by amplifying the quieter parts of your speech and compressing the louder parts, ensuring that more of your voice energy is transmitted.

How does a speech processor affect a single sideband phone signal? (G4D02)

A. It increases peak power
B. It increases average power
C. It reduces harmonic distortion
D. It reduces intermodulation distortion

The speech processor improves the average power output of the transmission, making the signal more readable, especially when using modes like single sideband. However, if set incorrectly, it can introduce distortion, affect bandwidth, or degrade the quality of the signal, so proper adjustment is key to its effectiveness.

The main benefit of a speech processor is that it increases the average power of a signal, particularly on single-sideband transmissions. This makes the voice signal stand out better and improves the overall readability of your transmission, which is crucial when dealing with weak signals or noisy band conditions.

What is the effect of an incorrectly adjusted speech processor? (G4D03)

A. Distorted speech
B. Excess intermodulation products
C. Excessive background noise
D. All these choices are correct

However, an incorrectly adjusted speech processor can cause significant issues. The most common problems include distorted speech, where the voice becomes unclear or garbled; excess intermodulation products, which result in unwanted signal mixing that can cause interference; and excessive background noise, making it harder for the receiving station to understand the transmission. These issues reduce the overall effectiveness of communication and may cause interference with other stations, so proper adjustment is needed.

In summary, while a speech processor can significantly enhance your signal, improper settings can lead to degraded transmission quality, affecting clarity and causing unwanted interference. Always ensure it's adjusted correctly for the best performance.

Chapter Summary

Transmitters, receivers, and transceivers are the core components of every ham radio station, transforming theoretical knowledge into practical communication. As a General Class operator, your expanded privileges demand a deeper understanding of these systems to confidently explore new bands and modes. With the right knowledge and tools, you can enhance your station's performance, minimize interference, and enjoy seamless communication across the airwaves. This solid foundation will prepare you for more advanced topics.

Chapter 10

Circuit Components and Design

Basic Electronic Components

ELECTRONIC CIRCUITS ARE COMPOSED of various fundamental components, each serving a specific function. Resistors, capacitors, and inductors are the primary passive components essential for controlling and manipulating electrical signals.

Resistors

Resistors introduce resistance to the flow of current, measured in ohms (Ω). They come in various types, including through-hole and surface-mount, each suited for different applications. Typical uses of resistors include current limiting, voltage division, and feedback circuits. Selecting the proper resistor involves considering its resistance value, power dissipation, and tolerance. These parameters ensure that the resistor can handle the required current without overheating or deviating from its specified value.

Wire-Wound Resistors

While wire-wound resistors are excellent for power handling and precision in certain circuits, they are not suitable for radio frequency (RF) circuits due to their inherent inductance. The wire-wound around the resistor's core can act like a small inductor, introducing inductive reactance at high frequencies. This inductive behavior can interfere with the resistor's performance, making the circuit unpredictable and less efficient in RF applications.

Why should wire-wound resistors not be used in RF circuits? (G6A06)

A. The resistor's tolerance value would not be adequate
B. The resistor's inductance could make circuit performance unpredictable
C. The resistor could overheat
D. The resistor's internal capacitance would detune the circuit

---

As RF circuits rely on precise tuning and performance, using a component that adds unintended inductance could severely affect signal quality or cause the circuit to malfunction.

Power Dissipation

Power dissipation refers to the process by which an electrical or electronic component converts electrical energy into heat or another form of energy that is not useful for the circuit's intended purpose. This happens whenever current flows through a component with resistance, such as a resistor, and energy is lost as heat. Power dissipation is an important concept in circuit design because excessive heat can damage components or reduce the system's efficiency.

Root Mean Square / RMS Voltage

An AC signal's RMS (Root Mean Square) value produces the same power dissipation in a resistor as a DC voltage of the same value. This is because the RMS value represents the effective or "average" voltage or current that would result in the same amount of heat or power loss as a steady DC signal. While the voltage in an AC signal constantly changes, the RMS value gives a way to measure how much actual power the signal delivers, just like a constant DC voltage would.

What value of an AC signal produces the same power dissipation in a resistor as a DC voltage of the same value? (G5B07)

A. The peak-to-peak value
B. The peak value
C. The RMS value
D. The reciprocal of the RMS value

The RMS (Root Mean Square) value expresses the practical or equivalent value of an alternating current or voltage. For AC signals, the current or voltage constantly changes. Hence, the RMS value represents the "average" level of power that would be equivalent to the same amount of power as a direct current signal.

Mathematically, the RMS value is the square root of the mean (average) of the squares of all the instantaneous values of the waveform over one complete cycle. For a sinusoidal AC signal, the RMS value is approximately 0.707 times the signal's peak value. The RMS value is important because it gives a practical measure of the power delivered by an AC signal, making it comparable to DC.

Math Diversion (not needed for the test)

The factor of 0.707 that relates the RMS voltage to the peak voltage comes from the mathematical definition of the Root Mean Square (RMS) value for a sine wave.

Here's a step-by-step explanation:

1. **Understanding RMS**: The RMS value is essentially the square root of the average of the squares of all instantaneous values of a waveform over one complete cycle. For an AC signal, the RMS value gives the equivalent DC value that would produce the same power dissipation in a resistor.

2. **The mathematical relationship**: For a sine wave, the RMS value is derived from integrating the square of the sine function over a complete cycle (from 0 to 2π radians) and then taking the square root of the average of those values. The detailed mathematical derivation shows that the RMS value of a sine wave is equal to the peak value (V_{peak}) divided by the reciprocal of the square root of 2.

3. The formula is:

$$V_{RMS} = \frac{V_{peak}}{\frac{1}{\sqrt{2}}} \approx V_{peak} \times 0.707$$

This factor of 0.707 comes from the fact that the square root of 2, divided by 2 gives approximately 0.707.

This is why, for a sine wave, the RMS value is approximately 70.7% of the peak value. This relationship is critical to understanding how AC voltages are compared to DC voltages in terms of power dissipation.

Example for Voltage

For an AC voltage with a peak value of V_{peak}:

$$V_{RMS} = V_{peak} \times 0.707$$

So, if the peak voltage is 100 volts, the RMS value is:

$$V_{RMS} = 100 \times 0.707 = 70.7 \; volts$$

The RMS value is used because it gives the equivalent voltage or current that would produce the same heating effect (or power) as a DC value in a resistive load.

Ok, let's try some exam questions now.

ELECTRICAL WAVEFORMS

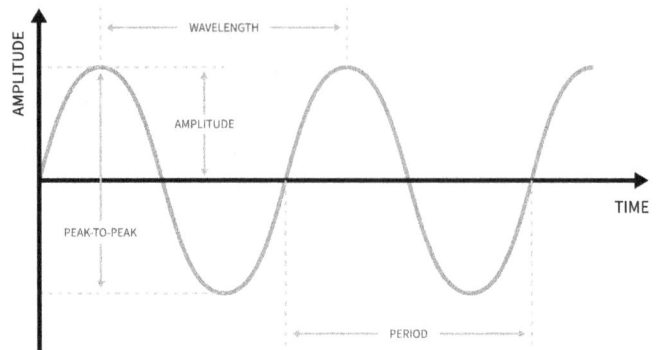

What is the peak-to-peak voltage of a sine wave with an RMS voltage of 120 volts? (G5B08)

A. 84.8 volts
B. 169.7 volts
C. 240.0 volts
D. 339.4 volts

To calculate the peak-to-peak voltage of a sine wave from its RMS voltage, you need to know the relationship between RMS voltage and peak voltage. For a sine wave, the RMS voltage is approximately 0.707 times the peak voltage. The peak-to-peak voltage is then twice the peak voltage.

Here's how you can calculate it:

Step 1: Start with the peak voltage formula.

$$V_{peak} = \frac{V_{RMS}}{0.707}$$

Step 2: Enter the variables from the question. RMS Voltage = 120 volts

$$V_{peak} = \frac{120}{0.707}$$

Step 3: Solve the math.

$$V_{peak} = \frac{120}{0.707} \approx 169.7 \; volts$$

Step 4: Find the peak-to-peak voltage using the formula.

$$V_{peak-to-peak} = 2 \times V_{peak}$$

Step 5: Enter the variables.

$$V_{peak-to-peak} = 2 \times 169.7$$

Step 6: Solve the math.

$$V_{peak-to-peak} = 2 \times 169.7 \approx 339.4 \; volts$$

So, for a sine wave with an RMS voltage of 120 volts, the peak-to-peak voltage is approximately 339.4 volts.

What is the RMS voltage of a sine wave with a value of 17 volts peak? (G5B09)

A. 8.5 volts
B. 12 volts
C. 24 volts
D. 34 volts

To find the RMS voltage of a sine wave when the peak voltage is known, we start with the same formula:

$$V_{peak} = \frac{V_{RMS}}{0.707}$$

But rearrange it to find RMS Voltage.

$$V_{RMS} = V_{peak} \times 0.707$$

In this case, the peak voltage is 17 volts. Applying the formula and solving.

$$V_{RMS} = 17 \times 0.707 \approx 12.02 \; volts$$

So, the RMS voltage of a sine wave with a peak voltage of 17 volts is approximately 12.02 volts... or 12 volts.

The RMS value gives the equivalent DC voltage that would produce the same power dissipation in a resistor as the AC signal – repetition is key!

RMS Voltage with Power

What is the RMS voltage across a 50-ohm dummy load dissipating 1200 watts? (G5B12)

A. 173 volts
B. 245 volts
C. 346 volts
D. 692 volts

Let's walk through how to find the RMS voltage across a 50-ohm dummy load dissipating 1200 watts based on the foundational formulas we know.

Step 1: Understand the relationship between power, voltage, and resistance = Power Formula

$$Power\ (P) = Current\ (I) \times Voltage\ (E)$$

Step 2: Relate current to voltage and resistance = Ohm's Law

$$Current\ (I) = \frac{Voltage\ (E)}{Resistance\ (R)}$$

Step 3: Substitute 'I' into the power formula

$$Using\ I = \frac{E}{R}\ substitue\ this\ into\ the\ Power\ Formula\ in\ Step\ \#1$$

$$Power\ (P) = \left(\frac{Voltage\ (E)}{Resistance\ (R)}\right) \times Voltage(V)$$

Step 4: Simplify the expression

$$P = \frac{E^2}{R}$$

Or, because we are talking about RMS Voltage, it can be written as

$$P = \frac{V_{RMS}^2}{R}$$

Where:

- P is the power (1200 watts),

- V_{RMS} is the RMS voltage, which we're solving for (and yes you will see V and E both used for voltage)!

- R is the resistance of the load (50 ohms).

Step 5: Rearranging the formula to solve for V_{RMS}

$$V_{RMS} = \sqrt{P \times R}$$

Step 6: Substituting the values and solve

$$V_{RMS} = \sqrt{1200 \times 50} = \sqrt{60,000} \approx 244.9\ volts$$

So, the RMS voltage across the 50-ohm dummy load dissipating 1200 watts is approximately 244.9 or simply 245 volts.

To remember this for the exam, start with the key formulas and recombine based on the variables you are given and what you are solving for.

Peak Envelope Power

Peak Envelope Power (PEP) is the maximum instantaneous power that a transmitter delivers during one cycle of the signal's modulation envelope, typically in an amplitude-modulated (AM), single-sideband (SSB), or other modulated signal. It represents the highest power level the signal reaches as it fluctuates over time due to modulation, which can be voice, data, or other types of information.

PEP is important in radio communication because it measures the highest power output point during transmission, particularly in modulated signals where the power level varies with the modulation (such as voice peaks). In an unmodulated carrier, PEP and average power are the same. Still, in a modulated signal, PEP can be much higher than the average power because it measures the peaks of the signal's envelope.

To sum it up, PEP is used to assess the transmitter's power output during the strongest part of the signal, giving a clear idea of the transmitter's maximum power capability during modulation. This is critical in ensuring transmissions stay within legal power limits while delivering strong and clear signals.

What is the PEP produced by 200 volts peak-to-peak across a 50-ohm dummy load? (G5B06)

A. 1.4 watts
B. 100 watts
C. 353.5 watts
D. 400 watts

Look at this question. What variables are in it? Power (PEP), Voltage (Peak to Peak), and Resistance (Ohm load). Like we did above, we can rearrange the Power Formula and Ohm's Law to get a formula with Power, Voltage and Resistance.

Step 1: Combined Power and Ohm's Law formula (see question above on how we can combine these two to get this formula).

$$P = \frac{V^2}{R}$$

Step 2: Rearrange Peak vs Peak-to-Peak Voltage

$$V_{peak-to-peak} = 2 \times V_{peak}$$

then rearrange to equal

$$V_{peak} = \frac{V_{p-p}}{2}$$

Step 3: Enter values and solve. Peak-to-peak voltage is the total voltage swing from the maximum positive to the maximum negative.

$$V_{peak} = \frac{V_{p-p}}{2}$$

substitue $V_{p-p} = 200\ V$

$$V_{peak} = \frac{200}{2} = 100\ V$$

Step 4: Convert Peak Voltage to RMS Voltage

$$RMS = V_{peak} \times 0.707$$

$$RMS = 100V \times 0.707$$

$$RMS = 70.7V$$

Step 5: Plug the RMS Voltage into the Power Formula from Step 1

$$P_{PEP} = \frac{RMS^2}{R}$$

substitue RMS = 70.7 V and R = 50 Ω

$$P_{PEP} = \frac{70.7^2}{50}$$

Step 6: Solve

$$P_{PEP} = \frac{4,998.5}{50} = 99.97\ W\ or\ simply\ 100\ Watts$$

Answer: The Peak Envelope Power (PEP) produced by 200 volts peak-to-peak across a 50-ohm dummy load is 100 watts.

To remember this for the exam, always convert the peak voltage to RMS before calculating power, as power is usually calculated using RMS voltage in AC circuits.

Do you have this straight? Let's try another one, just in case:

What is the output PEP of 500 volts peak-to-peak across a 50-ohm load? (G5B14)

A. 8.75 watts
B. 625 watts
C. 2500 watts
D. 5000 watts

It is similar to the steps we just walked through in the last question.

Step 1: Combined Power and Ohm's Law formula (see question above on how we can combine these two to get this formula).

$$P = \frac{V^2}{R}$$

Step 2: Rearrange Peak vs Peak-to-Peak Voltage

$$V_{peak-to-peak} = 2 \times V_{peak}$$

this formula rearranged equals

$$V_{peak} = \frac{V_{peak-to-peak}}{2}$$

Step 3: Enter values and solve

substitue $V_{peak-to-peak} = 500\ volts$

$$V_{peak} = \frac{500}{2} = 250\ volts$$

Step 4: Convert the peak voltage to RMS voltage

$$V_{RMS} = V_{peak} \times \frac{1}{\sqrt{2}}$$

substitue $V_{peak} = 250\ volts$

Step 5: Plug the Peak Voltage into the Power Formula from Step 1

the formula for Power when using RMS voltage is:

$$P = \frac{V_{RMS}^2}{R}$$

substitue RMS voltage of 176.75 volts and resistance of 50 ohms

$$P_{PEP} = \frac{176.75^2}{50}$$

Step 6: Solve

$$P_{PEP} = \frac{176.75^2}{50} = \frac{31,240}{50} = 625\ watts$$

The correct output PEP for 500 volts peak-to-peak across a 50-ohm load is 625 watts.

Capacitors

Electrolytic capacitors are known for their high capacitance relative to their physical size. They are ideal for applications requiring large amounts of charge storage in compact spaces. This makes them widely used in power supply circuits, where they help smooth out voltage fluctuations. The ability to store a large charge in a small volume is a defining characteristic of electrolytic capacitors. Still, they are also polarized, meaning they must be connected with the correct polarity in a circuit.

Which of the following is characteristic of an electrolytic capacitor? (G6A04)

A. Tight tolerance
B. Much less leakage than any other type
C. High capacitance for a given volume
D. Inexpensive RF capacitor

Electrolytic capacitors, such as power supplies, are key components in circuits requiring high capacitance in limited space. They are polarized, and connecting them the wrong way can cause failure.

Ceramic Capacitors

Low-voltage ceramic capacitors are known for their comparatively low cost. Due to their affordability, they are commonly used in electronics. These capacitors are ideal for high-frequency applications like decoupling and filtering in circuits. One of the advantages of ceramic capacitors is their stability, which makes them suitable for many different types of circuits, including radio frequency and power supply circuits. While they may not have the high capacitance of electrolytic capacitors, their cost-effectiveness and reliability make them a staple in many applications.

Which of the following is characteristic of low voltage ceramic capacitors? (G6A08)

A. Tight tolerance
B. High stability
C. High capacitance for given volume
D. Comparatively low cost

Low-voltage ceramic capacitors are cost-effective and commonly used in high-frequency applications. They offer stability at a relatively low cost compared to other capacitor types.

Inductor

When an inductor is operated above its self-resonant frequency, its behavior changes from inductive to capacitive. The self-resonant frequency is the point at which the inductance of the coil and the small amount of parasitic capacitance between the windings cancel each other out, effectively making the inductor resonate. Above this frequency, the capacitive effects dominate, and the inductor begins to behave like a capacitor rather than an inductor. This is due to the internal capacitance overtaking the inductance, shifting its overall impedance behavior.

What happens when an inductor is operated above its self-resonant frequency? (G6A11)

A. Its reactance increases
B. Harmonics are generated
C. It becomes capacitive
D. Catastrophic failure is likely

Inductors are typically designed to work below their self-resonant frequency. They lose their intended inductive properties when pushed beyond this frequency and act more like capacitors.

Diode

A diode is an electronic component that allows current to flow in one direction only, functioning as a one-way valve for electricity. It has two terminals: the anode (positive) and the cathode (negative). The diode allows current to pass through when the voltage applied to the anode is more positive than the cathode (a condition known as forward bias). However, the diode blocks current flow when the voltage is reversed (reverse bias).

Diodes are used in a wide variety of applications, including rectifying alternating current (AC) to direct current (DC), protecting circuits from voltage spikes, and in devices like LEDs (Light Emitting Diodes) for light emission.

Forward Biased

An LED (Light-Emitting Diode) emits light when it is forward-biased, meaning current flows from the anode (positive) to the cathode (negative). In this state, electrons recombine with holes in the LED's semiconductor material, releasing energy in the form of light. Forward biasing creates the necessary conditions for this electron-hole recombination, which is key to the light emission process in LEDs. Unlike regular diodes, LEDs are specially designed to emit visible light or infrared light when forward-biased.

How is an LED biased when emitting light? (G6B08)

A. In the tunnel-effect region
B. At the Zener voltage
C. Reverse biased
D. Forward biased

Forward biasing is essential for an LED to function and emit light. It allows current to flow through the device and enables the release of light energy. This concept is critical in understanding how LEDs operate in various applications, from indicators to communication systems.

LED vs LCD

Let's take a quick diversion here. Modern ham radio equipment often features either liquid crystal displays (LCDs) or light-emitting diode (LED) displays, each with its advantages depending on the operating environment.

LCDs are commonly found in handheld and mobile transceivers due to their ability to provide clear, high-contrast visuals in bright lighting (ambient) conditions, making them ideal for outdoor use.

On the other hand, LED displays are known for their bright and vivid colors, making them excellent for indoor or dimly lit environments. Understanding these differences helps operators select the right display type for their specific needs, ensuring clear readability during operation.

How does a liquid crystal display compare to an LED display? (G6B09)

A. Higher contrast in high ambient lighting
B. Wider dynamic range
C. Higher power consumption
D. Shorter lifetime

Think of it like reading a book versus looking at a phone screen outside—the book's print remains clear, while the phone's screen might glare or fade. This is why many portable and outdoor ham radios use LCD screens, as they provide higher contrast in high ambient lighting and ensure readability even in challenging conditions.

Germanium Diode

Germanium diodes have a lower forward threshold voltage than silicon diodes, approximately 0.3 volts. This lower voltage means that germanium diodes can begin conducting current at lower voltages than silicon diodes, making them useful in low-voltage applications.

> **What is the approximate forward threshold voltage of a germanium diode? (G6A03)**
>
> A. 0.1 volt
> **B. 0.3 volts**
> C. 0.7 volts
> D. 1.0 volts

Germanium diodes require about 0.3 volts to start conducting, making them suitable for low-voltage applications.

Silicon Junction Diode

Unlike germanium diodes, silicon junction diodes have a higher forward threshold voltage, typically around 0.7 volts. This means that in order for current to flow through a silicon diode, the applied voltage must be at least 0.7 volts – which is double germanium's. Silicon diodes are commonly used in general-purpose rectifiers and switching circuits.

> **What is the approximate forward threshold voltage of a silicon junction diode? (G6A05)**
>
> A. 0.1 volt
> B. 0.3 volts
> **C. 0.7 volts**
> D. 1.0 volts

Silicon junction diodes have a forward threshold voltage of around 0.7 volts. They are widely used in various electronic applications such as rectification and switching.

Transistors

Bipolar Transistors

A bipolar transistor used as a switch operates in two distinct operating points: saturation and cutoff. In the saturation region, the transistor is fully "on," meaning current flows freely from the collector to the emitter, allowing the transistor to conduct electricity. This state is similar to a closed switch, where the transistor provides a low-resistance path for current. On the other hand, in the cutoff region, the transistor is fully "off," meaning no current flows from the collector to the emitter, acting as an open switch with high resistance. The transistor can effectively control whether current flows by switching between these two states.

What are the operating points for a bipolar transistor used as a switch? (G6A07)

A. Saturation and cutoff
B. The active region (between cutoff and saturation)
C. Peak and valley current points
D. Enhancement and depletion modes

When used as switches, bipolar transistors operate in either saturation (fully on) or cutoff (fully off), enabling them to control current flow in circuits, similar to turning a switch on or off.

MOSFET

A MOSFET (Metal-Oxide-Semiconductor Field-Effect Transistor) is a type of transistor. Its construction involves a gate separated from the channel (the path through which current flows) by a thin insulating layer, typically made of silicon dioxide. This insulating layer allows the MOSFET to control the flow of current between the source and drain without a direct electrical connection, making it highly efficient and minimizing power loss. The gate voltage regulates the conductivity of the channel, allowing the MOSFET to act as a switch or amplifier.

Which of the following describes MOSFET construction? (G6A09)

A. The gate is formed by a back-biased junction
B. The gate is separated from the channel by a thin insulating layer
C. The source is separated from the drain by a thin insulating layer
D. The source is formed by depositing metal on silicon

MOSFETs are important to modern electronics because they can control large currents with minimal power. The insulated gate is a defining feature that enables them to efficiently switch and amplify signals in devices ranging from computers to radios.

Vacuum Tube

A vacuum tube, also known as a thermionic valve, is an electronic device used to control the flow of electrical current in a sealed glass or metal enclosure with no air inside. It works by heating a cathode (negative electrode) to release electrons into the vacuum, which are then attracted to the anode (positive electrode), creating a current. A control grid between the cathode and anode allows for regulating this flow, enabling the vacuum tube to amplify or modulate signals. Common types include diodes (converting AC to DC) and triodes (amplifying signals). Although largely replaced by transistors, vacuum tubes are still used in high-power RF transmitters, audiophile-grade audio equipment, and some military and scientific applications due to their reliability in extreme conditions and unique performance characteristics.

Control Grid

Which element of a vacuum tube regulates the flow of electrons between cathode and plate? (G6A10)

A. Control grid
B. Suppressor grid
C. Screen grid
D. Trigger electrode

In a vacuum tube, the control grid is the component responsible for regulating the flow of electrons between the cathode (which emits electrons) and the plate (which collects them). By applying a varying voltage to the control grid, you can control the amount of current that flows through the tube. This is the fundamental principle behind amplification in vacuum tubes. Essentially, the control grid allows small changes in voltage to create larger changes in current, enabling the tube to amplify signals effectively.

Vacuum Tubes

Screen Grid

What is the primary purpose of a screen grid in a vacuum tube? (G6A12)

A. To reduce grid-to-plate capacitance
B. To increase efficiency
C. To increase the control grid resistance
D. To decrease plate resistance

The screen grid serves a different function. It is placed between the control grid and the plate, and its primary purpose is to reduce grid-to-plate capacitance. This helps to stabilize the tube's operation at higher frequencies by preventing unwanted feedback and oscillations. By shielding the control grid from the plate, the screen grid ensures that the tube can operate more efficiently in RF applications.

The control grid manages the electron flow, making the tube act as an amplifier. The screen grid helps maintain stability by minimizing interference between the control grid and the plate, particularly in high-frequency applications.

Connectors

Connectors link different radio system components, such as transceivers, antennas, microphones, and power supplies. They ensure that signals are efficiently transmitted without interference, loss, or damage to the equipment.

SMA Connector

What is an SMA connector? (G6B11)

A. A type-S to type-M adaptor
B. A small threaded connector suitable for signals up to several GHz
C. A connector designed for serial multiple access signals
D. A type of push-on connector intended for high-voltage applications

A SMA connector (SubMiniature version A) is a small, threaded RF (Radio Frequency) connector commonly used for frequencies up to several GHz. It's designed for tight spaces and is often used in applications such as wireless communication systems, antennas, and test equipment. SMA connectors are known for their precision and durability, making them suitable for high-frequency use in ham radio and other RF applications.

For exam purposes, remember that an SMA connector is suitable for signals up to several GHz, making it ideal for high-frequency use. This is why they are popular for handheld transceivers, wireless devices, and even GPS antennas.

Connectors

Type N Connector

Which of the following describes a type N connector? (G6B07)

A. A moisture-resistant RF connector useful to 10 GHz
B. A small bayonet connector used for data circuits
C. A low noise figure VHF connector
D. A nickel-plated version of the PL-259

A Type N connector is a threaded RF connector designed to operate in frequencies up to 10 GHz. Its key feature is that it is moisture-resistant, making it suitable for outdoor applications where weatherproofing is essential. Type N connectors provide excellent low-loss performance, which is why they are commonly used in microwave and high-frequency applications like base stations and antenna setups for UHF and microwave bands.

The points to recall for the exam are that a Type N connector is moisture-resistant and can be used in frequencies as high as 10 GHz. Its durability makes it a preferred choice for outdoor installations and higher-frequency operations.

BNC Connector

The BNC connector (Bayonet Neill-Concelman) is a commonly used RF connector, but its performance is limited at higher frequencies. The typical upper-frequency limit for low SWR (Standing Wave Ratio) operation for 50-ohm BNC connectors is 4 GHz. While BNC connectors are popular for applications like oscilloscopes, network analyzers, and video transmission, they may cause higher losses or reflections at frequencies beyond 4 GHz.

What is a typical upper frequency limit for low SWR operation of 50-ohm BNC connectors? (G6B04)

A. 50 MHz
B. 500 MHz
C. 4 GHz
D. 40 GHz

The 4 GHz frequency is in the microwave spectrum with a wavelength of 7.5 cm.

For the exam, remember that BNC connectors are limited to low SWR operation up to 4 GHz, and beyond this point, they may not perform as well in maintaining signal integrity.

RCA Phono Connector

Which of these connector types is commonly used for low frequency or dc signal connections to a transceiver? (G6B12)

A. PL-259
B. BNC
C. RCA Phono
D. Type N

The RCA Phono connector is commonly used for low-frequency or DC signal connections to a transceiver. This type of connector, recognizable by its round shape with a central pin and a ring around it, is often seen in consumer electronics like audio and video equipment. In ham radio and other electronic systems, RCA connectors are typically used for connecting low-frequency audio signals or control signals, such as the connection between an audio output and an amplifier or a signal input into a transceiver.

RCA Phono connectors are widely used due to their simplicity, ease of use, and availability. They are reliable for low-frequency signals, making them a suitable choice for applications that don't require handling high-power or high-frequency RF signals. These connectors have been a standard in electronics for decades, primarily because they offer solid connections for various audio and control applications in transceiver setups.

Ferrite Core

A ferrite core is a type of magnetic core made from a ceramic compound composed of iron oxide mixed with metallic elements such as zinc, manganese, or nickel. These cores suppress high-frequency noise in electrical circuits, particularly transformers, inductors, and other electromagnetic devices. The high magnetic permeability of ferrite cores allows them to store magnetic energy effectively while minimizing electrical losses at high frequencies. In radio communication equipment, ferrite cores are often used to improve the performance of antennas and transformers.

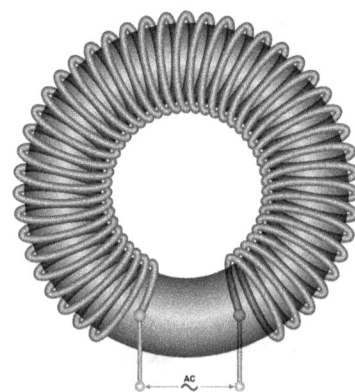

Ferrite Core Performance at Different Frequencies

Ferrite cores are widely used in electronics and radio frequency circuits to manage electromagnetic interference (EMI) and enhance the performance of inductors and transformers. The performance of a ferrite core is largely determined by the composition, or "mix," of materials used in its construction. Different mixtures of iron oxide and other metal oxides like zinc or manganese provide various electrical and magnetic properties, making specific ferrite cores more suited for particular frequency ranges. For instance, some ferrite mixes work better at high frequencies. In contrast, others are more effective at lower frequencies, allowing designers to choose the suitable core for optimal suppression of unwanted signals.

What determines the performance of a ferrite core at different frequencies? (G6B01)

A. Its conductivity
B. Its thickness
C. The composition, or "mix," of materials used
D. The ratio of outer diameter to inner diameter

Ferrite Core Toroidal Inductors

A ferrite core toroidal inductor offers several advantages in RF applications. One key benefit is that it allows for the creation of large inductance values while occupying relatively little space. The design of the toroid helps confine most of the magnetic field within the core itself, reducing electromagnetic interference (EMI) with nearby components. Additionally, the magnetic properties of the ferrite material can be optimized for a specific frequency range, making these inductors highly efficient and effective in RF circuits. This leads to better energy transfer and reduced losses.

What is an advantage of using a ferrite core toroidal inductor? (G6B05)

A. Large values of inductance may be obtained
B. The magnetic properties of the core may be optimized for a specific range of frequencies
C. Most of the magnetic field is contained in the core
D. All these choices are correct

Ferrite Bead

A ferrite bead, on the other hand, is a specific type of ferrite core that is shaped like a small cylindrical bead. It is placed around cables to suppress high-frequency electromagnetic interference (EMI) by acting as a passive filter. Ferrite beads are often found on USB cables, power cords, and signal lines, where they help to reduce noise and prevent interference from affecting the performance of connected devices. Ferrite beads work by increasing the impedance at high frequencies, dissipating unwanted energy such as heat, and thus blocking or filtering out noise.

How does a ferrite bead or core reduce common-mode RF current on the shield of a coaxial cable? (G6B10)

A. By creating an impedance in the current's path
B. It converts common-mode current to differential mode current
C. By creating an out-of-phase current to cancel the common-mode current
D. Ferrites expel magnetic fields

Both ferrite cores and beads help minimize interference in electronic systems and ensure smooth operation across a wide range of frequencies.

Integrated Circuit

An integrated circuit (IC) is a small electronic device made up of a collection of components such as transistors, resistors, diodes, and capacitors fabricated together on a single piece of semiconductor material, usually silicon. These components are connected to perform specific functions like amplification, switching, or digital data processing.

ICs are vital to modern electronics because they allow many components to be packed into a small chip, drastically reducing the size and cost of circuits while improving performance. Integrated circuits are used in a wide range of devices, including computers, smartphones, radios, and televisions.

The key advantage of ICs is that they integrate multiple components into a single chip, allowing for complex functionality in a compact and efficient form. This makes modern technology more powerful, efficient, and accessible.

Monolithic Microwave Integrated Circuit (MMIC)

What is meant by the term MMIC? (G6B02)

A. Multi-Mode Integrated Circuit
B. Monolithic Microwave Integrated Circuit
C. Metal Monolayer Integrated Circuit
D. Mode Modulated Integrated Circuit

An MMIC (Monolithic Microwave Integrated Circuit) is a type of integrated circuit specifically designed to operate at microwave frequencies, typically in the GHz range. Unlike traditional ICs, which work at lower frequencies, MMICs are used in applications like radar systems, satellite communication, and high-frequency wireless systems. The term "monolithic" refers to the fact that the circuit components, such as transistors and capacitors, are fabricated on a single semiconductor chip. The compactness of MMICs allows for efficient high-frequency performance, making them ideal for communication systems that require small, high-speed components.

CMOS vs TTL Integrated Circuits

Two of the most common families of ICs are CMOS (Complementary Metal-Oxide-Semiconductor) and TTL (Transistor-Transistor Logic). Each has unique characteristics, making one more suitable than the other depending on the application.

Which of the following is an advantage of CMOS integrated circuits compared to TTL integrated circuits? (G6B03)

A. Low power consumption
B. High power handling capability
C. Better suited for RF amplification
D. Better suited for power supply regulation

One significant advantage of CMOS over TTL is its low power consumption. CMOS circuits consume much less power at rest, as power is mainly drawn during switching, making them ideal for battery-operated devices. In contrast, TTL circuits consume more power continuously. Additionally, CMOS circuits tend to be more resistant to noise, which is another reason they are widely used in modern electronics like microcontrollers, smartphones, and digital logic devices.

Digital Circuits and Logic Gates

Digital circuits form the backbone of modern electronics, including ham radios. These circuits operate using logic gates, which process binary inputs to produce specific outputs based on set rules. Logic gates and binary counters are essential in controlling digital operations. At the same time, shift registers play a significant role in data handling and transmission. This section will help you understand how digital components work and why they're critical in ensuring smooth operations within your radio systems.

Which of the following describes the function of a two-input AND gate? (G7B03)

A. Output is high when either or both inputs are low
B. Output is high only when both inputs are high
C. Output is low when either or both inputs are high
D. Output is low only when both inputs are high

An AND gate is a basic digital logic gate that outputs only a "high" (or 1) signal when all its inputs are high. In the case of a two-input AND gate, this means that both inputs must be 1 for the output to also be 1. If either input is 0, the output will be 0. This type of gate is commonly used in circuits where multiple conditions must be met simultaneously, making it a vital building block in digital logic.

How many states does a 3-bit binary counter have? (G7B05)

A. 3
B. 6
C. 8
D. 16

A 3-bit binary counter has eight distinct states, counting from 000 to 111 in binary, which translates to decimal numbers 0 through 7. The number of states in a binary counter is determined by the number of bits it contains, and each additional bit doubles the number of possible states. These counters are vital in timing, counting, and control applications in digital electronics, including radios.

What is a shift register? (G7B06)

A. A clocked array of circuits that passes data in steps along the array
B. An array of operational amplifiers used for tri-state arithmetic operations
C. A digital mixer
D. An analog mixer

A shift register is a series of flip-flops connected in sequence that passes binary data from one flip-flop to the next at each clock cycle. It stores and transfers data in digital circuits. The clock pulses drive the transfer of data, and shift registers are often used in data handling applications like serial-to-parallel data conversion or for managing data transmission in communication systems.

Understanding how digital circuits, logic gates, and components like binary counters and shift registers work is key for mastering the digital aspects of modern radio systems. These elements enable the processing and handling of binary data, which is the foundation of all digital communication. By learning these basic principles, operators can ensure better control and efficiency in their radio operations.

Amplifiers And Oscillators

Amplifier Classes and Efficiency

Amplifiers are used by radios to increase the power of signals to suitable levels for transmission or processing. The different classes of amplifiers—Class A, B, AB, and C—vary in efficiency and performance, depending on the application. This section covers concepts such as neutralization to prevent unwanted oscillations and the comparative efficiencies of different amplifier classes.

What is the purpose of neutralizing an amplifier? (G7B01)

A. To limit the modulation index
B. To eliminate self-oscillations
C. To cut off the final amplifier during standby periods
D. To keep the carrier on frequency

Neutralizing an amplifier ensures that unwanted feedback doesn't cause the amplifier to generate oscillations on its own, known as self-oscillations. This is particularly important in RF amplifiers, where unintended oscillations can lead to poor signal quality or interference. Neutralization techniques adjust the amplifier's internal capacitances to counteract the feedback that could otherwise cause oscillations.

Which of these classes of amplifiers has the highest efficiency? (G7B02)

A. Class A
B. Class B
C. Class AB
D. Class C

Class C amplifiers are the most efficient, often reaching efficiencies above 70%. However, because of their nonlinear operation, they are not suitable for all modes of transmission. Class C amplifiers are typically used for frequency-modulated (FM) signals and other signals where signal distortion is acceptable since they amplify only a portion of the input signal.

For which of the following modes is a Class C power stage appropriate for amplifying a modulated signal? (G7B11)

A. SSB
B. FM
C. AM
D. All these choices are correct

Class C amplifiers are ideal for FM transmissions because their output is not directly related to the amplitude of the input signal. Frequency modulation relies on varying frequencies, so a Class C amplifier's tendency to distort amplitude signals does not affect the overall transmission quality.

In a Class A amplifier, what percentage of the time does the amplifying device conduct? (G7B04)

A. 100%
B. More than 50% but less than 100%
C. 50%
D. Less than 50%

A Class A amplifier conducts 100% of the time, meaning the amplifying device is always active during both halves of the signal. This makes Class A amplifiers highly linear and ideal for high-fidelity amplification. However, their continuous operation results in lower efficiency, typically around 25-30%, because they consume power even when no signal is being amplified.

How is the efficiency of an RF power amplifier determined? (G7B08)

A. Divide the DC input power by the DC output power
B. Divide the RF output power by the DC input power
C. Multiply the RF input power by the reciprocal of the RF output power
D. Add the RF input power to the DC output power

The efficiency of an RF amplifier is <u>calculated by dividing the RF power output by the DC input power</u>. This ratio indicates how effectively the amplifier converts the input energy into useful output signal power. Higher efficiency reduces heat dissipation and energy waste, particularly important in high-power applications.

Understanding amplifier classes, their efficiency, and their applications in radio is fundamental for selecting the right type for different transmission needs. Class C amplifiers offer high efficiency but are limited to specific uses like FM. In contrast, Class A amplifiers provide high linearity but at the cost of efficiency. Properly neutralizing an amplifier prevents unwanted oscillations, improving overall performance. In summary, the choice of amplifier type depends on the specific requirements of the radio signal and application.

Amplifier Characteristics

Amplifiers play a vital role in ensuring that the transmitted or processed signal is strong enough for effective communication. One key characteristic to consider when choosing an amplifier is whether it is linear or nonlinear. Linear amplifiers maintain the integrity of the signal's waveform throughout amplification, making them ideal for modes of communication where accuracy and signal fidelity are critical. This section will cover linear amplification and its significance in radio operations.

Which of the following describes a linear amplifier? (G7B10)

A. Any RF power amplifier used in conjunction with an amateur transceiver
B. An amplifier in which the output preserves the input waveform
C. A Class C high efficiency amplifier
D. An amplifier used as a frequency multiplier

<u>A linear amplifier accurately amplifies an input signal without altering its waveform</u>. This is crucial in applications such as SSB and AM, where the shape of the signal conveys essential information. A linear amplifier ensures that the amplified output is a scaled version of the input, preventing distortion that would degrade the signal quality. While linear amplifiers are less efficient than Class C amplifiers, their precision is critical for these communication modes.

Linear amplifiers ensure that signals are amplified without distortion, preserving the integrity of the waveform. This makes them indispensable for modes like AM and SSB, where even slight distortions can lead to a loss of important signal information. Radio operators can use the appropriate amplifier to ensure high-fidelity transmissions and efficient communications.

Oscillators and Frequency Determination

Oscillators are components in radio frequency circuits and generate the signal frequencies needed for transmission. These circuits convert DC into a periodic signal, often a sine wave, at a precise frequency. The frequency at which an oscillator operates is critical for maintaining clear communication. Key components, such as filters and amplifiers, stabilize and shape the signals. This section will cover the basic components of oscillators and the factors that determine their frequency.

Which of the following are basic components of a sine wave oscillator? (G7B07)

A. An amplifier and a divider
B. A frequency multiplier and a mixer
C. A circulator and a filter operating in a feed-forward loop
D. A filter and an amplifier operating in a feedback loop

A sine wave oscillator generates continuous waveforms, such as the sine wave. It relies on two key components: a filter and an amplifier. These two elements are arranged in a feedback loop, where the filter selects the correct frequency, and the amplifier increases the signal strength. The continuous feedback ensures that the oscillations remain stable and uninterrupted. Sine wave oscillators are used in various applications, including transmitters, signal generators, and clock circuits.

What determines the frequency of an LC oscillator? (G7B09)

A. The number of stages in the counter
B. The number of stages in the divider
C. The inductance and capacitance in the tank circuit
D. The time delay of the lag circuit

The frequency of an LC oscillator is determined by the inductance (L) and capacitance (C) values in its tank circuit. The tank circuit stores energy and sets the oscillation frequency. The larger the inductance or capacitance, the lower the frequency of oscillation. It helps operators fine-tune their transmitters to the desired frequency band.

Oscillators are the heart of signal generation in RF circuits, and their frequency is determined by key components such as inductors and capacitors in the tank circuit. Understanding the relationship between these components is crucial for setting precise frequencies and maintaining stable communication.

Direct Digital Synthesizer

Which of the following is characteristic of a direct digital synthesizer (DDS)? (G7C05)

A. Extremely narrow tuning range
B. Relatively high-power output
C. Pure sine wave output
D. Variable output frequency with the stability of a crystal oscillator

A direct digital synthesizer (DDS) allows for precise and stable frequency generation. Digital techniques can vary the output frequency while maintaining the stability of a crystal oscillator, making DDS useful in applications requiring accurate and stable frequency control, such as modern radio transceivers. DDS is preferred because of its ability to rapidly and precisely generate frequencies across a wide range, making it ideal for frequency-agile systems.

Operational Amplifiers

What kind of device is an integrated circuit operational amplifier? (G6B06)

A. Digital
B. MMIC
C. Programmable Logic
D. Analog

An operational amplifier (op-amp) is a versatile analog device commonly used in signal processing, filtering, or mathematical operations such as addition and integration. The op-amp is designed to amplify the difference in voltage between two input terminals and is highly useful in analog circuits. <u>Being an analog device</u> means that it processes continuous signals, making it crucial for applications like audio processing, control systems, and analog-to-digital conversion. Its high gain and flexibility make it a fundamental building block in a variety of electronic circuits.

Chapter Summary

This chapter has provided a detailed exploration of the critical components that form the foundation of circuit design and functionality in amateur radio. From understanding the behavior of resistors and capacitors to grasping the role of transistors, amplifiers, oscillators, and more, you've gained insight into the key elements that bring radio circuits to life.

Whether fine-tuning frequencies with LC oscillators, selecting the right diodes for rectification or choosing between CMOS and TTL integrated circuits, these principles are central to building and maintaining reliable radio systems. Armed with this knowledge, you're better prepared to troubleshoot, innovate, and optimize your equipment, empowering you to grow as a confident and capable amateur radio operator.

Chapter 11

Circuit Diagrams

CIRCUIT DIAGRAMS ARE THE blueprints of electronic circuits, offering a visual representation of how components are connected. They simplify complex circuits, making it easier to understand their functionality. For ham radio operators, these diagrams are indispensable tools for both designing new circuits and troubleshooting existing ones. By laying out each component and connection, circuit diagrams help you visualize the flow of current and the interaction between various elements. This clarity is crucial for diagnosing issues and ensuring that your designs function as intended. Moreover, circuit diagrams are a universal language for sharing designs with others, facilitating collaboration and innovation within the amateur radio community.

Reading and interpreting circuit diagrams involves understanding the symbols and layout that represent different components and their connections. You need to understand series and parallel connections. Components in series share a single path for the current flow, while parallel components provide multiple paths. Schematic notations and labels offer additional information, such as component values and specifications. These details are often annotated near the symbols to give clarity and context.

Using standard conventions, creating circuit diagrams involves translating your design ideas into a visual format. Circuit design software like KiCad and EasyEDA can be invaluable tools for this task. These programs offer libraries of component symbols and allow you to draw connections quickly. You can also simulate your designs to test their functionality before building them. For simpler circuits, hand-drawing techniques can be equally effective. Use a pencil and graph paper to sketch out your design, ensuring that each component is clearly labeled and that connections are neat and organized. Best practices for precise and accurate diagrams include keeping lines straight and avoiding unnecessary crossings, which can make the diagram confusing. Ensure that power and ground connections and group-related components are clearly marked to make the diagram easier to follow.

Schematic Symbols

This section will explore the common symbols used in schematics, their meanings, and how they help communicate complex electrical designs.

Exam Schematic

Here is the schematic that will be on the exam. Let's review the labeled components.

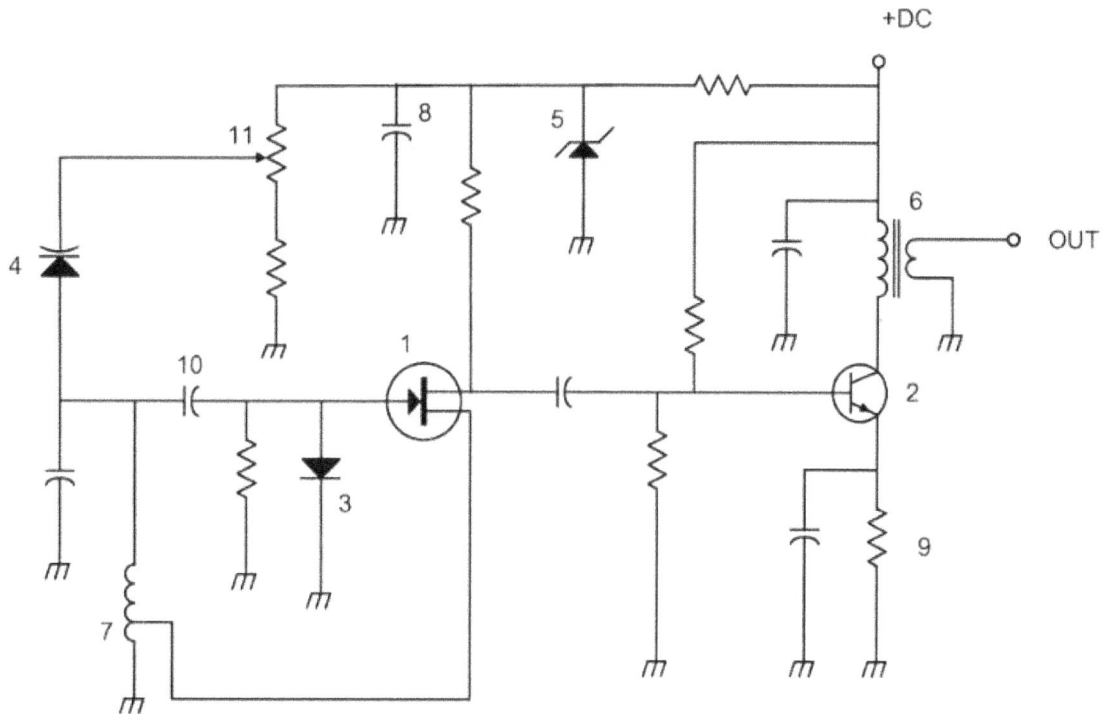

Figure G7-1

Let's walk through the labeled components in the schematic, which correspond to the numbered symbols one through eleven:

1. **Field-Effect Transistor (FET)**: A FET is a transistor used to control the flow of current with an electric field. It's widely used in radios for signal amplification due to its high input impedance, which allows for efficient control with little power. Fun Fact: FETs are used in radio because they can operate at very high frequencies, making them ideal for RF amplification.

2. **NPN Junction Transistor**: This is a Bipolar Junction Transistor (BJT) that allows current to flow when a small positive voltage is applied to its base. It amplifies weak RF or audio signals into stronger ones that can be transmitted or further processed. Interesting Fact: NPN transistors can switch on and off incredibly fast, making them key in digital signal processing.

3. **Diode/ Rectifier**: A diode is a component that allows current to flow in only one direction, making it helpful in converting alternating current (AC) to direct current (DC). Rectifiers, which are configurations of diodes, are used in powering circuits by turning AC into the DC that radios need. Diodes also protect circuits from voltage spikes and are important in voltage regulation.

4. **Varactor Diode**: A varactor diode, also known as a varicap diode, controls capacitance in a circuit. It acts like a variable capacitor, where the capacitance changes based on its voltage. Varactor diodes are often used in frequency tuning, allowing the radio to select different channels easily. They are key in voltage-controlled oscillators and other circuits requiring frequency adjustment.

5. **Zener Diode**: The Zener diode's ability to maintain a fixed voltage in reverse bias makes it a go-to component for voltage regulation in radio circuits. Zener diodes are commonly used in noise reduction circuits, which helps improve radio signal quality.

6. **Solid Core Transformer**: Transformers are used to step up or down the voltage in power supplies or radio circuits. The solid core transformer transfers energy between circuits through electromagnetic induction. In radio transmitters, transformers help match impedance between stages, ensuring efficient signal transfer.

7. **Tapped Inductor**: Tapped inductors are coils with an extra connection (tap), providing adjustable inductance. They're used for tuning and filtering. Tapped inductors are integral in antenna tuning, which helps radios receive clearer signals by optimizing the antenna's electrical length.

8. **Polarized Capacitor**: A polarized capacitor has a positive and negative terminal, meaning it must be connected in the correct direction in a circuit. Commonly used in power supply circuits, polarized capacitors store and release electrical energy. They are typically used in applications requiring larger capacitance, like filtering or smoothing signals. Connecting them with the wrong polarity can damage the capacitor or the circuit.

9. **Fixed Resistor**: A fixed resistor is a component that resists the flow of electrical current. Its resistance value is set and cannot be changed. It's used to control the amount of current in a circuit, helping to protect components or regulate voltage levels. Fixed resistors are widely used in all types of circuits for their simplicity and reliability in maintaining a consistent resistance value.

10. **Polarized Capacitor**: The same as #8.

11. **Potentiometer**: A potentiometer is a variable resistor that allows you to adjust the resistance within a circuit manually. It has three terminals: two connected to a resistive element and one connected to a movable wiper. Turning the wiper allows you to change the resistance between the terminals, effectively controlling voltage or current flow. Potentiometers are commonly used for volume controls in audio devices and for adjusting levels in various electronic circuits.

You have all those memorized, right? Nope? Well, don't worry too much. This is another one of those times when we use a little bit of exam strategy. While you will learn all the symbols and components over time, let's focus on the five you could see on the exam.

Exam Symbols

Which symbol in Figure G7-1 represents a field effect transistor? (G7A09)

A. Symbol 2
B. Symbol 5
C. Symbol 1
D. Symbol 4

A FET is a transistor that controls current flow using an electric field, making it ideal for amplifiers or switches due to its high input impedance and low power consumption.

Memory Trick: Think of the FET symbol as a **"gatekeeper" with a river flowing underneath.** The "gate" (the control pin) regulates the flow of electrons (the river), just like how the FET controls current. The straight line (drain) and arrow (source) show the flow direction.

Visual Cue: Picture a little toll booth on a bridge – only those allowed by the gatekeeper (the gate pin) can pass!

Which symbol in Figure G7-1 represents a Zener diode? (G7A10)

A. Symbol 4
B. Symbol 1
C. Symbol 11
D. Symbol 5

A Zener diode allows current to flow in the reverse direction when a specific voltage (the Zener voltage) is reached, making it essential for voltage regulation and protection circuits.

Memory Trick: The Zener diode's symbol looks like a regular diode but with a "zippy" zigzag at the tip. Think of it as a **"lightning bolt shield"** that activates when voltage gets too high, protecting your circuit.

Visual Cue: Imagine a superhero shield with a lightning bolt on it, standing guard to stop voltage surges!

Which symbol in Figure G7-1 represents an NPN junction transistor? (G7A11)

A. Symbol 1
B. Symbol 2
C. Symbol 7
D. Symbol 11

An NPN transistor is a type of bipolar junction transistor (BJT) that amplifies or switches electrical signals by controlling the current flow from the collector to the emitter, with a small current applied to the base.

Memory Trick: The NPN transistor has an arrow pointing **"Not Pointing iNward"** (NPN). This outward arrow represents the emitter pushing current out.

Visual Cue: Picture a rocket blasting off, with the arrow showing the current "blasting out" of the emitter.

Which symbol in Figure G7-1 represents a solid core transformer? (G7A12)

A. Symbol 4
B. Symbol 7
C. Symbol 6
D. Symbol 1

A transformer with a solid (non-laminated) core, typically used in applications where high-frequency signals need efficient coupling or where size and weight are critical.

Memory Trick: The transformer symbol has two coils (representing its windings) with a core between them. Think of it as a **"teleporter portal"** where electricity enters one side (primary) and exits the other (secondary), transformed!

Visual Cue: Imagine the core as the magic doorway linking the two coils like in a sci-fi movie.

Which symbol in Figure G7-1 represents a tapped inductor? (G7A13)

A. Symbol 7
B. Symbol 11
C. Symbol 6
D. Symbol 1

An inductor with one or more intermediate connections (taps) along its winding, enabling it to be used for impedance matching, filtering, or voltage step-up/step-down applications.

Memory Trick: The tapped inductor has extra lines (taps) poking out of the coil, like branches growing off a tree. Think of it as a **"tree of options,"** letting you pick different voltages or signals.

Visual Cue: Picture a winding staircase with little "landings" along the way where you can stop and "tap" into the energy.

Creating and interpreting circuit diagrams is fundamental for any ham radio operator. Whether you are designing a new project or troubleshooting an existing one, these diagrams offer invaluable insights into the workings of electronic circuits. They help you visualize connections, simplify complex designs, and communicate your ideas effectively.

Troubleshooting Common Circuit Issues

In electronic circuits, encountering issues is inevitable. Understanding typical problems and their symptoms is crucial for swift and effective troubleshooting.

A systematic approach to troubleshooting starts with a visual inspection. Begin by examining the circuit for obvious signs of damage, such as burnt components, broken wires, or loose connections. Look for any discoloration or physical deformities that might indicate overheating or stress. Continuity testing with a multimeter is the next step.

Multimeters and Voltmeters

Voltage measurements are taken to verify that expected voltage levels are present at key points in the circuit. Measure the voltage at different nodes using a multimeter set to the appropriate voltage range. Compare these readings with the expected values based on the circuit design. Discrepancies can reveal issues such as power supply problems or faulty components.

Because of that need to measure, multimeters are indispensable tools for measuring electrical properties like voltage, current, and resistance. They come in two main varieties: analog and digital.

When is an analog multimeter preferred to a digital multimeter? (G4B09)

A. When testing logic circuits
B. When high precision is desired
C. When measuring the frequency of an oscillator
D. When adjusting circuits for maximum or minimum values

An analog multimeter is preferred when adjusting circuits for maximum or minimum values because it provides a continuously moving needle. This visual feedback helps users make fine adjustments more smoothly than digital readouts, which can fluctuate or update less fluidly.

What is an advantage of a digital multimeter as compared to an analog multimeter? (G4B06)

A. Better for measuring computer circuits
B. Less prone to overload
C. Higher precision
D. Faster response

However, digital multimeters have their advantages, especially when precision is required. They offer higher accuracy and are better suited for exact measurements, providing clear numerical values without the interpretation needed for an analog display.

Voltmeters have high input impedance, which is another benefit. This reduces the loading effect on the circuit being measured, minimizing any impact the meter itself might have on the circuit's behavior. This feature makes them ideal for sensitive electronic components.

Why do voltmeters have high input impedance? (G4B05)

A. It improves the frequency response
B. It allows for higher voltages to be safely measured
C. It improves the resolution of the readings
D. It decreases the loading on circuits being measured

Multimeters—both analog and digital—each has its place. Analog versions shine in tasks where adjustments for maximum or minimum values are needed, while digital multimeters, due to their high input impedance, provide greater precision and reduce circuit loading. Understanding when to use each type is key to effective circuit troubleshooting and optimization.

Oscilloscope

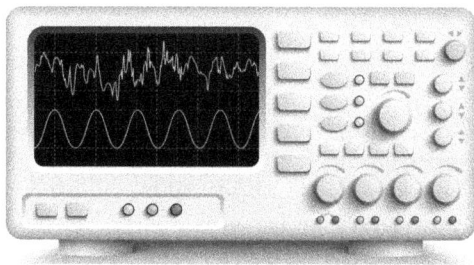

An oscilloscope is invaluable for visualizing waveforms and detecting issues that are not apparent through simple voltage measurements. An oscilloscope lets you see distortions, noise, and other anomalies by displaying the signal in real time. A signal generator can be used to inject test signals into the circuit, helping you trace and diagnose problems. Keeping a log of measurements and observations is also beneficial. Documenting your findings provides a reference for future troubleshooting and helps track changes over time.

The oscilloscope contains horizontal and vertical channel amplifiers, allowing it to display and measure electrical signals in real-time. The vertical channel represents the signal's voltage, while the horizontal channel represents time. By plotting the voltage against time, you can visualize waveforms and understand how a circuit is behaving. This makes the oscilloscope incredibly useful for diagnosing problems, monitoring signals, or observing complex waveforms that simpler tools like digital voltmeters cannot capture.

What item of test equipment contains horizontal and vertical channel amplifiers? (G4B01)

A. An ohmmeter
B. A signal generator
C. An ammeter
D. An oscilloscope

In addition to displaying basic voltage signals, oscilloscopes are particularly helpful in examining more complex waveforms, such as those generated by a CW (continuous wave) transmitter. When checking the keying waveform of a CW transmitter, an oscilloscope is the best tool for the job because it allows you to see the shape and timing of the keying pulses, ensuring that the transmitter is functioning correctly. This is especially important because distorted keying can lead to communication errors or signal interference.

Which of the following is the best instrument to use for checking the keying waveform of a CW transmitter? (G4B03)

A. An oscilloscope
B. A field strength meter
C. A sidetone monitor
D. A wavemeter

Another advantage of using an oscilloscope compared to a digital voltmeter is that it can display <u>complex waveforms that a voltmeter could not accurately measure.</u> While a voltmeter can only give you a numerical readout of voltage, the oscilloscope provides a visual representation, making it a superior tool for observing RF signals, such as when checking the RF envelope pattern of a transmitted signal by <u>connecting the attenuated RF output of the transmitter to the vertical input.</u>

Which of the following is an advantage of an oscilloscope versus a digital voltmeter? (G4B02)

A. An oscilloscope uses less power
B. Complex impedances can be easily measured
C. Greater precision
D. Complex waveforms can be measured

What signal source is connected to the vertical input of an oscilloscope when checking the RF envelope pattern of a transmitted signal? (G4B04)

A. The local oscillator of the transmitter
B. An external RF oscillator
C. The transmitter balanced mixer output
D. The attenuated RF output of the transmitter

Oscilloscopes are tools for visualizing electrical signals, particularly when measuring complex waveforms or monitoring the behavior of RF signals. They offer capabilities beyond simple voltage measurements, making them ideal for tasks like checking the keying waveform of a CW transmitter or examining RF envelope patterns.

Chapter Summary

Mastering circuit diagrams is an essential skill for any radio enthusiast, providing a clear roadmap for understanding, building, and troubleshooting the circuits that power a radio station. Recognizing schematic symbols and using diagnostic tools like multimeters, voltmeters, and oscilloscopes equips you to confidently address common issues, deepen your understanding of electronic circuits, and enhance your practical skills. By systematically tackling circuit challenges and employing effective troubleshooting techniques, you ensure the reliable operation of your ham radio equipment while strengthening your technical proficiency.

This solid foundation keeps your station running smoothly and prepares you for the challenges and opportunities that come with advancing your license.

Chapter 12

Mobile Stations

OPERATING A HAM RADIO station from a mobile setup offers unique opportunities and challenges. Mobile stations allow amateur radio operators to communicate on the go, such as from a car, boat, or other vehicles. This mobility is ideal for participating in emergency communications, DXing, or simply enjoying the flexibility of radio communication during travel. However, mobile setups often require a balance between convenience and performance.

One of the biggest challenges in mobile operations is the antenna. Unlike home-based stations, where operators can use full-sized antennas, mobile stations must use compact or shortened antennas due to vehicle space constraints. These shortened antennas are less efficient and provide limited bandwidth, which can reduce signal strength and require frequent tuning. Additionally, mobile stations often face interference from the vehicle's electrical systems, such as the battery charging system and onboard computers. Understanding and addressing these challenges is critical to maintaining effective communication while on the move.

Mobile Antennas

One of the most significant challenges in an HF mobile installation is the antenna's efficiency, mainly because mobile antennas tend to be electrically short. An electrically short antenna is one that is physically shorter than the optimal length needed for the frequency being used, which reduces its efficiency. The reduced efficiency happens because shorter antennas are less able to radiate energy effectively, resulting in weaker signal transmission and reception. In HF mobile setups, this limitation is critical since vehicles often require compact antennas. Therefore, the antenna's efficiency directly impacts the performance of the mobile station.

> **Which of the following most limits an HF mobile installation? (G4E05)**
>
> A. "Picket fencing"
> B. The wire gauge of the DC power line to the transceiver
> **C. Efficiency of the electrically short antenna**
> D. FCC rules limiting mobile output power on the 75-meter band

While convenient for vehicle installation, a shortened mobile antenna has a notable trade-off: a very limited operating bandwidth. This means the antenna is efficient only over a narrow range of frequencies, and tuning becomes necessary when switching between different parts of the HF spectrum. Full-sized antennas, by contrast, offer a much broader bandwidth, allowing them to cover a wider range of frequencies without requiring constant adjustments. For mobile users, this limitation can be frustrating, as it reduces the flexibility of the station's operation.

What is one disadvantage of using a shortened mobile antenna as opposed to a full-size antenna? (G4E06)

A. Short antennas are more likely to cause distortion of transmitted signals
B. Q of the antenna will be very low
C. Operating bandwidth may be very limited
D. Harmonic radiation may increase

Both of these questions highlight the inherent challenges of mobile HF operation, particularly focusing on antenna size and efficiency. Mobile antennas are often compromised in size, affecting their efficiency and bandwidth, which mobile operators must consider when setting up their stations.

Mobile Antenna Components

Mobile antennas have several components designed to optimize the performance of shortened antennas.

Capacitance Hat

A capacitance hat is a device used on a mobile antenna to "electrically lengthen" a physically short antenna. Since a full-sized HF antenna is often too long for practical vehicle use, a shortened one is used instead. The capacitance hat compensates for the lack of physical length by increasing the antenna's electrical length, making it behave more like a full-sized one. This improves the efficiency of the antenna, allowing it to radiate more power effectively. In simpler terms, it tricks the antenna into thinking it's longer than it really is, which helps it perform better despite its compact size.

What is the purpose of a capacitance hat on a mobile antenna? (G4E01)

A. To increase the power handling capacity of a whip antenna
B. To reduce radiation resistance
C. To electrically lengthen a physically short antenna
D. To lower the radiation angle

Hat

Corona Ball

The corona ball is another important component found on mobile antennas, particularly on HF antennas. It is a small, rounded metallic ball placed at the tip of the antenna. Its purpose is to reduce RF voltage discharge, or "corona discharge," from the sharp tip of the antenna while transmitting. Without the corona ball, high RF voltages can build up at the tip, which can lead to interference, energy loss, or even cause sparks in some cases. The corona ball smooths out this voltage, reducing the chance of unwanted discharge and improving transmissions.

Ball

What is the purpose of a corona ball on an HF mobile antenna? (G4E02)

A. To narrow the operating bandwidth of the antenna
B. To increase the "Q" of the antenna
C. To reduce the chance of damage if the antenna should strike an object
D. To reduce RF voltage discharge from the tip of the antenna while transmitting

The capacitance hat and corona ball help improve the functionality of mobile antennae despite their size and the challenges of operating in a mobile environment.

Power Connections in HF Mobile Installations

Power connections in HF mobile installations are critical for ensuring reliable, efficient, and safe operation. A 100-watt HF mobile transceiver draws significant current, often exceeding 20 amps during peak transmission. This power demand level means your power connection must be designed to handle high current without causing issues like voltage drops, overheating, or interference.

To achieve this, appropriately sized wires, typically at least 10-gauge or thicker, is needed to minimize resistance and avoid excessive heat buildup. Always connect directly to the vehicle's battery, as this provides the most stable and clean power source. Avoid using the vehicle's accessory power outlets, which often cannot handle the current required by an HF transceiver and may introduce electrical noise.

Properly fuse positive and negative leads as close to the battery as possible to protect against short circuits. Ensure all connections are secure, and use high-quality terminals and connectors to reduce the risk of power loss or sparking. Routing power cables away from other electrical components can minimize interference and maintain signal clarity.

By following these practices, you can ensure safe and reliable power delivery to your HF mobile radio, maximizing performance and minimizing potential issues during operation.

Direct, Fused Power Connection

The best method for powering a 100-watt HF mobile transceiver is to use a direct connection to the vehicle's battery with a heavy-gauge wire protected by a fuse. This approach ensures that the transceiver receives a stable supply of power without interference from other vehicle systems. The fuse is critical for protecting the equipment and vehicle in case of a short circuit or power surge. Heavy-gauge wire is necessary to handle the high current drawn by the transceiver during transmission, ensuring that the radio operates efficiently without causing a voltage drop, which could affect performance.

Which of the following direct, fused power connections would be the best for a 100-watt HF mobile installation? (G4E03)

A. To the battery using heavy-gauge wire
B. To the alternator or generator using heavy-gauge wire
C. To the battery using insulated heavy duty balanced transmission line
D. To the alternator or generator using insulated heavy duty balanced transmission line

---

Avoiding the Vehicle's Auxiliary Power Socket

Many mobile operators might be tempted to use the vehicle's auxiliary power socket (commonly the cigarette lighter socket) to supply power to their transceiver. However, this is not recommended for a 100-watt HF transceiver. The wiring in the auxiliary socket is typically not designed to handle the high current demands of the transceiver, leading to inadequate power delivery, potential overheating, or blown fuses. Instead, a dedicated power connection to the battery ensures the transceiver gets the power it needs for effective operation without compromising safety or performance.

Why should DC power for a 100-watt HF transceiver not be supplied by a vehicle's auxiliary power socket? (G4E04)

A. The socket is not wired with an RF-shielded power cable
B. The socket's wiring may be inadequate for the current drawn by the transceiver
C. The DC polarity of the socket is reversed from the polarity of modern HF transceivers
D. Drawing more than 50 watts from this socket could cause the engine to overheat

---

The direct power connection using the correct wire gauge and fuse ensures that your mobile HF radio setup will be both safe and reliable in your mobile operations.

Mobile Interference

Mobile installations present a unique set of interference challenges. Vehicles' internal electronics can severely degrade HF reception if proper precautions aren't taken.

Which of the following may cause receive interference to an HF transceiver installed in a vehicle? (G4E07)

A. The battery charging system
B. The fuel delivery system
C. The control computers
D. All these choices are correct

When installing a HF transceiver in a vehicle, various systems within the car can cause interference, affecting your ability to receive signals. The most common sources of interference include the vehicle's battery charging system, fuel delivery system, and control computers. Each of these systems generates electrical noise that can be picked up by the HF transceiver, leading to distorted or unclear signals. For example, alternators and fuel injectors can create pulsing noises, while onboard computers can generate broad-spectrum noise that interferes with your radio. Understanding and addressing these potential sources of interference is key to ensuring a clean and reliable mobile HF setup.

Chapter Summary

This chapter covered the exam questions on setting up and optimizing mobile and portable radio stations. From understanding the components of mobile antennas to learning about secure power connections and managing interference, each aspect is fundamental for achieving reliable, clear communication on the go. With this knowledge, you can be assured of strong signals and reliable operations wherever your mobile radio takes you.

Chapter 13

Feedlines

IN HAM RADIO AND other radio communications, feedlines are the cables or wires that carry radio frequency (RF) energy from the transmitter to the antenna and back to the receiver. They ensure efficient power transfer with minimal signal loss or interference.

Impedance Matching

Impedance matching ensures that the maximum amount of power is transferred from the transmitter to the antenna without unnecessary loss. This section will explore how impedance matching improves communication efficiency and clarity.

What is one reason to use an impedance matching transformer at a transmitter output? (G7C03)

A. To minimize transmitter power output
B. To present the desired impedance to the transmitter and feed line
C. To reduce power supply ripple
D. To minimize radiation resistance

An impedance matching transformer is used at a transmitter's output to match the transmitter's impedance to the feed line or antenna. This ensures efficient power transfer, minimizing signal loss due to reflection or mismatch. Without proper impedance matching, a portion of the signal would be reflected back to the transmitter, causing inefficiencies and potentially damaging equipment. The transformer helps balance the system for optimal performance.

Devices like a transformer, a Pi-network, and a length of transmission line can be used to adjust or match the impedance between components at radio frequencies, improving signal efficiency.

- **Transformers** work by adjusting the voltage and current ratios, allowing you to match different impedance levels between circuit stages.

- **Pi-networks** use a combination of capacitors and an inductor to tune the circuit and match impedances, which are commonly used in amplifiers or antenna tuners.

- **A length of transmission line** can also be used to match impedance by adjusting its length to a specific wavelength, which ensures that signals travel efficiently without reflections.

> **Which of the following devices can be used for impedance matching at radio frequencies? (G5A10)**
>
> A. A transformer
> B. A Pi-network
> C. A length of transmission line
> **D. All these choices are correct**

To remember this for the exam, consider these devices as "adjusters" that help optimize the connection between parts of a radio circuit, ensuring the signal flows smoothly and efficiently without losses due to mismatched impedance.

Feedlines

Feedlines come in different types, each with varying characteristics of impedance, attenuation (signal loss), and frequency capabilities. The most common types are:

1. **Coaxial Cable (Coax)**: The most widely used feedline, consisting of a central conductor surrounded by insulation, a shield (usually braided wire or foil), and an outer insulating jacket. It's easy to install and provides good shielding from external interference. Coaxial cables typically have impedances of 50 ohms or 75 ohms.

2. **Parallel Conductor Feedlines (Ladder Line or Window Line)**: This type consists of two parallel wires separated by spacers. These lines have much lower loss than coax, particularly at higher frequencies, but are more susceptible to environmental factors like moisture. Typical impedances are 300 ohms and 450 ohms.

The choice of feedline depends on factors like the operating frequency, the distance between the radio and the antenna, and the need for impedance matching to minimize the standing wave ratio (SWR), which affects signal transmission efficiency.

Feedlines bridge your transmitter/receiver and the antenna, ensuring the signal is carried with minimal loss. The right type of feedline ensures efficient operation by matching impedance and reducing losses.

Parallel Conductor Feedline

A parallel conductor feedline is a type of transmission line used to carry RF energy from a transmitter to an antenna. It consists of two parallel conductors, spaced apart by insulating material, which allows it to transfer signals efficiently over long distances.

Parallel conductor feedlines are ideal for efficient, multi-band antenna systems or situations where minimizing loss is crucial. They're a great choice for balanced antennas like dipoles, especially when paired with a tuner to handle impedance mismatches.

The characteristic impedance of a parallel conductor feed line is determined by two key factors: the distance between the centers of the two conductors and the radius of each conductor. Simply put, the farther apart the conductors are, the higher the impedance will be—likewise, the larger the conductor radius, the lower the impedance. The characteristic impedance is important because it helps determine how efficiently power is transferred from the transmitter to the antenna without creating reflections or losses.

Which of the following factors determine the characteristic impedance of a parallel conductor feed line? (G9A01)

A. The distance between the centers of the conductors and the radius of the conductors
B. The distance between the centers of the conductors and the length of the line
C. The radius of the conductors and the frequency of the signal
D. The frequency of the signal and the length of the line

For a parallel conductor feed line, the spacing between the conductors and the size of the conductors play a critical role in determining its characteristic impedance. Ensuring proper impedance matching can minimize transmission losses and maximize power transfer.

Window Line

Window line is a type of parallel conductor feedline used to carry RF signals between a transmitter and an antenna. It gets its name from the small "windows" or cutouts in the insulating material that separates the two conductors. These cutouts reduce weight and material while maintaining the proper spacing between the wires.

Window line is an excellent choice for operators seeking efficient feedline solutions, especially in high-frequency or multi-band setups where low loss and versatility are priorities. It's often paired with balanced antennas and antenna tuners to maximize performance. Its nominal characteristic impedance is 450 ohms.

What is the nominal characteristic impedance of "window line" transmission line? (G9A03)

A. 50 ohms
B. 75 ohms
C. 100 ohms
D. 450 ohms

The window line's 450-ohm nominal characteristic impedance makes it a popular feedline choice for HF communications. The open windows in the line help reduce losses while maintaining a manageable impedance level.

Attenuation and Frequency Effects

When transmitting signals from your radio to an antenna, it's essential to understand how energy losses occur in the feedline. Attenuation refers to the reduction of signal strength as it travels along the feedline, and this is influenced by factors such as the frequency of the signal and the type of feedline used. As the frequency increases, attenuation (or signal loss) generally increases, making it crucial to choose appropriate feedline types and lengths to minimize these losses. This section covers the effects of attenuation and frequency on feedlines and how RF feedline loss is measured and expressed.

How does the attenuation of coaxial cable change with increasing frequency? (G9A05)

A. Attenuation is independent of frequency
B. Attenuation increases
C. Attenuation decreases
D. Attenuation follows Marconi's Law of Attenuation

The attenuation of a coaxial cable refers to the amount of signal loss that occurs as the radio frequency signal travels through the cable. This loss becomes more pronounced as the frequency of the signal increases. This is because higher frequencies result in more energy being absorbed by the dielectric material inside the cable and more power being lost as heat. In practical terms, this means that if you're operating at higher frequencies, you may experience more significant signal degradation over long distances unless you use a low-loss cable specifically designed to minimize this effect.

Higher frequency signals experience greater attenuation, meaning more signal strength is lost over a given distance. This is an key factor to consider when designing a communication system, especially for high-frequency operations.

In what units is RF feed line loss usually expressed? (G9A06)

A. Ohms per 1,000 feet
B. Decibels per 1,000 feet
C. Ohms per 100 feet
D. Decibels per 100 feet

For practical purposes, RF feed line loss is typically measured in decibels (dB) over a given distance, commonly per 100 feet. The decibel is a logarithmic unit that quantifies the ratio of the power loss between two points along the feedline. A loss of 1 dB means that approximately 20% of the signal's power is lost, while a loss of 3 dB represents a 50% power loss. This measurement helps operators compare different types of feedlines and determine how much signal will be lost over a specific length of cable at a given frequency. The lower the dB value per 100 feet, the better the cable is at preserving signal strength over that distance.

Feed line loss is measured in decibels per 100 feet, which allows you to estimate the signal you will lose over a specific cable length.

Attenuation and signal loss are important considerations when planning your radio system. As frequency increases, attenuation becomes more significant, so choosing the right feedline with minimal loss is crucial for maintaining signal strength. Understanding how to measure loss in decibels per 100 feet gives you a practical way to select and optimize your feedline setup. In doing so, you can ensure that your signal reaches its intended destination with as little degradation as possible.

Chapter Summary

Feedlines are the critical link between your radio and antenna, ensuring efficient power transfer with minimal signal loss. Understanding the characteristics of different feedlines, such as coaxial cables and parallel conductor feedlines like window line, allows you to select the best option for your operating needs.

Key factors such as impedance, attenuation, and frequency effects influence feedline performance, making it important to choose and manage feedlines carefully to maintain signal strength. By recognizing how attenuation increases with frequency and measuring feedline loss in decibels per 100 feet, you can optimize your station for reliable and efficient communication. With the right feedline setup, your signals will travel farther and clearer, enhancing your overall operating experience.

Chapter 14

Antennas

YOUR ANTENNA IS THE gateway to global communication. Whether you're reaching out to local operators or making long-distance contacts, your antenna plays a pivotal role in your ability to transmit and receive signals. As you upgrade to General Class, having a deeper understanding of the different types of antennas and their specific characteristics becomes more important than it was with just a Technician's license.

Random Wire

A random wire HF antenna is a simple, versatile antenna made from a single length of wire, typically connected to an antenna tuner and used for transmitting and receiving on HF bands. It is called "random" because its length is not cut to a specific fraction of a wavelength for a particular frequency, making it usable across multiple bands with the help of a tuner.

What is a characteristic of a random-wire HF antenna connected directly to the transmitter? (G9B01)

A. It must be longer than 1 wavelength
B. Station equipment may carry significant RF current
C. It produces only vertically polarized radiation
D. It is more effective on the lower HF bands than on the higher bands

A random-wire antenna is an unbalanced, non-resonant antenna often used when space or other constraints limit antenna options. When connected directly to the transmitter without any matching or balancing network, significant RF currents may flow through the station's equipment, leading to potential RF interference or safety issues. This is due to the mismatched impedance between the antenna and the transmitter, causing RF currents to flow where they shouldn't, such as along the feed line or chassis of the equipment.

A random wire HF antenna is a versatile and cost-effective solution for multi-band HF operation, especially when simplicity and portability are key. It's easy to set up and ideal for operators with limited space or those operating in the field. While an antenna tuner and a good RF ground are essential for optimal performance, this antenna offers excellent flexibility across multiple bands. To ensure safe and efficient operation, it's crucial to properly match the impedance and establish a reliable ground to minimize RF currents in station equipment.

Monopole

A monopole antenna is a type of radio antenna consisting of a single conductive element—typically a rod or wire. It is called a "monopole" because it has only one pole or element, unlike a dipole antenna, which has two elements. Monopole antennas are popular due to their simplicity, compact size, and effectiveness in applications where vertical polarization and omnidirectional coverage are beneficial.

Quarter-wave

Quarter-waves are the most popular monopoles due to their balance of efficiency and simplicity. These antennas play a significant role in both amateur and professional radio setups.

Which of the following is a common way to adjust the feed point impedance of an elevated quarter-wave ground-plane vertical antenna to be approximately 50 ohms? (G9B02)

A. Slope the radials upward
B. Slope the radials downward
C. Lengthen the radials beyond one wavelength
D. Coil the radials

In a quarter-wave ground-plane vertical antenna, adjusting the angle of the radials (wires extending from the base of the antenna) can help match the antenna's feed point impedance to that of a standard 50-ohm coaxial cable. Slope the radials downward at an angle of around 45 degrees to lower the impedance to approximately 50 ohms, which is ideal for ensuring maximum power transfer between the transmitter and the antenna.

Adjusting the radials downward is a practical way to improve the impedance match of a quarter-wave vertical antenna, optimizing the antenna's efficiency and reducing signal loss.

Which of the following best describes the radiation pattern of a quarter-wave ground-plane vertical antenna? (G9B03)

A. Bi-directional in azimuth
B. Isotropic
C. Hemispherical
D. Omnidirectional in azimuth

A quarter-wave ground-plane vertical antenna radiates equally in all directions (omnidirectional in azimuth) along the horizontal plane, meaning the signal strength is uniform regardless of direction. This makes it ideal for applications like local communication, where signals need to reach receivers in all directions without favoring one particular angle over another.

The omnidirectional radiation pattern makes the quarter-wave vertical antenna versatile for various communication needs, particularly for covering a wide area equally.

Ok, I know what azimuth means. But can you explain it to me just one more time?

In the context of radio communication and antennas, azimuth refers to the horizontal angle or direction of an antenna's radiation pattern relative to a reference point, typically true north. It is measured in degrees, with 0° representing north, 90° east, 180° south, and 270° west. The azimuth angle helps to determine where the strongest signal is transmitted or received along the horizontal plane.

For example, in an omnidirectional antenna, the azimuth pattern would be a 360° circle, indicating equal radiation in all directions horizontally. In directional antennas, like a Yagi, the azimuth pattern might show a stronger signal in one specific direction, which is helpful for long-distance communications where focusing the signal is important.

In summary, azimuth helps operators understand how an antenna radiates or receives signals across the horizontal plane, aiding in positioning for optimal communication.

What is the approximate length for a 1/4 wave monopole antenna cut for 28.5 MHz? (G9B12)

A. 8 feet
B. 11 feet
C. 16 feet
D. 21 feet

Step 1: Use the Formula for Quarter-Wave Antenna Length

The formula to calculate the length of a quarter-wave antenna is:

$$Antenna\ Length\ (feet) = \frac{468\ feet}{Frequency\ (in\ MHz)}$$

Where:

- L = Length of the antenna in feet

- f = Frequency in megahertz (MHz)

Step 2: Plug in the Frequency & Solve

The given frequency is 28.5 MHz. Substituting into the formula:

$$Antenna\ Length = \frac{468}{28.5} \approx 16.4\ feet$$

$$16.4\ feet\ divided\ by\ two\ to\ get\ quarter\ wave\ = 8.2\ feet$$

L = 8.2 feet rounded equals 8 feet!

This calculation ensures that the antenna is the correct length for efficient resonance and transmission at that frequency. Knowing how to calculate the appropriate antenna length based on frequency is crucial for ensuring effective communication and minimizing signal loss.

Understanding how to adjust elements such as radial angles and length to match the operating frequency allows amateur radio operators to maximize their antenna's performance. Whether setting up for local or long-range communication, these foundational concepts in antenna design ensure reliable and clear signals.

Vertical

Vertical antennas are versatile and widely used, known for their compact design and effective performance.

Vertical antennas are often implemented as quarter-wave monopoles, combining the simplicity of the monopole design with vertical polarization for efficient omnidirectional coverage, making them ideal for local coverage and ground-wave propagation while also performing well for long-distance (DX) communication on lower HF bands.

Their vertically polarized signals minimize ground losses, and their straightforward design makes them suitable for setups with limited space or portable operations. Whether used in base stations or temporary field setups, vertical antennas offer a practical balance of simplicity, efficiency, and effectiveness for ham radio enthusiasts.

Where should the radial wires of a ground-mounted vertical antenna system be placed? (G9B06)

A. As high as possible above the ground
B. Parallel to the antenna element
C. On the surface or buried a few inches below the ground
D. At the center of the antenna

Radial wires in a ground-mounted vertical antenna system act as part of the ground system, helping to improve signal efficiency by creating a low-resistance path for the return current. These wires should be placed directly on the ground's surface or buried just a few inches below it. Burying them deeper isn't necessary and can make installation harder without providing added benefit. A well-laid radial system helps minimize ground losses, improving antenna performance, particularly for long-distance (DX) communication.

Which of the following is an advantage of using a horizontally polarized as compared to a vertically polarized HF antenna? (G9B09)

A. Lower ground losses
B. Lower feed point impedance
C. Shorter radials
D. Lower radiation resistance

Horizontal polarization typically results in lower ground losses compared to vertical polarization. This is because horizontally polarized waves interact with the ground in a way that reduces power absorption, allowing more energy to be radiated into the air. On the other hand, vertically polarized antennas tend to have higher ground losses since the signal must travel through the ground before radiating into the atmosphere. For HF communications, lower ground losses can significantly improve efficiency, making horizontally polarized antennas a better choice in many situations.

Dipole

Dipole antennas are among ham radio's most fundamental and widely used antennas. These antennas consist of two conductive elements, each half the length of the desired wavelength, making them versatile for various HF bands. The dipole antenna's simplicity makes it an excellent choice for beginners and experienced operators. It is straightforward to construct and requires basic materials like wire, insulators, and a center feed point. When installed at an appropriate height, typically half a wavelength above the ground, dipole antennas offer excellent local and medium-range communication performance. However, they require sufficient space for installation, which can be a limitation in urban or restricted environments.

Basic Characteristics of Dipole

What is the radiation pattern of a dipole antenna in free space in a plane containing the conductor? (G9B04)

A. It is a figure-eight at right angles to the antenna
B. It is a figure-eight off both ends of the antenna
C. It is a circle (equal radiation in all directions)
D. It has a pair of lobes on one side of the antenna and a single lobe on the other side

A dipole antenna exhibits a figure-eight radiation pattern when placed in free space. This means that most of the radiation is concentrated broadside to the antenna, while little to no radiation occurs off the ends. The pattern resembles a figure-eight, where the strongest signals are emitted at right angles to the antenna, making it ideal for long-range communication in those directions.

A dipole antenna's radiation pattern is highly directional, with maximum radiation occurring perpendicular to the antenna's wire. This figure-eight pattern is essential for focusing signals and optimizing communication range in the desired directions.

Due to their simplicity and efficiency, dipole and monopole antennas are widely used. Understanding how these antennas behave in terms of RF currents, radiation patterns, and impedance matching helps ensure safe and effective operation. Knowing the basics allows operators to make informed decisions about their antenna setup and optimize communication performance, whether using a random-wire HF antenna or a precisely tuned dipole.

Azimuthal Radiation Pattern

Dipole antennas are one of the most fundamental and widely used antennas in amateur radio due to their simplicity and efficiency. Understanding their characteristics, such as radiation patterns and impedance, helps in optimizing their performance for various frequencies and environments. The dipole's behavior, including how its height above ground and feed point location affect its impedance and radiation pattern, plays a crucial role in ensuring effective transmission and reception. These considerations are important for setting up reliable HF communications, especially for General Class operators.

How does antenna height affect the azimuthal radiation pattern of a horizontal dipole HF antenna at elevation angles higher than about 45 degrees? (G9B05)

A. If the antenna is too high, the pattern becomes unpredictable
B. Antenna height has no effect on the pattern
C. If the antenna is less than 1/2 wavelength high, the azimuthal pattern is almost omnidirectional
D. If the antenna is less than 1/2 wavelength high, radiation off the ends of the wire is eliminated

---

When a horizontal dipole antenna is placed closer to the ground (less than 1/2 wavelength high), its radiation pattern becomes nearly omnidirectional in the horizontal plane. This means it radiates signals in all directions more evenly. At higher elevation angles, the antenna sends energy more upward than outward. This is useful for short-range or "local" contacts where higher elevation angles are advantageous. For long-distance communication (DX), installing the dipole at a higher elevation is typically better for producing lower-angle radiation, favoring far-reaching signals.

Feed Point Impedance

Feed point impedance is the electrical resistance and reactance measured at the point where the transmission line connects to an antenna. It represents how much the antenna resists the flow of RF current from the feedline and is typically expressed in ohms. By understanding and matching the impedance, operators can minimize power losses, reduce SWR, and optimize antenna performance for better communication.

How does the feed point impedance of a horizontal 1/2 wave dipole antenna change as the antenna height is reduced to 1/10 wavelength above ground? (G9B07)

A. It steadily increases
B. It steadily decreases
C. It peaks at about 1/8 wavelength above ground
D. It is unaffected by the height above ground

As the height of a 1/2 wave dipole decreases relative to the ground (down to 1/10 wavelength), the impedance at the feed point drops significantly. A dipole placed higher above ground tends to have an impedance closer to the theoretical 73 ohms, but lowering it causes the antenna to interact more with the ground, <u>reducing its impedance.</u> This lower impedance can sometimes necessitate matching networks to prevent high SWR and improve the overall efficiency of the antenna.

How does the feed point impedance of a 1/2 wave dipole change as the feed point is moved from the center toward the ends? (G9B08)

A. It steadily increases
B. It steadily decreases
C. It peaks at about 1/8 wavelength from the end
D. It is unaffected by the location of the feed point

The feed point impedance of a 1/2 wave dipole is typically around 73 ohms when fed at the center. However, as you move the feed point towards the ends of the antenna, <u>the impedance increases,</u> reaching much higher values near the ends. This is because the current distribution along the dipole is highest in the center and drops towards the ends while the voltage increases. End-feeding a dipole results in very high impedance, often requiring impedance matching to avoid significant SWR issues.

An end-fed half-wave (EFHW) antenna is exactly what it sounds like—a half-wavelength long antenna fed at one of its ends instead of the center. Unlike a center-fed dipole, where the impedance at the center is around 50 to 75 ohms, the feed point impedance of an EFHW antenna is very high, typically in the range of several thousand ohms. This happens because the voltage at the end of a half-wave antenna is at a maximum while the current is very low, creating a high-impedance point.

What is the feed point impedance of an end-fed half-wave antenna? (G9D02)

A.   Very low
B.   Approximately 50 ohms
C.   Approximately 300 ohms
D.   **Very high**

To use an EFHW antenna effectively an impedance matching transformer is used to bring the impedance down to a more manageable level. Without proper matching, a radio connected directly to an EFHW would see excessive standing wave ratio (SWR), making the system inefficient and potentially damaging the transmitter. Remembering this concept will help you recognize why EFHW antennas require matching components and why their feed point impedance is so high.

Dipole Length

What is the approximate length for a 1/2 wave dipole antenna cut for 14.250 MHz? (G9B10)

A. 8 feet
B. 16 feet
C. 24 feet
D. 33 feet

To calculate the length of a 1/2 wave dipole, you can use the formula:

$$Length\ (in\ feet) = \frac{468\ feet}{Frequency\ (in\ MHz)}$$

Does this formula look familiar? It should be from the Technician's. It's also twice the length of what we just saw in the quarter-wave section, which makes sense going from quarter to half-wave!

Substitute into the formula and solve:

$$Length\ (in\ feet) = \frac{468\ feet}{14.250\ MHz} \approx 32.8\ or\ round\ to\ 33\ feet$$

The length of 14.250 MHz is approximately 33 feet. This formula helps determine the optimal size of the antenna to achieve resonance on the desired frequency band, ensuring efficient transmission and reception.

Try another.

What is the approximate length for a 1/2 wave dipole antenna cut for 3.550 MHz? (G9B11)

A. 42 feet
B. 84 feet
C. 132 feet
D. 263 feet

Using the same dipole formula:

$$Length\ (in\ feet) = \frac{468\ feet}{Frequency\ (in\ MHz)}$$

Substitute into the formula:

$$Length\ (in\ feet) = \frac{468\ feet}{3.550\ MHz} \approx 131.8\ or\ round\ to\ 132\ feet$$

This calculation gives a dipole length of around 132 feet. The required antenna length is much longer at lower frequencies, like 3.550 MHz (commonly in the 80-meter band). Operators may need to adjust the installation to accommodate the space required for such large antennas or use techniques like loading coils to shorten the physical length while maintaining electrical resonance.

Dipole antennas, whether used at high or low frequencies, present unique challenges regarding installation height, impedance matching, and overall effectiveness. Understanding the relationships between height, feed point location, and impedance is crucial for optimizing antenna performance. By knowing the correct calculations and characteristics, operators can ensure their antennas are tuned correctly for effective communication across various bands.

Inverted V

The inverted V antenna is a variation of the traditional dipole antenna with a single central support. Instead of being mounted horizontally, the antenna elements slope downwards from a central support, forming a "V" shape. This configuration is advantageous when space is limited, as it requires less horizontal area. The inverted V also offers a lower radiation angle than a standard dipole, which can enhance long-distance communication capabilities.

What is the common name of a dipole with a single central support? (G9D12)

A. Inverted V
B. Inverted L
C. Sloper
D. Lazy H

The inverted V dipole is a versatile and space-efficient antenna option for amateur radio operators, balancing ease of installation with solid all-around performance. Its flexibility makes it a popular choice for both fixed and portable operations.

Setting Up and Optimizing Your Antenna

Antenna Tuner

An antenna tuner is a device used in radio communications to improve the efficiency of power transfer from the transmitter to the antenna system through the feed line. In practical terms, the antenna tuner adjusts the impedance seen by the transmitter to match the feed line and antenna impedance. This is crucial because if the impedance of the antenna system is not correctly matched to the transmitter's output, a significant amount of power could be reflected back to the transmitter, causing poor performance and potential damage.

What is the purpose of an antenna tuner? (G4A06)

A. Reduce the SWR in the feed line to the antenna
B. Reduce the power dissipation in the feedline to the antenna
C. Increase power transfer from the transmitter to the feed line
D. All these choices are correct

Using an antenna tuner maximizes the amount of power radiated by the antenna rather than being reflected. This is particularly helpful when using antennas that operate on multiple frequencies, as the tuner helps match the impedance even when the antenna is not perfectly resonant at the desired operating frequency. Essentially, the tuner acts as a mediator between the transmitter and the antenna, ensuring efficient power transfer and better overall performance.

Antenna Analyzer

An antenna analyzer is a versatile tool amateur radio operators use to measure various parameters of antennas and transmission lines. One of the features of an antenna analyzer is its ability to measure the impedance of coaxial cables. Impedance is critical in ensuring efficient power transfer between the transmitter and antenna. By checking the impedance, an operator can verify that the feed line (coaxial cable) matches the antenna and transmitter correctly, reducing losses and preventing signal reflections. If there is a mismatch in impedance, power will be reflected back toward the transmitter, potentially causing inefficiencies and damage.

Which of the following can be measured with an antenna analyzer? (G4B13)

A. Front-to-back ratio of an antenna
B. Power output from a transmitter
C. Impedance of coaxial cable
D. Gain of a directional antenna

Impedance measurements with an antenna analyzer are necessary because they allow operators to fine-tune their transmission system for maximum performance. Coaxial cable impedance is usually specified (e.g., 50 ohms), and ensuring that this matches the antenna and transmitter prevents unnecessary power loss. Using an analyzer to measure impedance can also help diagnose issues like cable degradation, poor connections, or faulty components in the feed line.

An antenna analyzer is crucial for measuring the impedance of coaxial cables and ensuring that the entire transmission system is working efficiently. By matching the impedance of the cable, antenna, and transmitter, operators can optimize performance, reduce power loss, and avoid damaging their equipment.

Standing Wave Ratio (SWR)

In radio communication, SWR (Standing Wave Ratio) is a critical factor that determines how efficiently power is transferred from the transmitter to the antenna. SWR is affected by impedance mismatches between the feed line and the antenna. When this mismatch occurs, standing waves form on the feed line, causing signal reflections, which reduce the effectiveness of signal transmission. This section explores the causes and solutions for standing waves and discusses how transmission line losses can impact SWR measurements.

Which of the following can be determined with a directional wattmeter? (G4B10)

A. Standing wave ratio
B. Antenna front-to-back ratio
C. RF interference
D. Radio wave propagation

A directional wattmeter is a tool amateur radio operators use to measure the standing wave ratio (SWR). The SWR is a critical parameter that indicates how efficiently power is being transferred from the transmitter to the antenna through the feed line. A low SWR means that most of the power is reaching the antenna, while a high SWR suggests that some power is being reflected back toward the transmitter, which can lead to signal loss and even potential damage to equipment. Using a directional wattmeter helps operators monitor SWR and ensure efficient operation.

What must be done to prevent standing waves on a feed line connected to an antenna? (G9A07)

A. The antenna feed point must be at DC ground potential
B. The feed line must be an odd number of electrical quarter wavelengths long
C. The feed line must be an even number of physical half-wavelengths long
D. The antenna feed point impedance must be matched to the characteristic impedance of the feed line

Standing waves on a feed line occur when the impedance of the antenna and the feed line mismatch. To prevent these standing waves and ensure maximum power is transferred from the transmitter to the antenna, the impedance at the feed point must be matched to the characteristic impedance of the feed line (typically 50 ohms in most amateur radio setups). Using an antenna tuner or adjusting the antenna design can help achieve this impedance match, reducing the formation of standing waves and improving overall system performance.

Preventing standing waves is critical to ensuring efficient signal transmission. The most effective solution is to match the antenna's feed point impedance with the feed line's impedance.

What causes reflected power at an antenna's feed point? (G9A04)

A. Operating an antenna at its resonant frequency
B. Using more transmitter power than the antenna can handle
C. A difference between feed line impedance and antenna feed point impedance
D. Feeding the antenna with unbalanced feed line

Reflected power occurs when the impedance at the antenna's feed point does not match the impedance of the feed line. This mismatch causes some of the power sent from the transmitter to be reflected back along the feed line instead of being radiated by the antenna. The greater the mismatch, the more energy is reflected, leading to less efficient signal transmission and potential damage to the transmitter over time.

Reflected power results from impedance mismatches between the antenna and feed line, reducing the efficiency of your transmission.

What is the effect of transmission line loss on SWR measured at the input to the line? (G9A11)

A. Higher loss reduces SWR measured at the input to the line
B. Higher loss increases SWR measured at the input to the line
C. Higher loss increases the accuracy of SWR measured at the input to the line
D. Transmission line loss does not affect the SWR measurement

---

Transmission line losses occur as the RF signal travels through the feed line, especially over longer distances or at higher frequencies. These higher losses can reduce the SWR value measured at the transmitter's end of the line, even if there is an impedance mismatch at the antenna. This occurs because the line losses attenuate both the forward and reflected signals, making the SWR appear lower than it really is at the feed point.

Transmission line loss can mask a high SWR at the antenna, making it necessary to measure SWR as close to the antenna as possible for accuracy.

SWR plays a crucial role in efficient signal transmission, and managing standing waves through impedance matching is essential to maintain signal strength. Impedance mismatches cause reflected power, reducing efficiency and risking equipment damage. Transmission line losses can complicate SWR measurements, which is why monitoring SWR at the feed point is vital.

SWR Measurement and Effects

This section delves into how SWR is measured, its effects on transmission line losses, and the relationships between feed line impedance and resistive loads.

What is the relationship between high standing wave ratio (SWR) and transmission line loss? (G9A02)

A. There is no relationship between transmission line loss and SWR
B. High SWR increases loss in a lossy transmission line
C. High SWR makes it difficult to measure transmission line loss
D. High SWR reduces the relative effect of transmission line loss

The high SWR indicates a mismatch between the transmission line and the antenna, causing power to reflect back toward the transmitter. This reflected power in a lossy transmission line leads to even more signal loss as it travels back and forth along the line. The greater the SWR, the more signal is lost, which can reduce the overall performance of your radio system. The higher the frequency, the more pronounced this effect becomes, which is why keeping SWR low is important for minimizing losses and improving transmissions.

A higher SWR reduces signal transmission and increases transmission line loss, especially in lossy coaxial cables.

If the SWR on an antenna feed line is 5:1, and a matching network at the transmitter end of the feed line is adjusted to present a 1:1 SWR to the transmitter, what is the resulting SWR on the feed line? (G9A08)

A. 1:1
B. 5:1
C. Between 1:1 and 5:1 depending on the characteristic impedance of the line
D. Between 1:1 and 5:1 depending on the reflected power at the transmitter

Even though a matching network can adjust the SWR to 1:1 at the transmitter end, the SWR on the feed line remains unchanged. The matching network helps the transmitter operate more efficiently by presenting it with the desired impedance. Still, the impedance mismatch between the antenna and feed line remains. As a result, the standing waves continue to exist on the feed line, and the SWR on the line remains at 5:1.

A matching network at the transmitter end does not change the SWR on the feed line itself; it only makes the transmitter "see" a 1:1 SWR.

What standing wave ratio results from connecting a 50-ohm feed line to a 200-ohm resistive load? (G9A09)

A. 4:1
B. 1:4
C. 2:1
D. 1:2

SWR is calculated as the ratio between the load and feed line impedance. In this case, the 200-ohm load is four times the 50-ohm feed line impedance, resulting in an SWR of 4:1. This means that a significant portion of the power is being reflected back toward the transmitter. The system is not operating at peak efficiency.

A mismatch between the feed line and the load results in a high SWR, such as 4:1 when connecting a 50-ohm feed line to a 200-ohm load.

What standing wave ratio results from connecting a 50-ohm feed line to a 10-ohm resistive load? (G9A10)

A. 2:1
B. 1:2
C. 1:5
D. 5:1

In this case, the load impedance is much lower than the feed line impedance, creating a greater mismatch. SWR is calculated as the higher impedance ratio to the lower impedance. Therefore, dividing 50 ohms by 10 ohms results in an SWR of 5:1. This significant mismatch means that most of the power is reflected back to the transmitter, making the system highly inefficient.

A large impedance mismatch, such as connecting a 50-ohm line to a 10-ohm load, results in a high SWR, in this case, 5:1. High SWR values lead to increased transmission line loss and reduced system performance while matching the feed line impedance to the antenna can minimize standing waves optimizing the system's overall performance.

Antenna Analyzer for SWR Measurements

When measuring SWR with an antenna analyzer, it is essential that both the antenna and feed line are connected to the analyzer. The analyzer sends signals through the feed line to the antenna and measures how much of that signal is reflected back. This helps the operator adjust the antenna or feed line for optimal performance. However, strong signals from nearby transmitters can interfere with SWR readings on an antenna analyzer. These signals, picked up by the analyzer, can distort the measurements, leading to inaccurate readings and making it harder to fine-tune the system.

Which of the following must be connected to an antenna analyzer when it is being used for SWR measurements? (G4B11)

A. Receiver
B. Transmitter
C. Antenna and feed line
D. All these choices are correct

What effect can strong signals from nearby transmitters have on an antenna analyzer? (G4B12)

A. Desensitization which can cause intermodulation products which interfere with impedance readings
B. Received power that interferes with SWR readings
C. Generation of harmonics which interfere with frequency readings
D. All these choices are correct

Tools like directional wattmeters and antenna analyzers help ham radio operators measure SWR, ensuring that power is efficiently transferred to the antenna.

Chapter Summary

Antennas are key to any ham radio station, converting electrical signals into radio waves. From simple wire antennas to dipoles and verticals, each type has unique characteristics that affect performance. Understanding feed point impedance, radiation patterns, and tuning helps operators optimize communication. Mastering these basics improves station efficiency and sets the stage for exploring more advanced antenna systems.

Chapter 15

Directional Antennas

A DIRECTIONAL ANTENNA IS a general term for any antenna that focuses its signal in a specific direction rather than radiating equally in all directions (like an omnidirectional antenna). This focus increases the signal strength in the desired direction, improving performance for transmitting and receiving signals over long distances. Common directional antennas include Yagi, log-periodic, and parabolic antennas.

Yagi Antennas

Yagi antennas are prized for their high gain and directionality, making them ideal for long-distance communication. With a driven element, reflector, and directors arranged on a boom, Yagis focus signals in one direction, boosting strength and improving reception. This efficiency makes them a favorite for amateur radio operators seeking reliable performance over long distances. However, their setup requires space, sturdy support, and often a rotator for precise aiming. Understanding key concepts like element length, front-to-back ratio, and radiation lobes is crucial for optimizing their performance and maximizing range.

What is the approximate length of the driven element of a Yagi antenna? (G9C02)

A. 1/4 wavelength
B. 1/2 wavelength
C. 3/4 wavelength
D. 1 wavelength

The driven element of a Yagi antenna is responsible for receiving and transmitting the signal, and it is typically <u>around half the wavelength of the frequency it operates on</u>. This length allows for maximum signal capture and efficient energy radiation, which is why the driven element is precisely tuned to this length for optimal performance. For example, if you're operating on a 20-meter band, the driven element will be about 10 meters long.

How do the lengths of a three-element Yagi reflector and director compare to that of the driven element? (G9C03)

A. The reflector is longer, and the director is shorter
B. The reflector is shorter, and the director is longer
C. They are all the same length
D. Relative length depends on the frequency of operation

In a Yagi antenna, the reflector is slightly longer and placed behind the driven element. It helps to reflect energy forward, increasing the forward gain. The director is shorter and positioned in front of the driven element to guide the signal in the desired direction. This combination creates a directional beam, which makes Yagi antennas highly effective in targeting specific areas while reducing interference from other directions.

What does "front-to-back ratio" mean in reference to a Yagi antenna? (G9C07)

A. The number of directors versus the number of reflectors
B. The relative position of the driven element with respect to the reflectors and directors
C. The power radiated in the major lobe compared to that in the opposite direction
D. The ratio of forward gain to dipole gain

The front-to-back ratio is a measurement that compares the power radiated in the Yagi antenna's main direction (front) to the power radiated in the opposite direction (back). A higher front-to-back ratio means that the antenna focuses more power in the desired direction and minimizes power radiated in reverse, improving signal quality by reducing unwanted noise and interference from behind the antenna.

What is meant by the "main lobe" of a directive antenna? (G9C08)

A. The magnitude of the maximum vertical angle of radiation
B. The point of maximum current in a radiating antenna element
C. The maximum voltage standing wave point on a radiating element
D. The direction of maximum radiated field strength from the antenna

The "main lobe" of a Yagi antenna refers to the direction in which the antenna radiates the most power. This is the area of maximum signal strength, where communication will be strongest and most reliable. For Yagi antennas, the main lobe is directed forward toward the director elements, allowing the antenna to focus its energy in that direction and improving its effectiveness for long-distance communication.

Yagi antennas are powerful tools for directional communication, allowing radio operators to focus their transmissions in a specific direction while reducing interference from other areas. Understanding the role of each element—the driven element, reflector, and director—along with concepts like front-to-back ratio and the main lobe is key to optimizing antenna performance. These principles help radio enthusiasts use Yagi antennas for more effective and targeted communication.

Enhancing Performance

Yagi antennas are well-known for their ability to focus energy in a specific direction, which makes them excellent for long-distance communication. However, depending on your needs, several ways to improve their performance exist. Factors such as element diameter, boom length, the number of directors, and the spacing between elements can significantly impact a Yagi's bandwidth, gain, and front-to-back ratio. These adjustments allow users to fine-tune their Yagi antennas for better communication performance in varying conditions.

Which of the following would increase the bandwidth of a Yagi antenna? (G9C01)

A. Larger-diameter elements
B. Closer element spacing
C. Loading coils in series with the element
D. Tapered-diameter elements

Increasing the diameter of the elements in a Yagi antenna broadens the range of frequencies over which the antenna can operate efficiently. This is because larger elements lower the antenna's "Q factor," making it less sensitive to small changes in frequency. A lower Q factor means a wider bandwidth, allowing the antenna to transmit and receive signals over a broader range of frequencies without a significant drop in performance. This can be particularly useful when dealing with varying frequencies or when working in environments with a lot of signal interference.

What is the primary effect of increasing boom length and adding directors to a Yagi antenna? (G9C05)

A. Gain increases
B. Beamwidth increases
C. Front-to-back ratio decreases
D. Resonant frequency is lower

By increasing the boom length and adding more directors to a Yagi antenna, you focus the antenna's energy more narrowly in a specific direction. This results in an increase in antenna gain, meaning more signal strength is directed forward. The trade-off is that a longer boom and additional directors can make the antenna more directional, narrowing the beamwidth. Still, this enhanced focus is beneficial for reaching distant stations with greater signal strength.

Which of the following can be adjusted to optimize forward gain, front-to-back ratio, or SWR bandwidth of a Yagi antenna? (G9C10)

A. The physical length of the boom
B. The number of elements on the boom
C. The spacing of each element along the boom
D. All these choices are correct

Several factors can be adjusted to optimize the performance of a Yagi antenna. Increasing the boom length and adding more elements can improve forward gain, while careful spacing of the elements can fine-tune the front-to-back ratio, which measures how well the antenna rejects signals from the opposite direction. By adjusting these physical parameters, operators can achieve better signal reception and transmission tailored to their specific needs, whether that is increasing signal strength, improving selectivity, or widening bandwidth.

Optimizing a Yagi antenna's performance involves manipulating various physical characteristics like element size, boom length, and the number and spacing of directors. These adjustments allow radio operators to enhance the antenna's bandwidth, increase its gain, and fine-tune its directional properties. Understanding these factors is key to maximizing a Yagi antenna's potential, ensuring strong, focused communication over long distances with minimal interference.

Matching Techniques and Advanced Configurations

Matching techniques are used to ensure maximum power transfer from your transmitter to your antenna by minimizing reflected power. Different methods, such as the beta or hairpin match and the gamma match, are commonly used with Yagi antennas and other advanced configurations. These techniques help fine-tune impedance mismatches that can otherwise decrease efficiency and performance. By using specific matching methods, operators can optimize their antenna system for better transmission and reception, even when dealing with complex antenna setups.

What is a beta or hairpin match? (G9C11)

A. A shorted transmission line stub placed at the feed point of a Yagi antenna to provide impedance matching
B. A 1/4 wavelength section of 75-ohm coax in series with the feed point of a Yagi to provide impedance matching
C. A series capacitor selected to cancel the inductive reactance of a folded dipole antenna
D. A section of a 300-ohm twin-lead transmission line used to match a folded dipole antenna

The beta or hairpin match is a clever solution for impedance matching in Yagi antennas. Essentially, it is a shorted transmission line stub placed at the antenna's feed point. This stub helps transform the feed point impedance to a more desirable value—typically matching it to the feed line's impedance. This ensures that more of the signal energy is transmitted effectively, reducing losses due to mismatch. The simplicity and effectiveness of the beta match make it a popular choice for fine-tuning Yagi antennas, especially in high-performance settings.

Which of the following is a characteristic of using a gamma match with a Yagi antenna? (G9C12)

A. It does not require the driven element to be insulated from the boom
B. It does not require any inductors or capacitors
C. It is useful for matching multiband antennas
D. All these choices are correct

Gamma matching is another impedance technique often used with Yagi antennas. One of its advantages is that it doesn't require the driven element to be insulated from the boom, which simplifies the antenna construction. The gamma match adjusts the impedance by using a gamma rod connected partway along the driven element. This method allows the operator to match the antenna's impedance to the feed line without significantly modifying the boom or driven element, making it both a practical and efficient option.

Matching techniques like the beta and gamma match are used to fine-tune antennas, especially Yagi antennas, to ensure they perform optimally. These methods allow for efficient impedance matching without significant modifications to the antenna structure, making them ideal for both casual operators and advanced users looking to maximize their communication system's effectiveness. Understanding these configurations helps operators achieve better signal transmission, improving overall radio performance.

Advanced Configurations and Gain Comparisons

When optimizing the performance of Yagi antennas, especially in advanced configurations, operators often aim for increased gain and better signal directivity. Techniques like stacking Yagi antennas and understanding gain measurements (dBd vs. dBi) play a key role in fine-tuning antenna systems. These configurations allow for stronger signals and improved range, which is critical for long-distance communication and competitive performance in amateur radio.

In free space, how does the gain of two three-element, horizontally polarized Yagi antennas spaced vertically 1/2 wavelength apart typically compare to the gain of a single three-element Yagi? (G9C09)

A. Approximately 1.5 dB higher
B. Approximately 3 dB higher
C. Approximately 6 dB higher
D. Approximately 9 dB higher

Stacking two Yagi antennas vertically with a separation of 1/2 wavelength is a popular method for increasing antenna gain. By aligning the antennas in this way, the overall system increases its directional power and efficiency, resulting in approximately a 3 dB increase in gain compared to a single Yagi antenna. This is significant because every 3 dB increase in gain effectively doubles the power radiated in the desired direction, allowing for improved signal strength and coverage. This technique is often used when higher performance is needed in competitive or long-range communications.

How does antenna gain in dBi compare to gain stated in dBd for the same antenna? (G9C04)

A. Gain in dBi is 2.15 dB lower
B. Gain in dBi is 2.15 dB higher
C. Gain in dBd is 1.25 dBd lower
D. Gain in dBd is 1.25 dBd higher

Antenna gain can be expressed in two ways: dBd (relative to a dipole) and dBi (relative to an isotropic radiator). A dipole is a real-world antenna, while an isotropic radiator is a theoretical antenna that radiates equally in all directions. Gain in dBi is always 2.15 dB higher than in dBd because a dipole antenna has a natural gain of 2.15 dB over an isotropic radiator. Understanding this distinction helps operators accurately compare antenna types and determine their performance in specific configurations.

Advanced configurations such as stacking Yagi antennas and knowing the difference between dBi and dBd gain measurements are crucial for maximizing antenna performance. These strategies allow operators to boost signal strength, improve range, and fine-tune their setups for specific communication needs. By leveraging these techniques, radio operators can significantly enhance their transmission power and ensure more reliable communication over greater distances.

Vertically Stacking

Understanding antenna configurations is crucial for optimizing signal strength and range in ham radio operations. Different configurations, such as vertically stacked Yagi antennas or log-periodic antennas, offer unique advantages that can be tailored to specific communication needs. This section explores how these configurations affect antennas' gain, directivity, and bandwidth, providing insights into the technical details that make them efficient tools for operators.

What is an advantage of vertically stacking horizontally polarized Yagi antennas? (G9D05)

A. It allows quick selection of vertical or horizontal polarization
B. It allows simultaneous vertical and horizontal polarization
C. It narrows the main lobe in azimuth
D. It narrows the main lobe in elevation

Vertically stacking Yagi antennas is a technique used to improve the performance of an antenna system. By stacking two or more horizontally polarized Yagis one above the other, the main lobe of the radiation pattern narrows in elevation. This enhances the antenna's gain in the desired direction, focusing the signal and reducing wasted energy in other directions. This is particularly useful in DX communication, where a focused signal can improve reception at the target destination, especially when communicating at low elevation angles.

Log Periodic Antenna

A log-periodic antenna is characterized by its unique design, where the lengths of its elements and their spacing along the boom vary according to a logarithmic scale.

Log-periodic antennas differ from Yagi antennas in that they are designed to provide consistent performance across a wide frequency range rather than being optimized for a single frequency. This makes log-periodic antennas ideal for multi-band operation. Their ability to resonate at multiple frequencies comes from their unique design, where element lengths and spacing vary in a logarithmic progression, enabling broad bandwidth and versatility for various communication needs.

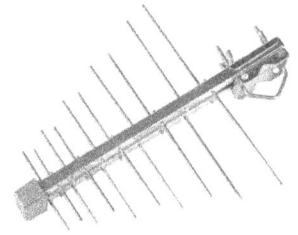

Which of the following describes a log-periodic antenna? (G9D07)

A. Element length and spacing vary logarithmically along the boom
B. Impedance varies periodically as a function of frequency
C. Gain varies logarithmically as a function of frequency
D. SWR varies periodically as a function of boom length

Which of the following is an advantage of a log-periodic antenna? (G9D06)

A. Wide bandwidth
B. Higher gain per element than a Yagi antenna
C. Harmonic suppression
D. Polarization diversity

A log-periodic antenna is designed to offer a <u>wide operational bandwidth</u>, meaning it can function efficiently across a broad range of frequencies. This characteristic makes log-periodic antennas highly versatile, as they can be used without needing to retune the antenna for different frequencies. This makes them ideal for applications that require communication on multiple bands, such as in amateur radio or other professional communication systems where versatility is vital.

Long-Path Communication

In amateur radio, long-path communication refers to a signal traveling the long way around the Earth to reach its destination, typically covering a greater distance than the direct short-path route. Signals can travel in both directions around the globe, and sometimes, the long path offers better propagation conditions, especially when short-path signals are weak or obstructed. Understanding how to use short and long paths can improve a ham radio operator's ability to reach distant stations under varying conditions.

How is a directional antenna pointed when making a "long-path" contact with another station? (G2D06)

A. Toward the rising sun
B. Along the gray line
C. 180 degrees from the station's short-path heading
D. Toward the north

<u>When making a long-path contact, the operator points their directional antenna 180 degrees opposite from the short-path heading.</u> For example, if the short path to a station is to the east, the operator would point their antenna west to reach the same station via the long path. This technique takes advantage of global propagation conditions, such as ionospheric reflection,

that might favor long-path transmission at certain times of day or during specific solar conditions. Knowing when to use long-path communication can help an operator maintain reliable contacts even when short-path signals are poor.

Chapter Summary

Directional antennas, such as Yagi and log-periodic designs, are essential tools for amateur radio operators looking to focus their signals and maximize performance. By concentrating energy in specific directions, these antennas enhance gain, improve signal clarity, and reduce interference from unwanted sources. Yagi antennas are prized for their high gain and precision, making them ideal for long-distance (DX) communication, while log-periodic antennas provide broad bandwidth for seamless operation across multiple bands.

Key concepts like element arrangement, front-to-back ratio, and main lobes help operators optimize their Yagi antennas for efficiency and reliability. Advanced configurations, such as stacking Yagis or employing impedance-matching techniques like the gamma or beta match, allow for fine-tuning performance based on operating conditions. Meanwhile, the log-periodic antenna stands out for its versatility, handling multi-band operation without frequent adjustments.

By mastering the principles and configurations of directional antennas, amateur radio enthusiasts can improve signal strength, minimize interference, and achieve reliable communication over long distances. Whether chasing DX or ensuring steady connections across multiple bands, these antennas empower operators to get the most out of their stations.

Chapter 16

Specialized Antennas

SPECIALIZED ANTENNAS SERVE UNIQUE purposes that go beyond the standard designs like dipoles, Yagis, or verticals, catering to specific operational needs and environments. These antennas include loop antennas for compact setups, helical antennas for circular polarization, and beverage antennas for long-distance reception. Each type offers distinct advantages, such as improved performance in challenging conditions, portability, or specific polarization patterns, making them invaluable tools for operators with unique communication goals. Understanding these specialized antennas expands your versatility as an amateur radio operator, enabling you to tackle everything from low-noise reception to satellite communication with confidence.

Near Vertical Incidence Skywave Antenna

A Near Vertical Incidence Skywave (NVIS) antenna is a specialized antenna designed to achieve reliable short-range HF communication (typically within 300-500 miles) by utilizing skywave propagation at near-vertical angles. Unlike traditional long-distance antennas, which aim to send signals at low angles toward the horizon, an NVIS antenna directs most of its energy upward at steep angles. These signals are then reflected back to Earth by the ionosphere, covering a broad area around the transmitter.

Which of the following antenna types will be most effective as a near vertical incidence skywave (NVIS) antenna for short-skip communications on 40 meters during the day? (G9D01)

A. A horizontal dipole placed between 1/10 and 1/4 wavelength above the ground
B. A vertical antenna placed between 1/4 and 1/2 wavelength above the ground
C. A horizontal dipole placed at approximately 1/2 wavelength above the ground
D. A vertical dipole placed at approximately 1/2 wavelength above the ground

---

A horizontal dipole placed relatively close to the ground, between 1/10 and 1/4 wavelength, is ideal for Near Vertical Incidence Skywave (NVIS) propagation. NVIS is a propagation method used for short-skip communications, where radio waves are reflected almost vertically by the ionosphere, allowing them to travel short distances effectively. This type of setup is commonly used on the 40-meter band for reliable daytime communications over a range of a few hundred miles, making it valuable for local and regional communication in emergency networks or casual contact.

A NVIS antenna is ideal when you need reliable short-range HF communication in areas where traditional line-of-sight methods, like VHF/UHF, fail due to obstructions such as mountains, forests, or urban environments. This makes NVIS antennas particularly valuable in emergency situations, where regional communication is critical, such as coordinating disaster relief efforts or maintaining contact across rugged terrain. They are also useful in rural areas with sparse infrastructure, providing dependable communication over distances of 300-500 miles. Additionally, NVIS antennas are commonly used by the military and emergency services for tactical operations, as they excel in covering local areas without the need for repeaters or other infrastructure. Their low deployment height and ability to work effectively on HF bands make them a practical choice for portable and field operations in challenging environments.

Beverage Antenna

The Beverage antenna is a long-wire <u>directional receiving antenna primarily used for medium-frequency (MF) and low-high-frequency (HF) bands</u>. Its main advantage is its excellent directional characteristics, which make it ideal for long-distance reception in the direction the antenna is pointed. Because of its low noise and ability to capture weak signals, the Beverage antenna is popular among operators who focus on long-range DXing, especially on bands like 160 meters, where noise can be a significant problem.

What is the primary use of a Beverage antenna? (G9D09)

A. Directional receiving for MF and low HF bands
B. Directional transmitting for low HF bands
C. Portable direction finding at higher HF frequencies
D. Portable direction finding at lower HF frequencies

The Beverage antenna is a specialized tool for amateur radio operators and shortwave listeners who prioritize weak signal reception, especially on the low bands. Its simplicity, effectiveness, and high SNR make it a favorite for low-frequency DXing and challenging listening environments.

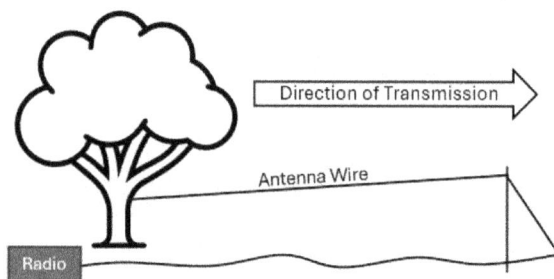

Halo Antenna

Understanding antenna directionality and radiation patterns is important for maximizing a radio station's performance. Some antennas, like the VHF/UHF "halo" antenna, are designed to radiate equally in all directions, while others, such as electrically small loop antennas, have specific directions of nulls or weaker signals. Knowing how these radiation patterns affect signal transmission and reception helps operators choose a suitable antenna for their needs, whether for local omnidirectional communication or long-distance directional contact.

In which direction is the maximum radiation from a VHF/UHF "halo" antenna? (G9D03)

A. Broadside to the plane of the halo
B. Opposite the feed point
C. Omnidirectional in the plane of the halo
D. On the same side as the feed point

A "halo" antenna is a horizontally polarized antenna, often used for VHF/UHF communications, with an <u>omnidirectional radiation pattern in the halo plane.</u> This means it radiates signals equally in all directions around the antenna, making it ideal for mobile or portable applications where communication is needed in all directions without requiring constant antenna repositioning. However, the signal strength is strongest in the horizontal plane, while the vertical radiation (above and below the antenna) is weaker, which aligns well with typical communication scenarios on these bands.

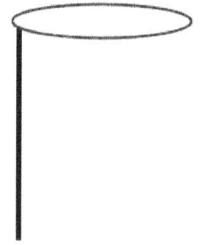

Loop Antenna

A loop antenna is a type of antenna formed by a loop or coil of conductive wire that is often used for both transmitting and receiving signals. Loop antennas are versatile and come in various sizes and designs, such as small magnetic loops for receiving or large full-wavelength loops for transmitting. Their compact size, directional properties, and efficiency make them popular among amateur radio operators, shortwave listeners, and professionals in RF communication.

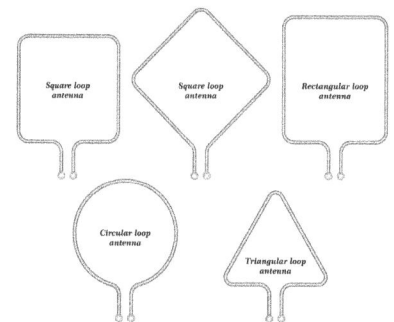

In which direction or directions does an electrically small loop (less than 1/10 wavelength in circumference) have nulls in its radiation pattern? (G9D10)

A. In the plane of the loop
B. Broadside to the loop
C. Broadside and in the plane of the loop
D. Electrically small loops are omnidirectional

An electrically small loop antenna, one whose circumference is less than 1/10 of a wavelength, has a characteristic radiation pattern with nulls, or points of minimal radiation, <u>broadside to the loop</u>. In simpler terms, the weakest signals are emitted directly perpendicular to the plane of the loop. This makes small loop antennas useful for direction-finding applications because the nulls can be used to pinpoint the direction of incoming signals by rotating the antenna until the signal fades. It also means that maximum radiation occurs along the axis of the loop, making it an effective antenna for specific directional applications.

Loop antennas offer a compact and efficient solution for operators seeking low-noise reception and stealth installations. These antennas can be magnetic loops or large wire loops, each with distinct characteristics. Magnetic loop antennas are particularly noted for their compact size and low noise reception, making them suitable for use in noisy environments. They are also highly portable, ideal for field operations or situations where a full-sized antenna is impractical. However, loop antennas generally have lower radiation efficiency and require precise tuning to operate effectively. This complexity in tuning can be a drawback for some operators.

Mobile Antenna

Mobile antennas are specifically designed for use on vehicles, enabling amateur radio operators to stay connected while on the move. These antennas are compact and rugged, built to withstand the challenges of mobile environments, such as vibration, weather exposure, and limited mounting space. Despite their smaller size compared to base station antennas, mobile antennas are engineered for efficient performance on popular bands, often using techniques like loading coils to achieve resonance. Whether you're engaging in local communication or participating in HF activities while traveling, mobile antennas are a critical part of ensuring reliable and flexible radio operation on the road.

How does a "screwdriver" mobile antenna adjust its feed point impedance? (G9D08)

A. By varying its body capacitance
B. By varying the base loading inductance
C. By extending and retracting the whip
D. By deploying a capacitance hat

A "screwdriver" mobile antenna is a type of adjustable antenna designed for use in vehicles, offering flexibility across different frequency bands. It adjusts its feed point impedance by <u>varying the base loading inductance.</u> This is accomplished by moving a coil or inductor up and down, effectively changing the antenna's electrical length. This feature allows the operator to tune the antenna across multiple frequency bands without needing to stop and manually adjust it, making it highly convenient for mobile ham radio operators who need quick and efficient impedance adjustments.

Trapped Antenna

Trapped antennas are versatile and efficient solutions for multi-band operation, designed to work seamlessly across multiple frequencies without requiring manual adjustments. These antennas use traps—coils and capacitors arranged in parallel—that act as resonant circuits to isolate specific sections of the antenna at different frequencies. By effectively "shortening" the antenna at higher frequencies while using its full length at lower ones, trapped antennas enable operators to cover multiple bands with a single design. Their convenience and simplicity make them a popular choice for those seeking efficient multi-band performance in limited space.

What is the primary function of antenna traps? (G9D04)

A. To enable multiband operation
B. To notch spurious frequencies
C. To provide balanced feed point impedance
D. To prevent out-of-band operation

Antenna traps are a key component used to enable multiband operation in antennas. These are resonant circuits that are placed along the length of an antenna, allowing it <u>to operate efficiently on multiple bands</u> without needing separate antennas for each band. When transmitting on a specific band, the trap effectively isolates parts of the antenna, allowing only the relevant portion to radiate the signal. This allows the same antenna to work across different frequency ranges, offering versatility and reducing the need for multiple antennas in a compact setup.

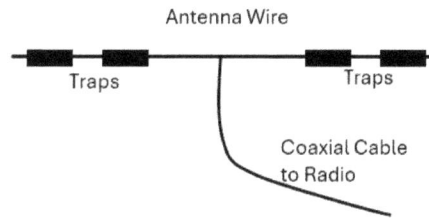

Antenna Wire

Traps   Traps

Coaxial Cable
to Radio

Multi-Band Antenna

A multi-band antenna is an antenna designed to operate efficiently on multiple frequency bands without requiring significant adjustments or additional components. These antennas allow amateur radio operators to communicate across various bands, making them versatile and convenient for both HF and VHF/UHF operation.

Which of the following is a disadvantage of multiband antennas? (G9D11)

A. They present low impedance on all design frequencies
B. They must be used with an antenna tuner
C. They must be fed with open wire line
D. They have poor harmonic rejection

While multiband antennas offer the convenience of operating on multiple frequency bands with a single antenna, one disadvantage is that they tend to have poor harmonic rejection. Harmonics are unwanted signals that are multiples of the fundamental frequency, and they can cause interference or reduce the clarity of communication. Due to their design, multiband antennas may allow these harmonics to radiate more easily. This can be problematic in environments where clear and precise signal transmission is crucial.

Multi-band antennas are an excellent choice for those who need flexibility and versatility in their setups. They offer reliable performance across multiple frequency ranges with minimal complexity.

Chapter Summary

Specialized antennas expand the capabilities of amateur radio operators, addressing unique challenges and requirements that standard antennas cannot. From the versatile NVIS antenna for short-range HF communication to the noise-reducing Beverage antenna for long-distance reception, each type serves a specific purpose. Loop antennas offer compact and efficient solutions for low-noise environments, while mobile and trapped antennas bring convenience to operators on the move or with limited space. Multi-band antennas add flexibility, enabling operation across multiple frequencies with a single setup.

By understanding the design, operation, and applications of these specialized antennas, operators can tailor their stations to achieve optimal performance in a variety of scenarios. Whether tackling challenging environments, enhancing portability, or maximizing versatility, these antennas empower amateur radio enthusiasts to broaden their horizons and elevate their communication capabilities.

Chapter 17

Power

Solar Energy

SOLAR POWER IS AN increasingly popular energy solution for ham operators.

In a typical ham radio solar setup, solar panels are connected to batteries through a charge controller to manage power storage. This enables the station to continue operating even when the sun is not shining. Key components such as photovoltaic cells, charge controllers, and series diodes work together to ensure that energy is harvested and stored safely and efficiently. By understanding the technical aspects of solar panel configurations, voltages, and protection mechanisms, radio operators can maintain reliable power systems, even in challenging environments.

This section will delve into the essentials of solar power, including proper setup, precautions, and the benefits of using solar energy for ham radio operations.

Solar Panel Configurations and Precautions

Individual photovoltaic (PV) cells are typically connected in a series-parallel configuration in solar panels. This setup combines the benefits of both series and parallel wiring. When cells are connected in series, their voltages add up, which helps produce a higher overall voltage. When connected in parallel, their currents add up, providing a higher current output. This configuration ensures that a solar panel can generate enough voltage and current to be useful for charging batteries or powering equipment, making it ideal for various applications, including setups in off-grid environments.

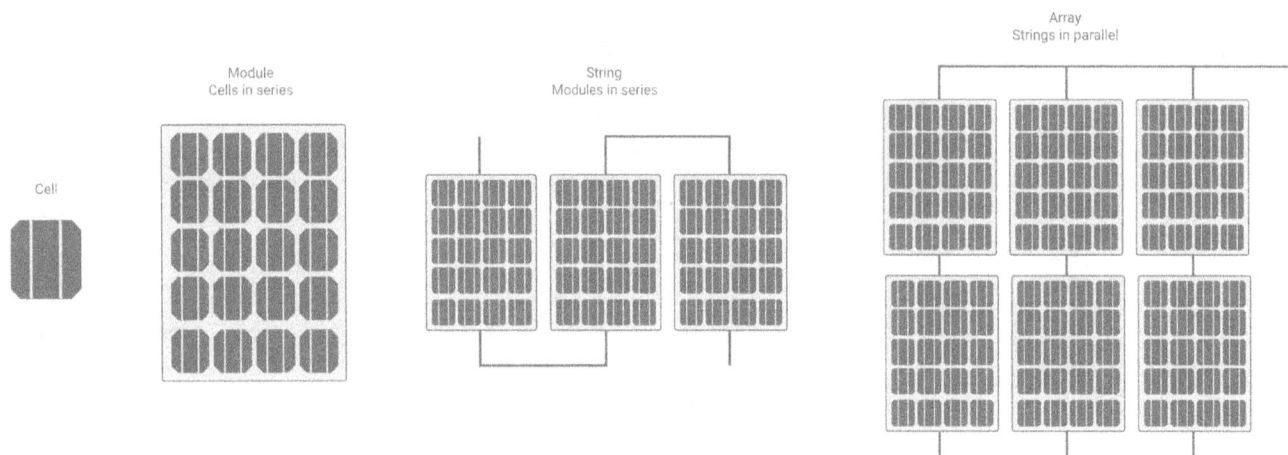

In what configuration are the individual cells in a solar panel connected together? (G4E08)

A. Series-parallel
B. Shunt
C. Bypass
D. Full-wave bridge

It is vital to use a charge controller when using a solar panel to charge a lithium-iron-phosphate battery. The charge controller regulates the voltage and current coming from the solar panel to prevent overcharging the battery, which could damage it or reduce its lifespan. Lithium iron phosphate batteries have specific charging requirements, and a charge controller helps manage these conditions, ensuring the battery is charged efficiently and safely without being overcharged.

What precaution should be taken when connecting a solar panel to a lithium iron phosphate battery? (G4E11)

A. Ground the solar panel outer metal framework
B. Ensure the battery is placed terminals-up
C. A series resistor must be in place
D. The solar panel must have a charge controller

Solar Panel Voltages and Protection

When fully illuminated, a single silicon photovoltaic (PV) cell generates an open-circuit voltage of approximately 0.5 volts DC (VDC). This measurement represents the maximum voltage the cell can produce when no current is flowing—essentially, when the circuit is open. The open-circuit voltage is a critical parameter for understanding the performance of a PV cell and designing solar power systems.

What is the approximate open-circuit voltage from a fully illuminated silicon photovoltaic cell? (G4E09)

A. 0.02 VDC
B. 0.5 VDC
C. 0.2 VDC
D. 1.38 VDC

Multiple PV cells are connected in series to achieve higher voltages suitable for practical applications. When connected this way, the voltages of individual cells are added together. However, it's important to note that the actual output voltage under load conditions (when current flows) will be slightly lower due to factors like internal resistance and power draw.

Diodes for Protection

A series diode is necessary when connecting a solar panel to a storage battery. The diode acts as a one-way valve, allowing current to flow from the solar panel to the battery during daylight while preventing the reverse flow of current at night or during low illumination. Without a diode, the battery could discharge through the panel, reducing its efficiency and potentially damaging the battery and solar panels over time.

Why should a series diode be connected between a solar panel and a storage battery that is being charged by the panel? (G4E10)

A. To prevent overload by regulating the charging voltage
B. To prevent discharge of the battery through the panel during times of low or no illumination
C. To limit the current flowing from the panel to a safe value
D. To prevent damage to the battery due to excessive voltage at high illumination levels

Solar panels are configured and protected to ensure they work efficiently and safely in various applications. Proper configurations and protective devices like diodes and charge controllers are crucial for maintaining performance and longevity, especially when using lithium iron phosphate batteries or other sensitive equipment.

Rectifiers

Rectifiers are essential components in power supply circuits. Their primary function is to convert alternating current into direct current, a necessary process for powering most electronic devices. Understanding how different rectifier types function allows radio operators to design power supplies that provide clean, consistent power. In this section, we'll explore full-wave and half-wave rectifiers, their characteristics, and how they impact the conversion of AC to DC in radio equipment.

What portion of the AC cycle is converted to DC by a half-wave rectifier? (G7A05)

A. 90 degrees
B. 180 degrees
C. 270 degrees
D. 360 degrees

In a half-wave rectifier, only one-half of the AC cycle (either the positive or the negative half) is used to produce DC. This means only <u>180 degrees</u> of the entire 360-degree AC waveform are converted into DC. During the other half of the cycle, no current flows. As a result, the output is less efficient and more "pulsed," creating a rougher DC signal that often requires additional filtering to be usable in sensitive circuits.

What is characteristic of a half-wave rectifier in a power supply? (G7A04)

A. Only one diode is required
B. The ripple frequency is twice that of a full-wave rectifier
C. More current can be drawn from the half-wave rectifier
D. The output voltage is two times the peak input voltage

A half-wave rectifier is simpler in design than a full-wave rectifier because it <u>requires only one diode</u> to block the negative portion of the AC signal, allowing only the positive half to pass through as DC. While this simplicity is an advantage in low-cost or low-power applications, the output is less efficient, with more ripple and less smooth DC than a full-wave rectifier.

Full-Wave Rectifiers

<u>A full-wave rectifier uses two diodes along with a center-tapped transformer</u> to convert the entire AC waveform (both positive and negative halves) into a usable DC output. The center tap allows the circuit to direct both halves of the AC cycle through the diodes, effectively doubling the frequency of the resulting DC pulses. This setup provides smoother power conversion than a half-wave rectifier, as it utilizes the full 360 degrees of the AC cycle. Full-wave rectifiers are more efficient and often used in higher-power applications requiring smoother DC.

Which type of rectifier circuit uses two diodes and a center-tapped transformer? (G7A03)

A. Full-wave
B. Full-wave bridge
C. Half-wave
D. Synchronous

A full-wave rectifier converts both the positive and negative halves of the AC cycle into DC, meaning it utilizes the entire 360-degree cycle. Compared to a half-wave rectifier, this results in a more continuous flow of current and less ripple in the output. The full-wave design ensures that the power supply delivers a more stable and efficient DC signal, which is particularly important for sensitive electronic equipment like radios.

What portion of the AC cycle is converted to DC by a full-wave rectifier? (G7A06)

A. 90 degrees
B. 180 degrees
C. 270 degrees
D. 360 degrees

When a full-wave rectifier is used without any filtering, the DC output is a series of pulses. These pulses occur at twice the frequency of the input AC signal because the full-wave rectifier inverts the negative half of the AC cycle, creating a positive pulse. This creates a pulsed DC signal, which can be smoothed using capacitors or inductors to create a more stable and continuous DC supply.

What is the output waveform of an unfiltered full-wave rectifier connected to a resistive load? (G7A07)

A. A series of DC pulses at twice the frequency of the AC input
B. A series of DC pulses at the same frequency as the AC input
C. A sine wave at half the frequency of the AC input
D. A steady DC voltage

Rectifiers are the backbone of power conversion in electronic systems, transforming AC into DC power. Understanding the difference between half-wave and full-wave rectification and the role of diodes and transformers helps ensure that the correct rectifier is chosen for specific applications. Full-wave rectifiers offer greater efficiency, while half-wave rectifiers are simpler but less effective. By mastering these concepts, operators can design more reliable power supplies that meet the needs of their radio equipment.

Power Supply

Switchmode

Power supplies are components that convert electrical energy into a form suitable for operating electronic devices. Two popular types of power supplies are switchmode and linear. Each has unique characteristics that make it ideal for different applications. Switchmode power supplies, known for their efficiency, utilize high-frequency operation, while linear power supplies offer a simpler design and lower noise but are generally less efficient. This section focuses on understanding the key differences between these two power supplies and how they impact performance and component selection in radio equipment.

Which of the following is characteristic of a switchmode power supply as compared to a linear power supply? (G7A08)

A. Faster switching time makes higher output voltage possible
B. Fewer circuit components are required
C. High-frequency operation allows the use of smaller components
D. Inherently more stable

Switchmode power supplies are highly efficient because they convert power by rapidly switching transistors on and off at high frequencies. This high-frequency operation allows the use of smaller components, such as transformers and capacitors, which reduces the overall size and weight of the power supply. Switchmode supplies also tend to generate less heat than linear supplies, making them more energy-efficient. However, because they operate at high frequencies, they can introduce more electrical noise, which may interfere with sensitive equipment unless adequately filtered.

Choosing between a switchmode and linear power supply depends on your equipment's needs. Switchmode power supplies offer significant efficiency and size advantages, particularly when high-frequency operation is necessary. However, they require proper noise filtering to avoid interference. Understanding these differences helps operators select the proper power supply for their radios, ensuring reliable and efficient performance in their setups.

Filtering

In a power supply, capacitors and inductors are components for smoothing out the ripple in the DC output after rectification. Capacitors store electrical energy and help stabilize voltage by filling in gaps in the DC signal while inductors resist changes in current. Together, they form a filter network that reduces unwanted AC ripple, ensuring the output is as close to pure DC as possible. This filtered DC is crucial for powering sensitive radio circuits without introducing noise or instability.

Which of the following components are used in a power supply filter network? (G7A02)

A. Diodes
B. Transformers and transducers
C. Capacitors and inductors
D. All these choices are correct

Protection

A power supply bleeder resistor is a safety component that ensures the capacitors in a power supply discharge their stored energy when the power is turned off. Capacitors can retain a charge for some time after the power is disconnected, which could be dangerous during maintenance or repair. The bleeder resistor slowly drains this stored energy, ensuring the system is safe to handle. Without a bleeder resistor, capacitors could deliver a shock if touched even after the equipment is turned off.

What is the function of a power supply bleeder resistor? (G7A01)

A. It acts as a fuse for excess voltage
B. It discharges the filter capacitors when power is removed
C. It removes shock hazards from the induction coils
D. It eliminates ground loop current

Chapter Summary

Powering your ham radio station effectively and safely is crucial for optimal performance, and understanding the diverse range of power sources and their components is key. From choosing between switchmode and linear power supplies to harnessing renewable energy like solar power, this chapter has equipped you with the knowledge to make informed decisions. We've explored the intricacies of rectifiers, filters, and power protection mechanisms, ensuring clean and stable energy for your setup. By mastering these concepts, you'll be well-prepared to maintain a reliable and efficient station, empowering your communication endeavors across all conditions.

Chapter 18

Electrical Safety

SAFETY HAS ALWAYS BEEN a cornerstone of amateur radio, emphasized heavily in the Technician's exam. Safety becomes even more critical as you progress to General Class, especially as you work with higher voltages, currents, and advanced setups. Electrical safety isn't just a precaution—protecting yourself, your equipment, and your surroundings is essential.

In ham radio, equipment often operates on mains electricity, presenting serious risks such as electric shock and potential fires if mishandled. Proper grounding is a cornerstone of safety, not only protecting your gear from electrical surges and static buildup but also reducing interference and ensuring a common electrical potential across all equipment. Using ground rods, copper wire, and lightning arrestors safeguards your station against these dangers.

Safe operation extends beyond grounding. Proper ventilation prevents overheating, fuses and circuit breakers protect against overloads, and a clean, organized workspace minimizes hazards. By following these best practices, you can enjoy amateur radio confidently and safely, knowing your setup is secure.

Power Safety

Emergency Generator Safety

Emergency generators provide backup power during outages; but they must be installed and operated safely. The primary concern is ensuring <u>the generator is used in a well-ventilated area</u> to prevent the buildup of dangerous exhaust gases like carbon monoxide. Carbon monoxide poisoning can occur quickly and without warning, so following safety guidelines is critical.

Which of the following is true of an emergency generator installation? (G0B09)

A. The generator should be operated in a well-ventilated area
B. The generator must be insulated from ground
C. Fuel should be stored near the generator for rapid refueling in case of an emergency
D. All these choices are correct

A generator should always be operated outdoors or in a well-ventilated area, never in enclosed spaces such as garages or basements. This prevents the accumulation of carbon monoxide, an odorless and lethal gas. By following this simple precaution, you can ensure that your generator provides power safely during an emergency without endangering the lives of those nearby.

Power Supply Interlock

Power supply interlocks are safety features designed to prevent accidental contact with dangerously high voltages. They ensure that the power is automatically disconnected when a device's cabinet is opened to avoid electrical shocks.

What is the purpose of a power supply interlock? (G0B12)

A. To prevent unauthorized changes to the circuit that would void the manufacturer's warranty
B. To shut down the unit if it becomes too hot
C. To ensure that dangerous voltages are removed if the cabinet is opened
D. To shut off the power supply if too much voltage is produced

A power supply interlock removes dangerous voltages when a power supply cabinet or similar device is opened. This safety mechanism prevents accidental exposure to high-voltage components, protecting operators from electric shock or injury. Always ensure that devices with interlocks are adequately maintained and functional to ensure safety during servicing.

National Electrical Code

Understanding the National Electrical Code (NEC) and Station Safety

The National Electrical Code (NEC) is a set of regulations and guidelines designed to ensure electrical safety in various installations, including amateur radio stations. Compliance with the NEC helps reduce the risk of electrical hazards, such as short circuits, overloads, and electrical fires. For ham radio operators, following these standards ensures that all electrical wiring, circuits, and connections in their station are safe and meet the required electrical codes. This is essential for personal safety and avoiding violations that could cause legal issues or lead to accidents.

Which of the following is covered by the National Electrical Code? (G0B06)

A. Acceptable bandwidth limits
B. Acceptable modulation limits
C. Electrical safety of the station
D. RF exposure limits of the human body

Minimum Wire Size for a 20-Amp Circuit (AWG 12)

The National Electrical Code specifies the minimum wire size for different circuit capacities to prevent overheating and fire hazards. For a 20-amp circuit, the minimum wire size allowed is AWG number 12. This gauge of wire is thick enough to handle the current safely without excessive heat buildup, ensuring that the wiring remains safe for prolonged use under load. Understanding wire sizes is crucial in amateur radio setups, especially when high-power equipment is used, as incorrect wiring can lead to failures or dangerous conditions.

According to the National Electrical Code, what is the minimum wire size that may be used safely for wiring with a 20-ampere circuit breaker? (G0B02)

A. AWG number 20
B. AWG number 16
C. AWG number 12
D. AWG number 8

AWG stands for American Wire Gauge, a standardized system used in North America to measure the thickness or diameter of electrically conductive wire. The AWG system assigns a number to the wire gauge, where higher numbers represent thinner wire and lower numbers represent thicker wire. For example, AWG 12 wire is thicker than AWG 18 wire.

The thickness of the wire affects its ability to carry current safely. Thicker wires (with lower AWG numbers) can carry more current without overheating, while thinner wires (with higher AWG numbers) are suitable for lower currents. This is important in electrical installations, including amateur radio setups, where different types of equipment may require wires of varying thicknesses to ensure safe operation.

Appropriate Fuse or Circuit Breaker Size for AWG 14 Wiring

In a circuit using AWG number 14 wiring, the maximum appropriate fuse or circuit breaker size is 15 amperes. This rating is selected to ensure that the wiring is not overloaded beyond its capacity, protecting the circuit from excessive current that could cause overheating. Using a fuse or breaker with a higher rating could allow dangerous current levels to flow through the wiring, leading to overheating or even fire. For amateur radio operators, properly sizing circuit protection devices is essential to maintain a safe operating environment.

> **Which size of fuse or circuit breaker would be appropriate to use with a circuit that uses AWG number 14 wiring? (G0B03)**
>
> A. 30 amperes
> B. 25 amperes
> C. 20 amperes
> **D. 15 amperes**

Hot Wires and Circuit Protection in a 240 VAC Circuit

In a 240 VAC circuit with four conductors, only the hot wires should be connected to fuses or circuit breakers. This ensures that if there is an overload or short circuit, the current flowing through the hot wires is interrupted, preventing damage to equipment or the risk of electrical shock. The neutral and ground wires should not be connected to fuses or breakers, as they are not intended to carry live current under normal operation. This understanding is important in maintaining safety when dealing with high-voltage equipment.

> **Which wire or wires in a four-conductor 240 VAC circuit should be attached to fuses or circuit breakers? (G0B01)**
>
> **A. Only the hot wires**
> B. Only the neutral wire
> C. Only the ground wire
> D. All wires

RF Safety

Amateur radio operators must ensure compliance with RF safety regulations to protect themselves and others from potential harm caused by radio frequency (RF) exposure. One key step is performing a routine RF exposure evaluation. This process assesses the amount of RF energy emitted by your station to determine whether it exceeds the Maximum Permissible Exposure (MPE) limits. If the evaluation reveals high-exposure areas, steps must be taken to prevent access to these areas, such as repositioning antennas, lowering power, or creating physical barriers.

Which of the following steps must an amateur operator take to ensure compliance with RF safety regulations? (G0A08)

A. Post a copy of FCC Part 97.13 in the station
B. Notify neighbors within a 100-foot radius of the antenna of the existence of the station and power levels
C. Perform a routine RF exposure evaluation and prevent access to any identified high exposure areas
D. All these choices are correct

RF Field Measurement

Operators use a calibrated field strength meter with a calibrated antenna to accurately measure RF field strength. This instrument helps measure the intensity of RF energy around the station, ensuring it remains within safe exposure limits. It is a crucial tool for verifying compliance with RF safety standards and identifying any concerns around the antenna.

What type of instrument can be used to accurately measure an RF field strength? (G0A09)

A. A receiver with digital signal processing (DSP) noise reduction
B. A calibrated field strength meter with a calibrated antenna
C. An SWR meter with a peak-reading function
D. An oscilloscope with a high-stability crystal marker generator

Actions for Excessive RF Exposure

Suppose an evaluation shows that RF energy from your station exceeds permissible limits for human absorption. In that case, you are required to take corrective action to prevent exposure. This might include limiting power output, increasing the distance between antennas and people, or adding warning signs. Failure to take such steps could result in harmful exposure to RF energy, which is regulated to prevent potential health risks.

What must you do if an evaluation of your station shows that the RF energy radiated by your station exceeds permissible limits for possible human absorption? (G0A05)

A. Take action to prevent human exposure to the excessive RF fields
B. File an Environmental Impact Statement (EIS-97) with the FCC
C. Secure written permission from your neighbors to operate above the controlled MPE limits
D. All these choices are correct

RF Exposure

Understanding RF Exposure and FCC Compliance

What stations are subject to the FCC rules on RF exposure? (G0A12)

A. All commercial stations; amateur radio stations are exempt
B. Only stations with antennas lower than one wavelength above the ground
C. Only stations transmitting more than 500 watts PEP
D. All stations with a time-averaged transmission of more than one milliwatt

RF exposure regulations ensure that amateur radio stations operate safely by preventing excessive exposure to radio frequency energy. According to the FCC rules, all stations transmitting more than one milliwatt of time-averaged power are subject to these RF exposure rules. This threshold applies broadly, covering most amateur radio stations, even those using relatively low power. Operators need to be aware of their station's power levels to ensure they comply with these regulations and minimize any potential risks associated with RF exposure.

If your station fails to meet the FCC's RF exposure exemption criteria, you are required to perform an RF Exposure Evaluation. This can be done following the FCC's OET Bulletin 65 guidelines, which provide detailed procedures for assessing RF safety. This evaluation ensures that your transmissions do not exceed the permissible limits for human exposure, protecting both yourself and others in the vicinity.

What must you do if your station fails to meet the FCC RF exposure exemption criteria? (G0A06)

A. Perform an RF Exposure Evaluation in accordance with FCC OET Bulletin 65
B. Contact the FCC for permission to transmit
C. Perform an RF exposure evaluation in accordance with World Meteorological Organization guidelines
D. Use an FCC-approved band-pass filter

How to Ensure Compliance with RF Exposure Regulations

Multiple ways exist to ensure that your station is FCC-compliant regarding RF exposure. These include calculations based on FCC OET Bulletin 65, computer modeling, or direct field strength measurements with calibrated equipment. By following these methods, you can verify that your station's RF emissions are within safe limits and avoid potential violations of FCC rules. Regular evaluations are necessary, especially when changing equipment or antenna setups.

How can you determine that your station complies with FCC RF exposure regulations? (G0A03)

A. By calculation based on FCC OET Bulletin 65
B. By calculation based on computer modeling
C. By measurement of field strength using calibrated equipment
D. All these choices are correct

RF Safety and Directional Antennas

When using a directional antenna, considering the potential for RF exposure to yourself and others, such as neighbors, is important. If an evaluation shows that your neighbor might experience more than the allowable RF exposure limit from the main lobe of your directional antenna, action must be taken. <u>The recommended precaution is to ensure that the antenna cannot be pointed in their direction when they are present.</u> This can involve physical restrictions on the rotation of the antenna or simply timing your transmissions when you know the area is clear.

What should be done if evaluation shows that a neighbor might experience more than the allowable limit of RF exposure from the main lobe of a directional antenna? (G0A10)

A. Change to a non-polarized antenna with higher gain
B. Use an antenna with a higher front-to-back ratio
C. Take precautions to ensure that the antenna cannot be pointed in their direction when they are present
D. All these choices are correct

Managing RF Exposure Risks

The FCC regulates RF exposure to protect individuals from excessive RF radiation, which can cause harmful effects like tissue heating. Directional antennas can concentrate RF energy in specific directions, potentially exposing people in those areas to higher-than-permissible levels. Therefore, amateur radio operators need to carefully evaluate and manage their antenna orientation to prevent unintentional exposure to others.

<u>One of the most significant effects of RF energy on the human body is its ability to heat body tissue.</u> Just like a microwave heats food, RF energy can cause tissue to heat up, potentially leading to burns or other damage if exposure is high or prolonged. This is why ensuring that exposure stays within regulated limits is critical for safety.

What is one way that RF energy can affect human body tissue? (G0A01)

A. It heats body tissue
B. It causes radiation poisoning
C. It causes the blood count to reach a dangerously low level
D. It cools body tissue

Understanding RF Exposure from a Transmitted Signal

Several factors must be considered to determine RF exposure from a transmitted signal. <u>These include duty cycle, frequency, and power density.</u> The duty cycle refers to the proportion of time the transmitter is actively sending a signal. For example, continuous transmissions have a higher duty cycle compared to intermittent transmissions. Frequency plays a role because RF exposure limits vary with different frequencies, as higher frequencies are typically more likely to penetrate human tissues. Power density, or the amount of RF energy in a given area, is another key factor, as higher power densities can increase exposure risk. Together, these factors help ensure that RF exposure stays within safe limits.

Which of the following is used to determine RF exposure from a transmitted signal? (G0A02)

A. Its duty cycle
B. Its frequency
C. Its power density
D. All these choices are correct

Modulation Duty Cycle and RF Exposure

The modulation duty cycle of a transmission also impacts RF exposure. <u>A lower duty cycle allows greater power levels to be transmitted while maintaining safe exposure limits.</u> Lower duty cycles mean the transmitter is off for a portion of the time, reducing overall exposure. Understanding how the duty cycle interacts with power levels helps operators maximize transmission efficiency while maintaining compliance with safety regulations.

> **What is the effect of modulation duty cycle on RF exposure? (G0A07)**
>
> **A. A lower duty cycle permits greater power levels to be transmitted**
> B. A higher duty cycle permits greater power levels to be transmitted
> C. Low duty cycle transmitters are exempt from RF exposure evaluation requirements
> D. High duty cycle transmitters are exempt from RF exposure requirements

Time Averaging in RF Radiation Exposure

<u>Time averaging refers to evaluating RF radiation exposure over a defined period</u>, rather than at a single instant. This approach smooths out fluctuations in exposure levels, considering both high and low power transmission times. For example, even if a high-power transmission occurs for a short time, if it is followed by periods of low-power or no transmission, the time-averaged exposure may remain within safe limits. This concept is essential for accurately assessing compliance with FCC guidelines, particularly for variable transmission power operations.

> **What does "time averaging" mean when evaluating RF radiation exposure? (G0A04)**
>
> A. The average amount of power developed by the transmitter over a specific 24-hour period
> B. The average time it takes RF radiation to have any long-term effect on the body
> C. The total time of the exposure
> **D. The total RF exposure averaged over a certain period**

Grounding & Bonding

Proper grounding and bonding are vital for both safety and performance. Grounding dissipates electrical surges, prevents static buildup, and shields equipment from lightning-induced damage. Ground rods should be driven deep into moist soil, and copper wire should connect all equipment to a common ground point. Bonding ensures all components share the same electrical potential, reducing the risk of shock and interference. Installing lightning arrestors adds an extra layer of protection, safeguarding your setup from harmful surges.

Grounding

Grounding refers to connecting the equipment to the earth, typically through a ground rod. Grounding provides a safe path for electrical energy to dissipate into the ground, like lightning strikes or electrical surges. The primary goal of grounding is safety, ensuring that any excess electrical charge doesn't remain on equipment and potentially harm operators.

Grounding all metal enclosures of station equipment is critical for safety. <u>The reason behind this is simple: it ensures that any hazardous voltages, such as those from power surges or faulty wiring, cannot build up on your equipment's chassis or metal surfaces.</u> If equipment isn't grounded correctly, there's a risk that dangerous voltages might appear on the metal case, which could lead to electrical shocks if touched. Proper grounding provides a path for those stray voltages to safely dissipate into the earth, reducing the risk of electrical hazards and improving the station's overall performance by reducing interference and hum.

Why must all metal enclosures of station equipment be grounded? (G4C12)

A. It prevents a blown fuse in the event of an internal short circuit
B. It prevents signal overload
C. It ensures that the neutral wire is grounded
D. It ensures that hazardous voltages cannot appear on the chassis

Consequences of Incorrect Grounding

What is a possible cause of high voltages that produce RF burns? (G4C05)

A. Flat braid rather than round wire has been used for the ground wire
B. Insulated wire has been used for the ground wire
C. The ground rod is resonant
D. The ground wire has high impedance on that frequency

A ground wire with high impedance can cause high voltages that produce RF burns at a specific frequency. When the ground wire has a high impedance, it does not effectively carry RF energy to the ground, allowing high voltage to accumulate on the surfaces of equipment enclosures. This can result in RF burns if you touch the equipment, especially at points where high RF voltage is present. The solution to this problem is to ensure that the ground wire is of low impedance at all operating frequencies by using proper grounding techniques, including shorter ground leads and bonding equipment together.

What is a possible effect of a resonant ground connection? (G4C06)

A. Overheating of ground straps
B. Corrosion of the ground rod
C. High RF voltages on the enclosures of station equipment
D. A ground loop

A resonant ground connection can also create problems by introducing high RF voltages onto the enclosures of station equipment. Resonance occurs when the length of the ground wire or system matches the wavelength of the operating frequency, creating a situation where RF energy is trapped or magnified in the system. This can lead to high RF voltages that are not safely dissipated, which increases the risk of RF burns and interference. To avoid this, ground connections should be made so they do not resonate with operating frequencies, ensuring safer and more stable station performance.

Lightening Ground Setup

A lightning ground setup is a system designed to safely direct lightning strikes into the ground. It prevents electrical surges from damaging equipment or posing a safety risk. This system is essential for protecting structures, antennas, and radio stations from lightning's destructive power.

Components of a Lightning Ground Setup:

1. **Lightning Rod or Air Terminal:** This is a metal rod installed at the highest point of a structure (e.g., the top of a radio tower or building). The purpose is to attract lightning strikes and provide a path for the energy to follow.

2. **Down Conductors:** Once the rod captures lightning, the down conductors carry the electrical energy from the rod to the ground. These conductors are typically made of copper or aluminum and are securely attached to the structure, ensuring a low-resistance path for the electricity to travel.

3. **Ground Rods (or Grounding Electrodes):** Ground rods direct energy into the ground. These rods are buried deep in the earth, providing a place for the electricity to dissipate safely. The rods are connected to the down conductor with strong mechanical clamps (not soldered connections) to ensure the joint doesn't fail under extreme heat conditions caused by lightning strikes.

4. **Bonding:** In the setup, all metal parts of the building and the station's equipment should be bonded or electrically connected to ensure that there are no potential differences between them. This prevents dangerous surges and reduces the chance of lightning damaging equipment.

A proper lightning grounding setup ensures that the electrical energy from a lightning strike is safely diverted into the earth, reducing the risk of fire, equipment damage, or harm to individuals in the area.

This setup is particularly critical for ham radio operators with tall antennas or towers, as these are prime targets for lightning strikes. Proper grounding protects equipment and ensures compliance with safety standards and local regulations.

Why should soldered joints not be used in lightning protection ground connections? (G4C07)

A. A soldered joint will likely be destroyed by the heat of a lightning strike
B. Solder flux will prevent a low conductivity connection
C. Solder has too high a dielectric constant to provide adequate lightning protection
D. All these choices are correct

When setting up a lightning protection system for a radio station, it is crucial to avoid soldered joints in ground connections. Soldered joints, while useful for electrical connections in many situations, are not suitable for lightning protection because of the intense heat generated by a lightning strike. The high temperature can melt or vaporize the solder, causing the joint to fail and compromising the entire grounding system. Instead, mechanical connections like clamps or compression fittings should be used, as they can withstand extreme conditions and remain intact during a lightning event.

Ground Fault Circuit Interrupter

A Ground Fault Circuit Interrupter (GFCI) is a critical safety device designed to protect individuals from electric shock. It monitors the flow of electricity through a circuit, specifically ensuring that the current flowing in the hot and neutral wires is balanced. If an imbalance occurs, such as when current flows from the hot wire to the ground, the GFCI immediately disconnects power to prevent injury or damage.

Which of the following conditions will cause a ground fault circuit interrupter (GFCI) to disconnect AC power? (G0B05)

A. Current flowing from one or more of the hot wires to the neutral wire
B. Current flowing from one or more of the hot wires directly to ground
C. Overvoltage on the hot wires
D. All these choices are correct

When the GFCI detects some electrical current flowing through an unintended path, such as a person or direct ground contact, it quickly interrupts the circuit, cutting off the electricity supply to prevent potential electric shock or harm. This is a vital safety feature for areas where electricity may come into contact with water, such as kitchens, bathrooms, and outdoor outlets.

Bonding

Bonding is the process of electrically connecting all metal components together to ensure they have the same electrical potential. This helps prevent differences in voltage between metal parts, which could otherwise result in sparks or dangerous electrical discharge. Bonding ensures that all pieces of equipment are at the same potential, reducing the risk of shock or interference.

Bond all equipment enclosures together to minimize RF "hot spots" in an amateur radio station. RF "hot spots" occur when uneven distributions of RF energy create areas of high radio frequency energy within the station, which can lead to interference or even electric shock. By bonding all equipment enclosures together, you ensure that all metal surfaces within the station are electrically connected. This helps to equalize the potential across all equipment, eliminating areas where stray RF energy could accumulate.

> **What technique helps to minimize RF "hot spots" in an amateur station? (G4C11)**
>
> A. Building all equipment in a metal enclosure
> B. Using surge suppressor power outlets
> **C. Bonding all equipment enclosures together**
> D. Placing low-pass filters on all feed lines

Ground Loop

A ground loop occurs when multiple ground paths in a system create a difference in potential between different points, leading to undesirable currents flowing through the system. This typically happens when equipment in an audio or radio setup is grounded at multiple points that are at different electrical potentials. The difference in potential causes current to flow through the ground connections, creating a loop. This can introduce hum, noise, or interference in the system, particularly in audio signals or transmitted radio waves.

Ground loops can interfere with transmitted and received signals in an amateur radio station or introduce audible "hum" in audio equipment. To avoid ground loops, bond all equipment to a single, common ground point and ensure that no multiple ground paths exist at different potentials.

By addressing ground loops, operators can maintain cleaner signals and reduce interference, ensuring better station performance.

> **How can the effects of ground loops be minimized? (G4C09)**
>
> A. Connect all ground conductors in series
> B. Connect the AC neutral conductor to the ground wire
> C. Avoid using lock washers and star washers when making ground connections
> **D. Bond equipment enclosures together**

A common technique for minimizing the effects of ground loops is bonding all the equipment enclosures together. By bonding, all the equipment is electrically connected to a single point, creating a unified ground reference. This eliminates differences in electrical potential between different devices, preventing unwanted currents from flowing between them and reducing the chance of interference. This technique ensures a clean and stable operating environment, improving both signal quality and equipment safety.

Making Repairs Safely

Lead-Tin Solder Safety

Lead-tin solder is commonly used in electronics, but mishandling it can lead to contamination. The primary danger from lead is its potential to contaminate food or surfaces, which can cause lead poisoning if ingested.

Which of the following is a danger from lead-tin solder? (G0B10)

A. Lead can contaminate food if hands are not washed carefully after handling the solder
B. High voltages can cause lead-tin solder to disintegrate suddenly
C. Tin in the solder can "cold flow," causing shorts in the circuit
D. RF energy can convert the lead into a poisonous gas

Lead-tin solder contains lead, a toxic substance that can be harmful if ingested. <u>After handling solder, it is crucial to wash your hands thoroughly before eating or touching food.</u> Failing to do so can cause lead particles to contaminate food, leading to long-term health risks like lead poisoning. Proper hygiene and cleanliness are essential to safely working with lead-based materials.

Chapter Summary

Safety is the foundation of every successful amateur radio operation, and this chapter has provided the essential knowledge and practices to ensure a secure environment for both you and your equipment. From grounding and bonding techniques to preventing RF exposure and managing high-voltage risks, these principles safeguard against accidents and maximize station efficiency. By adhering to safety standards, following the National Electrical Code, and performing regular evaluations, you can confidently explore the expanded opportunities that come with your General Class license. Prioritizing safety not only protects your station but also guarantees years of enjoyable, worry-free communication.

Chapter 19

Antenna Safety

The Technician Class exam introduced you to foundational antenna safety concepts, like grounding and basic lightning protection. As you advance to the General Class level, we'll take a deeper dive into these topics, expanding your knowledge to ensure you're prepared for more complex installations and operations. This chapter focuses on critical practices for antenna safety, including managing indoor transmitting setups, implementing advanced lightning protection systems, and optimizing grounding and bonding techniques. We'll also cover routine maintenance for high structures like towers, emphasizing the importance of preparation and proper equipment. By mastering these safety measures, you'll protect your station and yourself, equipping you to operate with confidence and expertise in any scenario.

Indoor Transmitting Antenna

When installing an indoor transmitting antenna, it's critical to consider safety, particularly concerning exposure to electromagnetic radiation. The Maximum Permissible Exposure (MPE) limits must not be exceeded in areas where people will be present. The FCC defines MPE limits to ensure individuals are not exposed to unsafe radio frequency (RF) energy levels. This is especially important indoors because there may be prolonged exposure to the RF fields generated by the antenna, which could be harmful if the levels are too high.

> **What precaution should be taken if you install an indoor transmitting antenna? (G0A11)**
>
> A. Locate the antenna close to your operating position to minimize feedline radiation
> B. Position the antenna along the edge of a wall to reduce parasitic radiation
> **C. Make sure that MPE limits are not exceeded in occupied areas**
> D. Make sure the antenna is properly shielded

Factors like the power output, frequency of operation, antenna placement, and distance from occupied areas must be considered to ensure MPE limits are not exceeded. Proper placement and minimizing transmission power are key ways to stay within safe exposure limits, protecting anyone in the room or nearby.

MPE is a crucial safety guideline for all amateur radio operations, particularly in confined spaces with a greater chance of high exposure. Indoor installations of transmitting antennas should always prioritize health and safety by ensuring RF exposure is below the limits set by regulatory authorities. These limits are designed to prevent long-term damage from exposure to RF energy, which can affect body tissues, especially during prolonged periods of use.

Lightning Protection

Lightning Protection Ground System

A key aspect of station safety is protecting equipment and operators from lightning strikes. System grounding is vital to safely direct any lightning strikes into the earth and away from station equipment.

Where should the station's lightning protection ground system be located? (G0B04)

A. As close to the station equipment as possible
B. Outside the building
C. Next to the closest power pole
D. Parallel to the water supply line

The station's lightning protection ground system should always be located outside the building to ensure that any potential strike is diverted away from interior equipment and occupants. This external ground system provides a path for the high-voltage energy to flow safely into the earth, minimizing the risk of electrical damage or fire inside the structure.

Lightning Arrestor Placement

Lightning arrestors are critical for protecting your equipment from voltage spikes caused by nearby lightning strikes. Correct placement of these arrestors can differentiate between a functional system and severe damage.

Where should lightning arrestors be located? (G0B13)

A. Where the feed lines enter the building
B. On the antenna, opposite the feed point
C. In series with each ground lead
D. At the closest power pole ground electrode

Lightning arrestors should be installed where the feed lines enter the building. This strategic placement ensures that any voltage surge due to a lightning strike is intercepted before it reaches indoor equipment, protecting your transceivers, amplifiers, and other sensitive devices from potential destruction.

Bonding Ground Rods for Lightning Protection

Proper bonding of ground rods is necessary to maintain a single, unified grounding system, ensuring safety and efficiency in lightning protection.

Which of the following is required for lightning protection ground rods? (G0B11)

A. They must be bonded to all buried water and gas lines
B. Bends in ground wires must be made as close as possible to a right angle
C. Lightning grounds must be connected to all ungrounded wiring
D. They must be bonded together with all other grounds

For effective lightning protection, <u>all ground rods must be bonded with all other grounds</u>, such as electrical system grounds and the station's RF grounds. This interconnected grounding system ensures that there are no voltage differences between grounds, minimizing the risk of damage from a lightning strike by allowing the energy to flow evenly and safely into the earth.

Routine Maintenance

Maintaining your antennas and feedlines is as critical as caring for your transceivers. Start by inspecting your antennas for any physical damage or corrosion. Look for signs of wear, such as rust on metal parts or cracks in plastic components. Ensuring that all elements are correctly aligned and tensioned is also critical. Misaligned elements can affect the performance of your antenna, reducing its efficiency. Use a level and tension gauge to check and adjust the alignment and tension as needed.

Cleaning and tightening connections and insulators is another vital task. Dirty or loose connections can lead to poor signal quality and increased SWR (Standing Wave Ratio). Use a wire brush or emery cloth to clean metal connectors and apply a small amount of dielectric grease to prevent corrosion. Tighten all bolts and screws to ensure a secure connection. Checking the insulators for cracks or dirt buildup is also necessary, as these can cause electrical leakage or arcing.

Climbing a Tower

Climbing towers, such as those used for ham radio antennas, require strict attention to safety to prevent accidents. An essential part is using a properly rated safety harness designed explicitly for tower work. <u>The harness should be inspected before each use to ensure it is within its service life, free from damage, and correctly rated for the climber's weight</u>. Proper use of a safety harness helps ensure that the climber is protected from severe injury in the event of a fall.

Which of these choices should be observed when climbing a tower using a safety harness? (G0B07)

A. Always hold on to the tower with one hand
B. Confirm that the harness is rated for the weight of the climber and that it is within its allowable service life
C. Ensure that all heavy tools are securely fastened to the harness
D. All these choices are correct

The harness must be able to support the climber's weight and remain in good condition. Harnesses can degrade over time, so confirming that they are within their service life ensures they will function properly if needed.

Lockout-Tagout Safety

The risk of accidental contact with electrical currents is high when climbing a tower that supports electrically powered devices. Before ascending the tower, it is crucial to follow lockout-tagout procedures, which involve <u>cutting off the electrical power supply and clearly tagging it to prevent accidental reactivation.</u> This eliminates the risk of electrocution while working on or near the tower.

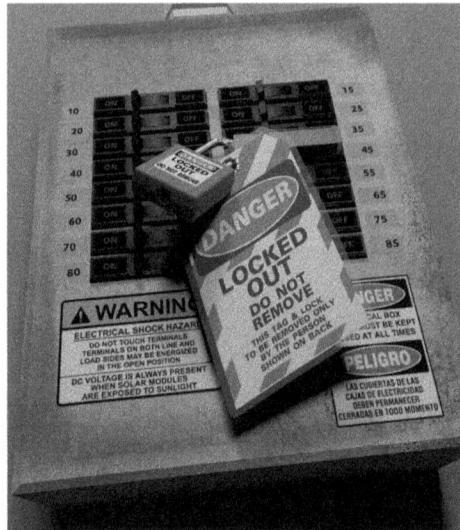

What should be done before climbing a tower that supports electrically powered devices? (G0B08)

A. Notify the electric company that a person will be working on the tower
B. Make sure all circuits that supply power to the tower are locked out and tagged
C. Unground the base of the tower
D. All these choices are correct

By ensuring all power circuits are disabled and tagged, tower climbers can safely work without encountering live electrical wires or components.

Chapter Summary

To conclude, ensuring antenna safety requires a careful balance of protective measures, regular maintenance, and adherence to established standards. From managing indoor antennas and preventing RF exposure to implementing robust lightning protection systems, each aspect of antenna safety is crucial for safeguarding the operator and equipment. Routine maintenance, especially when working with towers, demands meticulous attention to protocols to avoid accidents. By following these guidelines, radio operators can maintain a safe and reliable station that minimizes risk and maximizes performance, allowing them to focus on practical and enjoyable communication.

Operating Your Radio

MASTERING THE OPERATION OF your radio is key to becoming a skilled and confident ham operator. This section covers essential practices and knowledge, from proper operating procedures and understanding ham radio jargon to advanced CW operations and digital modes. You'll also learn how to manage interference, ensuring clear and reliable communication. Whether you're refining your skills or tackling new challenges, this section equips you with practical tools and techniques to ace your exam.

Chapter 20

Ham Radio Jargon

THIS CHAPTER BUILDS ON the Q codes you learned in the Technician's exam. Those were QRM and QSY. These three-letter codes act as shorthand to convey common phrases, making communication more efficient and precise. Q codes are understood universally in amateur radio, even across different languages, and they help streamline conversations, particularly in situations where speed and clarity are important. By mastering these signals, operators can enhance their CW proficiency and improve communication flow during their transmissions.

While all the codes can be useful, just like in the tech exam prep, let's focus on the ones you will need for the General exam. You will learn the rest over time through first-hand use.

Commonly Used Q Signals

What does the Q signal "QRL?" mean? (G2C04)

A. "Will you keep the frequency clear?"
B. "Are you operating full break-in?" or "Can you operate full break-in?"
C. "Are you listening only for a specific station?"
D. "Are you busy?" or "Is this frequency in use?"

The Q signal "QRL?" asks, "Are you busy?" or "Is this frequency in use?". It is a common courtesy for operators to send "QRL?" before starting a transmission to ensure they are not interfering with another ongoing communication. This helps prevent unintentional disruptions in radio communications, especially in busy bands.

What does the Q signal "QRV" mean? (G2C11)

A. You are sending too fast
B. There is interference on the frequency
C. I am quitting for the day
D. I am ready to receive

The Q signal "QRV" means "I am ready to receive." This signal is typically sent to indicate readiness to start receiving information from another station. It lets the operator know the receiver is standing by and prepared to copy the transmitted message.

What does the Q signal "QSL" mean? (G2C09)

A. Send slower
B. We have already confirmed the contact
C. I have received and understood
D. We have worked before

The Q signal "QSL" means "I have received and understood." This is an acknowledgment that the message has been successfully received. It's often used in live communication and the exchange of physical QSL cards to confirm contact between stations.

Troubleshooting and Adjustments Q Codes

What does the Q signal "QRN" mean? (G2C10)

A. Send more slowly
B. Stop sending
C. Zero beat my signal
D. I am troubled by static

The Q signal "QRN" means "I am troubled by static." This signal reports interference or noise in the receiver's environment, which may make it difficult to copy the signal clearly. Informing the other station of QRN can lead to adjusting transmission methods to overcome static or noise.

What should you do if a CW station sends "QRS?" (G2C02)

A. Send slower
B. Change frequency
C. Increase your power
D. Repeat everything twice

The Q signal "QRS?" is a request to "Send slower". This is used when the receiving station is having difficulty copying the transmission due to the speed of the code. It helps maintain clear and accurate communication by ensuring the message is understood.

Understanding the Q-codes helps with smooth and efficient communication. But you only need to memorize a few for the exam. Memorize those now, and the rest will come with use.

QRP Operation

What is QRP operation? (G2D10)

A. Remote piloted model control
B. Low-power transmit operation
C. Transmission using Quick Response Protocol
D. Traffic relay procedure net operation

QRP operation refers to low-power transmit operation in amateur radio. The term "QRP" comes from the Q code system, which originally meant "reduce power." QRP operators typically transmit with 5 watts or less for CW and digital modes or 10 watts or less for SSB. QRP aims to make efficient contacts using minimal power, challenging operators to improve their operating skills, antenna systems, and station efficiency.

This mode of operation is favored by many amateur radio operators who enjoy the challenge of making successful long-distance contacts with minimal power output. QRP enthusiasts often take pride in their ability to communicate over great distances with very little power, which requires skillful operating techniques, good propagation, and efficient antennas. The QRP community is a vibrant part of amateur radio, emphasizing efficiency and the thrill of low-power communication.

In contests and regular QSOs, QRP operators often note their low-power status, which adds to the challenge and sense of accomplishment when a contact is made under those conditions.

NATO Phonetic Alphabet

The NATO Phonetic Alphabet is an internationally recognized set of words used to clearly communicate letters over radio, telephone, or other audio transmissions. It helps reduce confusion, especially in noisy environments or when signals are weak. Each letter of the alphabet is represented by a distinct word, ensuring that the intended letters are still clear even if the transmission quality is poor. The military, aviation, emergency services, and amateur radio operators worldwide use the phonetic alphabet to enhance communication clarity.

In the NATO Phonetic Alphabet, each letter is assigned a word that sounds distinctly different from the others. For example, "A" is Alpha, "B" is Bravo, "C" is Charlie, and "D" is Delta. These specific words were chosen to minimize confusion and misunderstanding, especially in critical situations where clarity is paramount. This system allows radio operators to spell out words and call signs clearly, ensuring the message is accurately received regardless of signal interference or noise.

> **Which of the following are examples of the NATO Phonetic Alphabet? (G2D07)**
>
> A. Able, Baker, Charlie, Dog
> B. Adam, Boy, Charles, David
> C. America, Boston, Canada, Denmark
> **D. Alpha, Bravo, Charlie, Delta**

ALPHA—that's the key to this question. Remember that 'A' is alpha, and you have mastered this question.

Resources

Chapter Summary

In amateur radio, Q-codes and the NATO phonetic alphabet serve as valuable tools for clear and efficient communication. Q-codes simplify complex messages into easy-to-understand shorthand, while the NATO alphabet ensures accuracy in conveying critical information like call signs and locations. Mastering these tools enhances your operating skills and ensures you're understood in any situation, fostering better communication with operators worldwide.

Chapter 21

Voice Communication

Station Logs

KEEPING A STATION LOG is a common and recommended practice among amateur radio operators. A station log is a record of contacts made, equipment used, and important events related to the station's operation. While not always required by FCC rules, maintaining a log helps ensure that operators have a detailed account of their radio activities, which can be beneficial in a variety of circumstances, especially if questions arise about their transmissions or station operations.

Why do many amateurs keep a station log? (G2D08)

A. The FCC requires a log of all international contacts
B. The FCC requires a log of all international third-party traffic
C. The log provides evidence of operation needed to renew a license without retest
D. To help with a reply if the FCC requests information about your station

---

Many amateur radio operators keep a station log to have a detailed record of their communications. This can be especially useful if the FCC requests information about your station's activities. Suppose there's a need to verify that a transmission was made correctly. In that case, the log provides important details such as the frequency, mode, and time of communication. Additionally, keeping a log helps with personal record-keeping, tracking DX contacts, and organizing information for awards programs. A comprehensive log can also protect you in case of any dispute regarding interference or using specific frequencies.

Voice

When choosing a frequency to initiate a call, it is important to follow the voluntary band plan so much so that there is an exam question about it! Band plans are community-agreed guidelines that organize the amateur radio spectrum into specific sections based on usage, such as SSB, CW, or digital modes. By adhering to the band plan, operators help avoid interference with ongoing communications and ensure that they are operating in a manner that is respectful to others using the same spectrum. For example, calling a CQ on a frequency dedicated to digital modes with a voice transmission would cause unnecessary interference and go against commonly accepted practices.

Which of the following complies with commonly accepted amateur practice when choosing a frequency on which to initiate a call? (G2B07)

A. Listen on the frequency for at least two minutes to be sure it is clear
B. Identify your station by transmitting your call sign at least 3 times
C. Follow the voluntary band plan
D. All these choices are correct

Following the voluntary band plan also ensures that operators respect frequencies designated for specific purposes, like emergency communication or contesting. These guidelines help maintain an orderly use of the limited radio spectrum and make it easier for everyone to find and use the correct frequencies for their communication needs.

So, based on that info.

What is the voluntary band plan restriction for US stations transmitting within the 48 contiguous states in the 50.1 MHz to 50.125 MHz band segment? (G2B08)

A. Only contacts with stations not within the 48 contiguous states
B. Only contacts with other stations within the 48 contiguous states
C. Only digital contacts
D. Only SSTV contacts

The 50 MHz band, often referred to as the 6-meter band, is a unique section of the amateur radio spectrum that offers opportunities for long-distance communication, especially during periods of sporadic E propagation and solar activity. While the band is open to various types of communication, certain segments within it, such as the 50.1 MHz to 50.125 MHz range, have specific restrictions. These rules help ensure orderly and effective communication, particularly for DX (long-distance) contacts.

In the 50.1 MHz to 50.125 MHz band segment, US stations within the 48 contiguous states are limited to making contact only with stations outside the 48 contiguous states. This voluntary band plan restriction is designed to reserve this band portion for long-distance or DX communications, encouraging contacts with stations in Alaska, Hawaii, or other countries. By reserving this segment for DX contacts, operators avoid crowding the band with local communications, thus maximizing the chances for making rare, long-distance connections.

This restriction also helps reduce interference in a portion of the 6-meter band that is highly sought after for DX activity, especially during favorable propagation conditions. Remembering this for the exam is key, as it highlights how band segments are managed to ensure fair use and promote long-distance communication on specific frequencies.

Access to Frequencies

All operators share the same frequency spectrum, and there are no reserved or private frequencies for individual operators or groups. The principle of frequency sharing is fundamental to amateur radio, meaning that frequencies are generally available on a first-come, first-served basis. The only exception to this rule is during emergencies, where communication for emergency services takes precedence over regular operations.

When it comes to accessing frequencies, it is important to remember that, except during emergencies, no amateur station has priority access to any frequency. Any operator may use an available frequency, provided it is not currently in use. In the event that two stations wish to use the same frequency, it is customary to share or find another available frequency. However, if a frequency is used for an emergency or disaster-related communication, those transmissions take priority over all others.

Which of the following is true concerning access to frequencies? (G2B01)

A. Nets have priority
B. QSOs in progress have priority
C. Except during emergencies, no amateur station has priority access to any frequency
D. Contest operations should yield to non-contest use of frequencies

This principle ensures fair access for all operators and promotes cooperation within the amateur radio community. It is also an important guideline to recall for the exam, as understanding the rules for frequency access helps maintain smooth operations on the airwaves and prevents disputes between operators.

Receiver Bandwidth

Receiver bandwidth refers to the range of frequencies that a receiver can process at one time. Different operating modes like CW, SSB, and FM have unique bandwidth requirements. It's important to match the receiver's bandwidth to the operating mode's bandwidth for optimal performance. This ensures that the receiver only processes the necessary portion of the radio spectrum, helping to minimize noise and interference and allowing for clearer reception of the intended signal.

Matching the receiver bandwidth to the bandwidth of the operating mode is important because it results in the best signal-to-noise ratio (SNR). The signal-to-noise ratio measures how much the desired signal is received compared to background noise. If the receiver's bandwidth is too wide, it picks up unnecessary noise and interference from frequencies outside of the desired signal range, reducing the clarity of the transmission. By narrowing the bandwidth to just the amount needed for the mode you're using, you filter out unwanted noise, improving the clarity and overall quality of the received signal.

Why is it good to match receiver bandwidth to the bandwidth of the operating mode? (G8B09)

A. It is required by FCC rules
B. It minimizes power consumption in the receiver
C. It improves impedance matching of the antenna
D. It results in the best signal-to-noise ratio

To remember this for the exam, think of matching the bandwidth like "zooming in" on just the right portion of the signal, reducing noise, and making the signal stand out more clearly. This technique helps ensure the best listening experience with minimal interference.

Frequency Etiquette

Proper frequency etiquette is crucial to being a responsible amateur radio operator. Before transmitting on a frequency, operators should always check if the frequency is clear to avoid harmful interference with ongoing communications. Additionally, all amateur radio operators are expected to follow specific emergency protocols, such as prioritizing stations in distress. Furthermore, good amateur practice involves handling unexpected interference during communications and effectively managing nets to ensure smooth operation, even in poor conditions.

Let's review some radio etiquette that you will see on the exam.

How can you avoid harmful interference on an apparently clear frequency before calling CQ on CW or phone? (G2B06)

A. Send "QRL?" on CW, followed by your call sign; or, if using phone, ask if the frequency is in use, followed by your call sign
B. Listen for 2 minutes before calling CQ
C. Send the letter "V" in Morse code several times and listen for a response, or say "test" several times and listen for a response
D. Send "QSY" on CW or if using phone, announce "the frequency is in use," then give your call sign and listen for a response

To ensure that you are not interfering with ongoing communications, it is good practice to first check if the frequency is in use. For CW (Morse code), send the code "QRL?" followed by your call sign. On phone (voice), simply ask, "Is the frequency in use?" followed by your call sign. This helps prevent accidental interruptions of communications that may not be immediately audible.

What is the first thing you should do if you are communicating with another amateur station and hear a station in distress break in? (G2B02)

A. Inform your local emergency coordinator
B. Acknowledge the station in distress and determine what assistance may be needed
C. Immediately decrease power to avoid interfering with the station in distress
D. Immediately cease all transmissions

What is the first thing you should do if you are communicating with another amateur station and hear a station in distress break in? If a station in distress breaks in during your communication, the first and most important step is to acknowledge the station in distress. You should then determine what assistance may be required. In emergency situations, the distress signal takes priority, and all other communication should cease to allow assistance.

What is good amateur practice if propagation changes during a contact creating interference from other stations using the frequency? (G2B03)

A. Advise the interfering stations that you are on the frequency and that you have priority
B. Decrease power and continue to transmit
C. Attempt to resolve the interference problem with the other stations in a mutually acceptable manner
D. Switch to the opposite sideband

If propagation conditions change during a contact, leading to interference from other stations, it is good practice to attempt to resolve the interference issue in a mutually acceptable manner. This may involve moving to a new frequency or adjusting your equipment settings to minimize interference. Cooperation is key to ensuring all operators can continue communicating effectively.

Which of the following is good amateur practice for net management? (G2B10)

A. Always use multiple sets of phonetics during check-in
B. Have a backup frequency in case of interference or poor conditions
C. Transmit the full net roster at the beginning of every session
D. All these choices are correct

A key aspect of effective net management is having a backup frequency in case of interference or poor conditions. This ensures that the net can continue without interruption if the main frequency becomes unusable. Planning a backup frequency ensures successful and efficient net operation.

These principles and always prioritizing emergency communication help maintain the integrity and usefulness of the amateur radio spectrum for everyone.

Breaking Into a Phone Contact

It's common for operators to want to join or break into an ongoing conversation, also known as a QSO. It's important to follow established etiquette to avoid disrupting the conversation or causing interference when trying to do so. In phone (voice) contacts, there is a simple and polite method for breaking into a conversation, allowing you to join the QSO while maintaining good operating practices.

The recommended way to break into a phone contact is to say your call sign once. This brief and straightforward method alerts the operators in the ongoing conversation that someone wishes to join without interrupting their communication flow. By simply stating your call sign, you respect the conversation and give others the chance to invite you to join when they're ready.

What is the recommended way to break into a phone contact? (G2A08)

A. Say "QRZ" several times, followed by your call sign
B. Say your call sign once
C. Say "Breaker Breaker"
D. Say "CQ" followed by the call sign of either station

This practice ensures that you follow proper on-air etiquette and allows for smooth and courteous transitions when operators wish to join a QSO. Avoid saying "break" or other words that may be confusing or disruptive. Simply state your call sign clearly and wait for an acknowledgment.

And I appreciate the exam creator's sense of humor in including "breaker breaker!"

Another exam question similar to this:

Which of the following indicates that you are looking for an HF contact with any station? (G2D05)

A. Sign your call sign once, followed by the words "listening for a call" -- if no answer, change frequency and repeat
B. Say "QTC" followed by "this is" and your call sign -- if no answer, change frequency and repeat
C. Repeat "CQ" a few times, followed by "this is," then your call sign a few times, then pause to listen, repeat as necessary
D. Transmit an unmodulated carried for approximately 10 seconds, followed by "this is" and your call sign, and pause to listen -- repeat as necessary

---

Now, don't let the HF part of this question fool you. When you're open to contact with any station on any band, you send out a CQ call. "CQ" is a general call for communication, meaning you're looking for anyone to respond. This call is standard practice across the ham radio community and is crucial for starting conversations with fellow operators worldwide. It's an efficient way to reach out, especially when you're not looking for any particular station or operator.

To make a CQ call, the operator should follow a structured format: Repeat "CQ" a few times, followed by "this is," and then your call sign a few times. After sending the call, pause for a moment to listen for any responses. If no one replies, you can repeat the process until you establish contact with another operator. This method ensures your call is heard clearly and provides ample opportunity for others to respond. It's also important to ensure that the frequency is clear before making your call by asking, "Is this frequency in use?" to avoid interference.

DX Calls

In amateur radio, operators often seek to establish DX contacts, which means communicating with stations in distant locations, typically outside their own country or region. When an operator calls CQ DX, they are specifically requesting responses from stations that are far away, particularly outside of their immediate geographical area. Operators in the contiguous 48 states (the lower 48 in the US) are looking to contact stations outside this region.

When a station in the contiguous 48 states calls "CQ DX," only stations outside the lower 48 should respond. This is because the calling station is specifically seeking long-distance or international contacts, and responding from within the lower 48 would not fulfill the DX objective. Stations from Alaska, Hawaii, US territories, and other countries are typically the intended responders.

Generally, who should respond to a station in the contiguous 48 states calling "CQ DX"? (G2A11)

A. Any caller is welcome to respond
B. Only stations in Germany
C. Any stations outside the lower 48 states
D. Only contest stations

If you're in the lower 48, you should wait for another opportunity to call unless you are located outside of that region. This practice respects the operator's goal of seeking DX communication and follows proper on-air etiquette.

Single Sideband

Single Sideband (SSB) is a widely used voice communication mode in HF amateur bands. It is a variation of amplitude modulation (AM) that offers several advantages, including greater power efficiency and reduced bandwidth usage. Instead of transmitting both sidebands and the carrier as in traditional AM, SSB transmits only one sideband (upper or lower). In contrast, the other sideband and carrier are suppressed. This makes SSB an ideal mode for long-distance communication on crowded HF bands.

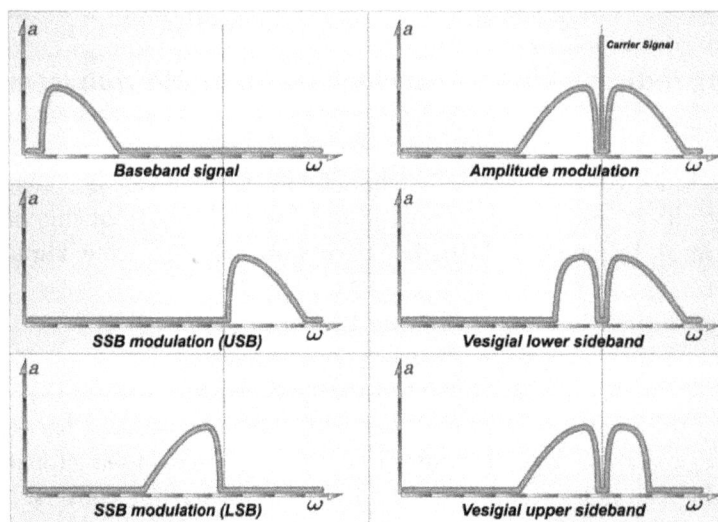

The most commonly used voice mode in HF amateur bands is single-sided band (SSB). SSB is preferred due to its power and bandwidth efficiency, making it ideal for long-distance contacts over the HF spectrum. By transmitting only one sideband, SSB allows for clearer communication with less interference.

Which mode of voice communication is most commonly used on the HF amateur bands? (G2A05)

A. Frequency modulation
B. Double sideband
C. Single sideband
D. Single phase modulation

The key characteristic of Single Sideband is that only one sideband is transmitted. In contrast, the other sideband and the carrier are suppressed. This selective transmission reduces the bandwidth used and focuses the power on a single sideband, allowing for more efficient use of the radio spectrum and minimizing interference with other signals.

Which of the following statements is true of single sideband (SSB)? (G2A07)

A. Only one sideband and the carrier are transmitted; the other sideband is suppressed
B. Only one sideband is transmitted; the other sideband and carrier are suppressed
C. SSB is the only voice mode authorized on the 20-, 15-, and 10-meter amateur bands
D. SSB is the only voice mode authorized on the 160-, 75-, and 40-meter amateur bands

One of the main advantages of SSB over other analog voice modes like AM or FM is that it uses less bandwidth and provides greater power efficiency. This means more of the transmitter's power is dedicated to the signal being sent, resulting in clearer, stronger signals over longer distances. Additionally, the reduced bandwidth allows more signals to fit into a given portion of the HF spectrum.

Which of the following is an advantage of using single sideband, as compared to other analog voice modes on the HF amateur bands? (G2A06)

A. Very high-fidelity voice modulation
B. Less subject to interference from atmospheric static crashes
C. Ease of tuning on receive and immunity to impulse noise
D. Less bandwidth used and greater power efficiency

SSB is the dominant mode for voice communication on HF amateur bands because it combines bandwidth conservation with power efficiency. Operators can communicate more effectively by suppressing the carrier and one sideband, especially in long-distance and crowded band conditions.

Automatic Level Control

In single sideband (SSB) transmissions, Automatic Level Control (ALC) ensures that the transmitter operates within safe power levels without causing distortion. ALC helps regulate the audio signal input to prevent overdriving the transmitter, which could result in signal distortion or interference. Adjusting the ALC properly helps maintain clear and clean transmissions, optimizing the performance of your SSB transceiver.

The transmit audio or microphone gain control is typically adjusted to properly set the ALC on a single sideband transceiver. This adjustment controls the audio level fed into the transmitter. The operator can ensure that the signal remains within the proper ALC range by increasing or decreasing the microphone gain. Too high a microphone gain can push the ALC beyond its limits, causing distortion, while too low a gain may result in weak transmissions.

What control is typically adjusted for proper ALC setting on a single sideband transceiver? (G2A12)

A. RF clipping level
B. Transmit audio or microphone gain
C. Antenna inductance or capacitance
D. Attenuator level

For optimal performance, the ALC should be adjusted to reach the recommended level during normal speech peaks, ensuring clear transmission and avoiding unnecessary over-modulation or distortion. This makes the transmit audio or microphone gain the key control to adjust for achieving the correct ALC settings.

Lower Sideband

Lower Sideband (LSB) is a type of Single Sideband (SSB) modulation that is widely used in voice communications on certain HF bands. In amateur radio, specific sidebands are typically used on different bands to standardize operations and avoid interference. For the 160-meter, 75-meter, and 40-meter bands, the lower sideband (LSB) is the preferred mode of voice communication. This practice ensures consistency and better signal management in the crowded HF bands, particularly for amateur operators.

The most commonly used mode for voice communication on the 160-, 75-, and 40-meter bands is Lower Sideband (LSB). In LSB transmission, only the lower sideband of the signal is transmitted, while the carrier and upper sideband are suppressed. This saves bandwidth and improves power efficiency, allowing for clearer communication, particularly over long distances.

Which mode is most commonly used for voice communications on the 160-, 75-, and 40-meter bands? (G2A02)

A. Upper sideband
B. Lower sideband
C. Suppressed sideband
D. Double sideband

And why is that?

Most amateur stations use Lower Sideband (LSB) on the 160-, 75-, and 40-meter bands because it is a commonly accepted amateur practice. This convention has been established to maintain consistency and order on the airwaves. By adhering to this practice, operators reduce confusion and interference when tuning into different bands, as they know what mode to expect for voice communications.

Why do most amateur stations use lower sideband on the 160-, 75-, and 40-meter bands? (G2A09)

A. Lower sideband is more efficient than upper sideband at these frequencies
B. Lower sideband is the only sideband legal on these frequency bands
C. Because it is fully compatible with an AM detector
D. It is commonly accepted amateur practice

Historically, Lower Sideband (LSB) was adopted for frequencies below 9 MHz, including the 160-, 75-, and 40-meter bands. Upper Sideband (USB) became the norm for frequencies above 9 MHz. This separation in sideband usage helps maintain efficiency and minimizes conflicts between operators on adjacent frequencies, making it easier to communicate clearly.

Lower Sideband (LSB) use on the 160-, 75-, and 40-meter bands is part of a long-standing tradition in amateur radio. By standardizing the use of sidebands, the amateur radio community has created a more predictable operating environment. This division between LSB on lower frequency bands and USB on higher frequency bands is part of accepted best practices, making it easier for operators to communicate effectively without unnecessary interference.

Now, some questions.

How close to the lower edge of a band's phone segment should your displayed carrier frequency be when using 3 kHz wide LSB? (G4D10)

A. At least 3 kHz above the edge of the segment
B. At least 3 kHz below the edge of the segment
C. At least 1 kHz below the edge of the segment
D. At least 1 kHz above the edge of the segment

When using LSB, especially near the lower edge of a band's phone segment, it's important to position your carrier frequency carefully. Since an LSB signal occupies frequencies below the carrier, your carrier frequency must be at least 3 kHz above the edge of the segment to ensure your signal doesn't extend outside the authorized band. For example, if the lower edge of the segment is 7.170 MHz, your carrier frequency should be no lower than 7.173 MHz when using a 3 kHz wide signal. This practice helps avoid transmitting outside the allocated band and ensures compliance with FCC regulations.

What frequency range is occupied by a 3 kHz LSB signal when the displayed carrier frequency is set to 7.178 MHz? (G4D08)

A. 7.178 MHz to 7.181 MHz
B. 7.178 MHz to 7.184 MHz
C. 7.175 MHz to 7.178 MHz
D. 7.1765 MHz to 7.1795 MHz

When you're transmitting on Lower Sideband (LSB), the signal occupies a range of frequencies below the displayed carrier frequency. For example, if your radio's carrier frequency is set to 7.178 MHz, a 3 kHz LSB signal will occupy a frequency range from 7.175 MHz to 7.178 MHz. This is because the LSB signal extends downward from the carrier frequency by 3 kHz. Understanding this relationship between your displayed frequency and the actual bandwidth occupied is critical for ensuring you do not interfere with signals on neighboring frequencies, especially near the band edges.

Both of these questions highlight the importance of understanding how LSB signals use bandwidth and how to properly position your carrier frequency on the band. To avoid interference or rule violations, you must maintain awareness of your signal's actual footprint, especially when operating near the band's edges.

Upper Sideband

In amateur radio, Upper Sideband (USB) is the most commonly used mode for Single Sideband (SSB) voice communication on VHF and UHF bands, as well as frequencies 14 MHz and higher. This convention arose to standardize operations, prevent interference, and maintain clear communication across the spectrum. By using USB on higher frequencies, operators ensure that their transmissions are compatible with standard practices, making communication easier and more reliable on these bands.

The standard mode for SSB voice communication on VHF and UHF bands is Upper Sideband (USB). USB is preferred for frequencies in these bands due to historical and technical reasons, including better signal clarity and ease of operation performance.

Which mode is most commonly used for SSB voice communications in the VHF and UHF bands? (G2A03)

A. Upper sideband
B. Lower sideband
C. Suppressed sideband
D. Double sideband

For voice communications on frequencies 14 MHz and higher, such as the 20-meter band, Upper Sideband is the most commonly used mode. This practice ensures uniformity and prevents confusion, as operators know that USB is the expected mode on these higher frequency bands.

Which mode is most commonly used for voice communications on frequencies of 14 MHz or higher? (G2A01)

A. Upper sideband
B. Lower sideband
C. Suppressed sideband
D. Double sideband

Upper Sideband is also the most common mode for voice communication on bands like 17 meters and 12 meters. This is part of the broader convention in amateur radio that USB is used on bands above 9 MHz, providing consistency across various frequencies.

Which mode is most commonly used for voice communications on the 17- and 12-meter bands? (G2A04)

A. Upper sideband
B. Lower sideband
C. Suppressed sideband
D. Double sideband

In amateur radio, the choice of sideband mode for voice communications is dictated by long-standing conventions. On frequencies 14 MHz and higher, including VHF and UHF bands, USB is the standard for SSB voice communication. This ensures that operators are transmitting in a predictable manner, reducing interference and improving the quality of communications. USB is particularly advantageous for these frequencies due to its more efficient use of bandwidth and clearer signal quality, especially for crowded bands.

By following the convention of using Upper Sideband on bands like 17 meters, 12 meters, and higher frequencies, operators maintain compatibility with the broader amateur radio community, ensuring that their transmissions are understood and that communication can happen smoothly worldwide.

Now, some questions.

How close to the upper edge of a band's phone segment should your displayed carrier frequency be when using 3 kHz wide USB? (G4D11)

A. At least 3 kHz above the edge of the band
B. At least 3 kHz below the edge of the band
C. At least 1 kHz above the edge of the segment
D. At least 1 kHz below the edge of the segment

When transmitting in USB, the signal extends upward from the carrier frequency. For a 3 kHz wide USB signal, the carrier frequency must be at least 3 kHz below the upper edge of the phone segment to prevent transmitting outside the authorized band. For example, if the upper edge of the band is 14.350 MHz, your carrier frequency should be no higher than 14.347 MHz. This ensures that the 3 kHz signal doesn't exceed the band limits, which helps you comply with regulations and avoid interfering with stations on adjacent frequencies.

What frequency range is occupied by a 3 kHz USB signal with the displayed carrier frequency set to 14.347 MHz? (G4D09)

A. 14.347 MHz to 14.647 MHz
B. 14.347 MHz to 14.350 MHz
C. 14.344 MHz to 14.347 MHz
D. 14.3455 MHz to 14.3485 MHz

A 3 kHz USB signal occupies a frequency range starting from the displayed carrier frequency and extending 3 kHz upward. For instance, if your radio is set to 14.347 MHz, the actual signal will occupy the range from 14.347 MHz to 14.350 MHz. It's important to understand how USB signals extend above the carrier frequency to avoid operating too close to the band edge and accidentally transmitting outside the allowed frequency range.

Both concepts relate to the importance of properly positioning your carrier frequency when using USB. Understanding the bandwidth of your signal ensures you stay within the band limits and avoid interfering with other stations.

Bandwidth

When selecting a frequency for SSB transmissions, it is important to maintain a separation of 2 kHz to 3 kHz from other stations. SSB signals typically use bandwidths of around 2.5 kHz, so maintaining this distance helps avoid interference. If stations are spaced too closely, signals can bleed into adjacent frequencies, causing distortion or difficulty hearing both transmissions.

When selecting an SSB transmitting frequency, what minimum separation should be used to minimize interference to stations on adjacent frequencies? (G2B05)

A. 5 Hz to 50 Hz
B. 150 Hz to 500 Hz
C. 2 kHz to 3 kHz
D. Approximately 6 kHz

Following this separation guideline, operators can transmit clean, undisturbed signals while minimizing the risk of interference for other stations. This is especially important in busy bands where multiple operators are transmitting simultaneously. Ensuring adequate separation is key to maintaining orderly and efficient communication on the airwaves.

Modulation and Sideband Selection

In radio communications, modulation allows a carrier signal to carry information by modifying its properties. Balanced modulators are integral to this process, especially when creating signals used in single-sideband transmission. When a signal is modulated, multiple "sidebands" are created, and selecting the appropriate one is crucial for efficient communication. This section explores how modulation circuits work and how sidebands are selected and extracted for effective transmission and reception.

What circuit is used to select one of the sidebands from a balanced modulator? (G7C01)

A. Carrier oscillator
B. Filter
C. IF amplifier
D. RF amplifier

A filter selects one of the sidebands from a balanced modulator. Balanced modulation generates both upper and lower sidebands around the carrier frequency. However, in single-sideband (SSB) transmissions, it's crucial to filter out one of these sidebands to conserve bandwidth and power. The filter allows only the desired sideband, either the upper or lower, to pass through and block the other sideband, ensuring cleaner transmission and reception.

What output is produced by a balanced modulator? (G7C02)

A. Frequency modulated RF
B. Audio with equalized frequency response
C. Audio extracted from the modulation signal
D. Double-sideband modulated RF

A balanced modulator creates a double-sideband (DSB) signal, meaning it produces two identical sidebands on either side of the carrier frequency. Importantly, the carrier itself is suppressed in a balanced modulator, allowing only the modulated sidebands to be transmitted. This makes it an efficient way to prepare signals for transmission, although further filtering (to extract just one sideband) is required for single-sideband operation.

How is a product detector used? (G7C04)

A. Used in test gear to detect spurious mixing products
B. Used in transmitter to perform frequency multiplication
C. Used in an FM receiver to filter out unwanted sidebands
D. Used in a single sideband receiver to extract the modulated signal

---

A product detector is a specialized circuit used in single-sideband receivers to demodulate signals. It works by mixing the incoming SSB signal with a locally generated signal to retrieve the original modulated information. This process allows the receiver to extract the voice or data information from the transmitted RF signal, making it essential for SSB and other types of amplitude-modulated communication systems.

Understanding how modulation circuits function, particularly in relation to sideband selection and signal extraction, is crucial for optimizing the efficiency of radio communications. Whether using filters to select the correct sideband or employing a product detector to recover transmitted data, each component ensures signals are transmitted clearly and efficiently.

Chapter Summary

This chapter has taken you through the critical aspects of voice communication in amateur radio, from selecting appropriate frequencies and understanding modulation to mastering sideband conventions and practicing proper on-air etiquette. Voice communication lies at the heart of the amateur radio experience, fostering global connections and ensuring clear, efficient exchanges. By adhering to band plans, respecting frequency etiquette, and understanding technical details like bandwidth and sideband usage, you not only improve your communication skills but also contribute to the smooth operation of HF bands. With these tools, you are well-equipped to navigate the airwaves confidently, ensuring your transmissions are effective, compliant, and a pleasure for others to hear.

Chapter 22

Continuous Wave

MORSE CODE, OR CONTINUOUS Wave (CW), remains one of amateur radio's most iconic and effective communication modes. Its simplicity, reliability, and ability to cut through noise make it a powerful tool for ham radio operators, especially in challenging conditions. For General Class operators, mastering CW goes beyond just learning the code—it's about understanding the techniques, protocols, and tools that make CW operation seamless and efficient.

In this chapter, we'll delve into key aspects of CW operation, from choosing the right frequency to understanding advanced practices like full break-in (QSK) operation and zero beating. You'll learn the meaning of common CW terms, the use of prosigns, and how to adjust your sending speed to match other operators. We'll also cover the role of equipment like electronic keyers and explore techniques to enhance signal clarity. Whether you're a seasoned CW operator or just beginning to explore its possibilities, this chapter will equip you with the knowledge and skills to confidently incorporate CW into your ham radio experience.

Basic Operations

Zero beat refers to the process of matching the transmit frequency to the frequency of a received signal. When two frequencies align, the tone becomes zero or minimal, ensuring both stations transmit on the same frequency, optimizing signal clarity.

What does the term "zero beat" mean in CW operation? (G2C06)

A. Matching the speed of the transmitting station
B. Operating split to avoid interference on frequency
C. Sending without error
D. Matching the transmit frequency to the frequency of a received signal

---

Zero beat is the process of tuning your receiver to match the frequency of a CW signal, allowing you to hear the clearest and most precise tone. This ensures your transmission is aligned with the station you're communicating with, improving signal clarity and minimizing interference.

As you tune into a CW signal, the pitch changes depending on how close you are to the correct frequency. The goal is to adjust your tuning until the tone matches a standard pitch, typically around 600-800 Hz, which corresponds to the beat frequency between your receiver's local oscillator and the incoming signal. Achieving zero beat is vital for clear communication, avoiding interference, and efficiently operating in crowded band conditions, such as during contests or DXing.

To achieve zero beat, you can manually adjust your receiver's tuning while listening for the tone to match, or use visual aids like bar graphs or indicators on modern transceivers. For transmitters without automatic offset for CW tones, aligning your transmitted signal with the receiving station's tone is necessary. Mastering zero beat is a fundamental skill in CW operation that enhances communication quality and demonstrates good operating practices.

Key Operating Practices

One key aspect of good amateur practice when operating Continuous Wave mode is selecting a transmitting frequency that minimizes interference with other stations. Adequate separation from adjacent frequencies is essential to avoid unwanted interference. Following the recommended spacing between CW signals, operators can maintain clear and effective communication without disrupting nearby stations.

When selecting a frequency to transmit CW, it is important to maintain a separation of 150 Hz to 500 Hz from other stations. This range of separation ensures that the narrow bandwidth of CW signals doesn't overlap with adjacent frequencies, reducing the risk of interference. The closer you are to another station's frequency, the more likely your signal will interfere with their communications.

When selecting a CW transmitting frequency, what minimum separation from other stations should be used to minimize interference to stations on adjacent frequencies? (G2B04)

A. 5 Hz to 50 Hz
B. 150 Hz to 500 Hz
C. 1 kHz to 3 kHz
D. 3 kHz to 6 kHz

This principle helps to keep CW transmissions clear and allows multiple stations to operate efficiently within a shared band. Operators are encouraged to monitor the band carefully, ensuring they are not too close to other signals before transmitting. Keeping this spacing in mind ensures smoother operation on the CW bands, helping you stay within the best practices of amateur radio.

Transmit Speed

The best speed to use when answering a CQ is the fastest speed you feel comfortable copying, but it should not exceed the speed of the station sending the CQ. This ensures smooth communication and prevents operators from struggling to understand or follow the message. When in doubt, match the speed of the other operator.

What is the best speed to use when answering a CQ in Morse code? (G2C05)

A. The fastest speed at which you are comfortable copying, but no slower than the CQ
B. The fastest speed at which you are comfortable copying, but no faster than the CQ
C. At the standard calling speed of 10 wpm
D. At the standard calling speed of 5 wpm

The best speed to use when answering a CQ in Morse code is to match the sending speed of the calling station. This ensures clear communication, avoids confusion, and shows courtesy by operating at a pace the other operator is comfortable with. Adapting your speed is key to effective CW operation.

Communication Protocols

When CW operators send "KN" at the end of a transmission, they listen only for a specific station or stations. This helps control communication flow by clarifying that the operator is not open to general calls.

These CW operational standards maintain clarity, precision, and etiquette, fostering smooth and respectful communication between operators.

What does it mean when a CW operator sends "KN" at the end of a transmission? (G2C03)

A. No US stations should call
B. Operating full break-in
C. Listening only for a specific station or stations
D. Closing station now

Prosigns

A prosign (short for procedural sign) is a combination of Morse code letters that represent a specific operational command or instruction during CW (Continuous Wave) or Morse code transmissions. Unlike regular characters or abbreviations, prosigns do not have distinct spacing between the letters; they are run together as a single entity. Prosigns are used to streamline communication and improve efficiency by providing standard, quick signals that convey particular actions or procedures.

What prosign is sent to indicate the end of a formal message when using CW? (G2C08)

A. SK
B. BK
C. AR
D. KN

The prosign "AR" signals the end of a formal message in CW. This lets the receiving station know that the transmission is complete and that there is no further message content.

Prosigns streamline communication by providing universally recognized signals that operators use to maintain clarity and professionalism during CW exchanges.

Proosign	Meaning
AR	End of message
SK	End of contact; signing off
KN	Go ahead, specific station only
BK	Break; inviting another station to respond
BT	Separator between parts of a message
CQ	Calling all stations; general call
AS	Wait; stand by
RR	All received; understood
K	Go ahead; inviting any station to transmit
SOS	Distress signal; emergency communication

These prosigns are sent as a single, continuous character (no spaces between dots and dashes) and are internationally recognized, making them important for efficient Morse code operation. Operators use these to convey instructions clearly and quickly during CW exchanges.

Signal Quality and Reporting

When sending CW, what does a "C" mean when added to the RST report? (G2C07)

A. Chirpy or unstable signal
B. Report was read from an S meter rather than estimated
C. 100 percent copy
D. Key clicks

When a "C" is added to the RST report, it indicates a chirpy or unstable signal. This helps inform the transmitting station of a frequency instability or issue with their equipment, enabling them to make necessary adjustments for a cleaner signal. This is often due to power supply instability or transmitter issues. While a chirp doesn't necessarily prevent communication, it can make the signal harder to copy and is considered a flaw. Adding "C" informs the sender about this issue, allowing them to check and adjust their equipment if needed.

Advanced Techniques

Which of the following describes full break-in CW operation (QSK)? (G2C01)

A. Breaking stations send the Morse code prosign "BK"
B. Automatic keyers, instead of hand keys, are used to send Morse code
C. An operator must activate a manual send/receive switch before and after every transmission
D. Transmitting stations can receive between code characters and elements

In full break-in (QSK) CW operation, transmitting stations can receive between code characters and elements, allowing them to hear the other station while transmitting. This makes for more efficient and conversational CW exchanges because operators can interrupt or respond more quickly, improving communication flow.

Reverse Sideband

What is the benefit of using the opposite or "reverse" sideband when receiving CW? (G4A02)

A. Interference from impulse noise will be eliminated
B. More stations can be accommodated within a given signal passband
C. It may be possible to reduce or eliminate interference from other signals
D. Accidental out-of-band operation can be prevented

Using the opposite or "reverse" sideband when receiving a continuous wave (CW) signal can help reduce interference from nearby signals. CW transmissions are narrowband; when receiving them, the radio's bandwidth plays a critical role. Interference might come from signals in the adjacent sideband (either upper or lower), which can make it harder to copy the intended signal.

By switching to the opposite sideband, you are effectively tuning in a slightly different part of the spectrum, often moving away from interfering signals that might be present on the current sideband. This allows for clearer reception of the desired signal, as it reduces the likelihood of overlapping interference from nearby stations. This simple technique can significantly improve signal clarity in crowded band conditions.

Equipment

Electronic Keyer

What is the function of an electronic keyer? (G4A10)

A. Automatic transmit/receive switching
B. Automatic generation of dots and dashes for CW operation
C. To allow time for switching the antenna from the receiver to the transmitter
D. Computer interface for PSK and RTTY operation

An electronic keyer is a device used in CW operation that automatically generates the correct sequence of dots (short signals) and dashes (long signals). It simplifies the process for operators by making sending code at consistent speeds easier. Rather than manually creating each dot or dash, the operator must press a paddle, and the keyer handles the timing.

This helps reduce operator fatigue and improves accuracy, especially at higher speeds. With the electronic keyer, sending CW becomes smoother and more efficient, as it helps maintain proper spacing between characters and reduces errors in manual timing. It's a useful tool for many CW operators, especially those participating in contests or long transmissions.

Chapter Summary

CW operation is a timeless and invaluable skill in amateur radio, offering reliability and effectiveness in various conditions. You can communicate more efficiently and avoid interference by mastering techniques like zero beating, choosing the right transmitting frequency, and adjusting your sending speed. Understanding advanced practices such as full break-in (QSK) operation and using electronic keyers further enhances your CW capabilities, making communication smoother and more enjoyable.

As you continue to develop your CW skills, remember that practice and patience are key to becoming a proficient operator. By applying the concepts and techniques covered in this chapter, you'll be well-prepared to make the most of this versatile mode and enjoy the unique opportunities it offers.

Chapter 23

Digital

IMAGINE SITTING AT YOUR ham radio station, but instead of speaking into a microphone, you are typing messages on a keyboard. Your computer converts these messages into audio tones that are transmitted over the airwaves. This is the essence of digital communication, a modern advancement in ham radio that has revolutionized how operators interact. Unlike traditional voice communication, digital modes offer improved signal clarity, enhanced data transmission capabilities, and efficient bandwidth use, making them indispensable in today's amateur radio operations.

Introduction to Digital Communications

Digital communication has revolutionized amateur radio by enabling efficient, reliable, and innovative ways to exchange information. Unlike analog modes, digital modes use software and precise modulation techniques to encode data, allowing hams to operate effectively in a variety of conditions, including low power and weak signals. Understanding digital communication starts with grouping modes by their underlying modulation techniques and practical applications, giving operators a solid foundation for making informed choices.

Group 1: Narrowband Efficiency: PSK31, or Phase Shift Keying at 31.25 baud, is a digital mode known for its efficiency and minimal bandwidth usage. Its narrow bandwidth of about 31 Hz allows multiple signals to coexist without interference, making it perfect for low-power (QRP) operations and casual QSOs. PSK31's simplicity and efficiency make it a favorite among operators focused on clear communication in crowded band conditions, particularly during digital contests or for general chit-chat with other hams.

Group 2: Text-Based Reliability: Radioteletype (RTTY) is one of the oldest digital modes that is still in active use. Utilizing Frequency Shift Keying (FSK), it transmits text messages with distinct frequencies representing binary data. RTTY's robustness makes it invaluable for long-distance and emergency communications, where reliability is paramount. Operating at a standard baud rate (often 45.45 baud), it handles challenging conditions with ease, bridging decades of tradition with modern functionality.

Group 3: Weak Signal Prowess: FT8 and JT65 are standout digital modes designed for weak-signal environments.

- **FT8** shines with its 15-second transmission cycles, enabling quick exchanges under poor propagation conditions. It's particularly popular for DXing, where operators aim to establish contacts over vast distances using minimal power. FT8's highly automated and efficient nature makes it the go-to choice for operators pushing the limits of their station capabilities.

- **JT65** operates with longer, 60-second transmission cycles and incorporates advanced error-correction algorithms to decode signals often inaudible to the human ear. This mode's slower pace makes it ideal for deliberate contacts and operators with lower power setups, maintaining reliable communication even in noisy environments.

Group 4: Versatile Voice Communication: Digital voice modes such as DMR (Digital Mobile Radio), D-STAR (Digital Smart Technologies for Amateur Radio), and System Fusion bring modern communication capabilities to amateur radio. These modes combine the clarity of digital audio with advanced networking features, enabling operators to link local repeaters to global networks seamlessly.

- **DMR**: Widely adopted in both amateur and commercial radio, DMR uses Time-Division Multiple Access (TDMA) technology to split a single frequency into two separate time slots, doubling its efficiency. DMR networks, such as BrandMeister, connect operators worldwide, making it a top choice for hams looking to explore international contacts.

- **D-STAR**: Developed specifically for amateur radio, D-STAR offers integrated voice and data capabilities, along with features like callsign-based routing. Operators can transmit text, GPS coordinates, or even files alongside their voice communication, making it a versatile mode for emergency communication and experimentation.

- **System Fusion**: Known for its ease of use, System Fusion allows operators to switch seamlessly between digital and analog communication. Yaesu's proprietary C4FM technology ensures high-quality audio and straightforward integration with analog repeaters, making it an excellent option for transitioning to digital modes without leaving analog users behind.

However, digital communication also presents certain challenges. One of the main hurdles is the need for additional equipment and software. Unlike traditional voice communication, which only requires a transceiver and microphone, digital modes necessitate a computer interface and specialized software. This setup can be complex for beginners, requiring a learning curve to understand and configure the various components. Additionally, troubleshooting digital communication issues can be more challenging, as problems can arise from both hardware and software configurations.

Despite these challenges, the benefits of digital communication make it a valuable addition to any ham radio operator's skill set. By embracing digital modes, you can enhance your communication capabilities, achieve reliable contacts under poor conditions, and explore new and exciting aspects of amateur radio. As you upgrade to your General Class license, understanding digital communication will open up a world of possibilities, allowing you to connect with operators around the globe using the latest advancements in radio technology.

Digital Modes and Frequencies

Digital modes are at the forefront of modern amateur radio, offering unique advantages in clarity, efficiency, and versatility. Whether you're experimenting with FT8's weak-signal prowess or exploring the bandwidth-efficient world of RTTY, understanding these modes is critical for today's ham operators. In this section, we'll examine the key digital voice modes and where you can commonly find them on the bands, setting the stage for deeper exploration of their applications and techniques.

DMR, D-STAR, and System Fusion are digital voice modes that enable high-quality voice transmission over radio frequencies. These systems convert voice into digital data before transmitting, allowing for clearer signals and added features like GPS tracking or text messaging. Each mode operates with slightly different technology and protocols, but all offer the benefits of digital transmission. DMR is popular for its widespread use in commercial systems, while D-STAR and System Fusion are designed specifically for amateur radio.

> **Which of the following provide digital voice modes? (G8C16)**
>
> A. WSPR, MFSK16, and EasyPAL
> B. FT8, FT4, and FST4
> C. Winlink, PACTOR II, and PACTOR III
> **D. DMR, D-STAR, and SystemFusion**

Frequencies

In what segment of the 20-meter band are most digital mode operations commonly found? (G2E08)

A. At the bottom of the slow-scan TV segment, near 14.230 MHz
B. At the top of the SSB phone segment, near 14.325 MHz
C. In the middle of the CW segment, near 14.100 MHz
D. Between 14.070 MHz and 14.100 MHz

On the 20-meter band, most digital mode operations are concentrated between 14.070 MHz and 14.100 MHz. This band portion is widely used for various digital modes, including PSK31, FT8, and other data transmissions. These modes are popular because they efficiently handle weak signals, making them ideal for long-distance communications under challenging conditions. Digital modes are less affected by noise and poor propagation, allowing operators to connect with stations worldwide even when conditions aren't ideal for voice transmission. Understanding the common digital mode frequencies can help operators quickly find contacts and take full advantage of these powerful communication tools.

Additionally, FT8 has specific frequency ranges where operators commonly use it, such as between 14.074 MHz and 14.077 MHz in the 20-meter band. Awareness of these frequencies helps you quickly tune in to FT8 activity and avoid interfering with other signals.

Which of the following is a common location for FT8? (G2E15)

A. Anywhere in the voice portion of the band
B. Anywhere in the CW portion of the band
C. Approximately 14.074 MHz to 14.077 MHz
D. Approximately 14.110 MHz to 14.113 MHz

FT8 is a low-power, highly efficient digital mode. Accurate computer timing is crucial for successful operation, and following good practices when selecting a transmission frequency helps ensure clear communication.

RTTY (Radio Teletype) is a digital mode used to transmit text-based messages over radio. One key element of RTTY is the frequency shift, which is the difference in frequency between the two tones used to represent digital data (known as mark and space). In amateur HF bands, the most common frequency shift is 170 Hz. This shift is optimal for standard RTTY operation because it balances signal robustness and bandwidth efficiency, allowing for clear communication without excessive spectrum.

What is the most common frequency shift for RTTY emissions in the amateur HF bands? (G2E06)

A. 85 Hz
B. 170 Hz
C. 425 Hz
D. 850 Hz

Understanding frequency shift is key for successful RTTY operation. In this mode, two frequencies are used to differentiate between binary signals. The most common shift used in amateur HF bands is 170 Hz, which helps ensure that RTTY signals remain narrow enough to avoid interference while being easily decoded by receiving stations.

Practical Operation and Troubleshooting

Once you've chosen a digital mode to operate, the next step is mastering its practical use. Whether it's fine-tuning your setup for FT8, understanding the nuances of RTTY, or troubleshooting a tricky signal, good operating practices are the foundation of successful communication. This section will guide you through the essentials, from selecting the right frequencies to diagnosing common decoding issues, ensuring you're ready to tackle any digital mode challenge.

FT8

FT8 is a popular digital mode known for its efficiency in low signal-to-noise ratio environments. To ensure accurate communication when using FT8, the operator's computer time must be synchronized within approximately one second. FT8 exchanges are time-sensitive, as each transmission is broken into alternating 15-second time slots. If your computer's clock is inaccurate, it can cause transmission errors or missed responses. This synchronization can be easily managed by using time-sync software to keep the computer's clock accurate.

Which of the following is required when using FT8? (G2E07)

A. A special hardware modem
B. Computer time accurate to within approximately 1 second
C. Receiver attenuator set to -12 dB
D. A vertically polarized antenna

When answering a CQ call on FT8, finding a clear frequency and transmitting during the alternate time slot from the calling station is considered good practice. This ensures that transmissions don't overlap, which can cause interference.

Which of the following is good practice when choosing a transmitting frequency to answer a station calling CQ using FT8? (G2E04)

A. Always call on the station's frequency
B. Call on any frequency in the waterfall except the station's frequency
C. Find a clear frequency during the same time slot as the calling station
D. Find a clear frequency during the alternate time slot to the calling station

Radio Teletype (RTTY)

What could be wrong if you cannot decode an RTTY or other FSK signal even though it is apparently tuned in properly? (G2E14)

A. The mark and space frequencies may be reversed
B. You may have selected the wrong baud rate
C. You may be listening on the wrong sideband
D. All these choices are correct

In RTTY and other Frequency Shift Keying (FSK) modes, decoding errors can happen even if the signal seems to be correctly tuned. Several factors could be at play. First, the mark and space frequencies—the two distinct tones representing binary 0s and 1s—might be reversed, which prevents accurate decoding. Another issue could be that you've selected the wrong baud rate, meaning the data transmission speed doesn't match between the sender and receiver. Listening on the wrong sideband(LSB vs. USB) could also distort the signal, as these modes require the correct sideband for proper reception.

Sideband Usage

Digital modes utilize different methods for transmitting data, such as AFSK (Audio Frequency Shift Keying), and they rely on specific sidebands depending on the type of transmission. Understanding the proper sideband to use ensures that your signals are transmitted correctly and can be easily received and decoded by other operators. In amateur radio, RTTY (Radio Teletype) signals and digital modes like FT8, JT65, and others have their own standard operating conventions regarding sideband usage.

Which mode is normally used when sending RTTY signals via AFSK with an SSB transmitter? (G2E01)

A. USB
B. DSB
C. CW
D. LSB

RTTY via AFSK with an SSB transmitter typically uses LSB (Lower Sideband). This is a common practice for sending RTTY signals, especially in the HF bands, where the lower sideband is the norm below 10 MHz. AFSK is a method where audio tones are modulated onto a carrier by using an SSB transmitter.

What is the standard sideband for JT65, JT9, FT4, or FT8 digital signal when using AFSK? (G2E05)

A. LSB
B. USB
C. DSB
D. SSB

For modes like JT65, JT9, FT4, or FT8, the standard sideband used when transmitting via AFSK is USB (Upper Sideband). This distinction is important because most modern digital modes on HF and higher frequencies use USB as the standard, even in frequency ranges where LSB would typically be used for voice communications. Knowing which sideband to use for each digital mode ensures compatibility with other stations and helps avoid unnecessary interference.

By understanding these sideband conventions, operators can ensure their signals are correctly transmitted and received, optimizing their chances of making successful digital mode contacts.

Advanced Digital Communication Protocols

Digital communication protocols like PACTOR, VARA, and Winlink offer hams powerful tools for reliable and efficient data transmission. These protocols excel in delivering text, email, and other data in both routine and emergency scenarios, even under challenging conditions. Here, we'll explore how these systems work, how to use them effectively, and their critical roles in modern amateur radio operations.

PACTOR

PACTOR is a digital communication mode designed for robust and error-free data exchange over HF radio. One of the key features of the PACTOR protocol is that it establishes a connection between only two stations at a time, meaning that joining an existing conversation or connection is not possible. This limitation ensures that the data transfer remains efficient and secure, with minimal interference or data loss. If you come across an active PACTOR session between two stations, the protocol does not allow for a third station to participate or interrupt the exchange.

How do you join a contact between two stations using the PACTOR protocol? (G2E09)

A. Send broadcast packets containing your call sign while in MONITOR mode
B. Transmit a steady carrier until the PACTOR protocol times out and disconnects
C. Joining an existing contact is not possible, PACTOR connections are limited to two stations
D. Send a NAK code

PACTOR is designed for reliable two-way communication between two stations only. This ensures secure, private, and interference-free data transfer, which makes it popular for applications like email transmission over HF or maritime communications. Because of this limitation, attempting to join an ongoing PACTOR connection is impossible, reinforcing this protocol's efficiency and focus.

Winlink

Winlink is a global messaging system that allows users to send and receive emails over HF, VHF, or UHF radio frequencies. It is particularly useful for emergency communications and off-grid scenarios where the internet is unavailable. Winlink can bridge traditional radio communication with the internet, allowing users to exchange messages with regular email addresses, even from remote locations.

Here's a detailed breakdown of how Winlink works and its key components:

How Winlink Works:

- **Radio-Based Email System**: At its core, Winlink allows users to send and receive emails over amateur radio frequencies using specialized software on the internet. These emails can be sent peer-to-peer via radio or routed to the internet via Radio Message Servers (RMS) stations.

- **Message Routing**: Messages (packets) sent over Winlink can either be relayed directly to another radio operator or sent to an RMS station connected to the Internet. The RMS server forwards the email to the recipient via the Internet.

- **Multiple Bands**: Winlink operates as a wireless network over HF, VHF, and UHF frequencies. On HF, long-distance communication is possible via ionospheric propagation, allowing users to connect to RMS stations that may be far away geographically.

- **Hybrid Network**: Winlink's hybrid network is unique. Even if the internet is down, Winlink can continue to function by relaying messages via radio from one RMS station to another. This makes it especially useful for emergency communication in disaster-prone areas.

- **How to Establish Contact:** To establish contact with a digital messaging system gateway station, such as those used for Winlink, you must transmit a "connect" message on the station's published frequency. Digital gateways are often used to send and receive messages in modes like PACTOR, VARA, or ARDOP. These gateways act as intermediaries, facilitating digital data transfer (such as emails) between radio operators and internet-linked servers. The "connect" message lets the gateway know that you are attempting to establish a communication link. The process is similar to dialing into an internet server in early internet systems, where the right frequency and digital protocol are crucial for a successful connection.

Based on that info, let's answer some exam questions.

Which of the following describes Winlink? (G2E12)

A. An amateur radio wireless network to send and receive email on the internet
B. A form of Packet Radio
C. A wireless network capable of both VHF and HF band operation
D. All of the above

If 'A,' 'B,' and 'C' are true, then it must be 'all of the above.'

Which of the following is a way to establish contact with a digital messaging system gateway station? (G2E10)

A. Send an email to the system control operator
B. Send QRL in Morse code
C. Respond when the station broadcasts its SSID
D. Transmit a connect message on the station's published frequency

---

Software and Hardware:

- **Winlink Software**: To use the system, users need software like Winlink Express, which can connect to the Winlink system through their computer. This software handles composing, sending, and receiving emails, similar to traditional email programs.

- **Modems**: Winlink uses modems (or sound cards) that convert digital email data into radio signals that can be transmitted over the airwaves. Examples include PACTOR and VARA modems for HF communication.

- **Winlink Remote Message Server**: These are the stations that provide the link between amateur radio frequencies and the internet. They act as relay points, sending messages from the radio network to the internet-based email systems and vice versa. They are also known as Gateways.

Question time!

What is another name for a Winlink Remote Message Server? (G2E13)

A. Terminal Node Controller
B. Gateway
C. RJ-45
D. Printer/Server

Winlink Modes:

- **Peer-to-Peer Mode**: In this mode, emails are sent directly from one amateur radio station to another without using the internet or RMS gateways.

- **RMS Mode**: This is the most common method. The message is sent via an RMS gateway, which forwards it to the recipient's internet-based email address.

- **Hybrid Mode**: In areas where internet connectivity is disrupted, the hybrid mode allows the system to relay messages entirely over the radio from one RMS to another until it reaches a station with internet access.

Winlink Applications:

- **Emergency Communications (EMCOMM)**: Winlink is an invaluable tool in disaster response. It can exchange messages, share forms (like ICS forms for incident management), and coordinate relief efforts when normal communication infrastructure is unavailable.

- **Off-Grid Communications**: Winlink allows users in remote areas without internet access to stay connected with the world via email.

- **Sailing and Remote Expeditions**: Winlink is popular among sailors and adventurers in remote locations because it allows them to send and receive emails even when they are far from any cellular network.

Winlink Protocols:

- **PACTOR** is one of the most efficient protocols used in Winlink. It is known for high-speed and reliable data transfer over HF radio. PACTOR modems can operate under noisy and weak signal conditions.

- **VARA**: VARA is a sound card digital protocol used with Winlink that offers fast transmission speeds and is an alternative to PACTOR, though PACTOR tends to offer higher reliability.

- **ARDOP**: This modern protocol designed specifically for Winlink offers a flexible and open-source solution for digital communication over HF.

- **AX.25 Packet Radio**: Used for VHF/UHF Winlink operation, this protocol is slower but efficient for shorter-range communications.

What is VARA? (G2E02)

A. A low signal-to-noise digital mode used for EME (moonbounce)
B. A digital protocol used with Winlink
C. A radio direction finding system used on VHF and UHF
D. A DX spotting system using a network of software defined radios

---

Benefits and Challenges:

- **Benefits**: Winlink is highly adaptable and resilient in disaster situations, providing a vital link for communication when other systems fail. Its ability to connect via HF radio makes it an excellent tool for long-distance communication, even without the internet.

- **Challenges**: Setting up Winlink requires some technical expertise, including understanding how to interface radios with modems and configuring the necessary software. Licensing and compliance with amateur radio rules are also essential for proper operation.

- **Interference**: When using digital modes like PACTOR or VARA, interference from other signals can cause various issues with your transmission. The most common symptoms include frequent retries or timeouts, which occur when the signal is disrupted to the point that the system must repeatedly try to send the data. You might also notice long pauses in message transmission as the system waits for a clear path to continue the transmission. Finally, severe interference may lead to the failure to establish a connection between stations entirely, preventing any communication. This happens because PACTOR and VARA are sensitive to signal clarity, and interference from overlapping frequencies or noise can significantly impact their effectiveness.

What symptoms may result from other signals interfering with a PACTOR or VARA transmission? (G2E03)

A. Frequent retries or timeouts
B. Long pauses in message transmission
C. Failure to establish a connection between stations
D. All these choices are correct

Winlink bridges radio communications and the internet, making it a versatile tool for emergency communication, off-grid messaging, and international coordination. It leverages both traditional radio frequencies and the internet to ensure messages get through under nearly any circumstances, making it a critical system for ham radio operators involved in emergency preparedness, remote expeditions, and disaster recovery.

Digital Networking and Emergency Applications

The integration of digital networking into amateur radio has revolutionized emergency communication and expanded the scope of what hams can achieve. Technologies like the Amateur Radio Emergency Data Network (AREDN) provide mesh networking solutions, while shared spectrum with unlicensed Wi-Fi enables innovative uses of existing equipment. This section will explore these exciting applications, highlighting their relevance to emergency preparedness and technical exploration.

Amateur Radio Emergency Data Network

An Amateur Radio Emergency Data Network (AREDN) mesh network is a high-speed data network designed to be deployed during emergencies or community events when regular communication systems are overloaded or unavailable. These networks use amateur radio frequencies and equipment to set up local area networks that can provide a variety of services, such as email, VoIP (Voice over IP) communications, file sharing, and even live video streaming. The network's decentralized and flexible nature allows for quick deployment, with each node in the network able to route data efficiently, even if some nodes go down. This makes it an invaluable tool for emergency responders and amateur radio operators involved in disaster relief.

What is the primary purpose of an Amateur Radio Emergency Data Network (AREDN) mesh network? (G2E11)

A. To provide FM repeater coverage in remote areas
B. To provide real-time propagation data by monitoring amateur radio transmissions worldwide
C. To provide high-speed data services during an emergency or community event
D. To provide DX spotting reports to aid contesters and DXers

---

The primary purpose of an AREDN mesh network is to deliver high-speed data services during emergencies when traditional communication networks fail. By using amateur radio frequencies, these networks can provide reliable and scalable communication options, ensuring critical data gets through.

Spectrum Use

Amateur radio operators share certain parts of the radio frequency spectrum with other services, including unlicensed users like Wi-Fi networks. The 2.4 GHz band is a shared range where licensed amateur radio and unlicensed services, such as Wi-Fi, Bluetooth, and other short-range devices, operate. Understanding which frequency bands are shared helps radio operators avoid interference with other services and ensures smooth operation for everyone using the spectrum.

On what band do amateurs share channels with the unlicensed Wi-Fi service? (G8C01)

A. 432 MHz
B. 902 MHz
C. 2.4 GHz
D. 10.7 GHz

Amateur radio operators share the 2.4 GHz band with unlicensed services, including Wi-Fi. The 2.4 GHz band is widely used for Wi-Fi networks because it offers a good balance between range and data-carrying capacity. While hams can legally transmit on this frequency, they must do so with caution to avoid causing interference to unlicensed users, as Wi-Fi and other devices also rely on this band for communication.

As an amateur radio operator, knowing about this shared usage is important to ensure harmonious operation with other users, such as Wi-Fi.

Software-Defined Radio (SDR)

Software-Defined Radio (SDR) represents a groundbreaking evolution in ham radio technology. Unlike traditional radios that rely on fixed hardware components, SDRs perform core functions like mixing, filtering, and modulating through software, offering unparalleled flexibility and adaptability. With the ability to update and reconfigure systems via software, SDRs are future-proof and versatile.

One of SDRs' most valuable features is real-time spectrum analysis and waterfall displays, which allow operators to visualize and analyze signals across multiple bands, from HF to UHF. These tools are indispensable for detecting interference, optimizing operating conditions, and experimenting with digital modes like FT8 and PSK31. SDRs also enhance signal processing through digital filters, noise reduction, and advanced automation, making them powerful tools for both beginners and experienced operators.

Setting up an SDR involves selecting compatible hardware, such as the SDRplay RSP1A or HackRF One, and pairing it with software like SDR# or HDSDR. Once configured, these systems open the door to monitoring faint signals, exploring new modes, and enhancing communication capabilities in ways that traditional radios cannot match.

By integrating cutting-edge technology with the spirit of amateur radio experimentation, SDRs empower operators to stay at the forefront of innovation, making them an essential tool for the modern ham.

Software-Defined Radios (SDR) and Modulation

Software-defined radios (SDR) represent a significant evolution in radio operation. They rely on software rather than solely on hardware components to process radio signals. This allows SDR systems to be much more flexible and adapt to different frequencies, modulation types, and other parameters through programming rather than physical changes. One key concept in SDR is I-Q modulation, where a signal's in-phase (I) and quadrature (Q) components are processed for efficient modulation and demodulation.

What is the phase difference between the I and Q RF signals that software-defined radio (SDR) equipment uses for modulation and demodulation? (G7C09)

A. Zero
B. 90 degrees
C. 180 degrees
D. 45 degrees

In software-defined radios, I-Q modulation is a method where two signals, one in-phase (I) and the other in quadrature (Q), represent different components of the radio signal. The key to this modulation technique is the 90-degree phase difference between the I and Q signals. This allows for the creation of complex modulation schemes that can be easily processed and demodulated by the SDR.

What is an advantage of using I-Q modulation with software-defined radios (SDRs)? (G7C10)

A. The need for high resolution analog-to-digital converters is eliminated
B. All types of modulation can be created with appropriate processing
C. Minimum detectible signal level is reduced
D. Automatic conversion of the signal from digital to analog

I-Q modulation offers SDRs a unique advantage in that virtually any modulation scheme can be generated with the right software processing. Whether it's AM, FM, SSB, or digital modes, SDRs can use I-Q modulation to create and manipulate signals, giving them an edge in flexibility and adaptability compared to traditional radios that require different hardware for each modulation type.

Which of these functions is performed by software in a software-defined radio (SDR)? (G7C11)

A. Filtering
B. Detection
C. Modulation
D. All these choices are correct

One of the main benefits of SDR technology is the ability to shift many functions—such as filtering, signal detection, and modulation—from hardware to software. This makes SDRs incredibly versatile and upgradeable, as software can be modified or updated to improve performance or adapt to new standards without changing the hardware.

Software-defined radios (SDRs) revolutionize radio technology by relying on software to perform functions traditionally managed by hardware. Using I-Q modulation techniques and advanced digital signal processing, SDRs provide unparalleled flexibility, allowing operators to easily switch between modulation modes and frequencies. As software improves, so do the capabilities of these radios, making them powerful tools for modern communications.

Chapter Summary

Incorporating digital communication into your operations requires careful planning and setup. By ensuring you have the right equipment, properly configuring your interfaces, and addressing any setup issues, you can enjoy the benefits of digital modes, enhance your communication capabilities, and connect with operators worldwide.

Chapter 24

Interference

INTERFERENCE IS AN INEVITABLE challenge in ham radio operations, affecting signal clarity and communication quality. From household devices to power lines, a wide range of sources can disrupt your transmissions, making it essential for operators to understand and address these issues effectively. Learning interference mitigation enhances your ability to communicate reliably and ensures compliance with regulations and respect for other operators on the airwaves.

In this chapter, we'll explore the most common types of interference, their causes, and practical solutions. You'll learn to identify specific interference patterns, use tools like bypass capacitors and ferrite chokes, and prevent issues like ground loops. Whether diagnosing distorted signals or eliminating RF noise from your setup, this chapter provides the tools and knowledge to keep your station operating at its best.

Several techniques and components can help reduce unwanted noise when dealing with RF interference in audio systems. One common solution is using a bypass capacitor. Bypass capacitors work by shunting high-frequency RF signals to the ground, preventing them from traveling along the audio path and causing interference. This is especially useful when the RF signals are coupled into sensitive circuits, such as amplifiers or microphones, where they can cause unwanted noise or distortion.

> **Which of the following might be useful in reducing RF interference to audio frequency circuits? (G4C01)**
>
> A. Bypass inductor
> **B. Bypass capacitor**
> C. Forward-biased diode
> D. Reverse-biased diode

Another potential source of RF interference is arcing at a poor electrical connection. Arcing can generate noise across a wide range of frequencies and is often caused by loose or corroded connections in electrical systems. This type of interference can be challenging to diagnose because it covers multiple frequencies, but addressing the faulty connection can eliminate the problem.

> **Which of the following could be a cause of interference covering a wide range of frequencies? (G4C02)**
>
> A. Not using a balun or line isolator to feed balanced antennas
> B. Lack of rectification of the transmitter's signal in power conductors
> **C. Arcing at a poor electrical connection**
> D. Using a balun to feed an unbalanced antenna

One effective solution for interference caused by common-mode currents on audio cables is to place a ferrite choke on the cable. A ferrite choke helps block high-frequency signals traveling along the outside of the cable, reducing interference without affecting the audio signal. This is a common solution when RF signals are being picked up by long audio cables acting as unintended antennas.

Which of the following would reduce RF interference caused by common-mode current on an audio cable? (G4C08)

A. Place a ferrite choke on the cable
B. Connect the center conductor to the shield of all cables to short circuit the RFI signal
C. Ground the center conductor of the audio cable causing the interference
D. Add an additional insulating jacket to the cable

To combat RF interference, use a bypass capacitor to filter out high-frequency noise, correct poor electrical connections that may be causing arcing, and add ferrite chokes to audio cables to eliminate common-mode currents. These steps help to maintain clear audio quality and prevent interference from external RF signals.

By combining these methods, amateur radio operators and audio engineers can reduce or eliminate the impact of RF interference on their audio equipment.

Interference Sounds

When operating a ham radio station, it's not uncommon to encounter different types of sounds and interference that can affect your transmission and reception quality. These sounds can be clues to specific types of interference, helping you diagnose and fix potential issues in your setup.

Each type of interference produces a unique audio signature, from distorted voices to humming or clicking sounds, making it easier to identify and address the underlying problems. This section will explore the different types of interference and associated characteristic sounds, providing a guide to recognizing and troubleshooting these common issues in ham radio setups.

What sound is heard from an audio device experiencing RF interference from a single sideband phone transmitter? (G4C03)

A. A steady hum whenever the transmitter is on the air
B. On-and-off humming or clicking
C. Distorted speech
D. Clearly audible speech

When RF interference from a single sideband (SSB) phone transmitter affects an audio device, it typically results in distorted speech. This occurs because the SSB signal, which carries voice information, is unintentionally picked up by the audio equipment, causing it to become distorted as the RF signal interferes with normal audio processing. This type of interference can make voices sound garbled or unclear, affecting communication quality in radio setups or audio systems.

CW Transmitter

What sound is heard from an audio device experiencing RF interference from a CW transmitter? (G4C04)

A. On-and-off humming or clicking
B. A CW signal at a nearly pure audio frequency
C. A chirpy CW signal
D. Severely distorted audio

In the case of CW (Continuous Wave) interference, the sound heard in an audio device is usually an on-and-off humming or clicking. This happens because CW transmissions are simple, on-and-off keyed RF signals, so when an audio device inadvertently picks them up, they manifest as repetitive pulses, clicks, or hums that follow the CW transmission pattern.

Ground Loop

What could be a symptom caused by a ground loop in your station's audio connections? (G4C10)

A. You receive reports of "hum" on your station's transmitted signal
B. The SWR reading for one or more antennas is suddenly very high
C. An item of station equipment starts to draw excessive amounts of current
D. You receive reports of harmonic interference from your station

A ground loop in an audio setup, which occurs when there are multiple ground paths between devices, can result in a persistent "hum" that is reported by other stations during transmission. This hum is caused by electrical interference being picked up and fed into the audio system, making the transmitted signal less clear.

Summary of Audio Device RF Interference:

- **SSB transmitter interference**: Results in distorted speech in audio devices.

- **CW transmitter interference**: Produces on-and-off humming or clicking sounds.

- **Ground loops**: Can lead to a persistent hum in transmitted signals.

By understanding these symptoms, operators can better diagnose and correct RF interference issues in their audio and radio setups.

Chapter Summary

Interference is an inevitable part of ham radio operations, but understanding its causes and solutions is key to maintaining a clear and reliable signal. By identifying the sources of interference—whether it's RF noise from household devices, arcing electrical connections, or common-mode currents—and applying targeted fixes such as bypass capacitors, ferrite chokes, or correcting ground loops, operators can significantly reduce its impact.

Recognizing the unique sounds and symptoms of interference further aids in swift diagnosis and resolution. Mastering these techniques not only enhances your operating experience but also ensures compliance with communication standards, allowing for more enjoyable and effective transmissions.

Rules & Regulations

IMAGINE A VAST, OPEN field where you can only explore a small corner. Almost the entire field is open to you with your General Class license. This newfound freedom brings exciting opportunities and responsibilities. As a General Class operator, you gain access to additional bands and frequencies that were previously out of reach. This chapter will guide you through your new frequency privileges and regulations, helping you make the most of them and understand the nuances that come with your upgraded license.

Chapter 25

Frequency Privileges and Band Plans

ITU Regions

THE INTERNATIONAL TELECOMMUNICATION UNION (ITU) divides the world into three regions to manage and regulate global radio frequency usage. These regions help coordinate radio communication practices and frequency allocations to ensure efficient and non-interfering use of the radio spectrum. Different regions have distinct frequency allocations, and understanding the applicable region is crucial for amateur operators to follow the correct regulations.

For radio amateurs operating in North and South America, the frequency allocations of ITU Region 2 apply. This region includes the Western Hemisphere, and its frequency allocations differ from those of Region 1 (Europe, Africa, and the Middle East) and Region 3 (Asia and the Pacific). Region 2 allocations help ensure consistent communication practices across North and South America while coordinating with the ITU to avoid user interference in other regions.

> **The frequency allocations of which ITU region apply to radio amateurs operating in North and South America? (G1E06)**
>
> A. Region 4
> B. Region 3
> **C. Region 2**
> D. Region 1

Frequency Privileges

With a General license, you access an array of high-frequency (HF) bands that significantly expand your operating capabilities compared to the Technician Class. These bands include 160, 80, 40, 20, 15, and 10 meters, each offering unique propagation characteristics and opportunities for long-distance communication.

Beyond these HF bands, General operators also enjoy expanded privileges on VHF and UHF bands. While Technician Class licensees have access to these bands, the General Class license offers additional segments and power limits, enhancing your ability to communicate over longer distances and through various propagation modes. This broader access increases your reach and opens up new avenues for experimenting with different modes and frequencies.

The differences between Technician and General Class privileges are noteworthy. As a General Class operator, your power limits increase significantly on certain bands, allowing you to transmit with up to 1500 watts PEP on most HF bands. This is a substantial upgrade from the 200-watt limit on the Technician bands. Moreover, the General Class license grants you access to expanded band segments for voice (SSB), CW, and digital modes. This expanded access is particularly advantageous for international communication, enabling you to participate in DX contacts and join international nets and contests.

Bands

60-Meter

The 60-meter band is a unique allocation in the amateur radio spectrum, designed to operate on five specific channels (or frequencies) rather than a continuous frequency range like other HF bands. Because it is shared with federal and other government users, the FCC has implemented additional rules and restrictions to minimize interference and ensure coexistence between amateur radio and other services.

For General class licensees, the 60-meter band offers a rare opportunity to operate within a fixed channelized structure, which differs from other bands' open, frequency-sweeping operations. Here are the key points to know:

- **Channelized Operation**: The five channels are 5330.5, 5346.5, 5357.0, 5371.5, and 5403.5 kHz, and operators must ensure that their signals remain centered on these frequencies.

- **Power Limits**: Operators are limited to 100 watts PEP Effective Radiated Power (ERP). This ensures that amateur transmissions remain within acceptable limits to prevent interference with primary users on the band.

- **Antenna Regulations**: If you use an antenna other than a dipole, you are required to keep a record of its gain. The ERP calculation is based on a dipole's performance, and using a more directional or high-gain antenna can exceed the power limit without proper documentation.

- **Operating Modes**: USB (Upper Sideband) is the primary mode for voice transmissions, and other digital and CW modes are allowed as long as they fit within the channel bandwidth restrictions.

Overall, the 60-meter band provides valuable opportunities for communications, particularly in emergency and disaster response situations where operators may need to coordinate with other services. The band's unique constraints make it an interesting and important part of the General class operating privileges.

Channelized Bands

A channel in radio communication refers to a specific, pre-assigned frequency or set of frequencies within a broader radio band that is designated for communication. Instead of allowing operators to use any frequency within a range, channels are fixed, clearly defined points on the spectrum where communication is permitted. Each channel has a specific center frequency and bandwidth, ensuring that multiple operators can share the same band without interference.

In these cases, operators must tune their radios to the exact frequency of the channel and follow the band's specific operating rules.

Key Points:

- **Fixed frequencies**: A channel is a specific frequency within a band, not a range.

- **Shared use**: Channels help organize communication, especially in shared bands, by preventing users from interfering with each other.

- **Efficient spectrum use:** predefined channels ensure that users use the available spectrum efficiently and fairly.

The 60-meter band is unique in that amateur operators are limited to transmitting on specific channels rather than a continuous frequency range. Five channels are designated for amateur use in the United States, and these channels must be used exactly as specified, with no deviation. This channelization is important because 60 meters is a shared band, meaning both amateurs and government services use it. Channelization helps avoid interference and ensures that both users can coexist without disrupting communications.

> **Which amateur bands are restricted to communication only on specific channels rather than frequency ranges? (G1A04)**
>
> A. 11 meters
> B. 12 meters
> C. 30 meters
> **D. 60 meters**

Special Regulations on the Exam

The 60-meter band is unique in amateur radio as it operates on specific channels and is shared with federal and other users. The FCC imposes additional regulations to ensure responsible use, including specific power limits and antenna guidelines. These rules help manage interference and ensure that the band is used efficiently and fairly by both amateur and non-amateur services.

When operating on the 60-meter band, if you are using an antenna other than a dipole, the FCC requires that you keep a record of the gain of your antenna. This is because power limits on the 60-meter band are measured in terms of Effective Radiated Power (ERP) relative to a dipole antenna. Keeping a record of the antenna gain ensures that you remain within the 100-watt ERP limit set for this band, even when using more complex or directional antennas that can increase your signal strength.

> **Which of the following is required by the FCC rules when operating in the 60-meter band? (G1C04)**
>
> **A. If you are using an antenna other than a dipole, you must keep a record of the gain of your antenna**
> B. You must keep a record of the date, time, frequency, power level, and stations worked
> C. You must keep a record of all third-party traffic
> D. You must keep a record of the manufacturer of your equipment, and the antenna used

This helps operators stay compliant with power restrictions and prevents interference with other users on the 60-meter band.

Bandwidth

The FCC enforces strict rules on how much bandwidth an amateur transmission can occupy to avoid interference. This ensures that while taking advantage of the 60-meter channels, amateur operators do not unintentionally interfere with other services that also use this band. The rules apply particularly to USB (Upper Sideband) voice transmissions, the primary mode used on this band.

On the 60-meter band, the maximum bandwidth permitted for amateur stations transmitting on USB frequencies is 2.8 kHz. This limitation ensures that the signals remain narrow enough to fit within the allocated channels without spilling into adjacent frequencies, which could interfere with other communications. The 2.8 kHz limit accommodates typical voice transmissions while preventing over-modulation or excessive bandwidth usage.

What is the maximum bandwidth permitted by FCC rules for amateur radio stations transmitting on USB frequencies in the 60-meter band? (G1C03)

A. 2.8 kHz
B. 5.6 kHz
C. 1.8 kHz
D. 3 kHz

To remember this for the exam, think of 2.8 kHz as the standard bandwidth for voice transmissions in the USB mode on this shared band, ensuring that communication remains efficient and interference-free.

15-Meter

The 15-meter band is a valuable resource for General class licensees, offering excellent long-distance communication opportunities, especially during periods of high solar activity. As a General class operator, you gain access to a significant portion of this band, specifically from 21.275 kHz to 21.450 kHz. This range provides ample space for voice, CW (Morse code), and digital modes, allowing General licensees to make global contacts, particularly during daylight hours.

And now, for an exam question.

The frequency of 21.300 kHz falls within the General class portion of the 15-meter band. General class operators are allowed to transmit in the range from 21.275 kHz to 21.450 kHz on the 15-meter band. This portion provides general class licensees with a substantial range of communication capabilities. In contrast, Amateur Extra licensees have additional access to a smaller segment of the band. General class operators must ensure they are transmitting within this authorized range to avoid operating outside of their privileges.

Which of the following frequencies is within the General class portion of the 15-meter band? (G1A09)

A. 14250 kHz
B. 18155 kHz
C. 21300 kHz
D. 24900 kHz

Remember, 21,300 kHz is part of the "middle ground" in the 15-meter band, where General-class operators have full access, while certain lower frequency segments are reserved for Amateur Extra operators.

10-Meter

The 10-meter band is unique and versatile, offering exciting opportunities for General class licensees. It sits at the upper end of the HF spectrum. It provides the potential for both local and long-distance (DX) communication, particularly during periods of high solar activity. The band is well-suited for various modes, including voice (SSB), CW, digital modes, and even repeater use. General class operators have broad access to the 10-meter band, allowing them to experiment with different modes and take advantage of favorable propagation conditions to make contacts around the world, especially when the band is open during the peak of the solar cycle.

Continuous Wave

CW (Continuous Wave) is a mode of communication used primarily for Morse code transmissions. Different license classes have specific privileges on various frequency bands, but CW operation is often permitted across broader segments because of its narrow bandwidth and efficient spectrum use. The 10-meter band is a portion of the HF spectrum that offers a wide range of privileges for General class licensees.

On the 10-meter band, stations with a General class control operator may transmit CW emissions across the entire band. Unlike some other bands where specific frequency ranges are reserved for higher license classes, General class operators have full access to the 10-meter band for CW transmissions. This makes the 10-meter band a valuable resource for General license holders who wish to use Morse code for long-distance communication.

On which amateur frequencies in the 10-meter band may stations with a General class control operator transmit CW emissions? (G1A07)

A. 28.000 MHz to 28.025 MHz only
B. 28.000 MHz to 28.300 MHz only
C. 28.025 MHz to 28.300 MHz only
D. The entire band

Repeaters

On the 10-meter band, repeaters play a crucial role in providing long-distance communication, particularly during periods of favorable propagation. Unlike VHF or UHF repeaters, which are commonly used for local communication, 10-meter band repeaters allow for extended reach, sometimes spanning hundreds or even thousands of miles.

The portion of the 10-meter band available for repeater use is above 29.5 MHz. This section of the band is specifically allocated for repeater operation, enabling General and higher-class licensees to take advantage of repeater networks for long-distance communication. Repeaters in this frequency range can provide clear, reliable communications over large distances, particularly when the band is open due to favorable propagation conditions.

What portion of the 10-meter band is available for repeater use? (G1A10)

A. The entire band
B. The portion between 28.1 MHz and 28.2 MHz
C. The portion between 28.3 MHz and 28.5 MHz
D. The portion above 29.5 MHz

Everything above this frequency is available for repeaters, helping to extend communication range beyond what is possible with direct line-of-sight radio operation.

12-Centimeters

The 12-centimeter band is part of the SHF (Super High Frequency) spectrum, spanning frequencies from 2.3 GHz to 2.45 GHz. Amateur radio operators primarily use frequencies close to 2.4 GHz. This band is shared with many other services, including Wi-Fi, Bluetooth, and other unlicensed consumer technologies that operate under Part 15 of the FCC rules.

Amateur radio operators use the 12 cm band for various communications, including short-range, line-of-sight transmissions and experiments with digital and spread spectrum techniques. One key aspect of this band is that it falls into the microwave frequency range, allowing for high data rate communications and making it ideal for fast-scan television and various digital modes.

Due to its shared nature, amateur operators must be mindful of potential interference with unlicensed devices and use proper coordination and care when operating on the 2.4 GHz portion of the band. The relatively short wavelength of about 12.5

centimeters makes it suitable for highly directional antenna designs, allowing for focused signal transmission and reception over short distances.

Amateur radio operators in the 2.4 GHz band must be mindful that this portion of the spectrum is shared with unlicensed services like Wi-Fi. While both services occupy this frequency range, amateur stations are not permitted to communicate with non-licensed Wi-Fi stations. The FCC rules keep amateur operations separate from commercial and consumer applications, ensuring each service operates within its designated scope without causing interference or blending the two services.

In what part of the 2.4 GHz band may an amateur station communicate with non-licensed Wi-Fi stations? (G1E07)

A. Anywhere in the band
B. Channels 1 through 4
C. Channels 42 through 45
D. No part

Amateur stations are not allowed to communicate with non-licensed Wi-Fi stations in any part of the 2.4 GHz band. Even though both amateur radio and Wi-Fi operate within this shared spectrum, they serve different purposes, and the rules prohibit cross-communication between them. Wi-Fi is regulated under Part 15 of the FCC rules, while amateur radio is governed by Part 97. These rules prevent service interference and maintain the integrity of amateur operations and consumer networks like Wi-Fi.

Restrictions

Amateur Extra Licensees

Higher-level licenses, like the Amateur Extra class, come with greater operating privileges. While General licensees have access to most of the HF bands, certain portions of these bands are reserved exclusively for Amateur Extra licensees. These reserved segments allow more experienced operators to enjoy less crowded operating conditions and prioritize using these critical frequencies, especially in busy bands.

The 80-meter, 40-meter, 20-meter, and 15-meter HF bands have segments that are exclusively allocated to Amateur Extra licensees. These segments are set aside to give Extra class operators exclusive access to certain frequencies within these bands, which means that General and Technician class operators cannot transmit in these areas. These reserved segments are typically at the lower end of each band, offering more experienced operators a quieter, less congested space for communication.

Which HF bands have segments exclusively allocated to Amateur Extra licensees? (G1A08)

A. All HF bands
B. 80 meters, 40 meters, 20 meters, and 15 meters
C. All HF bands except 160 meters and 10 meters
D. 60 meters, 30 meters, 17 meters, and 12 meters

To remember this for the exam, think of the 80, 40, 20, and 15-meter bands as shared spaces, but with exclusive "VIP sections" reserved for Amateur Extra operators, providing them with additional operating privileges.

Primary and Secondary Users

Certain frequency bands are shared between different services, which are classified as primary or secondary users. Primary users have priority on the band, and their communications take precedence. Secondary users, such as amateur radio operators on some bands, are allowed to operate but must ensure that they do not interfere with the primary users. Additionally, secondary users must be prepared to accept any interference caused by primary users.

When the FCC designates amateur radio operators as secondary users on a band, it means that they must not cause any harmful interference to primary users and must accept any interference from those primary users. Primary users, such as government or military services, have the right to use the frequency without disruption. Suppose a secondary user, like a ham operator, is causing interference to a primary user. In that case, the amateur operator must reduce power, change frequency, or stop transmitting.

Which of the following applies when the FCC rules designate the amateur service as a secondary user on a band? (G1A06)

A. Amateur stations must record the call sign of the primary service station before operating on a frequency assigned to that station
B. Amateur stations may use the band only during emergencies
C. Amateur stations must not cause harmful interference to primary users and must accept interference from primary users
D. Amateur stations may only operate during specific hours of the day, while primary users are permitted 24-hour use of the band

---

Partial Band Restrictions

General class licensees have significant access to the HF bands, but on some bands, they are restricted from using the entire voice (phone) segment. These restrictions help manage band usage and ensure that more experienced operators, such as Amateur Extra class licensees, have exclusive access to certain frequencies. General class operators are typically allowed to use the upper portion of the voice segment on these restricted bands, enabling them to participate in voice communication while adhering to FCC regulations.

When General class licensees are not permitted to use the entire voice portion of a band, they are allowed to operate in the upper-frequency portion of the voice segment. This means that while Amateur Extra operators may have access to the entire voice section, General class operators are limited to the higher end of that portion. These upper segments are still well-suited for voice communication and offer ample contact opportunities.

When General class licensees are not permitted to use the entire voice portion of a band, which portion of the voice segment is available to them? (G1A11)

A. The lower frequency portion
B. The upper frequency portion
C. The lower frequency portion on frequencies below 7.3 MHz and the upper portion on frequencies above 14.150 MHz
D. The upper-frequency portion on frequencies below 7.3 MHz and the lower portion on frequencies above 14.150 MHz

---

While we can't operate on the lower part of the voice range reserved for Extra class, the upper portion provides plenty of room for our communication needs.

HF and MF Band Restrictions

General licensees are granted access to a large portion of the HF and MF bands used for long-distance communication. However, some restricted segments within these bands are reserved for higher license classes.

On the 80-meter, 40-meter, 20-meter, and 15-meter bands, there are specific portions where General class licensees cannot transmit, as these segments are reserved for Amateur Extra class operators. These restrictions give more experienced operators exclusive access to certain parts of the spectrum. General class operators must be aware of these limits to avoid operating outside their privileges.

On which HF and/or MF amateur bands are there portions where General class licensees cannot transmit? (G1A01)

A. 60 meters, 30 meters, 17 meters, and 12 meters
B. 160 meters, 60 meters, 15 meters, and 12 meters
C. 80 meters, 40 meters, 20 meters, and 15 meters
D. 80 meters, 20 meters, 15 meters, and 10 meters

These restricted segments are designed to ensure that higher-class license holders have their own space on the bands.

Frequency Privileges for General Class Licensees

General class licensees have access to a large portion of the HF bands, but they are not permitted to operate as control operators in certain frequency ranges. These restricted segments are reserved for operators with higher-level licenses. Knowing which frequencies are off-limits is important to ensure compliance with FCC regulations.

General class licensees are prohibited from operating as control operators on frequencies between 7.125 MHz and 7.175 MHz. This range falls within the 40-meter band, but only Amateur Extra license holders are allowed to operate in this portion of the band. General license holders must stay outside this segment when operating as control operators to avoid operating beyond their legal privileges.

On which of the following frequencies are General class licensees prohibited from operating as control operator? (G1A05)

A. 7.125 MHz to 7.175 MHz
B. 28.000 MHz to 28.025 MHz
C. 21.275 MHz to 21.300 MHz
D. All of the above

Phone Restrictions

Some bands are restricted to certain modes of operation, meaning that specific activities, like voice (phone) transmissions, may be prohibited. These restrictions help manage spectrum use and avoid interference, ensuring that different modes of communication can coexist efficiently. Understanding which bands allow or restrict certain types of operation is needed to comply with FCC regulations.

The 30-meter band is one of the HF bands where phone operation (voice communication) is prohibited. This band is reserved for digital modes and CW (Morse code) transmissions. The prohibition of phone operation on this band helps avoid congestion and interference, especially since 30 meters is often used for long-distance digital communication under challenging propagation conditions.

On which of the following bands is phone operation prohibited? (G1A02)

A. 160 meters
B. 30 meters
C. 17 meters
D. 12 meters

The 30-meter band is a "digital-only" band where operators can use CW and data modes but not voice communication. This restriction ensures efficient band use for specific communication modes that benefit from quieter, less crowded conditions.

Image Restrictions

In amateur radio, different frequency bands have specific regulations regarding which modes and types of communication are allowed. One of these modes is image transmission, which includes sending pictures or visual data, typically through modes like SSTV (Slow Scan Television). Some bands, however, have restrictions on image transmission to maintain their designated use for other communication types, such as digital modes and CW (Morse code). Understanding where image transmission is prohibited helps operators stay within legal guidelines and use the spectrum efficiently.

Image transmission is prohibited on the 30-meter band. This band is designated primarily for digital modes and CW, and voice and image transmissions are not allowed. The prohibition on image transmission helps keep the band clear for its intended uses, particularly for long-distance CW and data communications, which require a less congested environment to function optimally.

On which of the following bands is image transmission prohibited? (G1A03)

A. 160 meters
B. 30 meters
C. 20 meters
D. 12 meters

Again, the exam is asking about the 30-meter band. This band is unknown as a "quiet" band reserved for Morse code and data communication, where both phone and image transmissions are prohibited to avoid interference with the more narrow-bandwidth modes that rely on this space.

Chapter Summary

This chapter has explored the diverse frequency privileges and band plans available to General class licensees, highlighting the opportunities and responsibilities that come with expanded access to the amateur radio spectrum. By understanding ITU regions, channelized bands, and specific regulations for bands like 60 meters, you gain insight into the broader scope of your operating privileges. We've also examined critical aspects like restrictions, modulation types, and the shared use of certain frequencies.

Band Plan

Chapter 26

Adhering to FCC Rules

COMPLIANCE WITH FCC RULES is a legal obligation and a fundamental aspect of being a responsible ham radio operator. Following these regulations ensures that every operator can use the spectrum fairly. The radio spectrum is a limited resource, and equitable access is crucial for all operators to enjoy their hobby without causing or experiencing interference. Adhering to these rules contributes to a well-organized and respectful environment where everyone can communicate effectively.

A lot of the basic and universal rules were covered on the Technician's exam.

As a General Class licensee, you must be familiar with specific FCC rules that apply to your operating privileges. Frequency allocations and band limits are defined to ensure that operators use designated frequencies responsibly. General Class operators can access a wider range of frequencies, including portions of the HF bands, which allow for long-distance communications. Knowing the exact frequencies and band segments you are authorized to use is necessary to avoid interfering with other services.

Power limits and restrictions vary across different bands. While the General Class license allows for increased power output, the best practice is to use the minimum power necessary to maintain effective communication. This practice helps reduce interference with other operators and ensures efficient use of the spectrum. Familiarize yourself with the power limits for each band and adhere to them strictly.

The FCC also clearly defines authorized modes of operation and emission types. As a General Class operator, you can use various modes, including CW, SSB, AM, FM, and digital modes like PSK31 and FT8. Each mode has specific emission types and bandwidth requirements that must be followed. Understanding these parameters ensures that your transmissions are within legal limits and do not cause undue interference.

Operating Rules

Operators must follow specific regulations outlined in Part 97 of the FCC rules and adhere to good engineering and amateur practice principles. This concept refers to conducting operations in a way that ensures the best technical standards are met, promotes efficient spectrum use, and avoids interference with other users. These guidelines extend beyond the letter of the law, promoting responsible behavior among amateur operators.

The FCC determines what constitutes good engineering and good amateur practice for amateur radio operators. While many aspects of amateur radio operations are clearly defined in Part 97, these rules do not cover some situations. In such cases, the FCC expects operators to follow established best practices in their technical operations and conduct. This includes ensuring that their transmissions are clean, clear of interference, and within the proper power limits, even when specific details are not laid out in the regulations.

> **Who or what determines "good engineering and good amateur practice," as applied to the operation of an amateur station in all respects not covered by the Part 97 rules? (G1B11)**
>
> **A. The FCC**
> B. The control operator
> C. The IEEE
> D. The ITU

Permitted Transmissions

Amateur radio operators are generally restricted from retransmitting third-party communications. Still, there are exceptions for specific types of valuable public information. One of these exceptions is the occasional retransmission of weather and propagation forecasts from authorized US government stations. This rule allows amateur radio operators to share important information that benefits the wider amateur community, particularly when it comes to safety or enhancing radio communication effectiveness.

All amateur stations are permitted to occasionally retransmit weather and propagation forecast information from US government stations. This exception allows hams to share important forecasts, such as severe weather alerts or propagation reports, that can affect communication or public safety. While amateur operators are generally not allowed to retransmit content from other services, weather, and propagation information from official government sources is an important exception.

> **Which of the following transmissions is permitted for all amateur stations? (G1B04)**
>
> A. Unidentified transmissions of less than 10 seconds duration for test purposes only
> B. Automatic retransmission of other amateur signals by any amateur station
> **C. Occasional retransmission of weather and propagation forecast information from US government stations**
> D. Encrypted messages, if not intended to facilitate a criminal act

One-Way Transmissions

Hopefully, you remember from the Technician's exam that amateur radio transmissions are generally intended to be two-way communications between operators. However, there are certain exceptions where one-way transmissions are allowed. One key exception is for transmissions that assist with learning and practicing the International Morse code. These transmissions can be valuable for operators who are working to improve their Morse code proficiency, which remains a popular and important mode in amateur radio.

One-way transmissions that are specifically allowed under FCC rules include those made to assist with learning the International Morse code. This means that operators can transmit Morse code practice messages or educational content to help other hams develop their skills without needing a return signal. These transmissions support the amateur radio community by fostering skills in CW (continuous wave) operation, which is still widely used today.

> **Which of the following one-way transmissions are permitted? (G1B05)**
>
> A. Unidentified test transmissions of less than 10 seconds in duration
> **B. Transmissions to assist with learning the International Morse code**
> C. Regular transmissions offering equipment for sale, if intended for amateur radio use
> D. All these choices are correct

It's one of the few exceptions to the general rule that amateur radio transmissions should be two-way.

Harmful Interference

Radio operators are responsible for ensuring that their transmissions do not cause harmful interference to other users or services. The FCC has specific guidelines to ensure that operators take extra precautions when transmitting under certain conditions, such as near FCC monitoring stations, on bands where the amateur service is secondary, or when using spread spectrum emissions. These rules protect the shared use of the radio spectrum and ensure cooperation among different services and users.

A licensed amateur radio operator must take specific steps to avoid harmful interference under three key conditions:

1. **Operating within one mile of an FCC Monitoring Station**: These stations monitor the radio spectrum for unauthorized transmissions and interference. To avoid interfering with their work, operators must ensure their signals are clean and well-controlled when operating nearby.

2. **Using a band where the Amateur Service is secondary**: In some frequency bands, amateur radio is designated as a secondary service, meaning it must not interfere with primary users, such as government or commercial services. Operators must be especially careful in these cases, ensuring they don't interfere with primary users.

3. **Transmitting spread spectrum emissions**: Spread spectrum transmissions cover a wider bandwidth and have the potential to interfere with other users. Extra care must be taken to minimize interference, as these transmissions can overlap with other signals.

Which of the following conditions require a licensed amateur radio operator to take specific steps to avoid harmful interference to other users or facilities? (G1E04)

A. When operating within one mile of an FCC Monitoring Station
B. When using a band where the Amateur Service is secondary
C. When a station is transmitting spread spectrum emissions
D. All these choices are correct

Consider these situations as special cases where additional care is required to avoid interference, protecting both the amateur service and other radio spectrum users.

Transmitted Communications

Operators often use abbreviations and procedural signals (such as Q-codes or prosigns) to simplify and speed up their transmissions. These shorthand signals help reduce the time spent transmitting while conveying commonly understood meanings. However, the use of such signals is regulated to ensure that the intended meaning of a message remains clear and understandable to other operators.

Abbreviations and procedural signals are allowed in amateur radio, but they must not obscure the message's meaning. This means that while using shorthand is helpful for efficient communication, it is important that these signals do not make the message difficult to understand. For instance, Q-codes like "QSL" (acknowledge receipt) or "QRZ" (who is calling me?) are widely accepted, but using them in a way that confuses the listener or hides the message's intent is not permitted.

What are the restrictions on the use of abbreviations or procedural signals in the amateur service? (G1B07)

A. Only "Q" signals are permitted
B. They may be used if they do not obscure the meaning of a message
C. They are not permitted
D. They are limited to those expressly listed in Part 97 of the FCC rules

Remember back to the Technician's exam when we learned that only space station or radio-controlled craft communications are allowed to have messages that obscure their meaning? So, always ensure you don't cloud the meaning of your message.

International Communications

Amateur radio provides operators the exciting opportunity to communicate with other hams worldwide. However, these international communications are subject to specific rules and agreements governed by the FCC and the International Telecommunication Union (ITU). While most countries support amateur radio communication, there are a few exceptions where certain governments have objected to allowing such communication across their borders. Understanding these rules is essential for General class operators who want to make contacts beyond the U.S.

Communicating with amateur stations in countries outside the areas administered by the FCC is permissible as long as those countries have not notified the ITU that they object to such communications. In other words, most international contacts are allowed unless a country explicitly requests to restrict amateur radio communications with other nations. This rule ensures that international amateur radio contacts comply with global agreements and respect the laws of other countries.

When is it permissible to communicate with amateur stations in countries outside the areas administered by the Federal Communications Commission? (G1B08)

A. Only when the foreign country has a formal third-party agreement filed with the FCC
B. When the contact is with amateurs in any country except those whose administrations have notified the ITU that they object to such communications
C. Only when the contact is with amateurs licensed by a country which is a member of the United Nations, or by a territory possessed by such a country
D. Only when the contact is with amateurs licensed by a country which is a member of the International Amateur Radio Union, or by a territory possessed by such a country

To remember this for the exam, think of it as a general rule: you can talk to hams worldwide, except in countries whose governments have told the ITU that they do not want such communications—as of this writing, only North Korea and Yemen. This keeps international amateur radio contacts within legal boundaries and respects international agreements.

International Remote Control

With technological advancements, operating a US amateur radio station remotely has become increasingly common. However, when an operator controls a US-based station from outside the country, certain licensing requirements apply to ensure compliance with FCC rules. These regulations ensure that remote operations are conducted by individuals authorized under US law, even when controlling the station from international locations.

When operating a US station by remote control from outside the country, the operator must hold a valid US operator/primary station license. This means that even though you may be physically located in another country while remotely controlling a US-based station, you must still hold the appropriate FCC-issued license to operate legally. The rule is in place to ensure that all remote operations fall under US jurisdiction and are held to the same standards as any local operation within the United States.

When operating a US station by remote control from outside the country, what license is required of the control operator? (G1D05)

A. A US operator/primary station license
B. Only an appropriate US operator/primary license and a special remote station permit from the FCC
C. Only a license from the foreign country, as long as the call sign includes identification of portable operation in the US
D. A license from the foreign country and a special remote station permit from the FCC

However, the regulations for controlling a station remotely depend on the country where the remote station is located. This ensures that operations comply with the laws of the station's location, regardless of where the control operator is physically present.

For example, when operating a station in South America remotely from the US, the operator must follow the regulations of the remote station's country. This means that the rules governing the station's operation are set by the country in which the station is physically located, not by the FCC or US regulations. The operator must ensure they are familiar with and comply with that specific country's licensing and operational standards, even if controlling the station from a location outside its borders, such as from the United States.

When operating a station in South America by remote control over the internet from the US, what regulations apply? (G1D12)

A. Those of both the remote station's country and the FCC
B. Those of the remote station's country and the FCC's third-party regulations
C. Only those of the remote station's country
D. Only those of the FCC

Third-Party Transmissions

Third-party communication involves relaying a message between two amateur radio operators with a third, non-licensed person participating in the exchange. The FCC sets rules to ensure that third-party communication follows amateur radio guidelines, especially regarding licensing status and international agreements. These rules aim to maintain the integrity and proper use of amateur radio frequencies while allowing for important personal or emergency-related communications.

A third party is disqualified from participating in an amateur radio message if their amateur license has been revoked and not reinstated. This rule ensures that only properly authorized individuals can participate in amateur radio communication, maintaining compliance with FCC standards.

Which of the following would disqualify a third party from participating in sending a message via an amateur station? (G1E01)

A. The third party's amateur license has been revoked and not reinstated
B. The third party is not a US citizen
C. The third party is speaking in a language other than English
D. All these choices are correct

Messages sent to a third party in a country with which the U.S. has a Third-Party Agreement must be limited to amateur radio-related topics, personal remarks, or communications related to emergencies or disaster relief. This restriction ensures that the content aligns with the purpose of amateur radio and avoids commercial or inappropriate use of the frequencies.

What are the restrictions on messages sent to a third party in a country with which there is a Third-Party Agreement? (G1E05)

A. They must relate to emergencies or disaster relief
B. They must be for other licensed amateurs
C. They must relate to amateur radio, or remarks of a personal character, or messages relating to emergencies or disaster relief
D. The message must be limited to no longer than 1 minute in duration, and the name of the third party must be recorded in the station log

---

Finally, third-party messages can be transmitted via remote control under any circumstances in which the FCC permits them. This includes situations that follow the rules for third-party communication, whether the station is operated remotely or in person.

When may third-party messages be transmitted via remote control? (G1E12)

A. Under any circumstances in which third-party messages are permitted by FCC rules
B. Under no circumstances except for emergencies
C. Only when the message is intended for licensed radio amateurs
D. Only when the message is intended for third parties in areas where licensing is controlled by the FCC

By adhering to these rules, amateur radio operators ensure that third-party communication is conducted properly, maintaining the purpose and spirit of amateur radio.

Cross-Band Operations

Repeaters are commonly used to extend the range of communication by receiving a signal on one frequency and retransmitting it on another. Cross-band repeaters, such as those operating between the 10-meter and 2-meter bands, allow signals to be received on one band and transmitted on another. However, the operator's license class controlling the repeater affects what frequencies can be legally used for retransmission, as FCC rules are designed to ensure that higher-class licensees are responsible for operations on bands that require more advanced privileges.

A 10-meter repeater can only retransmit a 2-meter signal from a Technician class control operator if the control operator of the 10-meter repeater holds at least a General class license. This is because transmitting on the 10-meter band requires General class privileges or higher. In contrast, Technician class operators have limited privileges on HF bands. The General or higher class repeater control operator ensures compliance with FCC regulations by taking responsibility for the proper operation on the 10-meter band.

When may a 10-meter repeater retransmit the 2-meter signal from a station that has a Technician class control operator? (G1E02)

A. Under no circumstances
B. Only if the station on 10-meters is operating under a Special Temporary Authorization allowing such retransmission
C. Only during an FCC-declared general state of communications emergency
D. Only if the 10-meter repeater control operator holds at least a General class license

The repeater operator is responsible for ensuring legal operation on the higher frequency.

Digital Station Control

Digital stations are often operated under automatic control, which means the station can function without direct operator intervention. However, strict rules apply when conducting digital communications outside designated automatic control band segments, especially to avoid interference and ensure proper control. Operators must maintain local or remote control to ensure compliance with FCC regulations when operating outside these segments.

<u>To conduct communications with a digital station under automatic control outside the designated automatic control segments, the station initiating the contact must be under local or remote control.</u> This ensures an operator can intervene and take responsibility for the communication, avoiding unintended interference. Automatic control is generally allowed only within certain segments of the band, but outside of these areas, active control by the operator is required.

What is required to conduct communications with a digital station operating under automatic control outside the automatic control band segments? (G1E03)

A. The station initiating the contact must be under local or remote control
B. The interrogating transmission must be made by another automatically controlled station
C. No third-party traffic may be transmitted
D. The control operator of the interrogating station must hold an Amateur Extra class license

---

<u>Automatically controlled stations that transmit RTTY (Radio Teletype) or data emissions can communicate with other automatically controlled digital stations on 6-meter or shorter wavelength bands and in limited segments of some HF bands.</u> These specific allocations are designed to minimize interference while still allowing for automatic communication on certain frequencies.

On what bands may automatically controlled stations transmitting RTTY or data emissions communicate with other automatically controlled digital stations? (G1E11)

A. On any band segment where digital operation is permitted
B. Anywhere in the non-phone segments of the 10-meter or shorter wavelength bands
C. Only in the non-phone Extra Class segments of the bands
D. Anywhere in the 6-meter or shorter wavelength bands, and in limited segments of some of the HF bands

---

To summarize both questions, it is important to follow FCC rules carefully when operating digital stations under automatic control. Outside of the designated segments, operators must have local or remote control to ensure proper oversight. Automatically controlled digital stations are restricted to certain bands, including the 6-meter and shorter wavelength bands and certain parts of the HF bands for RTTY and data emissions. These rules are in place to ensure the responsible use of automatic control in the amateur radio service.

New Digital Protocols

Digital modes have expanded rapidly, and operators often experiment with new digital communication protocols. However, these new protocols must comply with FCC rules to ensure transparency and interoperability across the amateur community. Before putting a new digital protocol on the air, certain steps must be taken to ensure its technical characteristics are available for public scrutiny and avoid interference or confusion with other established modes.

<u>Before using a new digital protocol on the air, operators must publicly document its technical characteristics.</u> This rule ensures that other operators and regulators can understand how the protocol works and verify that it complies with FCC regulations.

Public documentation helps maintain transparency, allowing the amateur community to use and experiment with the new protocol without causing interference or violating rules.

> **What must be done before using a new digital protocol on the air? (G1C07)**
>
> A. Type-certify equipment to FCC standards
> B. Obtain an experimental license from the FCC
> **C. Publicly document the technical characteristics of the protocol**
> D. Submit a rule-making proposal to the FCC describing the codes and methods of the technique

Make sure everyone knows how the new digital protocol operates before it's used on the airwaves. By documenting the protocol publicly, the amateur radio community ensures responsible and compliant experimentation and innovation.

Power Regulations

The FCC regulates the maximum power amateur radio stations can transmit to ensure that operators use the radio spectrum responsibly and avoid interference with other services. The measurement used to regulate this maximum power is called PEP (Peak Envelope Power), which refers to the highest level of power output during a transmission. Understanding this measurement helps amateur operators comply with FCC power limits and maintain good amateur practice.

The FCC specifies that maximum power limits are measured in terms of PEP output from the transmitter. PEP represents the highest instantaneous power level in the radio signal's envelope, typically measured during voice or data transmission peaks. The FCC restricts operators to a maximum of 1500 watts PEP for most amateur bands. This measurement ensures that the transmitted signal remains within safe and legal limits, preventing interference with other users on the band.

> **What measurement is specified by FCC rules that regulate maximum power? (G1C11)**
>
> A. RMS output from the transmitter
> B. RMS input to the antenna
> C. PEP input to the antenna
> **D. PEP output from the transmitter**

The FCC imposes specific power regulations across different bands for General class licensees. Here's an overview of the key power regulations:

1. **Maximum Power Limit**: General class licensees are permitted to transmit with up to **1500 watts PEP** (Peak Envelope Power) on most amateur HF, VHF, and UHF bands. The FCC allows the highest power for amateur radio operations on most frequencies.

2. **Exceptions and Lower Power Limits**:

   - **30-meter band (10.100 to 10.150 MHz)**: The maximum power limit is **200 watts PEP**. This band is shared with other services, and the lower power limit helps minimize interference.

   - **60-meter band (5 channels between 5.3305 and 5.3715 MHz)**: The maximum power is **100 watts PEP**, and transmissions must not exceed the equivalent of 100 watts PEP relative to an isotropic radiator (EIRP).

   - **Certain VHF/UHF bands**: Power limits on the 2200-meter band (135.7-137.8 kHz) and 630-meter band (472-479 kHz) are typically lower, restricted to **1 watt EIRP** or less.

3. **Restrictions on Specific Modes**: Certain digital and weak signal modes may also have specific power limitations to prevent interference, but these are rare. In general, using the **minimum power necessary** to maintain commu-

nication is recommended.

With that overview, let's review some exam questions.

What is the maximum transmitter power an amateur station may use on 10.140 MHz? (G1C01)

A. 200 watts PEP output
B. 1000 watts PEP output
C. 1500 watts PEP output
D. 2000 watts PEP output

The 30-meter band is a unique part of the HF spectrum, shared with other radio services, which necessitates stricter regulations on power limits. As a result, the FCC imposes lower power restrictions on this band to avoid interference with other users. While General class licensees have the privilege to operate on this band, they must adhere to these power limitations.

On 10.140 MHz, which is part of the 30-meter band, amateur stations are restricted to a maximum transmitter power of 200 watts PEP (Peak Envelope Power). The 30-meter band has stricter regulations than other HF bands because it is shared with other services, including international broadcasts and fixed services. This power limit ensures that amateur radio operators do not interfere with these other critical services while still allowing effective communication within the amateur community.

What is the maximum transmitter power an amateur station may use on the 12-meter band? (G1C02)

A. 50 watts PEP output
B. 200 watts PEP output
C. 1500 watts PEP output
D. An effective radiated power equivalent to 100 watts from a half-wave dipole

---

The 12-meter band is part of the HF spectrum allocated for amateur radio use. It offers excellent opportunities for long-distance communication, particularly during periods of high solar activity.

You will notice that the exceptions above don't mention this band. Therefore, for the 12-meter band, an amateur station's maximum power is 1500 watts PEP (Peak Envelope Power). This is the standard maximum power limit for most HF bands under the FCC regulations, and it allows operators to transmit with sufficient power for long-distance communication. This power level is particularly useful on the 12-meter band during times of favorable propagation, such as when the solar cycle is active, enabling effective global communication.

What is the limit for transmitter power on the 28 MHz band for a General Class control operator? (G1C05)

A. 100 watts PEP output
B. 1000 watts PEP output
C. 1500 watts PEP output
D. 2000 watts PEP output

Again, we don't see this frequency called out in the exceptions. So, on the 28 MHz band (also known as the 10-meter band), General class control operators are allowed to use up to 1500 watts of PEP (Peak Envelope Power) output. This power limit enables operators to maximize their signal strength, which is particularly beneficial when conditions on the band are less favorable. The high power limit on this band makes it easier for hams to achieve long-distance contacts, especially during periods of sunspot activity, when the band can support global communications.

What is the limit for transmitter power on the 1.8 MHz band? (G1C06)

A. 200 watts PEP output
B. 1000 watts PEP output
C. 1200 watts PEP output
D. 1500 watts PEP output

The 160-meter band, centered around 1.8 MHz, is often referred to as the "top band" and is the lowest-frequency HF band available to amateur radio operators. This band offers a unique challenge for long-distance communication, especially at night when atmospheric conditions allow for better propagation. The FCC permits high power levels on this band to help overcome the challenges of lower-frequency communication.

Have you found the pattern yet?

If not called out as an exception, then on the 1.8 MHz band, the maximum allowable transmitter power for a General class operator is 1500 watts PEP). This high power limit is particularly important for 160 meters, as lower frequencies tend to experience more signal attenuation and require more power to maintain effective communication. The combination of the high power limit and the unique propagation characteristics of this band provides excellent opportunities for long-distance communication, especially during nighttime hours when the band tends to open up.

And our last question on power restrictions.

What is the maximum power limit on the 60-meter band? (G1C09)

A. 1500 watts PEP
B. 10 watts RMS
C. ERP of 100 watts PEP with respect to a dipole
D. ERP of 100 watts PEP with respect to an isotropic antenna

First, you see, it is an exception. So, special rules apply.

The 60-meter band is unique in amateur radio as it operates on five specific channels rather than across a frequency range, and it is shared with other users, such as federal agencies. Because of this shared usage, the FCC imposes stricter power limits and other regulations to ensure that amateurs do not interfere with other services. Power on the 60-meter band is measured in terms of Effective Radiated Power (ERP), which considers the antenna's gain relative to a dipole.

On the 60-meter band, the maximum allowable power is 100 watts PEP (Peak Envelope Power) ERP, with respect to a dipole antenna. This means that the transmitted power must not exceed the equivalent of 100 watts as a standard dipole antenna would radiate it. This restriction is important because the 60-meter band is shared with government and non-amateur services, and limiting power helps to prevent interference with these primary users.

Spread Spectrum Transmissions

Spread spectrum is a transmission technique used in radio that spreads the signal over a wider bandwidth than necessary. It provides benefits such as resistance to interference, secure communication, and reduced potential for signal interception. While spread spectrum is often used in commercial and military applications, it is also permitted in the amateur radio service under certain restrictions, particularly regarding power limits.

The maximum Peak Envelope Power (PEP) output allowed for spread spectrum transmissions in amateur radio is 10 watts. The FCC sets this lower power limit to minimize the potential for interference with other users in the band, given that spread spectrum signals occupy a wider bandwidth. The 10-watt limit ensures that spread spectrum transmissions remain low-power while still taking advantage of their interference-resistant properties.

What is the maximum PEP output allowed for spread spectrum transmissions? (G1E08)

A. 100 milliwatts
B. 10 watts
C. 100 watts
D. 1500 watts

To remember this for the exam, associate spread spectrum with its unique ability to handle interference and the requirement to operate within a 10-watt PEP power limit to avoid disrupting shared frequencies.

Antenna Structures

Height Regulations

Antennas are crucial in optimizing communication, and their height can significantly impact performance. However, there are federal regulations in place to ensure that antenna structures do not interfere with air traffic. For antennas not located near public-use airports, operators must still be mindful of height limits set by the FAA (Federal Aviation Administration) and FCC (Federal Communications Commission). These regulations help ensure safety in air navigation while allowing amateur operators to enjoy the benefits of elevated antenna installations.

The maximum height for an antenna structure without requiring notification to the FAA and registration with the FCC is 200 feet above ground level, provided it is not near a public-use airport. This limit is set to avoid potential hazards to air navigation. Any structure exceeding this height must be reported to the FAA and registered with the FCC to ensure that it does not pose a risk to aircraft. Antenna height can greatly improve signal reach, but safety regulations must always be followed.

What is the maximum height above ground for an antenna structure not near a public-use airport without requiring notification to the FAA and registration with the FCC? (G1B01)

A. 50 feet
B. 100 feet
C. 200 feet
D. 250 feet

Two hundred feet is the safe height limit for antennas in areas not close to public airports. Anything taller requires official clearance to ensure that it does not interfere with air traffic.

Local Restrictions

While HAM radio is governed primarily by federal regulations, state and local governments also have some authority to regulate antenna structures. However, such regulations must ensure that amateur radio operators are reasonably accommodated in their communication ability. This balance allows local authorities to address safety, aesthetic, or zoning concerns while ensuring these regulations do not overly burden or prevent amateur radio operations.

State and local governments are permitted to regulate amateur radio antenna structures under certain conditions, but they must ensure that the regulations allow for amateur service communications to be reasonably accommodated. This means that regulations cannot unnecessarily restrict an operator's ability to set up and use antennas. The minimum regulations must also be necessary to serve a legitimate purpose, such as safety or zoning requirements. For example, a local government might limit antenna heights for safety reasons, but they cannot outright ban antennas without a valid reason.

Under what conditions are state and local governments permitted to regulate amateur radio antenna structures? (G1B06)

A. Under no circumstances, FCC rules take priority
B. At any time and to any extent necessary to accomplish a legitimate purpose of the state or local entity, provided that proper filings are made with the FCC
C. Only when such structures exceed 50 feet in height and are clearly visible 1,000 feet from the structure
D. Amateur Service communications must be reasonably accommodated, and regulations must constitute the minimum practical to accommodate a legitimate purpose of the state or local entity

---

To remember this for the exam, consider the regulations as needing to strike a balance: They must protect legitimate local interests while ensuring that hams can operate their stations effectively. Regulations must be the least restrictive necessary to achieve their purpose.

Beacon Stations

A beacon station is an amateur radio station that continuously transmits a signal on a set frequency to provide information about propagation conditions or serve as a reference signal for other operators. Beacon stations typically transmit a simple, repetitive signal, often including the station's call sign and other identifying information, such as location or operating frequency.

According to FCC rules, a beacon station's primary purpose is to observe propagation and reception. This means that beacon stations are set up to help operators monitor how radio waves travel across different regions, which bands are open for communication, and how conditions change over time. Operators can use the information provided by beacon stations to decide which frequencies might be optimal for their communication.

Which of the following is a purpose of a beacon station as identified in the FCC rules? (G1B03)

A. Observation of propagation and reception
B. Automatic identification of repeaters
C. Transmission of bulletins of general interest to amateur radio licensees
D. All these choices are correct

Beacon stations are valuable for determining how well radio waves are propagating over long distances at a given time. By listening for beacon signals on different frequencies, other operators can assess the current state of the ionosphere and make informed decisions about which frequencies might be optimal for communication.

Key Points:

- **Propagation testing**: Beacons help operators check signal propagation conditions.

- **Repetitive signal**: Beacons usually send a simple, repeating transmission with their call sign.

- **Frequency reference**: Beacons provide a stable signal that operators can tune to for testing equipment or making adjustments.

Beacon Station Restrictions

Since beacon stations are continuously active, specific regulations ensure they do not interfere with other operations or unnecessarily congest the bands.

One important regulation for beacon stations is that no more than one beacon station may transmit on the same band from the same station location. This rule helps prevent multiple beacons from creating interference or congestion on a single frequency band. By limiting a location to one beacon per band, the FCC ensures that the spectrum is used efficiently and that operators can rely on clean, clear signals from each beacon station.

With which of the following conditions must beacon stations comply? (G1B02)

A. No more than one beacon station may transmit in the same band from the same station location
B. The frequency must be coordinated with the National Beacon Organization
C. The frequency must be posted on the internet or published in a national periodical
D. All these choices are correct

Multiple beacons on the same band from the same place could cause confusion and interfere with the station's purpose of providing propagation information.

Automatically Controlled Beacon Stations

In the HF bands, beacons can either be manually controlled or automatically controlled, meaning they transmit without an operator needing to be physically present. However, automatic control of beacons is only permitted on specific frequency ranges to prevent interference with other communications.

Automatically controlled beacon stations are permitted on HF frequencies ranging from 28.20 MHz to 28.30 MHz, which falls within the 10-meter band. This 100 kHz segment is reserved for beacons to transmit signals continuously, helping operators monitor propagation conditions on this band. Automatic control allows beacons to transmit without the need for constant operator input, ensuring they remain on air for extended periods.

On what HF frequencies are automatically controlled beacons permitted? (G1B09)

A. On any frequency if power is less than 1 watt
B. On any frequency if transmissions are in Morse code
C. 21.08 MHz to 21.09 MHz
D. 28.20 MHz to 28.30 MHz

Remember this for the exam: 28.20 MHz to 28.30 MHz as the dedicated beacon segment on the 10-meter band, where beacons can run automatically, providing useful propagation data for operators tuning in.

Power Limits

Beacon stations must operate under specific power limitations to prevent interference with other spectrum users and ensure that they can coexist efficiently with other amateur services.

The power limit for beacon stations is 100 watts of PEP (Peak Envelope Power) output. This means that beacons are restricted to a maximum of 100 watts of output power during their transmissions. This limitation helps ensure that beacons do not overpower other stations on the band and allows for efficient use of the spectrum, particularly on shared frequency ranges.

What is the power limit for beacon stations? (G1B10)

A. 10 watts PEP output
B. 20 watts PEP output
C. 100 watts PEP output
D. 200 watts PEP output

Transmission Restrictions

Amateur operators should avoid transmitting on 14.100 MHz, 18.110 MHz, 21.150 MHz, 24.930 MHz, and 28.200 MHz because a system of propagation beacon stations uses these frequencies. These beacons, part of an international network, continuously transmit signals that help operators assess propagation conditions across various parts of the radio spectrum. By listening to these beacons, amateur operators can determine which bands are open for long-distance communication. Transmitting on these frequencies would interfere with the beacons' signals, disrupting their use for propagation analysis.

Why should an amateur operator normally avoid transmitting on 14.100, 18.110, 21.150, 24.930 and 28.200 MHz? (G1E10)

A. A system of propagation beacon stations operates on those frequencies
B. A system of automatic digital stations operates on those frequencies
C. These frequencies are set aside for emergency operations
D. These frequencies are set aside for bulletins from the FCC

Avoiding transmission on these frequencies ensures that these valuable tools remain functional and interference-free for all amateur operators.

Chapter Summary

To stay informed about FCC rules and regulations, utilize various resources available to ham radio operators. The official FCC website offers comprehensive information on current regulations, updates, and bulletins. Subscribing to ham radio newsletters and magazines can provide regular updates and insights on regulatory changes and best practices. Joining ham radio organizations like the American Radio Relay League (ARRL) offers access to valuable resources, training, and community support.

Staying updated on FCC rules is necessary for maintaining compliance and ensuring a positive operating experience.

Chapter 27

Rules for the General Exam

AFTER PASSING A LICENSING exam, amateur operators receive a Certificate of Successful Completion of Examination (CSCE). This certificate grants temporary operating privileges for the new license class until the official FCC upgrade is processed. For Technician class operators, this means they can start using General class privileges while waiting for their license upgrade, expanding the range of frequencies they can operate on.

Suppose you are a Technician class operator with an unexpired CSCE for General class privileges. In that case, you are allowed to operate on any General or Technician class band segment. This means that as soon as you pass the General exam, you can immediately start using both General and Technician frequency allocations, even before your official license is upgraded in the FCC database. However, these privileges are temporary and last for 365 days or until your upgrade is officially processed; after 365 days, your CSCE expires.

On which of the following band segments may you operate if you are a Technician class operator and have an unexpired Certificate of Successful Completion of Examination (CSCE) for General class privileges? (G1D03)

A. Only the Technician band segments until your upgrade is posted in the FCC database
B. Only on the Technician band segments until you have a receipt for the FCC application fee payment
C. On any General or Technician class band segment
D. On any General or Technician class band segment except 30 meters and 60 meters

---

And

How long is a Certificate of Successful Completion of Examination (CSCE) valid for exam element credit? (G1D09)

A. 30 days
B. 180 days
C. 365 days
D. For as long as your current license is valid

Temporary Operating Privileges

When a Technician class licensee passes the General class exam, they are granted temporary operating privileges before the upgrade is processed in the FCC database. These privileges allow the operator to use General class frequency bands. Still, specific identification rules must be followed during this interim period. The requirement to use a special identifier ensures that other operators know the licensee is temporarily operating under General privileges, which are not yet reflected in the official FCC records.

Until the General class upgrade is shown in the FCC database, a Technician class licensee must identify with "AG" (short for Authorized General) after their call sign whenever they are using General class frequency privileges. This is a way of informing others on the air that the operator is operating with General privileges based on a CSCE (Certificate of Successful Completion of Examination), but the upgrade has not yet been processed. When operating on Technician class bands, the "AG" is not required, but it must be used every time the operator transmits on General class frequencies.

Until an upgrade to General class is shown in the FCC database, when must a Technician licensee identify with "AG" after their call sign? (G1D06)

A. Whenever they operate using General class frequency privileges
B. Whenever they operate on any amateur frequency
C. Whenever they operate using Technician frequency privileges
D. A special identifier is not required if their General class license application has been filed with the FCC

It's a way of signaling your new privileges to other operators while waiting for the database update.

Renewing an Expired Amateur License

Amateur radio licenses are valid for 10 years. If a license expires, there is a two-year grace period during which the operator can renew their license without taking another exam. However, reinstating the license becomes more complex once this grace period has passed. Operators who want to regain their license privileges after this period must meet certain requirements, including taking parts of the licensing exam again.

Suppose an operator's General class license has expired and the two-year grace period has passed. In that case, the applicant must provide proof of their expired license and pass the current Element 2 exam (the Technician class exam). Element 2 is the foundational exam required for all amateur operators, and passing it reinstates the individual's privileges, including General class, if they can show proof of holding a previously valid General class license. The Element 2 requirement ensures that the operator is up-to-date on basic rules and regulations, while their previous experience is credited for the more advanced privileges.

What action is required to obtain a new General class license after a previously held license has expired and the two-year grace period has passed? (G1D11)

A. They must have a letter from the FCC showing they once held an amateur or commercial license
B. There are no requirements other than being able to show a copy of the expired license
C. Contact the FCC to have the license reinstated
D. The applicant must show proof of the appropriate expired license grant and pass the current Element 2 exam

---

To remember this for the exam, think of it as a two-step process. After the grace period, you must pass the Technician class exam (Element 2) and show proof of your expired General license to regain your operating privileges.

Partial Credit for Expired Licenses

Amateur radio operators who once held higher-class licenses, such as General, Advanced, or Amateur Extra, may be eligible to regain their privileges even if their license has expired. The FCC allows former licensees to avoid retaking all licensing exams by allowing them to receive partial credit for their previously held license. This helps operators re-enter the hobby more easily, recognizing their past achievements while ensuring they still meet current requirements.

Anyone who can demonstrate that they once held an FCC-issued General, Advanced, or Amateur Extra class license, which was not revoked, may receive partial credit for the exam elements that were previously passed. This means that while the operator will need to take the Element 2 exam (Technician), they are not required to retake the General or Extra class exams, as their previous license covers those elements. This provision allows former licensees to restore their privileges without starting from scratch, provided they can show proof of their expired license.

Who may receive partial credit for the elements represented by an expired amateur radio license? (G1D01)

A. Any person who can demonstrate that they once held an FCC-issued General, Advanced, or Amateur Extra class license that was not revoked by the FCC
B. Anyone who held an FCC-issued amateur radio license that expired not less than 5 and not more than 15 years ago
C. Any person who previously held an amateur license issued by another country, but only if that country has a current reciprocal licensing agreement with the FCC
D. Only persons who once held an FCC issued Novice, Technician, or Technician Plus license

---

Think of it as the FCC's way of honoring past licensees' achievements: if your General, Advanced, or Extra license expired but wasn't revoked, you can get credit for the elements you passed, making it easier to get back on the air.

Volunteer Examiners and Examiner Coordinators

The Volunteer Examiner (VE) program is fundamental in administering amateur radio licensing exams. The FCC authorizes qualified amateur operators to serve as VEs, allowing them to administer licensing exams for new and upgrading operators. These VEs are accredited by Volunteer Examiner Coordinators (VECs), organizations recognized by the FCC to manage and coordinate VEs. The VEs ensure that exams are administered fairly and according to FCC regulations and enable the smooth operation of the amateur radio licensing process.

To administer a Technician class license examination, at least three Volunteer Examiners (VEs) of General class or higher must observe and oversee the exam. This ensures the integrity of the examination process, as multiple VEs confirm the accuracy and fairness of the test administration.

Who must observe the administration of a Technician class license examination? (G1D04)

A. At least three Volunteer Examiners of General class or higher
B. At least two Volunteer Examiners of General class or higher
C. At least two Volunteer Examiners of Technician class or higher
D. At least three Volunteer Examiners of Technician class

Volunteer Examiners are accredited by a Volunteer Examiner Coordinator (VEC), which oversees the accreditation process and ensures that VEs are properly qualified to administer exams. There are several VECs across the United States, with the ARRL VEC being the largest.

Volunteer Examiners are accredited by what organization? (G1D07)

A. The Federal Communications Commission
B. The Universal Licensing System
C. A Volunteer Examiner Coordinator
D. The Wireless Telecommunications Bureau

To qualify as a Volunteer Examiner, <u>one must be at least 18 years old</u> and hold an FCC-granted amateur radio license of General class or above. This rule applies to US and <u>non-US citizens, with the latter requiring a valid FCC-issued General class license or higher.</u>

What is the minimum age that one must be to qualify as an accredited Volunteer Examiner? (G1D10)

A. 16 years
B. 18 years
C. 21 years
D. There is no age limit

Which of the following criteria must be met for a non-US citizen to be an accredited Volunteer Examiner? (G1D08)

A. The person must be a resident of the US for a minimum of 5 years
B. The person must hold an FCC-granted amateur radio license of General class or above
C. The person's home citizenship must be in ITU region 2
D. None of these choices is correct; a non-US citizen cannot be a Volunteer Examiner

A VE holding a General class license may administer exams <u>only for the Technician class</u>. Higher-class exams, such as General or Extra, require Volunteer Examiners with a higher level of licensure.

What license examinations may you administer as an accredited Volunteer Examiner holding a General class operator license? (G1D02)

A. General and Technician
B. None, only Amateur Extra class licensees may be accredited
C. Technician only
D. Amateur Extra, General, and Technician

This system ensures that exams are fair, properly monitored, and in compliance with FCC regulations, allowing qualified individuals to enter the amateur radio hobby.

Chapter Summary

This chapter highlights key elements of amateur radio licensing that directly prepare you for the exam. Understanding the privileges granted by the Certificate of Successful Completion of Examination (CSCE) ensures you know what operating rights you gain immediately after passing an exam. Familiarizing yourself with the roles of Volunteer Examiners and the structure of licensing tests provides clarity on the process and requirements. By mastering these rules and procedures, you'll be better equipped to approach the exam with confidence.

Further Expanding Your Ham Experience

AMATEUR RADIO IS A gateway to a world of opportunities, blending technical expertise with real-world applications that go beyond simply operating a station. In this section, we delve into some of the most rewarding and impactful ways to expand your journey in amateur radio. Whether you're stepping up to provide critical communications during emergencies, supporting the integrity of the airwaves through volunteer monitoring, or pushing your skills and equipment to the limit in HF contesting, amateur radio offers avenues for both service and personal growth. By exploring these activities, you'll uncover the true depth of this hobby—one that fosters community, resilience, and innovation, all while keeping the thrill of discovery alive.

Chapter 28

The Responsibility and Excitement of Amateur Radio

AMATEUR RADIO IS MORE than just a hobby—it's a versatile tool that connects people, communities, and even entire regions during critical times. This chapter dives into three essential areas that showcase the depth and impact of amateur radio operations.

First, we'll explore the Radio Amateur Civil Emergency Service (RACES), a vital program that supports disaster relief and emergency communications. Next, we'll examine the Volunteer Monitor Program, a unique initiative that empowers radio operators to help maintain order and compliance on the airwaves. Finally, we'll delve into the dynamic world of HF contesting, where operators test their skills and equipment under fast-paced, competitive conditions.

Together, these topics reveal amateur radio's multifaceted nature and the important roles operators play in public service and personal achievement.

Radio Amateur Civil Emergency Service

The Radio Amateur Civil Emergency Service (RACES) is an important part of amateur radio operations, designed to assist with disaster relief and emergency communications. It is a partnership between amateur radio operators and emergency management agencies to provide communications during times of crisis. RACES operations are subject to specific rules and regulations, especially concerning who may operate a station under these circumstances. This ensures that only qualified and licensed individuals are responsible for critical emergency communications.

In RACES operations, only a person holding an FCC-issued amateur operator license can act as the station's control operator. This means that during disaster relief efforts, the responsibility for controlling the station lies exclusively with licensed amateurs whom the FCC authorizes. The control operator ensures that all transmissions comply with the rules and that communications aid in emergency or relief operations.

Who may be the control operator of an amateur station transmitting in RACES to assist relief operations during a disaster? (G2B09)

A. Only a person holding an FCC-issued amateur operator license
B. Only a RACES net control operator
C. A person holding an FCC-issued amateur operator license or an appropriate government official
D. Any control operator when normal communication systems are operational

---

This restriction guarantees that RACES stations are operated by qualified individuals who understand the importance of proper communication protocols during critical situations, ensuring smooth and effective disaster response efforts.

How often may RACES training drills and tests be routinely conducted without special authorization? (G2B11)

A. No more than 1 hour per month
B. No more than 2 hours per month
C. No more than 1 hour per week
D. No more than 2 hours per week

RACES training drills and tests can be conducted no more than 1 hour per week without special authorization. This limit allows RACES operators to maintain their skills and be ready for emergencies while ensuring that frequencies are not unnecessarily occupied for extended periods. The 1-hour weekly limit is sufficient for routine practice and maintaining operational readiness. Still, special authorization from the FCC would be necessary if more frequent or longer training sessions are required.

This regulation helps to strike a balance between maintaining emergency preparedness and ensuring that amateur frequencies remain available for general use. By adhering to these rules, operators can be ready to assist during real emergencies while respecting the broader amateur radio community's need for shared frequency access.

Volunteer Monitor Program

The Volunteer Monitor Program is an initiative established by the FCC and managed by the ARRL (American Radio Relay League) that empowers amateur radio operators to help maintain compliance with FCC rules and regulations. These volunteers monitor amateur radio transmissions and help identify and address rule violations. The program fosters a sense of self-regulation within the amateur radio community, allowing operators to take an active role in keeping the airwaves free from harmful interference and illegal activity.

Explanation of the Volunteer Monitor Program

The Volunteer Monitor Program consists of amateur radio operators who are formally enlisted to observe and report on airwaves activity, ensuring that transmissions comply with FCC rules. The volunteers are trained and authorized to monitor the bands for violations, such as improper use of frequencies or intentional interference. The goal is to encourage self-regulation among amateur radio operators and foster a culture of compliance.

What is the Volunteer Monitor Program? (G2D01)

A. Amateur volunteers who are formally enlisted to monitor the airwaves for rules violations
B. Amateur volunteers who conduct amateur licensing examinations
C. Amateur volunteers who conduct frequency coordination for amateur VHF repeaters
D. Amateur volunteers who use their station equipment to help civil defense organizations in times of emergency

Which of the following are objectives of the Volunteer Monitor Program? (G2D02)

A. To conduct efficient and orderly amateur licensing examinations
B. To provide emergency and public safety communications
C. To coordinate repeaters for efficient and orderly spectrum usage
D. To encourage amateur radio operators to self-regulate and comply with the rules

The program also includes a technique for locating problem stations. For example, if a continuous carrier holds a repeater open inappropriately, Volunteer Monitors can use direction-finding techniques. <u>By comparing beam headings on the repeater input from different locations, they can triangulate the source of the signal and assist in addressing the issue.</u>

What procedure may be used by Volunteer Monitors to localize a station whose continuous carrier is holding a repeater on in their area? (G2D03)

A. Compare vertical and horizontal signal strengths on the input frequency
B. Compare beam headings on the repeater input from their home locations with that of other Volunteer Monitors
C. Compare signal strengths between the input and output of the repeater
D. All these choices are correct

This collaborative monitoring effort ensures that the amateur radio bands remain a reliable and well-regulated resource for all operators.

Contesting

Ham radio contesting is a fast-paced activity where operators compete to make as many contacts as possible within a set timeframe, testing their skills and endurance. It fosters community and camaraderie while pushing the limits of amateur radio capabilities. Contests vary widely, from DX contests emphasizing long-distance contacts to VHF/UHF contests challenging operators with line-of-sight propagation, CW contests highlighting Morse code skills, and digital mode contests like RTTY or FT8 requiring digital expertise.

Each contest has unique rules for exchanging information, such as call signs, signal reports, or grid locators, with accurate logging essential for scoring. Points are awarded per contact, with multipliers for different regions or bands and bonuses for achievements like working rare stations. Contests typically last hours to weekends, sometimes requiring mandatory breaks for fairness.

Success in contesting often depends on specialized tools, like logging software (e.g., N1MM Logger+) to track contacts and calculate scores. Advanced transceivers with noise reduction, high-gain antennas like Yagis, and accessories like headsets and foot switches can improve performance. Operators must always follow FCC rules, including proper station identification, to ensure legal and transparent operation.

<u>Operators must identify their station according to normal FCC regulations when participating in a HF frequency contest.</u> This means that despite the rapid pace of contest exchanges, you must still announce your call sign at the required intervals, just as you would during any other communication. This rule helps prevent confusion on the bands and ensures that all operators follow the appropriate identification protocols.

Which of the following is required when participating in a contest on HF frequencies? (G2D09)

A. Submit a log to the contest sponsor
B. Send a QSL card to the stations worked, or QSL via Logbook of The World
C. Identify your station according to normal FCC regulations
D. All these choices are correct

<u>Additionally, many operators use an azimuthal projection map during contests or DXing to help determine the true bearings and distances from their location to other stations.</u> These maps are centered on a specific point, usually the operator's station, and display the correct direction and distance to any location on Earth. This tool is particularly useful for pointing directional antennas to improve signal strength during long-distance communications.

Which of the following describes an azimuthal projection map? (G2D04)

A. A map that shows accurate land masses
B. A map that shows true bearings and distances from a specific location
C. A map that shows the angle at which an amateur satellite crosses the equator
D. A map that shows the number of degrees longitude that an amateur satellite appears to move westward at the equator with each orbit

By combining the awareness of FCC identification rules with practical tools like azimuthal projection maps, contest participants can operate more efficiently while adhering to legal requirements.

Strategies for Winning Contests

Preparation and planning are the bedrock of successful contesting. To maximize your performance, start by researching each contest's specific rules and strategies. Every contest has unique requirements and nuances, so familiarize yourself with the contest's official rules, which can typically be found on the organizing body's website. Understanding the scoring system, exchange requirements, and any special bonuses or multipliers will give you a strategic advantage.

Maximizing your contact rates during a contest requires a combination of effective techniques. Running a frequency, where you hold a frequency and call CQ continuously, can attract a steady stream of contacts. This method is particularly effective when you have a strong signal and a clear frequency. However, choose a frequency that is not too crowded to avoid becoming a target of interference. Alternatively, the search and pounce method involves actively tuning across the band to find and contact other stations. This technique is beneficial when the band is crowded, allowing you to capitalize on every available opportunity.

Chapter Summary

In this chapter, we explored key aspects of amateur radio operations, including the critical roles of RACES in disaster communications, the Volunteer Monitor Program's importance in maintaining compliance, and the excitement and strategy involved in HF contesting. From understanding the unique rules and responsibilities of emergency operations to leveraging tools like azimuthal projection maps during contests, each topic highlights the diverse ways amateur radio operators contribute to public service and personal achievement. Whether preparing for emergencies, fostering self-regulation, or competing in contests, these skills and practices embody the spirit of amateur radio—combining technical expertise, community involvement, and a commitment to operating responsibly and effectively.

Keeping the Hobby Alive & Growing

Keeping the Hobby Alive & Growing

Now that you've taken the step to upgrade to your General Class license, you're in a prime position to explore even more of what ham radio has to offer. But this journey isn't just about you—it's about keeping the spirit of amateur radio alive for future generations.

Would you help someone you've never met, even without getting credit for it? That person might be where you were before you picked up this book—curious about ham radio, eager to expand their knowledge, and looking for guidance. Your honest review could encourage them to take their next step.

At Morse Code Publishing, we aim to make ham radio accessible and exciting for everyone. But to continue growing the hobby and reaching new operators, we need your help. A good book cover catches the eye, but reviews are what truly help others make decisions. By taking just 60 seconds to leave a review, you could:

- Help a beginner discover the joy and appreciate the intricacies of ham radio.

- Help other Technicians prepare for and pass their General Class exam.

Your review costs nothing but could make all the difference in someone else's journey to becoming a licensed operator or upgrading to General.

Thank you for your support! Ham radio grows and thrives when we share our knowledge and enthusiasm. With your help, we can continue to inspire the next generation of operators.

Welcome to the community of General Class operators!

Your biggest fan,

Morse Code Publishing

PS – Helping others succeed is one of the greatest rewards in this hobby. If you believe this book can benefit someone, consider sharing it with them. Let's keep the hobby thriving!

Conclusion

THE JOURNEY THROUGH THIS book was designed with a singular purpose: to provide you, a dedicated ham radio operator, with the knowledge and tools necessary to upgrade from a Technician to a General class license. This guide aims to expand your understanding and capabilities in amateur radio, ensuring you are well-prepared to ace the FCC exam and enjoy the enhanced privileges of the General class license.

From the outset, my vision has been clear—to make complex information accessible and easy to retain. This book has been structured to break down intricate topics into manageable sections, using practical tips and examples to facilitate learning. This approach ensures the material is understandable and memorable, enabling you to apply your newfound knowledge confidently.

The book was organized into four key sections, each addressing crucial aspects of the General class license exam.

We began with the **Science of Radio** section and explored the fascinating principles of radio wave propagation, the types of antennas and their uses, and the intricacies of modulation techniques. This scientific understanding is vital for optimizing your communication range and quality, allowing you to maximize your operating privileges.

Next, in the **Physical Ham Station** section, we covered the technical aspects of setting up and maintaining your ham radio station. From choosing the right equipment to understanding transmitters and receivers and ensuring proper grounding and power supply management, this section provided the practical knowledge necessary to create a functional and efficient station.

In **Operating Procedures**, where you learned the proper techniques for making and answering calls, handling emergency communications, and participating in ham radio nets. These procedures are essential for effective and respectful communication on the airwaves.

Lastly, **Rules & Regulations**, where we explored the role of the FCC, the specific regulations that General class licensees must follow, and the importance of operating legally and ethically. This section laid the foundation for understanding the responsibilities that come with greater privileges.

Key takeaways from this book include a comprehensive understanding of FCC regulations, mastery of operating procedures, the ability to set up and maintain a physical ham station, and a deep knowledge of radio science. These insights will help you pass the General class license exam and enhance your overall ham radio experience.

I encourage you to utilize this guide's study techniques and tips as you prepare for the exam. Practice with the included exams (take all 16!), and remember to manage your time effectively during the test. Your dedication to studying and passion for ham radio will undoubtedly lead you to success.

I wish you the best of luck in your exam and future ham radio endeavors. Keep exploring, learning, and, most importantly, enjoying the incredible world of ham radio.

Jared Johnson (KF0RTU/AG) @ Morse Code Publishing

Exam Question Index

References

- *47 CFR Part 97 -- Amateur Radio Service* https://www.ecfr.gov/current/title-47/chapter-I/subchapter-D/part-97

- *General Class Frequency Privileges in Ham Radio* https://www.dummies.com/article/technology/digital-audio-radio/ham-radio/general-class-frequency-privileges-in-ham-radio-164187/

- *1068. Violation of FCC Regulations—47 U.S.C. § 502* https://www.justice.gov/archives/jm/criminal-resource-manual-1068-violation-fcc-regulations-47-usc-502#:~:text=Under%2047%20U.S.C.,on%20which%20a%20violation%20occurs.

- *Frequency Allocations* http://www.arrl.org/frequency-allocations

- *Amateur Radio Service Enforcement - AMAT* https://www.fcc.gov/tags/amateur-radio-service-enforcement-amat

- *Upgrading to a General License* http://www.arrl.org/upgrading-to-a-general-license

- *General Class Frequency Privileges in Ham Radio* https://www.dummies.com/article/technology/digital-audio-radio/ham-radio/general-class-frequency-privileges-in-ham-radio-164187/

- *How to Stop Radio Frequency Interference | PCB Design Blog* https://resources.pcb.cadence.com/blog/2022-how-to-stop-radio-frequency-interference

- *Ethics and Operating Procedures for the Radio Amateur* https://www.arrl.org/files/file/DXCC/Eth-operating-EN-ARRL-CORR-JAN-2011.pdf

- *Choosing a Field Radio: How to find the perfect transceiver ...* https://qrper.com/2022/07/choosing-a-field-radio-how-to-find-the-perfect-transceiver-for-your-outdoor-radio-activities/

- *Improving Receiver Reception: Tips and Tricks* https://www.hamradiostore.co.uk/blog/improving-receiver-reception-tips-and-tricks#:~:text=IF%20Bandwidth%20Adjustment,neighboring%20signals%20and%20background%20noise.

- *Choosing a Power Supply for Your Station * https://www.onallbands.com/choosing-a-power-supply-for-your-station%EF%BB%BF/

- *Ham Radio Tech: Choosing a Battery System for Portable ...* https://www.onallbands.com/ham-radio-tech-choosing-a-battery-system-for-portable-operations/

- *Ham Radio Antennas: Types, Differences, and Pros & Cons* https://strykerradios.com/ham-radios/ham-radio-antenna-types-differences-pros-cons/

- *Amateur Radio Station Grounding and Lightning Protection* https://www.bwcelectronics.com/articles/WP30A190.pdf

- *Basic Amateur Radio - HF Propagation* https://www.arrl.org/files/file/Technology/tis/info/pdf/8312011.pdf

- *Basic Electronic Components - Sierra Circuits* https://www.protoexpress.com/kb/basic-components-overview/#:~:text=Some%20of%20the%20most%20commonly,switches%2C%20ICs%2C%20and%20connectors.

- *How to Read a Schematic - SparkFun Learn* https://learn.sparkfun.com/tutorials/how-to-read-a-schematic/all

- *Loads of Modes - Ham Radio School* https://www.hamradioschool.com/post/loads-of-modes#:~:text=Modulation%20Mode%20%E2%80%93%20A%20modulation%20mode,voice%20into%20the%20radio%20transmission.

- *How to Demodulate an FM Waveform | Radio Frequency ...* https://www.allaboutcircuits.com/textbook/radio-frequency-analysis-design/radio-frequency-demodulation/how-to-demodulate-an-fm-waveform/

- *How to Call and Answer a CQ* https://www.ailunce.com/blog/How-to-Call-and-Answer-a-CQ

- *How to Participate in Ham Radio Nets* https://www.dummies.com/article/technology/digital-audio-radio/ham-radio/how-to-participate-in-ham-radio-nets-160474/

- *Amateur Radio Digital Communications - SARCNET* https://www.sarcnet.org/amateur-radio-digital-communications.html

- *RTTY / FSK Configuration - Fldigi Users Manual* http://www.w1hkj.com/FldigiHelp/rtty_fsk_configuration_page.htm

- *Amateur Radio Station Grounding* https://w7aia.org/files/operating_files/RCC_Mar_2007_N7ZXP_Amateur_Radio_Station_Grounding.pdf

- *ARRL Handbook: The ARRL Handbook is a comprehensive guide to amateur radio that covers everything from basic electronics to advanced operating techniques.* https://hamradioprep.com/ham-radio-study-guide/

- *Contest Operating Tips, originated by K1AR* https://www.mapability.com/ei8ic/contest/k1ar.php

- *Practical Signal Reports* https://www.hamradioschool.com/post/practical-signal-reports

- *The Ionosphere | Center for Science Education* https://scied.ucar.edu/learning-zone/atmosphere/ionosphere

- *How to Use Meteor Scatter Communications in Ham Radio* https://www.electronics-notes.com/articles/ham_radio/amateur-propagation/meteor-scatter-burst-communications.php

- *How to Troubleshoot ham radio problems* https://brara.org/BLOG/2018/11/06/new-hams-how-to-troubleshoot-ham-radio-problems-radio-wave-frequency-and-power/

- *Test equipment every amateur radio operator should have* http://forums.radioreference.com/threads/test-equipment-every-amateur-radio-operator-should-have.453318/

- *The four pieces of radio test equipment you really need* https://vk3ye.com/gateway/noaug00.htm

- *Ethics and Operating Procedures for the Radio Amateur* https://www.arrl.org/files/file/DXCC/Eth-operating-EN-ARRL-CORR-JAN-2011.pdf

- *Interference   Resolution | Federal   Communications   Commission* https://www.fcc.gov/enforcement/areas/interference-resolution#:~:text=Consumers%20or%20members%20of%20the,using%20the%20Consumer%20Complaint%20Center.

- *The Trouble with Practice Exams - HamTestOnline* https://www.hamradiolicenseexam.com/trouble-with-practice-exams.htm#:~:text=Practice%20exams%20do%20serve%20a,they%20should%20not%20replace%20education.